Feature Papers in Materials Simulation and Design

Feature Papers in Materials Simulation and Design

Editors

Michele Bacciocchi
Abbas S. Milani

MDPI • Basel • Beijing • Wuhan • Barcelona • Belgrade • Manchester • Tokyo • Cluj • Tianjin

Editors

Michele Bacciocchi
Dipartimento di Economia,
Scienze e Diritto (DESD)
University of the Republic
of San Marino
Dogana
San Marino

Abbas S. Milani
School of Engineering
University of British Columbia
Kelowna
Canada

Editorial Office
MDPI
St. Alban-Anlage 66
4052 Basel, Switzerland

This is a reprint of articles from the Special Issue published online in the open access journal *Materials* (ISSN 1996-1944) (available at: www.mdpi.com/journal/materials/special_issues/feature_simulation_design).

For citation purposes, cite each article independently as indicated on the article page online and as indicated below:

LastName, A.A.; LastName, B.B.; LastName, C.C. Article Title. *Journal Name* **Year**, *Volume Number*, Page Range.

ISBN 978-3-0365-7577-3 (Hbk)
ISBN 978-3-0365-7576-6 (PDF)

Contents

About the Editors

Michele Bacciocchi

Michele Bacciocchi is Associate Professor at the University of the Republic of San Marino (San Marino). He obtained his PhD in Civil, Chemical, Environmental, and Materials Engineering at the University of Bologna (Italy). He was a research-visiting scholar at Texas A&M University (United States). His research interests are the mechanics of solids, elasticity, and numerical modeling of structures. He co-authored more than 60 international peer-reviewed journal papers, 6 books, and more than 15 abstracts in international and national conferences. He organized, as Conference Co-Chair, several international events on composite materials and structures.

Abbas S. Milani

Abbas S. Milani is Professor at the School of Engineering of the University of British Columbia (Canada). He is Director of the Materials and Manufacturing Research Institute (MMRI) at the same university. His primary expertise falls into advanced modeling, simulation, and multi-criteria design optimization of composite/bio-composite materials, structures, and manufacturing processes. His research to date has resulted in over 300 publications, including six books, and his projects are well supported by government and numerous industrial organizations.

Preface to "Feature Papers in Materials Simulation and Design"

The title of the current Special Issue, "Feature Papers in Materials Simulation and Design", has identified the aims of this collection since its opening: the gathering of research works and comprehensive review papers that advance the understanding and prediction of material behavior at different scales, from atomistic to macroscopic, through innovative modeling and simulation.

In this context, interdisciplinary researches that tackled challenging and complex material problems where the governing phenomena might span different scales of material behavior, with an emphasis on the development of quantitative approaches to explain and predict experimental observations, have been collected. Similarly, particular attention has been given to homogenization techniques for the evaluation of the mechanical properties of new materials and multi-phase composites. Innovative numerical approaches for the mechanical analysis, highlighting their accuracy, reliability, and stability features, have been also welcomed. Significant space has been also given to advanced and sustainable technologies. From the aforementioned topics, it is evident that this Special Issue represents the ideal forum for disseminating excellent research findings, as well as sharing innovative ideas in this significant field.

Throughout the several months of activity, this Special Issue has been able to attract much interesting research. Its success is proven by the sixteen papers collected and published, which have passed through the rigorous review process carried out by experts in the field, who should be gratefully acknowledged for their efforts. A heartfelt thanks should be also given to all the authors, coming from many different countries of the world, who contributed to the success of the collection. This important achievement could not be reached without the constant and kind support given by the Section Managing Editor, Ms. Fay Liu, who should be gratefully thanked for her dedication and commitment. Finally, the Editors-in-Chief of *Materials* should be also mentioned for the fantastic opportunity of managing this successful Special Issue.

Michele Bacciocchi and Abbas S. Milani
Editors

materials

Editorial

Special Issue: "Feature Papers in Materials Simulation and Design"

Michele Bacciocchi [1,*] and Abbas S. Milani [2,*]

1 Dipartimento di Economia, Scienze e Diritto (DESD), University of San Marino, Via Consiglio dei Sessanta, 47891 Dogana, San Marino
2 School of Engineering, The University of British Columbia, Kelowna, BC V1V 1V7, Canada
* Correspondence: michele.bacciocchi@unirsm.sm (M.B.); abbas.milani@ubc.ca (A.S.M.)

The title of the current Special Issue, "Feature Papers in Materials Simulation and Design", has identified the aims of this collection since its opening: the gathering of research works and comprehensive review papers that advance the understanding and prediction of material behavior at different scales, from atomistic to macroscopic, through innovative modeling and simulation.

In this context, interdisciplinary researches that tackled challenging and complex material problems where the governing phenomena might span different scales of material behavior, with an emphasis on the development of quantitative approaches to explain and predict experimental observations, have been collected. Likewise, peculiar attention has been given to homogenization techniques for the evaluation of the mechanical properties of new materials and multi-phase composites. Innovative numerical approaches for the mechanical analysis, highlighting their accuracy, reliability, and stability features, have been also welcomed. Significant space has been also given to advanced and sustainable technologies. From the aforementioned topics, it is evident that this Special Issue has represented the ideal forum for disseminating excellent research findings, as well as sharing innovative ideas in this significant field.

Throughout the several months of activity, this Special Issues has been able to attract many interesting researches. Its success is proven by the sixteen papers collected and published, which have passed through the rigorous review process carried out by experts in the field, who should be gratefully acknowledged for their efforts. An heartfelt thanks should be also given to all the authors, coming from many different countries of the world, who contributed to the success of the collection. This important achievement could be barely reached without the constant and kind support given by the Section Managing Editor, Ms. Fay Liu, who should be gratefully thanked for her dedication and commitment. Finally, the Editors-in-Chief of Materials should be also mentioned for the fantastic opportunity of managing this successful Special Issue.

A brief review of the papers published in the collection is now presented as a proof of the advancements and innovations achieved in the field of materials simulation and design. Nurek et al. [1] investigated the possibility of using forest logging residues as an alternative to plant biomass of various origins. In particular, they concluded that shredded logging residues can be used to produce briquettes. A series of test has been presented to discuss the density and compaction of the briquette, providing useful results for manufacturers.

The paper by Tulska et al. [2] has been focused on the kinematics of cone opening in the European larch (*Larix decidua Mill.*) during a four-step seed extraction process and to determine optimum process time on that basis, describing also the microscopic cellular structure of scales in cones for different values of moisture content. Their results have highlighted the conditions for the automation of this process.

De Schryver et al. [3] proved that numerical finite volume simulations are a powerful means in the context of the rheological assessment of concrete. Their results should be used by engineers to build confidence on the reliability of numerical simulations. In addition,

Citation: Bacciocchi, M.; Milani, A.S. Special Issue: "Feature Papers in Materials Simulation and Design". *Materials* **2023**, *16*, 1900. https://doi.org/10.3390/ma16051900

Received: 9 February 2023
Accepted: 13 February 2023
Published: 24 February 2023

they stated that to improve bias due to uncertain rheology, a rheological configuration close to the engineer's aimed application should be taken into account without overlooking important phenomena.

The paper by Chen et al. [4] aimed at studying the thermal properties and thermoelectric performance of imidazole-graphyne (ID-GY) by combining first principle calculations with the Boltzmann transport theory. A detailed analysis of the harmonic and anharmonic properties has been presented to prove that the low lattice thermal conductivity can be attributed to the low Young's modulus, low Debye temperature, and high Grüneisen parameter. Finally, the authors demonstrated that it is possible to reduce the lattice thermal conductivity and enhance the thermoelectric performance of carbon-based materials by changing structural units from hexagonal to pentagonal

On the other hand, the paper by Ursini and Collini [5] has been focused on the Fused Deposition Modeling (FDM) additive technology, emphasizing that the analysis of the functional behavior of FDM parts is still a topic of great interest. They investigated experimentally and numerically the effects of different phenomena that could affect the mechanical features of 3D printed lattice structures.

Bragov et al. [6] have discussed the influence of specimen geometry and loading conditions on the mechanical properties of porous brittle media. Their results proved that the structure of the material should be taken into account so that the size of the specimen (in terms of both length and diameter) exceeds the size of the internal fractions of the material by at least five times, when the geometry of specimens of brittle porous media is defined.

A fully autonomous computational workflow to identify light-harvesting materials for water splitting devices based on properties has been presented by Ludvigsen et al. [7]. In particular, they discussed relevant features such as stability, size of the band gap, position of the band edges, and ferroelectricity, proving that ferroelectric materials represents a possible solution to enhance the efficiency of solar cells and photoelectrocatalytic devices in light-harvesting applications.

Chomka et al. [8] investigated, instead, the possibility of using a high-pressure water-ice jet as a new method to remove a worn-out paint coating from the surface of metal parts (including those found in means of transportation) and to prepare the base surface for the application of renovation paint coating. The results have been supported by an experimental campaing.

Vahed et al. [9] highlighted the need of mathematical predictive models to overcome the complex and non-linear nature of material properties evolution during 3D printing, especially when Fused Deposition Modeling (FDM) is considered. Once the process parameters have been discussed, artificial neural networks have been trained to predict both the storage and loss moduli of the samples. An optimization of the process parameters through the Particle Swarm Optimization (PSO) has been also performed.

The topic of green composites has been discussed in the paper by Jiao-Wang et al. [10], developing fully biodegradable composites made of microbially degradable polymers reinforced with natural fibers. In particular, they enhanced numerical models to predict the damage of these structures with the aim to extend their employment in industrial applications of structural responsibility.

Civalek et al. [11] developed an efficient eigenvalue algorithm for the axial vibration analysis of embedded short-fiber-reinforced micro-/nano-composite rods subjected to arbitrary boundary conditions. Their investigations have been carried out within the framework provided by the nonlocal elasticity theory needed to capture the size effect. The influence of elastic spring boundary conditions on the axial vibration frequencies and mode shapes has been also discussed.

A beam model for thermal buckling analysis of a bimetallic box beam has been developed by Simonetti et al. [12], considering Euler–Bernoulli–Vlasov beam theory and including large rotations but small strains. An updated Lagrangian formulation is used to discuss the nonlinear stability analysis. The effects of different boundary conditions, beam lengths and material thickness ratios on the critical buckling temperature and post-buckling responses have been discussed through several numerical tests.

The results presented in the paper by Esencan Türkaslan et al. [13] proved that the efficiency of the photocatalytic activity of the nanocomposite depends on different synthesis methods and the morphology resulting from the changed method. To this aim, tungsten trioxide/graphene oxide nanocomposites have been successfully synthesized using in situ and ex situ chemical approaches.

A series of experimental and numerical studies concerning the influence of temperature on mode II fracture of a T800/epoxy unidirectional laminates has been presented by Gong et al. [14]. The failure mechanism has been also determined by using a scanning electron microscope. The validity of the numerical model has been confirmed by the experimental tests. Therefore, the results represent an helpful guidance for the design of composite laminates.

Dastjerdi et al. [15] developed a novel nonlinear elasticity theory for annular and circular plates made of hyperelastic materials. The effect of viscosity has been taken into account to obtain the long-term structural response. The partial differential equations have been solved through a semi-analytical method, and have been confirmed by the successful comparison with the results found in other sources.

The last paper by Cheng and Vescovini [16] presented instead an accurate and efficient numerical method for the analysis and design of Variable Stiffness (VS) laminates. In particular, a ps-version of the Finite Elements Method (ps-FEM) has been developed for the global/local analysis through different refinement approaches. Peculiar attention has been given to the implementation aspects to illustrate the potential of the methodology.

The Guest Editors would like to congratulate all the authors for the remarkable results presented in these papers.

Author Contributions: Conceptualization, M.B. and A.S.M.; writing—original draft preparation, M.B. and A.S.M.; writing—review and editing, M.B. and A.S.M.; visualization, M.B. and A.S.M.; supervision, M.B. and A.S.M. All authors have read and agreed to the published version of the manuscript.

Funding: This research received no external funding.

Conflicts of Interest: The authors declare no conflict of interest.

References

1. Nurek, T.; Gendek, A.; Dąbrowska, M. Influence of the Die Height on the Density of the Briquette Produced from Shredded Logging Residues. *Materials* **2021**, *14*, 3698. [CrossRef] [PubMed]
2. Tulska, E.; Aniszewska, M.; Gendek, A. The Kinematics of Scale Deflection in the Course of Multi-Step Seed Extraction from European Larch Cones (*Larix decidua* Mill.) Taking into Account Their Cellular Structure. *Materials* **2021**, *14*, 4913. [CrossRef] [PubMed]
3. De Schryver, R.; El Cheikh, K.; Lesage, K.; Yardimci, M.Y.; De Schutter, G. Numerical Reliability Study Based on Rheological Input for Bingham Paste Pumping Using a Finite Volume Approach in OpenFOAM. *Materials* **2021**, *14*, 5011. [CrossRef] [PubMed]
4. Chen, Y.; Sun, J.; Kang, W.; Wang, Q. Phonon Transport and Thermoelectric Properties of Imidazole-Graphyne. *Materials* **2021**, *14*, 5604. [CrossRef] [PubMed]
5. Ursini, C.; Collini, L. FDM Layering Deposition Effects on Mechanical Response of TPU Lattice Structures. *Materials* **2021**, *14*, 5645. [CrossRef] [PubMed]
6. Bragov, A.M.; Lomunov, A.K.; Igumnov, L.A.; Belov, A.A.; Eremeyev, V.A. The Influence of Specimen Geometry and Loading Conditions on the Mechanical Properties of Porous Brittle Media. *Materials* **2021**, *14*, 7144. [CrossRef] [PubMed]
7. Ludvigsen, A.C.; Lan, Z.; Castelli, I.E. Autonomous Design of Photoferroic Ruddlesden-Popper Perovskites for Water Splitting Devices. *Materials* **2022**, *15*, 309. [CrossRef] [PubMed]
8. Chomka, G.; Chodór, J.; Kukiełka, L.; Kasperowicz, M. The Use of a High-Pressure Water-Ice Jet for Removing Worn Paint Coating in Renovation Process. *Materials* **2022**, *15*, 1168. [CrossRef] [PubMed]
9. Vahed, R.; Zareie Rajani, H.R.; Milani, A.S. Can a Black-Box AI Replace Costly DMA Testing?-A Case Study on Prediction and Optimization of Dynamic Mechanical Properties of 3D Printed Acrylonitrile Butadiene Styrene. *Materials* **2022**, *15*, 2855. [CrossRef] [PubMed]
10. Jiao-Wang, L.; Loya, J.A.; Santiuste, C. On the Numerical Modeling of Flax/PLA Bumper Beams. *Materials* **2022**, *15*, 5480. [CrossRef] [PubMed]

11. Civalek, Ö.; Uzun, B.; Yaylı, M.Ö. Nonlocal Free Vibration of Embedded Short-Fiber-Reinforced Nano-/Micro-Rods with Deformable Boundary Conditions. *Materials* **2022**, *15*, 6803. [CrossRef] [PubMed]
12. Simonetti, S.K.; Turkalj, G.; Banić, D.; Lanc, D. Bimetallic Thin-Walled Box Beam Thermal Buckling Response. *Materials* **2022**, *15*, 7537. [CrossRef] [PubMed]
13. Esencan Türkaslan, B.; Çelik, A.K.; Dalbeyler, A.; Fantuzzi, N. The Effect of Different Morphologies of WO3/GO Nanocomposite on Photocatalytic Performance. *Materials* **2022**, *15*, 8019. [CrossRef] [PubMed]
14. Gong, Y.; Jiang, L.; Li, L.; Zhao, J. An Experimental and Numerical Study of the Influence of Temperature on Mode II Fracture of a T800/Epoxy Unidirectional Laminate. *Materials* **2022**, *15*, 8108. [CrossRef] [PubMed]
15. Dastjerdi, S.; Akgöz, B.; Civalek, Ö. Viscoelasticity in Large Deformation Analysis of Hyperelastic Structures. *Materials* **2022**, *15*, 8425. [CrossRef] [PubMed]
16. Yan, C.A.; Vescovini, R. Application of the ps-Version of the Finite Element Method to the Analysis of Laminated Shells. *Materials* **2023**, *16*, 1395. [CrossRef] [PubMed]

materials

MDPI

Article

Influence of the Die Height on the Density of the Briquette Produced from Shredded Logging Residues

Tomasz Nurek, **Arkadiusz Gendek** * and **Magdalena Dąbrowska**

Department of Biosystems Engineering, Institute of Mechanical Engineering, Warsaw University of Life Sciences–SGGW, Nowoursynowska 164, 02-787 Warsaw, Poland; tomasz_nurek@sggw.edu.pl (T.N.); magdalena_dabrowska@sggw.edu.pl (M.D.)

* Correspondence: arkadiusz_gendekk@sggw.edu.pl

Abstract: An alternative to plant biomass of various origins are forest logging residues. They differ significantly from other, previously used plant materials. This difference is due to the heterogeneous composition and relatively large size of individual particles. This research on the compaction of this type of shredded material was aimed at determining the influence of the die height on the density and relaxation of briquettes. This parameter is crucial for the proper construction of compaction devices. The measurements were carried out for the same fractional composition of the shredded logging residues, with variable input parameters of the material and process. It was found that the briquette density and relaxation are influenced by the die height, as well as the material moisture content and process temperature. The highest density at maximum compaction pressure (1.40 $g \cdot cm^{-3}$) was obtained at a moisture content of 16%, temperature of 80 °C, and the lowest die height (195 mm). In the case of the briquette density after ejection from the die, the best results were obtained at the same temperature and die height but at a moisture content of 9%. The tests confirmed that, regardless of the process temperature and material moisture, the briquette density increases as the die height is reduced. The relaxation coefficient of compacted logging residues ranges from 21.7% to 50.1% and depends mainly on the material moisture content and the temperature of the process. The lowest value of the relaxation coefficient (21.7 ± 1.61) was obtained at 9% moisture content, 60 °C temperature, and 220 mm die height.

Keywords: die height; biomass compaction; relaxation coefficient; briquette density; logging residues; mechanical engineering

Citation: Nurek, T.; Gendek, A.; Dąbrowska, M. Influence of the Die Height on the Density of the Briquette Produced from Shredded Logging Residues. *Materials* **2021**, *14*, 3698. https://doi.org/10.3390/ma14133698

Academic Editors: Abbas S. Milani and Michele Bacciocchi

Received: 26 May 2021
Accepted: 28 June 2021
Published: 1 July 2021

Publisher's Note: MDPI stays neutral with regard to jurisdictional claims in published maps and institutional affiliations.

1. Introduction

Logging residues are characterized by a heterogeneous composition. The content of wood, bark, and pine needles makes their chemical composition different from other traditional raw materials. Therefore, it becomes necessary to investigate and, consequently, develop new parameters for the compaction process of this type of biomass, as well as to develop design guidelines for compaction devices. Particular attention must be paid to the height of the compacting die, which affects the height of the biomass compacted using a piston in a closed die, or in the case of briquetting machines, in an open die.

For several years, energy policies pursued in the world have resulted in an increasing demand for plant biomass, which is intended for direct combustion or for processing into another type of fuel, including refined fuels. There are numerous scientific studies related to this topic that seek optimal process and material parameters, e.g., a reduction of processing costs, an increase in durability, and the energy obtained per unit volume of solid fuels in the form of briquettes or pellets.

Based on scientific reports, it can be concluded that tests of physical properties [1–6] (bulk density, moisture content, particle size), chemical properties [7–18] (carbon, hydrogen, nitrogen, ulphur oxygen and ash contents), and energy properties (gross and net calorific values) of plant biomass and other materials have been studied by multiple scientists.

The increasing demand for solid fuels in the form of briquettes and pellets is conducive to market development. In the case of the Polish market, there is a relatively large number of enterprises with a production of 1–5 thousand tons of briquettes or pellets per year, intended mainly for private use or for individual users on the local market [4]. Such enterprises have a low chance of purchasing sawdust for briquette production due to competition from large factories. Therefore, an opportunity for them to use other sources of raw material supply exists in shredded logging residues.

The by-products of the logging process are residues in the form of branches, treetops, and small trees and shrubs after the removal of undergrowth and saplings. This material, in its primary form or after shredding, can be used as a renewable energy source for direct combustion, or after processing, it can be used for the production of refined fuels in the form of briquettes [19]. Conversely, in the literature, there are multiple descriptions of research on the production of briquettes from plant materials of uniform composition and particle size [15,20,21]. The authors of the literature considered the parameters of the material and the process influencing the quality of the final product, such as moisture, particle size, temperature, pressure, and others.

Taulbee [20] indicates that the increase in process temperature causes an increase in the mechanical strength of briquettes. Higher temperatures cause plasticization of particles and activation of natural binders in the plant material [22]. Increased temperature during agglomeration is a desirable phenomenon. This causes hardening of the briquette, increasing its density, which is also confirmed by Kaliyan and Morey [23,24]. These researchers produced durable and dense briquettes from plant materials at a temperature of 65–100 °C. Similar conclusions were obtained by Wang et al. [25]; however, they conducted agglomeration at a temperature of 120 °C, while obtaining shorter compaction time.

An important factor that determines the mechanical properties of the briquette and its quality is particle size. Considering the use of shredded logging residues to briquette production, particle size depends on the type and construction of the shredder, the type of wood, the part of the tree that will be shredded, and the angle of setting and sharpening of the knife [7,26,27]. Barontini et al. [28] found that by shredding the wood with a blunt knife, smaller particles are obtained and the proportion of finer particles is increased. The fine particles during briquette production cause the briquettes to obtain higher density and durability [29,30].

During the agglomeration of plant materials, an important factor influencing the quality of the obtained briquette is moisture content. According to the conclusions presented by Kaliyan et al. [23,31], Wang et al. [25], and Gurdil and Melki [32], material moisture content should be within a range of 15–20%. An increase in moisture causes deterioration of the briquette quality [33,34] and adversely affects the compaction process [35]. Conversely, and based on their obtained results, Mani et al. [36] and Shaw et al. [37] stated that in order to obtain greater durability and density, agglomeration should be carried out at 5–10% moisture. However, Gendek et al. [38] produced briquettes from shredded cones with a moisture content of approx. 8%, achieving mechanical durability of briquettes at a level of 0–27%, and by increasing the moisture to approx. 15–20%, the briquette durability was above 87%.

The density of the final product depends on its relaxation. Wongsiriamuay and Tippayawong [39] found that a decrease in temperature causes an increase in relaxation of the pellets. For a temperature of 30 °C, relaxation was 16–30%, while at a temperature of 60–80 °C, relaxation did not exceed 5%. The influence of temperature on the size of briquette relaxation was also confirmed by Shaw and Tabil [37]. Kaliyan and Morey [29] obtained an average relaxation of briquettes composed of switchgrass within a range of 15–32%, which depended on particle size and material moisture. High moisture, large particles, and low agglomeration temperature results in a greater relaxation of the briquette.

The studies mentioned above were carried out mainly for the compaction of materials with a homogeneous composition, e.g., sawdust, crushed various parts of maize or energy crops, etc. Nevertheless, there is not enough research describing the possibilities of producing a briquette from shredded logging residues and the properties of the obtained briquette [19,40,41]. Authors indicated the optimal fractional composition of shredded logging residues, material moisture, and process temperature to obtain a briquette of appropriate density and durability. Depending on the assumed process parameters, they achieved a density ranging from approx. 800 kg·m^{-3} to approx. 1200 kg·m^{-3} and a durability coefficient of up to approx. 50–72%.

Taking into consideration the above information, the next stage of the authors' research is to consider the changes in the height of the biomass bed by changing the die height to indicate how this height affects briquetting. Therefore, the aim of the research is to determine the effect of die height (biomass bed height) on the density of the produced briquette and its relaxation at a constant fractional composition, while considering the variable input parameters of the material and the briquetting process.

2. Materials and Methods

2.1. Research Stand

Biomass compaction was carried out in a closed die (Figure 1). A Veb Thüringer Industriewerk Rauenstein (TIRA GmbH, Schalkau, Germany) testing machine with a maximum pressure force of 100 kN was used for the tests. The stand was equipped with a measuring system, enabling the recording of results (HBM Catman v.2.1 program—Hottinger Baldwin Messtechnik GmbH, Darmstad, Germany). The temporary compaction force was recorded with a strain gauge placed on the piston with an accuracy of ±1 N, and the displacement of the piston was recorded using a displacement sensor with an accuracy of ±0.01 mm.

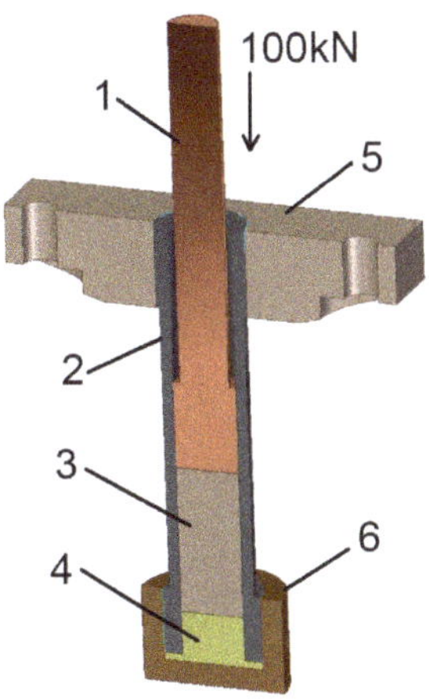

Figure 1. Construction and cross-section of the stand for biomass compaction. 1: piston; 2: bushing of compaction die; 3: compacted biomass; 4: replaceable spacer for adjustment of die height; 5: mounting for die; 6: die cover.

The parameters of the die enabled a maximum compaction pressure equal to p = 65 MPa. The total volume of the die was V_k = 448.3 cm^3. Its diameter was d = 45 mm and the maximum die height was h = 295 mm. The following values of the die height, identical to the bed height of the compacted biomass, were assumed: h_z = 295, 270, 245, 220 and 195 mm. The effect of changing the bed height was achieved by filling the die with cylindrical spacers with heights of 25, 50, 75 and 100 mm composed of S235 steel. The speed of compaction was constant at 3.2 mm·s^{-1}, and the compaction time, depending on the bed height, was within the range of 50–80 sec, while the total time of biomass remaining in the die was 3–4 min.

Two micanite band heaters with a total power of 600 W (ϕ60x25/230V300W, Selfa GE S.A., Bydgoszcz, Poland) were mounted on the outer surface of the die. The temperatures of heating the die set at T = 60, 80, 100, 120 °C were maintained with a range of ±1 °C by the EMKO ESM-3710 controller (EMKO Elektronik A.S., Bursa, Turkey).

2.2. Material

The investigated material was shredded logging residues. This biomass was collected directly from the forest area located in the Rajgród Forest District, Poland (GPS: N 53.4909, E 22.5825) in the form of branches left after harvesting the wood of *Pinus sylvestris* at the age of 86 years. It contained white wood, a considerable amount of pine needles, and small non-woody shoots. The collected branches and needles were shredded in a BT13HP-90 mm hammer shredder (Grupa REDMET sp.j., Dębica, Poland).

The particle size distribution of the shredded biomass was tested using the sieving method. The chips were divided into fractions using a sieve separator in accordance with ISO 17827-1: 2016 standard [42]. Screens produced in accordance with ISO 3310-2: 2013 [43] and ISO 565: 2000 standards [44] were used for separation. Detailed research methodology and construction of the stand were described by Lisowski et al. [45,46]. Based on the cited ISO 17827 standard, the value d_{50} = 5.86 mm was determined, i.e., the average size of the sieve through which 50% of the particles were passed.

2.3. Compaction Process

Compaction was carried out for two biomass moisture contents MC = 9% and 16%. Moisture content was controlled by the drying-weighing method according to the EN13183-1:2004 standard. Samples of 100 ± 0.5 g were placed in the laboratory drier UF55 plus (Memmert, Schwabach, Germany). The drying process at a temperature of 105 °C was carried out until the dry substance was obtained. The mass of samples and loss in moisture content was controlled with an accuracy of ±0.01 g using laboratory scales RADWAG WTC 600 (Radwag, Radom, Poland).

Moisture (MC, %) for shredded logging residues was determined using the formula:

$$MC = \frac{m_{bw} - m_{bs}}{m_{bw}} \cdot 100, \quad (1)$$

where: m_{bw} is a mass of wet sample (g), m_{bs} is a mass of dry sample (g).

For two moistures, the bulk densities of tested chips were ρ_9 = 0.22 g·cm^{-3} and ρ_{16} = 0.24 g·cm^{-3}.

For individual measurement series (different die heights), to maintain the same initial conditions of compaction, i.e., the same density for different volumes, samples were prepared with the masses calculated from the following formula:

$$m_b = \rho_w \cdot V_k, \quad (2)$$

where: m_b is a mass of tested sample (g), ρ_w is a bulk density of wood chips for a specific moisture (g·cm^{-3}), and V_k is the total volume of compacting die (cm^3).

After filling the compacting die at a predetermined temperature with a suitable mass of chips, the testing machine was started. The temporary value of the compaction force and the piston displacement were recorded automatically every 0.1 s until the maximum pressure P = 65 MPa was obtained. After stopping, the piston was retracted from the compacting die, and after opening the die, the briquette was pushed out of the bushing.

2.4. Determination of the Volume and Density of the Briquette

During the measurements, the briquette height was measured twice. The height at maximum compaction pressure (P = 65 MPa) h_0 was measured when the briquette was inside the compaction die. This height was determined with an accuracy of 0.1 mm based on the reading from the apparatus used to measure piston displacement. The second measurement of the briquette height h_1 was rendered using an electronic caliper with an accuracy of 0.5 mm immediately after the briquette was ejected from the die. Measurements were constructed in two perpendicular planes and the result was averaged.

Briquette volume (V_i) and density (ρ_i) were determined based on the measurement of briquette height in the die (h_0) and its height after rejection from the die (h_1), and the mass of a single briquette (m_a) was determined every time by weighing the briquette using the formulas:

$$V_i = \frac{\pi d^2 h_i}{4}, \quad (3)$$

$$\rho_i = \frac{m_a}{V_i}, \quad (4)$$

where: 'i' is the index (0 for briquette in the die, 1 for briquette ejected from the die), V_i is the briquette volume (cm^3), d is the die diameter (cm), and h_i is the briquette height (cm).

Considering briquette density in the die (ρ_0) and after ejection from the die (ρ_1) the relaxation coefficient was determined, λ:

$$\lambda = \frac{\rho_1 - \rho_0}{\rho_1} \cdot 100, \quad (5)$$

For each of the combinations of input parameter values, a minimum of 10 repetitions was performed.

2.5. Statistical Analysis

The parameters were analyzed using the Statistica v.13 program (TIBCO Software Inc., Palo Alto, Santa Clara, CA, USA). Analyses of variances (ANOVA) were performed at a significance level of $p = 0.05$.

3. Results and Discussion

3.1. Briquette Density in the Die ρ_0

Briquettes obtained from a biomass with a higher moisture content (MC = 16%) were characterized by a slightly higher density than those obtained at a lower moisture content (Table 1). For the temperature T = 80 °C and the die height h_z = 195 mm, density was ρ_0 = 1.40 g·cm^{-3}, while for the biomass with lower moisture MC = 9% it was on average ρ_0 = 1.22 g·cm^{-3}. The graph for the moisture content of MC = 16% (Figure 2a) is irregular, although it shares similarity with the graph for the moisture content MC = 9% (Figure 2b). In both cases, the density increased with increasing temperature, although for MC = 9%, it is more visible and consistent with previous studies by other authors [23–25,47]. For moisture MC = 16%, changes in the density values were not high. The maximum density of 1.40 g·cm^{-3} was obtained for the temperature T = 80 °C and the die height h_z = 195 mm, and the lowest density ρ_0 = 1.31 g·cm^{-3} was obtained for the temperature T = 60 °C and die height h_z = 220 mm. The inclination of the graph plane showed a tendency of increasing briquette density composed of biomass with a moisture content MC = 16% with an increasing temperature. The effect of changes in the die height on the density was less visible than the effect of temperature.

Table 1. The results of changes in briquette densities.

T, °C	hz, mm	ρ_0, g·cm^{-3}					ρ_1, g·cm^{-3}					λ, %				
		Mean	SD	−95%	+95%	SE	Mean	SD	−95%	+95%	SE	Mean	SD	−95%	+95%	SE
MC = 16%																
60	195	1.36	0.07	1.29	1.43	0.03	0.77	0.02	0.75	0.79	0.01	42.8	3.63	39.03	46.64	1.48
	220	1.31	0.04	1.27	1.36	0.02	0.75	0.04	0.72	0.79	0.01	42.4	4.06	38.18	46.69	1.66
	245	1.33	0.04	1.29	1.37	0.02	0.75	0.02	0.72	0.77	0.01	43.9	2.74	41.04	46.79	1.12
	270	1.32	0.02	1.30	1.34	0.01	0.74	0.02	0.72	0.76	0.01	43.9	2.17	41.62	46.18	0.89
	295	1.31	0.03	1.28	1.35	0.01	0.74	0.02	0.73	0.76	0.01	43.2	2.73	40.32	46.04	1.11
80	195	1.40	0.04	1.36	1.45	0.02	0.76	0.01	0.75	0.78	0.00	45.6	1.96	43.58	47.69	0.80
	220	1.34	0.03	1.31	1.37	0.01	0.75	0.01	0.74	0.75	0.00	44.3	1.27	43.00	45.66	0.52
	245	1.34	0.02	1.32	1.36	0.01	0.72	0.03	0.70	0.75	0.01	45.8	2.33	43.31	48.19	0.95
	270	1.34	0.02	1.32	1.35	0.01	0.73	0.03	0.70	0.76	0.01	45.5	2.56	42.76	48.14	1.05
	295	1.33	0.03	1.29	1.36	0.01	0.74	0.02	0.72	0.75	0.01	44.5	2.28	42.06	46.84	0.93
100	195	1.39	0.05	1.34	1.44	0.02	0.74	0.02	0.72	0.76	0.01	46.5	1.02	45.46	47.60	0.42
	220	1.33	0.02	1.31	1.35	0.01	0.73	0.02	0.71	0.75	0.01	45.0	1.18	43.72	46.21	0.48
	245	1.34	0.07	1.34	1.42	0.03	0.72	0.02	0.70	0.73	0.01	46.7	3.33	43.17	50.16	1.36
	270	1.34	0.04	1.32	1.38	0.02	0.72	0.02	0.70	0.74	0.01	46.3	2.84	43.36	49.31	1.16
	295	1.34	0.03	1.34	1.37	0.01	0.71	0.03	0.69	0.74	0.01	46.9	1.94	44.87	48.93	0.79
120	195	1.39	0.04	1.34	1.43	0.02	0.71	0.03	0.68	0.74	0.01	48.9	1.30	47.50	50.23	0.53
	220	1.35	0.04	1.31	1.39	0.02	0.69	0.01	0.68	0.70	0.00	48.8	2.02	46.69	50.94	0.83
	245	1.38	0.04	1.34	1.42	0.02	0.69	0.02	0.67	0.71	0.01	50.1	1.49	48.55	51.68	0.61
	270	1.35	0.03	1.32	1.38	0.01	0.69	0.03	0.67	0.72	0.01	48.6	2.9	45.52	51.61	1.18
	295	1.35	0.02	1.34	1.37	0.01	0.69	0.02	0.67	0.71	0.01	49.2	1.80	47.32	51.08	0.73
MC = 9%																
60	195	1.11	0.01	1.10	1.12	0.00	0.86	0.01	0.85	0.87	0.00	22.5	1.23	21.70	23.27	0.36
	220	1.11	0.01	1.10	1.12	0.00	0.87	0.02	0.86	0.88	0.01	21.7	1.61	20.62	22.68	0.47
	245	1.13	0.02	1.12	1.14	0.01	0.86	0.01	0.85	0.86	0.00	24.2	1.50	23.26	25.17	0.43
	270	1.13	0.02	1.12	1.14	1.01	0.85	0.01	0.84	0.86	0.00	24.8	1.25	23.97	25.55	0.36
	295	1.10	0.02	1.09	1.11	0.01	0.83	0.01	0.83	0.84	0.00	24.5	1.41	23.65	25.44	0.41
80	195	1.22	0.02	1.20	1.23	0.01	0.86	0.02	0.84	0.87	0.01	29.6	2.05	28.25	30.86	0.59
	220	1.19	0.03	1.17	1.21	0.01	0.86	0.02	0.85	0.87	0.00	28.0	2.38	26.50	29.53	0.69
	245	1.15	0.03	1.13	1.17	0.01	0.85	0.02	0.84	0.96	0.00	26.4	2.21	24.94	27.76	0.64
	270	1.12	0.01	1.12	1.13	0.00	0.85	0.01	0.84	0.86	0.00	24.3	1.61	23.30	25.34	0.46
	295	1.10	0.01	1.10	1.11	0.00	0.83	0.01	0.83	0.84	0.00	24.4	1.12	23.71	25.14	0.32
100	195	1.27	0.05	1.24	1.30	0.01	0.85	0.01	0.84	0.86	0.00	33.1	2.53	31.49	34.71	0.73
	220	1.23	0.05	1.21	1.26	0.01	0.84	0.02	0.83	0.85	0.00	31.9	3.45	29.70	34.08	1.00
	245	1.21	0.02	1.20	1.22	0.01	0.85	0.02	0.84	0.86	0.00	29.9	1.53	28.88	30.82	0.44
	270	1.18	0.02	1.16	1.19	0.01	0.84	0.01	0.83	0.85	0.00	28.5	1.56	27.55	29.53	0.45
	295	1.13	0.02	1.12	1.14	0.01	0.83	0.01	0.83	0.84	0.00	26.3	1.44	25.35	27.18	0.42
120	195	1.31	0.02	1.30	1.33	0.01	0.85	0.02	0.84	0.86	0.00	35.4	1.90	34.14	36.56	0.55
	220	1.28	0.02	1.27	1.30	0.01	0.85	0.02	0.84	0.86	0.00	33.4	1.61	32.41	34.46	0.47
	245	1.25	0.01	1.24	1.26	0.00	0.84	0.02	0.83	0.86	0.01	32.4	1.81	31.28	33.81	0.52
	270	1.21	0.02	1.20	1.22	0.00	0.85	0.01	0.84	0.86	0.00	29.8	1.52	28.87	30.81	0.44
	295	1.16	0.01	1.15	1.17	0.00	0.83	0.01	0.82	0.84	0.00	28.0	1.61	27.01	29.06	0.47

Note: SD: standard deviation; +95% and −95%: confidence interval; SE: standard error.

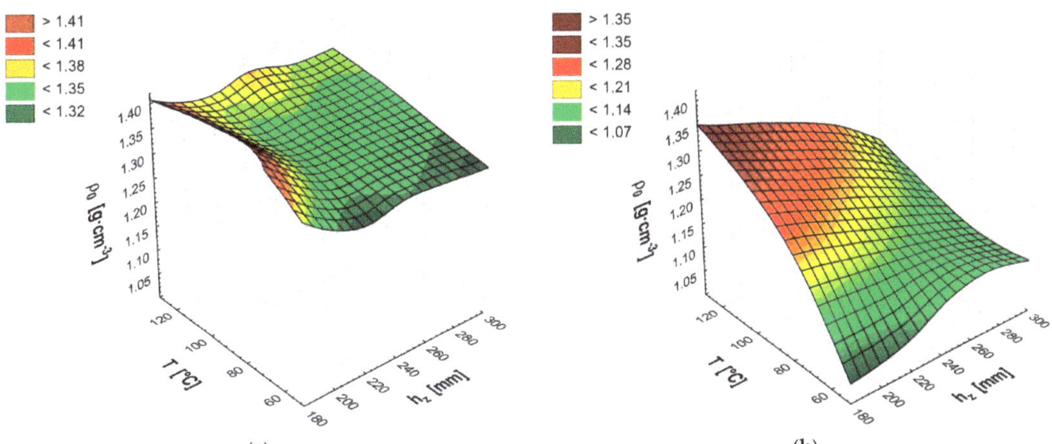

Figure 2. Briquette density inside the die at maximum pressure ρ_0 for moisture content 16% (**a**) and moisture content 9% (**b**).

Compacting the chips with a moisture content of 9% resulted in more regular changes in the briquette density (Figure 2b, Table 1). The maximum value of density ρ_0 = 1.31 g·cm^{-3} was obtained for temperature T = 120 °C and die height h_z = 195 mm. The density decreased with reduction in the temperature of the compaction process to T = 80 °C and an increase to the die height. The compaction of biomass at the lowest of the assumed temperatures (T = 60 °C) resulted in a significant reduction in density by about 15%, but negligible for a maximum die height h_z = 295 mm of about 5%. This was related to the effect of "amortization" pressure in the upper layer of the biomass. The temperature T = 60 °C did not lead to full plasticization of resins and other substances contained in the biomass. The matter retained elastic properties, which resulted in the aforementioned effect of amortization and reduction of the degree of compaction.

The discussed briquette densities (1.2–1.4 g·cm^{-3}) inside the die at a maximum pressure of approx. 63 MPa were similar to the values obtained by Gendek et al. [38] for briquettes from shredded spruce cones (1.1 g·cm^{-3}). Similar densities (>1.0 g·cm^{-3}) were obtained, among others, by Borowski [48] for briquettes composed from a mixture of coal with biomass, and Gürdil and Demirel [21] for briquettes composed from walnut shells. Such densities are characteristic for pellets composed from various plant materials [49–51], but in all cases the agglomerated particles were approx. 1 mm in size, and measurements were composed for briquettes and pellets outside the die.

Statistical analysis (ANOVA) showed the presence of the influence of all input parameters on the briquette density inside the die for the moisture content MC = 9%. In the case of moisture content MC = 16%, only for the combination of input parameters (bed height and temperature), was it indicated that they did not affect the tested parameter ($p > 0.05$), Table 2.

3.2. Briquette Density after Ejection from the Die ρ_1

The briquette density after ejection from the die changed to a small extent. In the case of biomass moisture content MC = 16%, the lowest density ρ_1 = 0.69 g·cm^{-3} was obtained for the temperature T = 120 °C in four out of five die heights. Only for the lowest die height h_z = 195 mm was the density higher, only by 3%. The highest density ρ_1 = 0.77 g·cm^{-3} was obtained for the temperature T = 60 °C and the small die height h_z = 195 mm. Even less significant changes were observed for compaction of biomass with a moisture content MC = 9% (Figure 3, Table 1). On average, it was about 12% higher than the density for biomass with moisture MC = 16% and ranged from ρ_1 = 0.83 to ρ_1 = 0.87 g·cm^{-3}. As shown in the graph for briquette density from biomass of moisture content MC = 9% (Figure 3b), there is no clear, unequivocal trend of changes. There is a weak tendency to decrease the density for the maximum die height. This confirms the results obtained in the study of the height of compacted biomass. The graph for the moisture content MC = 16% (Figure 3a) shows a lower briquette density obtained for higher temperatures. This is also confirmed by the results of the research on the die height (compacted biomass). Statistical analysis for both values of moisture (Table 3) showed the existence of the influence of temperature and die height on the briquette density ($p < 0.05$). An Analysis of the simultaneous influence of these parameters on the briquette density showed no such influence ($p > 0.05$).

Table 2. Results of ANOVA analysis for density of briquettes in the die ρ_0 at two moisture contents.

	df	SS	MS	F	*p*-Value
MC = 16%					
Die height (A)	4	0.0457	0.0114	7.9	0.000
Temperature (B)	3	0.0215	0.0072	5.0	0.003
Interaction (A × B)	12	0.0085	0.0007	0.5	0.915
MC = 9%					
Die height (A)	4	0.3069	0.0767	150.9	0.000
Temperature (B)	3	0.5617	0.1872	368.2	0.000
Interaction (A × B)	12	0.1276	0.0106	20.9	0.000

Note: df: degrees of freedom; SS: sum of squares, MS: mean square, F: F test value; *p*-value: significance level.

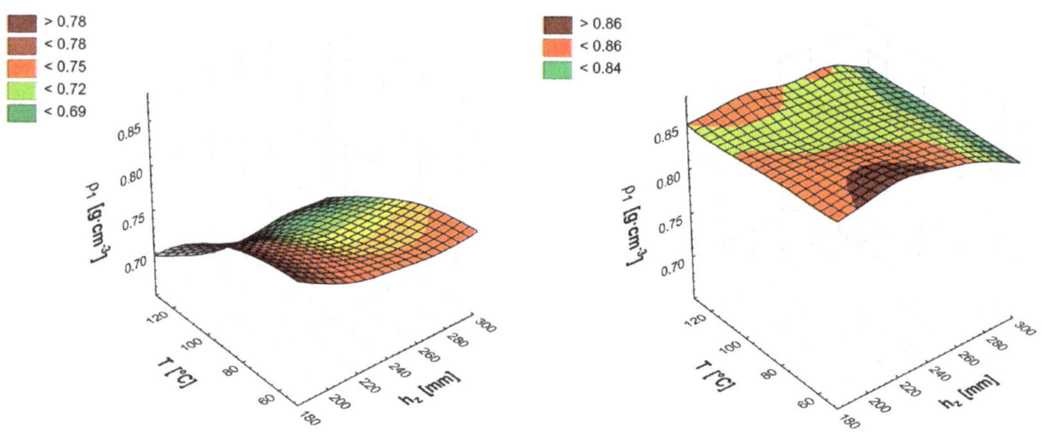

(**a**) (**b**)

Figure 3. Briquette density after ejection from the die ρ_1 for moisture content 16% (**a**) and moisture content 9% (**b**).

Table 3. Results of ANOVA analysis for density of briquette after ejection from the die ρ_1 at two moisture contents.

	df	SS	MS	F	*p*-Value
MC = 16%					
Die height (A)	4	0.0145	0.0036	8.0	0.000
Temperature (B)	3	0.0567	0.0189	41.6	0.000
Interaction (A × B)	12	0.0016	0.0001	0.3	0.991
MC = 9%					
Die height (A)	4	0.0142	0.0036	17.4	0.000
Temperature (B)	3	0.0032	0.0011	5.2	0.002
Interaction (A × B)	12	0.0043	0.004	1.7	0.061

The obtained density (0.7–0.87 g·cm^{-3}) of the briquette immediately removed from the die was lower than the density of the briquette composed of shredded forest cones (0.94–1.1 g·cm^{-3}) [38], and partially within the range or slightly below the density (0.8–1.0 g·cm^{-3}) for briquettes composed of rice straw [25] and wood sawdust [32]. It is satisfactory that this density was higher than that of a briquette composed of tropical wood sawdust (0.62–0.72 g·cm^{-3}) obtained by Mitchual et al. [52]. In every case, the authors compacted the shredded plant materials to particles of approx. 1 mm in size. However, referring to the ISO 17225 standard [53] describing the classification of briquettes, briquettes composed of shredded wood chips do not meet the requirements of this standard.

3.3. The Relaxation Coefficient λ

By comparing of values of the relaxation coefficient presented in Figure 4 (Table 1), it can be concluded that in each of the considered cases, (combination of temperature and die height) a much more favorable situation (less relaxation) occurred for the moisture content $MC = 9\%$. For moisture content $MC = 16\%$ (Figure 4a), the relaxation coefficient ranged from $\lambda = 42.4\%$ for the temperature T = 60 °C and the die height $h_z = 220$ mm to $\lambda = 50.1\%$ for the temperature T = 120 °C and die height $h_z = 245$ mm. Moreover, for moisture content $MC = 9\%$, the minimum relaxation $\lambda = 21.7\%$ was obtained by the briquette produced at T = 60 °C and $h_z = 220$ mm, and the maximum relaxation $\lambda = 35.4\%$ was obtained by the briquette produced at T = 120 °C and $h_z = 195$ mm. For moisture content $MC = 16\%$, a significant increase in the relaxation coefficient with increasing temperature was observed, which is the opposite of the phenomenon found by Kaliyan and Morey [29] during switchgrass briquetting and by Wongsiriamuay and Tippayawong [39] testing pellets composed of various parts of maize. However, these authors created briquettes and pellets from materials characterized by homogeneity and much smaller particle sizes. The increase in the value of the relaxation coefficient along with the increase in temperature had a negative effect, which may be linked with a strong, excessive liquefaction of the resins contained in the biomass and the resulting loss of the possibility of binding (sticking) the biomass immediately after leaving the compacting die [54]. Moreover, during compaction, an intense evaporation of water and other substances contained in the biomass occurred. After compaction and ejection, the briquette rapidly expanded by gases produced in the process of vaporization and its linear dimensions increased. Less significant changes in the relaxation coefficient occurred with a change in the die height. However, less relaxation was observed at a lower height for all tested temperatures. This change was around one percentage point. For moisture content $MC = 16\%$, the statistical analysis showed that there was no influence of the die height and the interaction of temperature and the die height on the value of the relaxation coefficient. In both cases, the p-value was less than 0.05.

As noted for moisture content $MC = 9\%$ (Figure 4b), the relaxation coefficient was lower for all the values of the input parameters. In this case, the influence of the die height on the value of the relaxation coefficient was visible. In the case of high temperatures, reducing the die height had a negative effect (increasing the relaxation coefficient), while for lower temperatures, the effect was positive. The lowest value of the relaxation coefficient $\lambda = 21.7\%$ was obtained for the temperature T = 60 °C and the die height $h_z = 220$ mm. An increase in temperature caused an increase in the value of the relaxation coefficient for each die height. This increase was clearer for the lower die height, from the value of $\lambda = 22.5\%$ for $h_z = 195$ mm to $\lambda = 35.4\%$ for $h_z = 195$ mm. For the die height $h_z = 295$ mm, the relaxation coefficient varied from 24.5% to 28%.

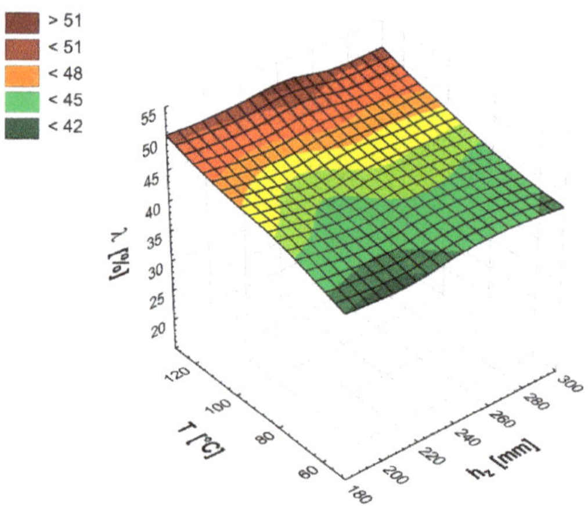

(a)

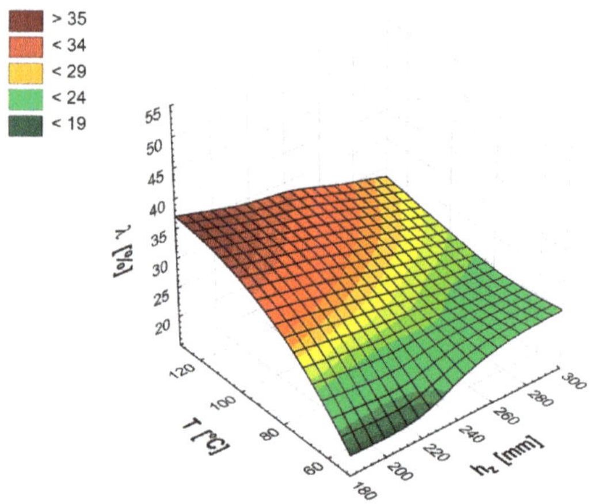

(b)

Figure 4. The changes of relaxation coefficient values [%] for moisture content 16% (**a**) and moisture content 9% (**b**).

The statistical analysis for moisture content *MC* = 9% confirmed the influence of all input parameters on the value of the relaxation coefficient (Table 4).

Table 4. Results of ANOVA analysis for the relaxation coefficient for two moistures.

	df	SS	MS	F	*p*-Value
MC = 16%					
Die height (A)	4	26.7	6.7	1.14	0.341
Temperature (B)	3	542.1	180.7	30.93	0.000
Interaction (A × B)	12	17.3	1.4	0.25	0.995
MC = 9%					
Die height (A)	4	5355	1339	39.32	0.000
Temperature (B)	3	24,249	808.3	237.41	0.000
Interaction (A × B)	12	5603	46.7	13.71	0.000

It can be assumed that for lower compaction temperatures up to T = 80 °C, the more advantageous design of compacting devices is at the lower die height. For high compaction temperatures less frequently used, the die height does not significantly affect the value of the relaxation coefficient.

4. Summary and Conclusions

Based on these studies, it can be concluded that shredded logging residues can be used to produce briquettes, but it is difficult to obtain briquettes of appropriate size and density over time. The density of the briquette obtained in the tests is lower than that required by the relevant standards ISO 17225-3 [53].

Studies have shown that better briquetting effects (higher density) are achieved at a lower temperature and for a small die height. In the case of moisture content *MC* = 16%, temperature T = 60 °C, and die height h_z = 195 mm, the density was equal to $\rho_1 = 0.77$ g·cm^{-3}. However, for moisture content *MC* = 9%, the highest density $\rho_1 = 0.86$ g·cm^{-3} was obtained for temperature T = 60 °C and die height h_z = 220 mm. For die height h_z = 195, the obtained density was only slightly lower (by 0.01 g·cm^{-3}). In all test series, a decrease in density was observed with increasing die height. The obtained results are a direct indication for briquette manufacturers. It was found that for the type of tested biomass, it is beneficial to compact a portion of raw material. This design solution results in greater compaction and a lower degree of relaxation.

The research showed unequivocally that the compaction of biomass with a moisture content of 9% leads to briquettes of higher density, while increasing the temperature to 100–120 °C does not improve the compaction effects. An important result of the conducted research is the determination of the relaxation coefficient. The smallest change in density was observed for a moisture content of 9%, temperature of 60 °C, and die height of 220 mm. This indicates that increasing the process temperature has no benefit on improving the density or lowering energy consumption of the process. High temperatures may lead to a reduction in the bonding effect of the briquette, for example, causing scorched resin immediately after the briquette has been withdrawn from the die. Increased relaxation at high temperatures may also be the result of rapid evaporation of water and other substances contained in the compacted biomass.

Author Contributions: Conceptualization, T.N. and A.G.; methodology, T.N. and A.G.; formal analysis, T.N., A.G. and M.D.; writing—original draft preparation, T.N. and A.G.; writing—review and editing, T.N., A.G. and M.D.; supervision T.N., A.G. and M.D. All authors have read and agreed to the published version of the manuscript.

Funding: This research did not receive any specific grant from funding agencies in the public, commercial, or not-for-profit sectors.

Institutional Review Board Statement: Not applicable.

Informed Consent Statement: Not applicable.

Data Availability Statement: Not applicable.

Conflicts of Interest: The authors declare no conflict of interest.

References

1. Phanphanich, M.; Mani, S. Drying Characteristics of Pine Forests Residues. *BioResources* **2009**, *5*, 108–121. [CrossRef]
2. Jensen, P.D.; Hartmann, H.; Böhm, T.; Temmerman, M.; Rabier, F.; Morsing, M. Moisture Content Determination in Solid Biofuels by Dielectric and NIR Reflection Methods. *Biomass Bioenergy* **2006**, *30*, 935–943. [CrossRef]
3. Sultana, A.; Kumar, A. Optimal Configuration and Combination of Multiple Lignocellulosic Biomass Feedstocks Delivery to a Biorefinery. *Bioresour. Technol.* **2011**, *102*, 9947–9956. [CrossRef]
4. Stolarski, M.J.; Szczukowski, S.; Tworkowski, J.; Krzyżaniak, M.; Gulczyński, P.; Mleczek, M. Comparison of Quality and Production Cost of Briquettes Made from Agricultural and Forest Origin Biomass. *Renew. Energy* **2013**, *57*, 20–26. [CrossRef]
5. Chaloupková, V.; Ivanova, T.; Krepl, V. Particle Size and Shape Characterization of Feedstock Material for Biofuel Production. *Agron. Res.* **2019**, *17*, 1861–1873. [CrossRef]
6. Wróbel, M.; Jewiarz, M.; Mudryk, K.; Knapczyk, A. Influence of Raw Material Drying Temperature on the Scots Pine (*Pinus Sylvestris* L.) Biomass Agglomeration Process—A Preliminary Study. *Energies* **2020**, *13*, 1809. [CrossRef]
7. Friedl, A.; Padouvas, E.; Rotter, H.; Varmuza, K. Prediction of Heating Values of Biomass Fuel from Elemental Composition. *Anal. Chim. Acta* **2005**, *544*, 191–198. [CrossRef]
8. Munalula, F.; Meincken, M. An Evaluation of South African Fuelwood with Regards to Calorific Value and Environmental Impact. *Biomass Bioenergy* **2009**, *33*, 415–420. [CrossRef]
9. Reva, V.; Fonseca, L.; Lousada, J.L.; Abrantes, I.; Viegas, D.X. Impact of the Pinewood Nematode, Bursaphelenchus Xylophilus, on Gross Calorific Value and Chemical Composition of Pinus Pinaster Woody Biomass. *Eur. J. For. Res.* **2012**, *131*, 1025–1033. [CrossRef]
10. Wielgosiński, G.; Łechtańska, P.; Namiecińska, O. Emission of Some Pollutants from Biomass Combustion in Comparison to Hard Coal Combustion. *J. Energy Inst.* **2017**, *90*, 787–796. [CrossRef]
11. Zhao, D.; Kane, M.; Teskey, R.; Markewitz, D.; Greene, D.; Borders, B. Impact of Management on Nutrients, Carbon, and Energy in Aboveground Biomass Components of Mid-Rotation Loblolly Pine (*Pinus Taeda* L.) Plantations. *Ann. For. Sci.* **2014**, *71*, 843–851. [CrossRef]
12. Chandrasekaran, S.R.; Hopke, P.K.; Rector, L.; Allen, G.; Lin, L. Chemical Composition of Wood Chips and Wood Pellets. *Energy Fuels* **2012**, *26*, 4932–4937. [CrossRef]
13. Sulaiman, S.A.; Roslan, R.; Inayat, M.; Yasin Naz, M. Effect of Blending Ratio and Catalyst Loading on Co-Gasification of Wood Chips and Coconut Waste. *J. Energy Inst.* **2018**, *91*, 779–785. [CrossRef]
14. Akhmedov, S.; Ivanova, T.; Abdulloeva, S.; Muntean, A.; Krepl, V. Contribution to the Energy Situation in Tajikistan by Using Residual Apricot Branches after Pruning as an Alternative Fuel. *Energies* **2019**, *12*, 3169. [CrossRef]
15. Ivanova, T.; Mendoza Hernández, A.H.; Bradna, J.; Fernández Cusimamani, E.; García Montoya, J.C.; Armas Espinel, D.A. Assessment of Guava (*Psidium Guajava* L.) Wood Biomass for Briquettes' Production. *Forests* **2018**, *9*, 613. [CrossRef]

16. Mudryk, K.; Jewiarz, M.; Wróbel, M.; Niemiec, M.; Dyjakon, A. Evaluation of Urban Tree Leaf Biomass-Potential, Physico-Mechanical and Chemical Parameters of Raw Material and Solid Biofuel. *Energies* **2021**, *14*, 818. [CrossRef]
17. Bożym, M.; Siemiątkowski, G. Characterization of Composted Sewage Sludge during the Maturation Process: A Pilot Scale Study. *Environ. Sci. Pollut. Res.* **2018**, *25*, 34332–34342. [CrossRef] [PubMed]
18. Bożym, M.; Siemiątkowski, G. Assessment of Composition Changes, Stability Degree and the Potential of Biogas Formation of Sewage Sludge Composts During Maturation Process. *Waste Biomass Valorization* **2020**, *11*, 4081–4091. [CrossRef]
19. Nurek, T.; Gendek, A.; Roman, K.; Dąbrowska, M. The Impact of Fractional Composition on the Mechanical Properties of Agglomerated Logging Residues. *Sustainability* **2020**, *12*, 6120. [CrossRef]
20. Taulbee, D.; Patil, D.P.; Honaker, R.Q.; Parekh, B.K. Briquetting of Coal Fines and Sawdust Part I: Binder and Briquetting-Parameters Evaluations. *Int. J. Coal Prep. Util.* **2009**, *29*, 1–22. [CrossRef]
21. Gürdίl, G.A.K.; Demίrel, B. Effect of Moisture Content, Particle Size and Pressure on Some Briquetting Properties of Hazelnut Residues. *Anadolu Tarım Bilim. Derg.* **2020**, *35*, 330–338. [CrossRef]
22. Lisowski, A.; Dąbrowska-Salwin, M.; Ostrowska-Ligęza, E.; Nawrocka, A.; Stasiak, M.; Świętochowski, A.; Klonowski, J.; Sypuła, M.; Lisowska, B. Effects of the Biomass Moisture Content and Pelleting Temperature on the Pressure-Induced Agglomeration Process. *Biomass Bioenergy* **2017**, *107*, 376–383. [CrossRef]
23. Kaliyan, N.; Morey, R.V. Densification Characteristics of Corn Cobs. *Fuel Process. Technol.* **2010**, *91*, 559–565. [CrossRef]
24. Kaliyan, N.; Morey, R.V. Factors Affecting Strength and Durability of Densified Biomass Products. *Biomass Bioenergy* **2009**, *33*, 337–359. [CrossRef]
25. Wang, Y.; Wu, K.; Sun, Y. Effects of Raw Material Particle Size on the Briquetting Process of Rice Straw. *J. Energy Inst.* **2018**, *91*, 153–162. [CrossRef]
26. Spinelli, R.; Hartsough, B.R.; Magagnotti, N. Testing Mobile Chippers for Chip Size Distribution. *Int. J. For. Eng.* **2005**, *16*, 29–36. [CrossRef]
27. Nati, C.; Spinelli, R.; Fabbri, P. Wood Chips Size Distribution in Relation to Blade Wear and Screen Use. *Biomass Bioenergy* **2010**, *34*, 583–587. [CrossRef]
28. Barontini, M.; Scarfone, A.; Spinelli, R.; Gallucci, F.; Santangelo, E.; Acampora, A.; Jirjis, R.; Civitarese, V.; Pari, L. Storage Dynamics and Fuel Quality of Poplar Chips. *Biomass Bioenergy* **2014**, *62*, 17–25. [CrossRef]
29. Kaliyan, N.; Morey, R.V. Constitutive Model for Densification of Corn Stover and Switchgrass. *Biosyst. Eng.* **2009**, *104*, 47–63. [CrossRef]
30. Jewiarz, M.; Wróbel, M.; Mudryk, K.; Szufa, S. Impact of the Drying Temperature and Grinding Technique on Biomass Grindability. *Energies* **2020**, *13*, 3392. [CrossRef]
31. Kaliyan, N.; Morey, R.V. Strategies to Improve Durability of Switchgrass Briquettes. *Trans. ASABE* **2009**, *52*, 1943–1953. [CrossRef]
32. Gürdίl, G.A.K.; Melki, S. Determining Briquetting Parameters for Peach Tree Pruning Residues For Biofuel. *Fresenius Environ. Bull.* **2018**, *27*, 9083–9090.
33. Ganesan, V.; Muthukumarappan, K.; Rosentrater, K.A. Flow Properties of DDGS with Varying Soluble and Moisture Contents Using Jenike Shear Testing. *Powder Technol.* **2008**, *187*, 130–137. [CrossRef]
34. Zou, Y.; Brusewitz, G.H. Flowability of Uncompacted Marigold Powder as Affected by Moisture Content. *J. Food Eng.* **2002**, *55*, 165–171. [CrossRef]
35. Landillon, V.; Cassan, D.; Morel, M.-H.; Cuq, B. Flowability, Cohesive, and Granulation Properties of Wheat Powders. *J. Food Eng.* **2008**, *86*, 178–193. [CrossRef]
36. Mani, S.; Tabil, L.G.; Sokhansanj, S. Effects of Compressive Force, Particle Size and Moisture Content on Mechanical Properties of Biomass Pellets from Grasses. *Biomass Bioenergy* **2006**, *30*, 648–654. [CrossRef]
37. Shaw, M.D.; Tabil, L.G. *Compression and Relaxation Characteristics of Selected Biomass' Grinds*; ASABE Annual International Meeting, Technical Papers, Paper Number 076183; American Society of Agricultural and Biological Engineers: Minneapolis, MN, USA, 2007.
38. Gendek, A.; Aniszewska, M.; Malaťák, J.; Velebil, J. Evaluation of Selected Physical and Mechanical Properties of Briquettes Produced from Cones of Three Coniferous Tree Species. *Biomass Bioenergy* **2018**, *117*, 173–179. [CrossRef]
39. Wongsiriamnuay, T.; Tippayawong, N. Effect of Densification Parameters on the Properties of Maize Residue Pellets. *Biosyst. Eng.* **2015**, *139*, 111–120. [CrossRef]
40. Nurek, T.; Gendek, A.; Roman, K. Forest Residues as a Renewable Source of Energy: Elemental Composition and Physical Properties. *BioResources* **2019**, *14*, 6–20. [CrossRef]
41. Nurek, T.; Gendek, A.; Roman, K.; Dąbrowska, M. The Effect of Temperature and Moisture on the Chosen Parameters of Briquettes Made of Shredded Logging Residues. *Biomass Bioenergy* **2019**, *130*, 105368. [CrossRef]
42. International Organization for Standardization. *ISO 17827-1:2016—Solid Biofuels—Determination of Particle Size Distribution for Uncompressed Fuels—Part 1: Oscillating Screen Method Using Sieves with Apertures of 3,15 Mm and Above*; International Organization for Standardization: Geneva, Switzerland, 2016.
43. International Organization for Standardization. *ISO 3310-2:2013—Test Sieves—Technical Requirements and Testing—Part 2: Test Sieves of Perforated Metal Plate*; International Organization for Standardization: Geneva, Switzerland, 2013.
44. Polish Committee for Standardization. *PN-ISO 565:2000—Test Sieves—Metal Wire Cloth, Perforated Metal Plate and Electroformed Sheet—Nominal Sizes of Openings*; Polish Committee for Standardization: Warsaw, Poland, 2000.

45. Lisowski, A.; Świątek, K.; Kostyra, K.; Chlebowski, J. Methods for Evaluation of Breaking up of Maize Chaff Separated on the Sieve Separator. *Ann. Wars. Univ. Life Sci. SGGW Agric.* **2008**, *52*, 23–30.
46. Lisowski, A.; Sar, Ł.; Świątek, K.; Kostyra, K. Sieve separator to analysis of chaff length distribution. *Tech. Rol. Ogrod. Leśna* **2008**, *2*, 17–19.
47. Chen, W.-H.; Kuo, P.-C. A Study on Torrefaction of Various Biomass Materials and Its Impact on Lignocellulosic Structure Simulated by a Thermogravimetry. *Energy* **2010**, *35*, 2580–2586. [CrossRef]
48. Borowski, G. *Metody Przetwarzania Odpadów Drobnoziarnistych Na Produkty Użyteczne [Methods of Fine-Grained Waste Processing into Useful Products]*; Monografie—Politechnika Lubelska; Politechnika Lubelska: Lublin, Poland, 2013; ISBN 978-83-63569-43-3.
49. Biswas, A.K.; Yang, W.; Blasiak, W. Steam Pretreatment of Salix to Upgrade Biomass Fuel for Wood Pellet Production. *Fuel Process. Technol.* **2011**, *92*, 1711–1717. [CrossRef]
50. Garcia-Maraver, A.; Zamorano, M.; Fernandes, U.; Rabaçal, M.; Costa, M. Relationship between Fuel Quality and Gaseous and Particulate Matter Emissions in a Domestic Pellet-Fired Boiler. *Fuel* **2014**, *119*, 141–152. [CrossRef]
51. Brunerová, A.; Müller, M.; Gürdil, G.A.K.; Šleger, V.; Brožek, M. Analysis of the Physical-Mechanical Properties of a Pelleted Chicken Litter Organic Fertiliser. *Res. Agric. Eng.* **2020**, *66*, 131–139. [CrossRef]
52. Mitchual, S.J.; Frimpong-Mensah, K.; Darkwa, N.A. Effect of Species, Particle Size and Compacting Pressure on Relaxed Density and Compressive Strength of Fuel Briquettes. *Int. J. Energy Environ. Eng.* **2013**, *4*, 30. [CrossRef]
53. International Organization for Standardization. *ISO 17225-3:2014—Solid Biofuels—Fuel Specifications and Classes—Part 3: Graded Wood Briquettes*; International Organization for Standardization: Geneva, Switzerland, 2014.
54. Styks, J.; Knapczyk, A.; Łapczyńska-Kordon, B. Effect of Compaction Pressure and Moisture Content on Post-Agglomeration Elastic Springback of Pellets. *Materials* **2021**, *14*, 879. [CrossRef] [PubMed]

materials

MDPI

Article

The Kinematics of Scale Deflection in the Course of Multi-Step Seed Extraction from European Larch Cones (*Larix decidua* Mill.) Taking into Account Their Cellular Structure

Ewa Tulska, Monika Aniszewska and Arkadiusz Gendek *

Department of Biosystems Engineering, Institute of Mechanical Engineering, Warsaw University of Life Sciences, Nowoursynowska 164, 02-787 Warsaw, Poland; ewa_tulska@sggw.edu.pl (E.T.); monika_aniszewska@sggw.edu.pl (M.A.)

* Correspondence: arkadiusz_gendek@sggw.edu.pl

Abstract: The objective of the study was to elucidate the kinematics of cone opening in the European larch (*Larix decidua* Mill.) during a four-step seed extraction process and to determine optimum process time on that basis. Each step lasted 8 h with 10 min of water immersion between the steps. The study also described the microscopic cellular structure of scales in cones with a moisture content of 5% and 20%, as well as evaluated changes in cell wall thickness. The obtained results were compared with the structural investigations of scales conducted using scanning electron microscopy (SEM) of characteristic sites on the inner and outer sides of the scales. The greatest increment in the scale opening angle was noted on the first day of the process (34°) and in scales from the middle cone segment (39°). In scales with a moisture content of 5% and 20%, the greatest changes in cell wall thickness were recorded for large cells (57%). The inner and outer structure of scales differed in terms of the presence and size of cells depending on the moisture content of the cones (5%, 10%, or 20%). The study demonstrated that the moisture content of cones was the crucial determinant of the cellular structure and opening of scales in larch cones. The scale opening angle increased with decreasing moisture content but did not differ significantly for various segments of cones or various hours of the consecutive days of the process. This finding may lead to reducing the seed extraction time for larch cones. The internal and external structure of scales differed depending on moisture content, which also determined the size and wall thickness of cells.

Keywords: scale opening mechanics; seed extraction; morphological structure

Citation: Tulska, E.; Aniszewska, M.; Gendek, A. The Kinematics of Scale Deflection in the Course of Multi-Step Seed Extraction from European Larch Cones (*Larix decidua* Mill.) Taking into Account Their Cellular Structure. *Materials* **2021**, *14*, 4913. https://doi.org/10.3390/ma14174913

Academic Editors: Michele Bacciocchi and Abbas S. Milani

Received: 16 July 2021
Accepted: 26 August 2021
Published: 29 August 2021

1. Introduction

Seed extraction from the cones of various forest tree species has been described in the research literature since the 1950s [1,2] as a complex process [3] determined by taxonomic characteristics [4]. Most publications on the subject tend to analyze pine seed extraction due to the fact that the seeds of that species are in greatest demand [5–7]. Conifer seeds constitute a valuable propagation material needed for forest regeneration either via natural processes or for the needs of nurseries [8].

On average, between 200 and 500 kg of cones and 17.5 to 28.0 kg of seeds can be harvested per 1 ha of larch stands in Poland [9,10]. In Polish conditions, approximately 10,000 kg of larch cones were harvested annually between 2010 and 2020. In years of low harvest, there is a significant proportion of empty cones, which may be attributed to pests or diseases [11], or even climate change [12,13]. Furthermore, the total area of larch stands in Poland is to be reduced by 20% [9]. In view of these factors, it seems important to add to the understanding of the process of seed extraction from larch cones to maximize the amount and quality of seeds and enable their long-term storage.

Post-extraction cones constitute waste, which can be briquetted [14] or torrefied [15] and used together with damaged seeds for power generation purposes [16–18].

In practice, seeds are obtained from larch cones in two ways: via thermal extraction (involving alternating drying and moistening of cones) or thermal-mechanical extraction (long-term drying with additional mechanical crushing of scales) [19–21]. In the first method, moistening treatments extend the seed extraction time considerably, up to 60 h [22]. In addition, seed shaking in devices manufactured by BCC (Sweden), Nomeko (Sweden), or OTL Jarocin (Poland) is carried out between drying stages, directly before moistening [23–25]. Seeds obtained by thermal extraction are easier to clean, and it is possible to obtain nearly 100% purity. The second method, in turn, carries the risk of damaging the coat of the obtained seeds by the grinding elements of cone crushing equipment [20]. As reported by Suszka [21], mechanical extraction of seeds from larch cones was attempted by Drachal and Tyszkiewicz using a self-developed device, TD Mechanical Seed Extractor. The separation of seeds from a mixture of dust and cone debris makes the method difficult to implement [19].

In Poland, seeds from larch cones are extracted using pine and spruce extraction programs in seed extraction cabinets using two-step extraction programs with variable drying temperature [26] to prevent thermal damage to the seeds [1]. Researchers seek new devices and technological solutions to make the process more effective, for example, by microwave irradiation of cones in the initial stage of seed extraction [27,28].

The structure and properties of cell layers may affect the mechanical movement of scales [29]. The humidity of air surrounding the cones has a significant impact on moisture absorption and transpiration of the water vapor contained in the scale cells, which, due to changes in the temperature of the drying air, expand and contract anisotropically in a direction perpendicular to scale tissue orientation [30,31]. Periodic changes in the moisture content of larch cones after reaching the preliminary dry state lead to the contraction and relaxation of scale cells, causing scale movement and outward displacement of the seeds [20]. The process is gradual, and the seeds are released from the cones only after several instances of cone opening and closing [19]. Under natural conditions, approx. three weeks after the beginning of spring the upper parts of seed wings begin to project by approx. 2–3 mm outside the scales in cones on trees. Subsequently, following a slight decrease in cone moisture the scales are gradually deflected and the seeds fall out. Partially displaced seeds do not slide back to their initial positions, even after cone moistening. This is due to the fact that the space under the scale is the narrowest at the cone rachis (where the seed was originally located) and becomes wider in the outward direction [20]. Specific mechanisms of scale opening and closing are linked to plant evolution and survival strategy, which enables conifers to release seeds to greater distances on sunny and dry days [32].

In addition to a publication by Aniszewska [33,34], the available literature provides some other studies on the cone structure and the scale opening process [29,35–37], but these do not concern European larch cones.

The research problem addressed in this paper concerns difficulties with seed extraction from larch cones associated with their scale structure. Thus, the study evaluates the kinematics of scale deflection caused by changes in moisture content in the cones, the cellular structure of scales, and the resulting changes in cell wall thickness during scale opening. It also examines the scale surface in the process of seed extraction.

2. Materials and Methods

2.1. Provenance and Characterization of the Material

The study involved European larch cones (MP/3/41001/05) collected at the beginning of December 2019 from the seed orchard at the Grabowiec Nursery, division 282 k, Bielsk Podlaski municipality, Podlaskie Province (GPS: 52°41′0 N, 23°60′ E). The cones were transferred to the laboratory of the Department of Biosystems Engineering, Warsaw University of Life Sciences; divided into batches; and stored in an LKexv 3600 laboratory refrigerator (Liebherr, Bulle, Switzerland) at 2 ± 1 °C until examination. The length and thickness of all cones were measured (length—*h* and diameter—*d*) using a Silverline 677,256 electronic

Vernier caliper (Silverline Tools, Yeovil, UK) with an accuracy of ±0.1 mm; their initial weight m_0 was determined using a WPS210S laboratory balance (Radwag, Radom, Poland) with an accuracy of ±0.001 g.

2.2. Provenance and Characterization of the Material

The mechanics of scale deflection from the rachis were examined throughout the process of seed extraction. Individual closed cones were cut in half along the axis using an originally developed blade with holder [38] mounted in a modified 10 T screw press (Cormak, Siedlce, Poland). Each cone was placed on a special base, bottom side to the baffle. Subsequently, the turn of the lever lowered the blade that cut the cone from top to bottom, perpendicularly to its axis.

In the resulting half cones, three reference points were marked on selected scales (Figure 1a): one at the junction of the scale with the cone rachis (1), another one on the curve of the scale (2), and the last one (3) at the distal end of the scale.

After marking the reference points, the prepared cone halves were placed in the holder of the purpose-developed stand to examine the opening angle of the scales (Figure 1b).

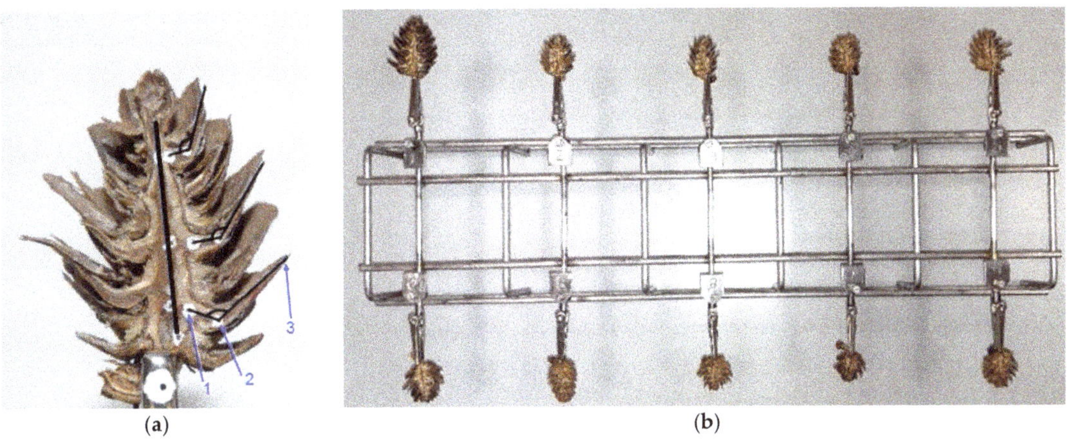

Figure 1. Cone half with marked reference points and axis (**a**), and stand for examining the scale opening angle in ten cone samples (**b**), where: 1—junction of the scale with the cone rachis, 2—point on the scale curve, and 3—scale apex.

Subsequently, the cones were placed in a Heraeus UT612 circulating air oven (Kendro Laboratory Products GmbH, Hanau, Germany). The drying air temperature was set to 35 °C for the first two hours and then increased to 50 °C for another six hours. Every hour throughout the process, the stand was taken out of the oven and individual cones were photographed using a Nikon D3000 camera (Nikon, Tokyo, Japan) with an AF-S DX NIKKOR 18–105 mm f/3.5–5.6G ED VR lens. The cones on the stand were photographed against a white background with a Modeco MN 85-001 manual Vernier caliper (Modeco Expert, Wrocław, Poland), which served as a measurement reference for scaling. Images acquired at a focal length of 105 mm and an aperture of f = 5.6 were saved in JPG format at a resolution of 3872 × 2592 pixels. The distance between the cones and the lens was 350 mm.

The other half of each cone was placed on a glass disc with a diameter of $\Phi = 0.90$ mm (Chemland, Stargard, Poland) in the oven next to the cone stand. After acquiring images of the first half, the other half on the glass disc was removed from the oven and weighed on WPS 210S laboratory scales (Radwag, Radom, Poland) with an accuracy of 0.001 g.

After 8 h of seed extraction and taking nine photographs of each cone half, the halves were immersed in distilled water at approx. 25 °C in laboratory beakers (Chemland, Stargard, Poland) for approx. 10 min, after which they were removed and left to soak for 14 h. The cycle was repeated over the next four days.

After the completion of seed extraction, the other halves were dried at 105 °C for 24 h to constant weight.

After the end of examination, the acquired images were analyzed using MultiScan Base v. 18.03 software (Computer Scanning System, Warsaw, Poland). In the images, three reference points on scales were connected by lines to determine the scale opening angle, α, with an accuracy of $\pm 0.01°$ in each hour of the process (Figure 1a). Analysis involved scales from three cone regions: apex, middle, and base.

The methodology for investigating the scale opening angle was described by Dawson et al. [35], who studied *Pinus radiata* cones, and by Aniszewska [34], who studied *Pinus sylvestris, Picea abies* and *Larix deciduas* cones. It was also followed by Bae and Kim [29] in their investigation of the scale opening angle in pine cones (*Pinus*).

It was assumed that for each of the halved cones the absolute moisture content of one half mounted in a holder for photographing was the same as that of the other half on the glass disc. Therefore, moisture content in each cone was estimated on the basis of weighing its half on a glass disc and determining its dry matter content; that moisture content was then assigned to the scale opening angle at the time of measurement.

2.3. Cellular Structure of Cone Scales

Scales for cellular structure examination were taken from the middle segment of cones with a moisture content of 5% and 20%. Cross-sections of the middle region of the scales were prepared as microscope slides (Figure 2a).

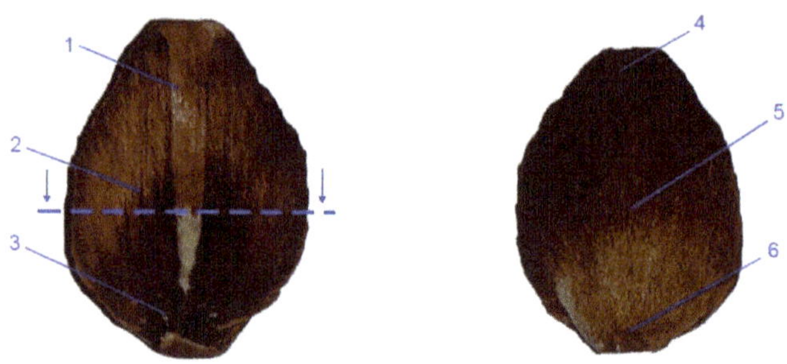

(a) (b)

Figure 2. View of larch scale from the middle part of the cone: (**a**)—inner side (with the cross-section area marked): 1—wing area margin in the distal part of the scale, 2—wing area in the middle part of the scale, and 3 – seed depression in the proximal part of the scale; (**b**)—outer side: 4—distal part, 5—middle part, and 6—proximal part of the scale.

Samples of scales with a moisture content of 5% were taken using an NT Cutter BA-170 blade (NT Incorporated, Tokyo, Japan) with a WSL-lab microtome (Swiss Federal Research Institute WSL, Zürich, Switzerland). Samples of scales with a moisture content of 20% were taken using a Leica 22 C blade (Leica, Wetzlar, Germany) with a MC 2 u4.2 microtome (Moscow, Russia). The slides were observed at magnifications of ×40, ×100, and ×400. Cross-sections from scales with a moisture content of 5% were examined using an Olympus BX61 (ZEISS, Oberkohen, Germany) biological microscope coupled to an Asion 556 camera (ZEISS, Oberkohen, Germany). Cross-sections of scales with a moisture content of 20% were examined using a Nikon Alphaphot–2 YS2 biological microscope (Nikon, Tokyo, Japan) coupled to a Panasonic GP—KR222E camera (Panasonic, Kadoma, Japan). This measurement method was used for spruce cones by Aniszewska [34] and for pine cones by Bae and Kim [29].

Prior to the preparation of scale slides from cones with a moisture content of 5%, the collected scales were immersed for 15 min in plant glycerin to decrease their brittleness and enable microtome cutting; in the case of scales with a moisture content of 20%, such a treatment was not necessary.

The acquired microscopic images were analyzed by means of MultiScan Base v.18.03 (Computer Scanning System, Warsaw, Poland) and ZEN v. 2.3 software (ZEISS, Oberkohen, Germany) to measure the distance between the outer cell margin and lumen termed "cell wall thickness", with an accuracy of 0.0001 ± µm.

2.4. Surface Structure of Scales under an Electron Scanning Microscope

The surface structure of scales was examined under a SEM 200 electron scanning microscope (Quanta, FEI, Europe). Scales for examination were taken from the middle region of whole cones used for cellular studies. Characteristic areas were examined both on the inner surfaces (to which seeds with wings are attached, Figure 2a) and outer surfaces of scales (Figure 2b) at magnifications of ×50 and ×500. Photographs of the inner side involved the following regions: wing area margin (1), wing area (2), and seed depression area (3), while the outer areas were the distal part of the scale (4), the middle part, adjoined by a lower scale (5), and the proximal part of the scale (6). The acquired SEM images were analyzed using MultiScan Base v. 18.03 software (Computer Scanning System, Warsaw, Poland) to measure the dimensions of the structural elements of scales with moisture contents of 5%, 10%, and 20%. The SEM-based method for determining the surface structure of scales or other plant materials is part of public domain and was described by, inter alia, Aniszewska et al. [39], Dawson et al. [35], Bae and Kim [29], and Berthlott et al. [40].

2.5. Statistical Analysis

The parameters were analyzed using the Statistica v.13 program (TIBCO Software Inc., Palo Alto, Santa Clara, CA, USA). Analyses of variances (ANOVA) were performed at a significance level of $\alpha = 0.05$. The differences were statistically significant for $p < 0.05$.

3. Study Results

Table 1 presents mean values with standard deviations, as well as minimum and maximum values, ranges, and coefficients of variation for the entire set of studied cones.

Table 1. Characteristic parameters of the studied cones.

Data	Mean ± SD	Min.	Max.	Range	Coefficient of Variation
Length h, mm	31.2 ± 1.0	30.0	33.1	3.1	3.2
Thickness d, mm	16.3 ± 0.6	15.5	17.1	1.6	3.7
Initial mass m_0, mm	1.247 ± 0.398	0.725	2.052	1.327	31.9
Mass of dry cone m_s, g	0.946 ± 0.306	0.546	1.568	1.022	32.3
Number of scales l_w, [pcs]	53 ± 5	45	61	16	10

Note: SD is standard deviation.

The studied cones had a length of 30.0 to 33.1 mm with a mean of 31.2 ± 1.0 mm and a diameter of 15.5 to 17.1 mm with a mean of 16.3 ± 0.6 mm. The mean initial weight of the cones was 1.247 ± 0.398 g, while their mean initial dry weight was 0.946 ± 0.306 g. The number of scales per cone ranged from 45 to 61, with a mean of 53 ± 5.

3.1. Changes in the Scale Opening Angle during Seed Extraction from Larch Cones

Figure 3 presents images of the opening states of an individual cone on the first day of seed extraction, as well as at the beginning and 8 h into the process over the next days of extraction.

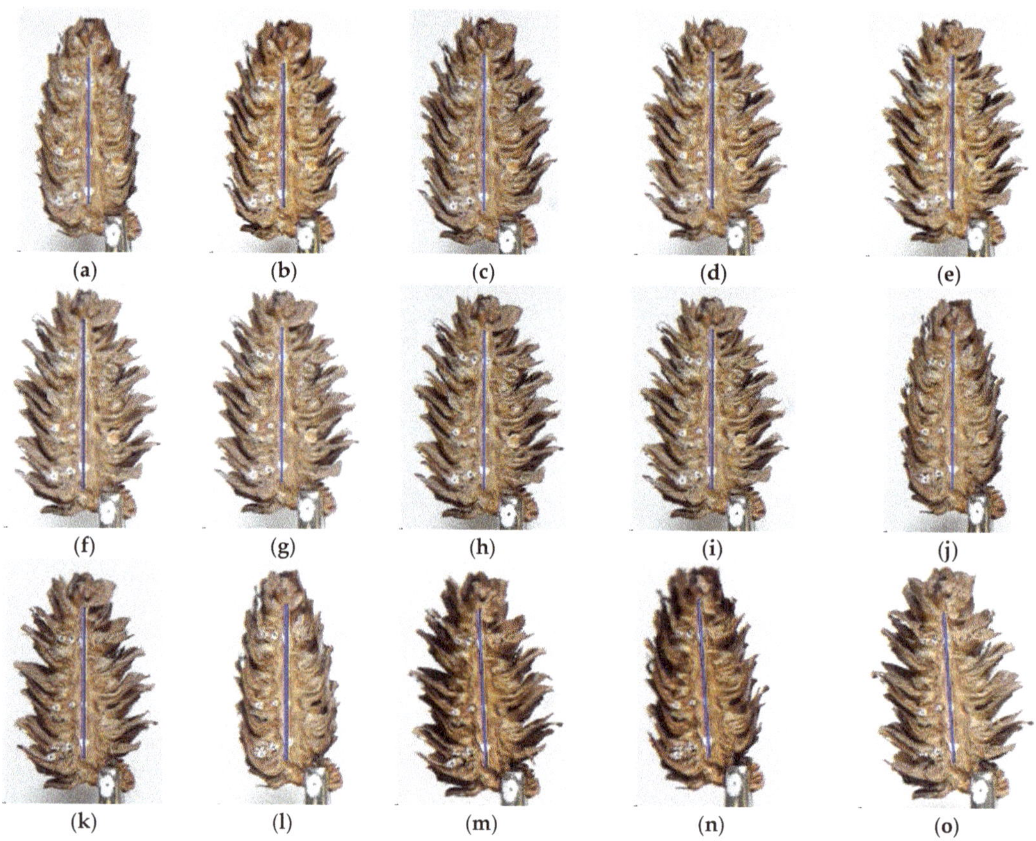

Figure 3. Opening states of an individual cone on the first day of seed extraction as well as at the beginning and 8 h into the process over the following days; day 1: (**a**)—initial state, (**b**)—at 1 h, (**c**)—at 2 h, (**d**)—at 3 h, (**e**)—at 4 h, (**f**)—at 5 h, (**g**)—at 6 h, (**h**)—t 7 h, and (**i**)—at 8 h; day 2: (**j**)—initial state, (**k**)—at 8 h; day 3: (**l**)—initial state, (**m**)—at 8 h, and day 4: (**n**)—initial state, and (**o**)—at 8 h.

At the beginning of the process (day 1 initial state) the cones were fully closed; then, they gradually opened throughout the day with the greatest angles of scale deflection from the rachis being reached after 8 h. Subsequently, the cones were immersed in water for 10 min and left to absorb he moisture for 14 h. As a result, the cones closed, leading to a more intensive opening process the following day. Throughout the seed extraction process, changes in scale deflection angles were most pronounced during the first 2–3 h of extraction at the lower temperature and soon after increasing the temperature setting to 50 °C. In subsequent hours (from 4 to 8 h), the changes were imperceptible to the human eye but the opening angle continued to increase.

Table 2 shows mean moisture content values u_1–u_4 and scale opening angles α_1–α_4 together with standard deviations, measured in scales from the bottom, middle, and top cone segments for each hour of the studied seed extraction process.

Table 2. Mean moisture content in cones and the corresponding opening angles of scales in the bottom, middle, and top cone segments over four consecutive days of measurement.

Time	Day 1		Day 2		Day 3		Day 4	
[h]	u_1	α_1	u_2	α_2	u_3	α_3	u_4	α_4
Scale from the bottom of the cone segments								
0	0.326 ± 0.016	100.98 ± 10.28	0.346 ± 0.117	104.78 ± 10.50	0.377 ± 0.085	106.79 ± 10.83	0.364 ± 0.106	111.53 ± 9.80
1	0.260 ± 0.016	116.06 ± 13.35	0.286 ± 0.088	117.51 ± 12.59	0.282 ± 0.111	118.70 ± 10.68	0.232 ± 0.091	124.20 ± 10.07
2	0.197 ± 0.015	121.46 ± 12.87	0.242 ± 0.064	122.84 ± 10.72	0.232 ± 0.070	122.91 ± 11.65	0.174 ± 0.069	127.73 ± 10.38
3	0.156 ± 0.010	126.64 ± 11.60	0.135 ± 0.043	129.05 ± 10.30	0.112 ± 0.039	129.45 ± 11.18	0.098 ± 0.042	131.31 ± 10.54
4	0.104 ± 0.014	129.23 ± 11.01	0.103 ± 0.024	131.92 ± 9.52	0.095 ± 0.024	131.29 ± 10.18	0.078 ± 0.023	132.68 ± 10.56
5	0.090 ± 0.008	130.53 ± 10.88	0.091 ± 0.013	132.90 ± 9.68	0.080 ± 0.012	132.55 ± 10.28	0.070 ± 0.012	134.42 ± 10.70
6	0.083 ± 0.004	131.74 ± 10.81	0.081 ± 0.006	133.81 ± 9.70	0.076 ± 0.008	133.40 ± 10.17	0.064 ± 0.008	135.18 ± 10.68
7	0.081 ± 0.004	132.93 ± 11.09	0.078 ± 0.005	134.56 ± 9.89	0.071 ± 0.011	134.51 ± 10.44	0.063 ± 0.008	136.10 ± 10.89
8	0.077 ± 0.004	133.88 ± 10.95	0.075 ± 0.004	135.13 ± 9.79	0.069 ± 0.012	135.42 ± 10.39	0.062 ± 0.008	136.88 ± 10.64
Scale from the middle of the cone segments								
0	0.329 ± 0.018	106.40 ± 9.86	0.387 ± 0.091	108.63 ± 10.72	0.421 ± 0.069	108.37 ± 11.73	0.425 ± 0.113	113.23 ± 11.15
1	0.265 ± 0.019	125.93 ± 9.93	0.306 ± 0.062	124.81 ± 10.47	0.301 ± 0.078	124.52 ± 11.10	0.270 ± 0.083	127.76 ± 11.39
2	0.198 ± 0.017	131.90 ± 9.91	0.251 ± 0.046	129.99 ± 11.11	0.230 ± 0.046	129.85 ± 11.39	0.192 ± 0.059	130.92 ± 11.27
3	0.156 ± 0.013	138.66 ± 10.06	0.135 ± 0.035	137.28 ± 11.53	0.106 ± 0.030	138.02 ± 11.46	0.104 ± 0.034	137.53 ± 11.78
4	0.103 ± 0.013	140.87 ± 10.33	0.101 ± 0.021	139.80 ± 11.52	0.093 ± 0.018	139.66 ± 12.02	0.078 ± 0.018	140.14 ± 11.84
5	0.091 ± 0.010	141.89 ± 10.38	0.089 ± 0.013	140.92 ± 11.38	0.080 ± 0.009	141.07 ± 11.91	0.069 ± 0.010	141.66 ± 11.66
6	0.085 ± 0.008	143.16 ± 10.58	0.081 ± 0.009	142.55 ± 11.55	0.076 ± 0.006	141.96 ± 12.06	0.064 ± 0.007	142.71 ± 11.60
7	0.082 ± 0.008	144.03 ± 10.69	0.078 ± 0.008	143.48 ± 11.56	0.073 ± 0.006	142.88 ± 11.97	0.063 ± 0.007	143.47 ± 11.80
8	0.079 ± 0.008	144.99 ± 10.88	0.076 ± 0.008	144.25 ± 11.53	0.073 ± 0.005	143.84 ± 12.00	0.062 ± 0.007	144.02 ± 11.74
Scale from the top of the cone segments								
0	0.334 ± 0.018	108.97 ± 5.81	0.377 ± 0.113	110.96 ± 6.15	0.416 ± 0.104	111.09 ± 5.97	0.415 ± 0.118	114.65 ± 6.95
1	0.267 ± 0.020	122.41 ± 7.91	0.296 ± 0.079	122.40 ± 8.33	0.291 ± 0.105	122.19 ± 6.76	0.263 ± 0.098	125.28 ± 7.30
2	0.196 ± 0.022	127.59 ± 8.94	0.242 ± 0.060	126.50 ± 9.03	0.225 ± 0.066	128.20 ± 7.33	0.186 ± 0.070	129.93 ± 8.54
3	0.152 ± 0.014	132.98 ± 8.06	0.129 ± 0.043	133.00 ± 8.69	0.102 ± 0.039	133.48 ± 8.01	0.098 ± 0.040	134.17 ± 7.80
4	0.101 ± 0.017	135.03 ± 8.21	0.099 ± 0.025	135.56 ± 8.53	0.091 ± 0.023	135.86 ± 7.17	0.075 ± 0.022	136.16 ± 7.97
5	0.090 ± 0.012	136.71 ± 8.09	0.088 ± 0.015	136.76 ± 8.44	0.078 ± 0.012	136.22 ± 8.23	0.066 ± 0.012	137.83 ± 7.88
6	0.084 ± 0.010	137.81 ± 7.75	0.079 ± 0.010	137.64 ± 8.38	0.074 ± 0.008	137.05 ± 8.23	0.062 ± 0.008	139.09 ± 8.25
7	0.082 ± 0.009	138.67 ± 7.74	0.077 ± 0.009	138.79 ± 8.39	0.070 ± 0.010	138.35 ± 8.14	0.061 ± 0.008	139.89 ± 8.42
8	0.078 ± 0.009	139.14 ± 7.59	0.075 ± 0.008	139.64 ± 8.25	0.069 ± 0.011	139.55 ± 8.17	0.060 ± 0.008	140.65 ± 8.48

Note: u_1–u_4 is mean moisture content ± SD [$kg_{water} \cdot kg_{dry\ weight}^{-1}$]; α_1–α_4 is mean angle opening angle ± SD [°].

The smallest mean scale opening angle at the cone base was 100.98°; it was found at the beginning of the process, at a mean cone moisture content of 0.326 $kg_{water} \cdot kg_{dw}^{-1}$. The greatest opening angle was recorded on the fourth day at 8 h; it was 136.88° at the lowest mean cone moisture content in the process (0.062 $kg_{water} \cdot kg_{dw}^{-1}$).

The mean scale opening angle at the cone base increased with each day of the process: from 100.98° to 133.88° (by 32.91°) on the first day, from 104.78° to 135.13° (by 30.35°) on the second day, from 106.79° to 135.42° (by 28.63°) on the third day, and from 111.53° to 136.88° (by 25.35°) on the fourth day.

The lowest mean scale opening angle in the middle cone segment was found at the beginning of the process (106.40°). The highest mean opening angles were recorded on the first day at 8 h (144.99°) at a mean moisture content of 0.079 $kg_{water} \cdot kg_{dw}^{-1}$ and on the second day at 8 h (144.25°) at a mean moisture content of 0.076 $kg_{water} \cdot kg_{dw}^{-1}$. At the lowest moisture content, on the fourth day at 8 h (0.062 $kg_{water} \cdot kg_{dw}^{-1}$), the opening angle was 144.02°.

The mean scale opening angles in the middle cone segment increased from 106.40° to 144.99 (by 38.59°, the highest increment) on the first day, from 108.63° to 144.25° (by 35.63°) on the second day, from 108.37° to 143.84° (by 35.46°) on the third day, and from 113.23° to 144.02° (by 30.80°) on the fourth day.

The mean scale opening angle at the cone apex was the lowest at the beginning of the process (108.97°) and the highest on the fourth day at 8 h (140.65°). The mean initial scale opening angle at the cone apex increased with each day of seed extraction: from 108.97° to 139.14° (by 30.17°) on the first day, from 110.96° to 139.64° (by 28.68°) on the second day, from 111.09° to 139.55° (by 28.46°) on the third day, and from 114.65° to 144.65° (by 26.01°) on the fourth day.

The relationships between the opening angle and moisture content over the consecutive days and times are shown in Figure 4. For example, for scales from the middle cone segment the relationship was described with a polynomial Equation (1) and a linear Equation (2) for the first day and linear equations for the remaining days (3)–(5).

$$\text{Day 1} \quad \alpha_1 = -469.53u_1^2 + 50.082u_1 + 142.26 \; (R = 0.991;\; t_{crit} = 0.156), \quad (1)$$

$$\text{Day 1} \quad \alpha_1 = -131.68u_1 + 155.62 \quad (R = 0.965;\; t_{crit} = 0.125), \quad (2)$$

$$\text{Day 2} \quad \alpha_2 = -97.52u_2 + 150.92 \quad (R = 0.976;\; t_{crit} = 0.125), \quad (3)$$

$$\text{Day 3} \quad \alpha_3 = -91.31u_3 + 149.21 \quad (R = 0.990;\; t_{crit} = 0.125), \quad (4)$$

$$\text{Day 4} \quad \alpha_4 = -79.835u_4 + 147.5 \quad (R = 0.992;\; t_{crit} = 0.125), \quad (5)$$

where t_{crit} is the critical value of the simple or multiple correlation coefficient at $\alpha = 0.05$.

For the first day, also a linear function was calculated (2) with the following opening angles: $\alpha_{oh} = 112.30°$, $\alpha_{1h} = 120.72°$, $\alpha_{2h} = 129.55°$, $\alpha_{3h} = 135.08°$, $\alpha_{4h} = 142.06°$, $\alpha_{5h} = 143.64°$, $\alpha_{6h} = 143.64°$, $\alpha_{7h} = 144.82°$, and $\alpha_{8h} = 145.22°$. The opening angle increased rapidly from the initial state up to 4 h on the first day (by 29.76°), and then slowly from 4 h to 8 h—on average by 3.16°.

The greatest increments in the mean scale opening angle at the base, middle, and apex were recorded on the first day. With increasing moisture content, the opening angle of scales in those three cone regions increased by 33.89° on the first day, 31.55° on the second day, and 30.85° on the third and fourth days.

The statistical analysis (Tukey HSD (honestly significant difference) test for unequally sized samples), which compared the scale opening angle in different sections of the cone (base, middle, and apex) for different hours on consecutive days of the process, revealed no significant differences ($p < 0.05$). This seems to indicate that the scale position in the cone does not influence the opening angle. Furthermore, the analysis demonstrated that scale opening remained similar on subsequent days, which implies that the process duration (in days) has no statistically significant impact on the scale opening angle ($p < 0.05$).

The initial moisture content of cones increased with each day, while the final moisture content continued to decrease. Moisture content changes in scales from the middle segment were 0.251 $kg_{water} \cdot kg_{dw}^{-1}$ on the first day, 0.312 $kg_{water} \cdot kg_{dw}^{-1}$ on the second day, 0.349 $kg_{water} \cdot kg_{dw}^{-1}$ on the third day, and 0.362 $kg_{water} \cdot kg_{dw}^{-1}$ on the fourth day of the process.

It was found that in scales from the middle cone segment, a decrease in moisture content of 0.01 $kg_{water} \cdot kg_{dw}^{-1}$ was associated with a mean increase in the opening angle of 1.3° on the first day and, on average, 0.90° on the following days (0.98°, 0.91°, and 0.80° on days 2, 3, and 4, respectively).

Figure 5 presents the movement of a scale in the middle cone segment throughout the four-day seed extraction process by showing the location of reference points 1, 2, and 3 at consecutive process times.

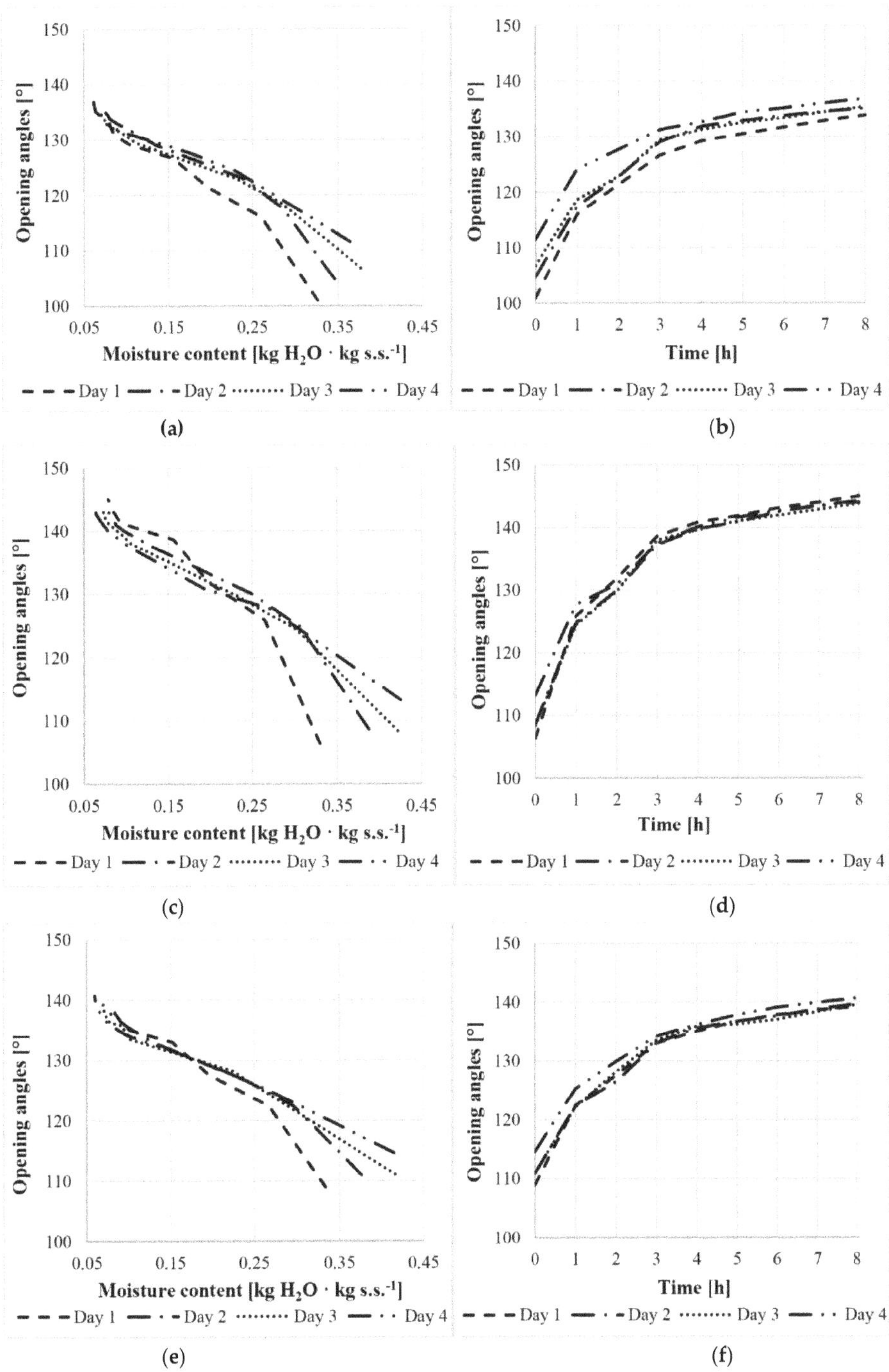

Figure 4. Relationship between the scale opening angle and moisture content/process duration for scales obtained from the base (**a**,**b**), middle (**c**,**d**), and apex (**e**,**f**) of cones.

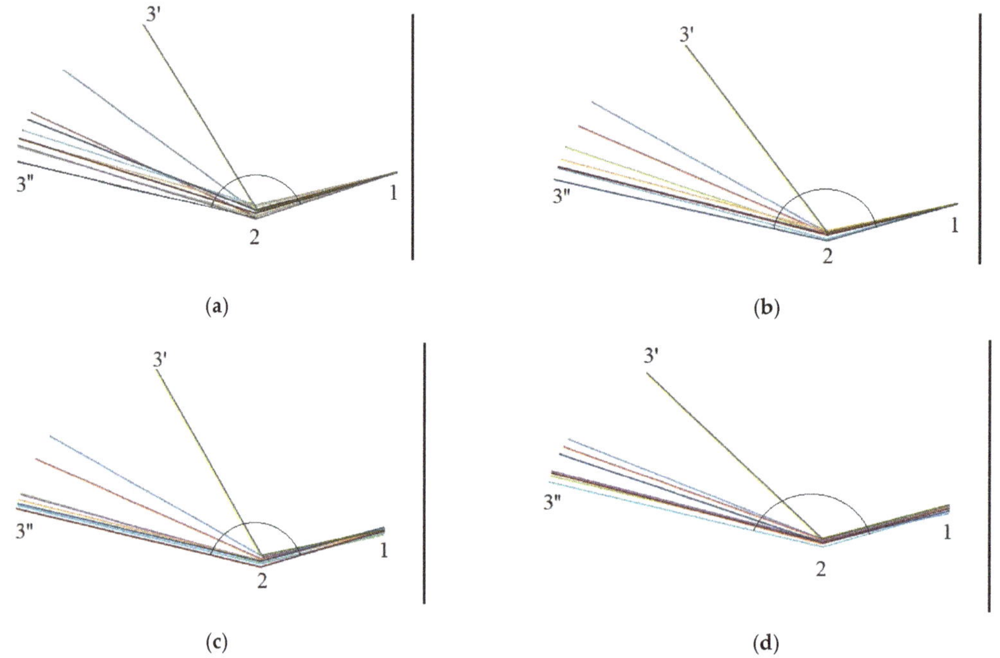

Figure 5. Changes in the opening angle of a scale from the middle cone segment at consecutive measurement times on (**a**)—day 1, (**b**)—day 2, (**c**)—day 3, and (**d**)—day 4(3′—start day; 3′′—end day).

The mean opening angle of scales in the middle cone segment increased from 102.08° to 150.03° (by 47.95°, the largest increment) on the first day, from 113.30° to 150.22° (by 36.92°) on the second day, from 111.54° to 149.84° (by 38.3°) on the third day, and from 122.23° to 149.82° (by 27.59°) on the fourth day.

3.2. The Cellular Structure of Scales

Larch scales consist of three major cell types:

- Small cells with thick cell walls occur in bundles in the central layer,
- Medium-sized cells with thick walls in the outer and inner epidermal layers,
- Large cells with thin walls and large lumina in the central layer.

Cross-sections of scales with a moisture content of 5% are shown in Figure 6.

Figure 6a show medium cells of the outer epidermis with a mean cell wall thickness of 2.750 ± 0.530 μm as well as inner epidermal cells with a mean wall thickness of 7.007 ± 1.376 μm. The darker regions in the large cell layer probably represent bundles of cells.

As can be seen from Figure 6b, the epidermal cells on the outer side of the sale are packed more tightly than those on the inner side. Due to loss of water, the cells on the inner side contract, causing scale deflection from the rachis in larch cones (albeit the deflection is less pronounced than in spruce and pine cones).

Figure 6c presents a cross-section involving a scale margin on which glycerin particles penetrated into empty intercellular spaces that emerged as a result of moisture loss during seed extraction. It should also be noted that loss of moisture led to cell deformation.

Figure 6d shows large cells with a cell wall thickness of 3.497 ± 0.946 μm and epidermal cells on the inner side of the scale; their lumina were larger than those on the outer side.

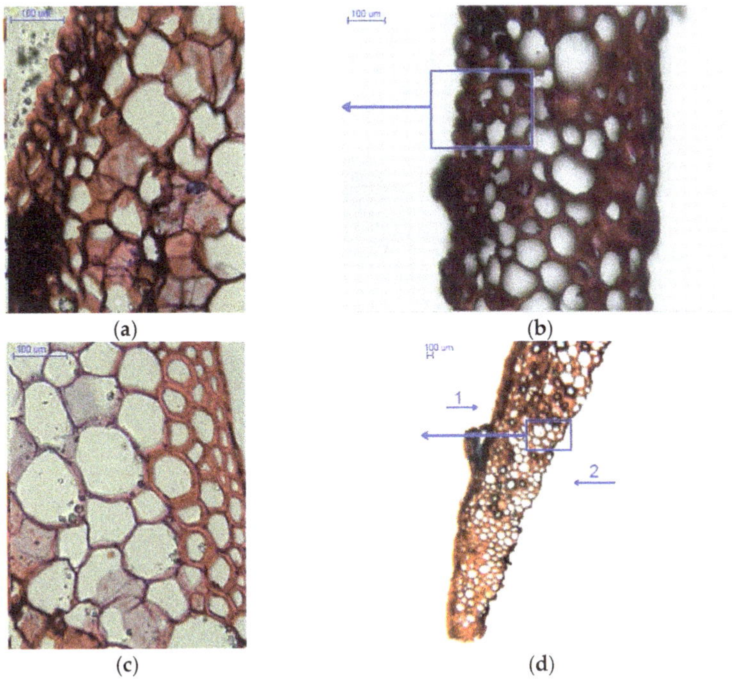

Figure 6. Cross-sections of scales with a moisture content of 5%: (a) outer epidermal cell layer (400×); (b) scale with visible cell layers (100×); (c) scale with visible cell layers on the inner side (400×); and (d) scale with visible cell layers in the marginal region (40×), 1—outer side of the scale, and 2—inner side of the scale.

Figure 7 presents cross-sections of scales with a moisture content of 20%. Figure 7a shows cross-sections of outer and inner epidermal cells, while large cells and cells in bundles are shown in Figure 7c. Outer and inner epidermal cells had mean wall thicknesses of 3.670 ± 0.561 μm and 9.458 ± 2.335 μm, respectively. The difference in cell wall thickness between the inner and outer scale sides could be the reason why the opening process of larch cones is slower than that of Scots pine and Norway spruce cones.

In Figure 7b, in the middle of the cross-section one can see large cells and cells in bundles with a mean wall thickness of 8.206 ± 1.482 μm and 3.313 ± 0.599 μm, respectively. It was found that the layer of large cells narrows down towards the scale margin (Figure 7d).

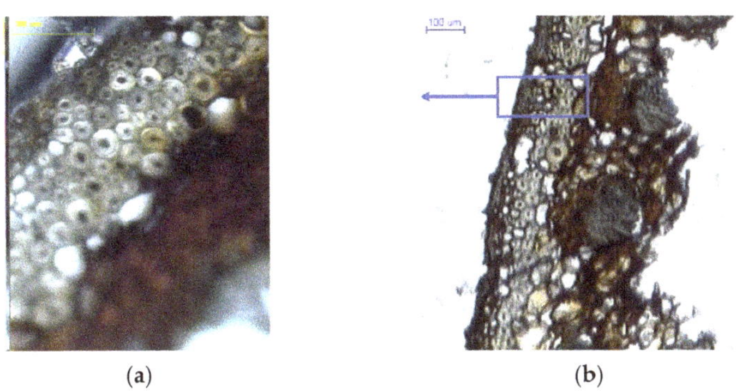

Figure 7. *Cont.*

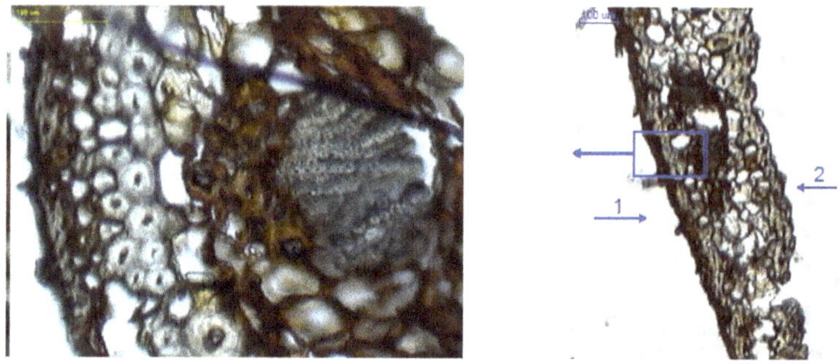

(c) (d)

Figure 7. Cross-sections of scales with a moisture content of 20%: (**a**) outer epidermal cell layer (100×); (**b**) scale with visible cell layers on the inner side (100×); (**c**) cell layer in bundles between large cells (100×); (**d**) scale with visible cell layers in the marginal region (40×), 1–outer side of the scale, and 2–inner side of the scale.

A comparison of mean cell wall thickness in scales with a moisture content of 5% and 20% indicates that in medium-sized cells it increased by 25% (outer side) and by approx. 18% (inner side), as compared to approx. 57% for large cells. Analysis of variance revealed significant wall thickness differences between large cells (p = 0.00) and medium-sized cells both on the outer side (p = 0.00) and inner side of the epidermis ($p < 0.05$) in dry scales (5% moisture) and wet scales (20% moisture).

3.3. Results of Structural Examinations of Cone Scales

Figures 8–13 present the results of structural examinations of the inner and outer sides of larch cones with moisture contents of 5%, 10%, and 20%.

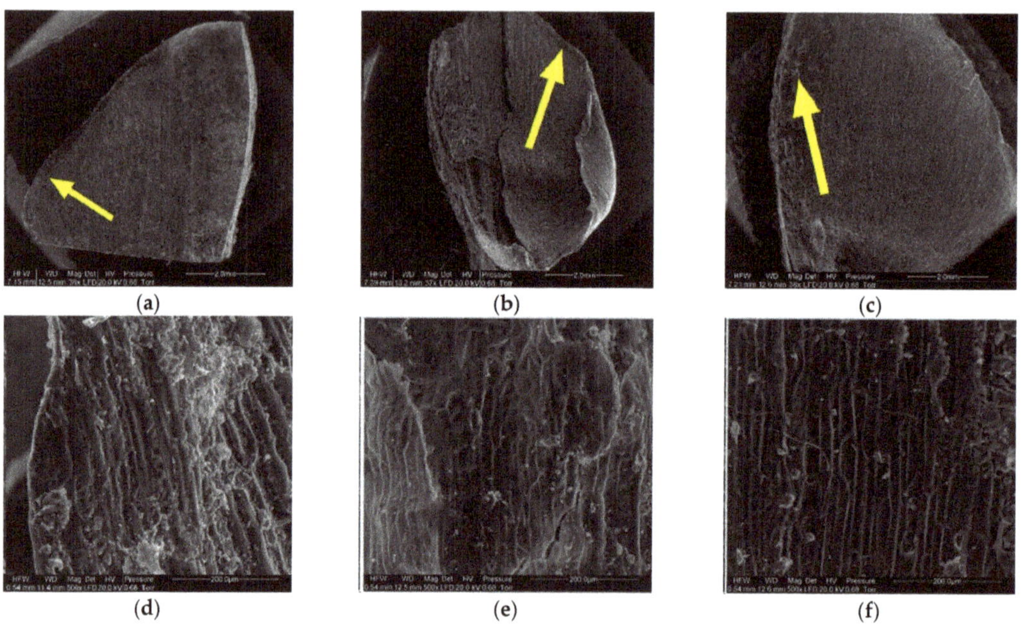

Figure 8. Inner side of larch scale. Scale margin outside of the wing area: (**a**) MC = 20%, zoom 50×; (**b**) MC = 10%, zoom 50×; (**c**) MC = 5%, zoom 50×; (**d**) MC = 20%, zoom 500×; (**e**) MC = 10%, zoom 500×; and (**f**) MC = 5%, zoom 500×.

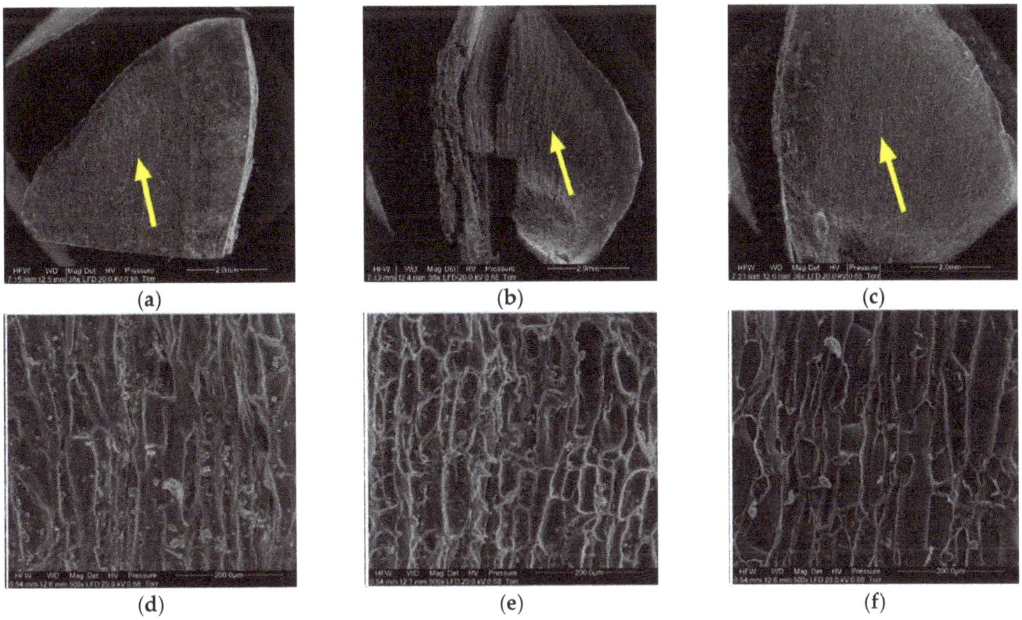

Figure 9. Inner side of larch scale. Wing area: (**a**) MC = 20%, zoom 50×; (**b**) MC = 10%, zoom 50×; (**c**) MC = 5%, zoom 50×; (**d**) MC = 20%, zoom 500×; (**e**) MC = 10%, zoom 500×; and (**f**) MC = 5%, zoom 500×.

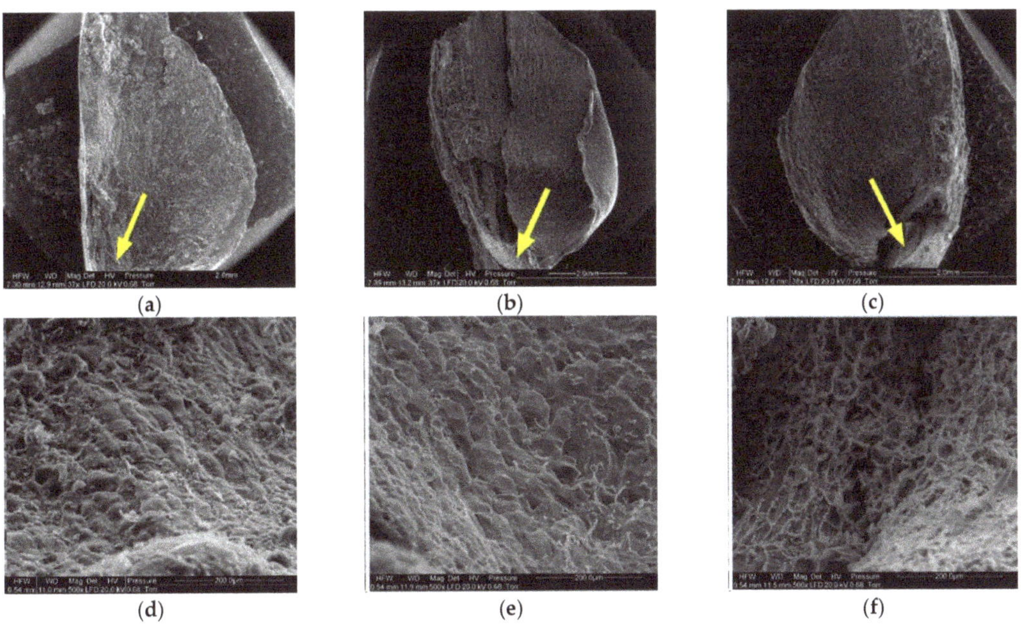

Figure 10. Inner side of larch scale. Seed depression area: (**a**) MC = 20%, zoom 50×; (**b**) MC = 10%, zoom 50×; (**c**) MC = 5%, zoom 50×; (**d**) MC = 20%, zoom 500×; (**e**) MC = 10%, zoom 500×; and (**f**) MC = 5%, zoom 500×.

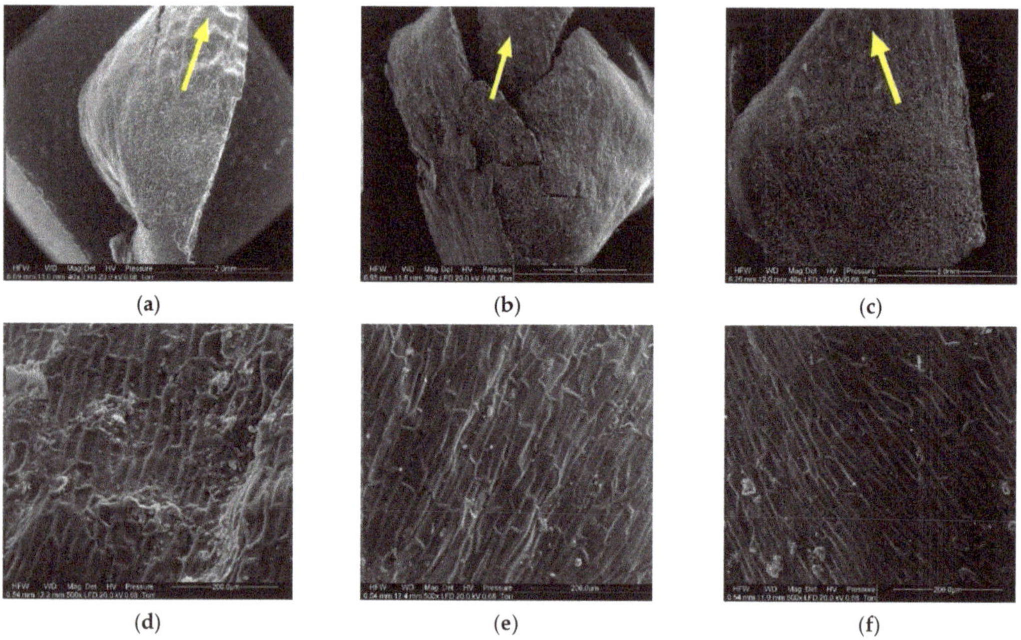

Figure 11. Outer side of larch scale. Distal scale area: (**a**) MC = 20%, zoom 50×; (**b**) MC = 10%, zoom 50×; (**c**) MC = 5%, zoom 50×; (**d**) MC = 20%, zoom 500×; (**e**) MC = 10%, zoom 500×; and (**f**) MC = 5%, zoom 500×.

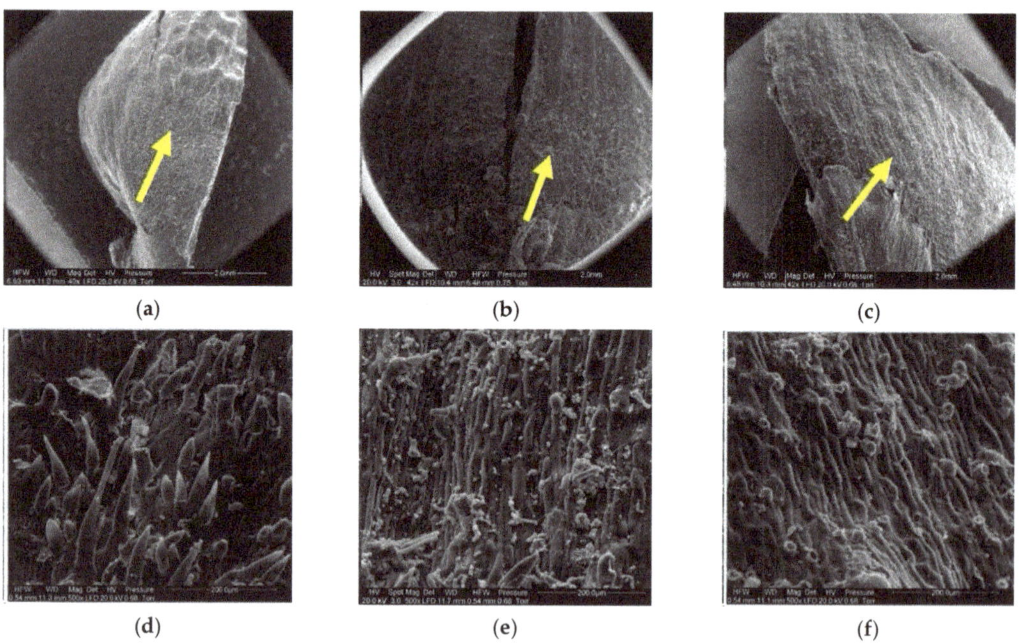

Figure 12. Outer side of larch scale. Middle scale area, adjoin by a lower scale: (**a**) MC = 20%, zoom 50×; (**b**) MC = 10%, zoom 50×; (**c**) MC = 5%, zoom 50×; (**d**) MC = 20%, zoom 500×; (**e**) MC = 10%, zoom 500×; and (**f**) MC = 5%, zoom 500×.

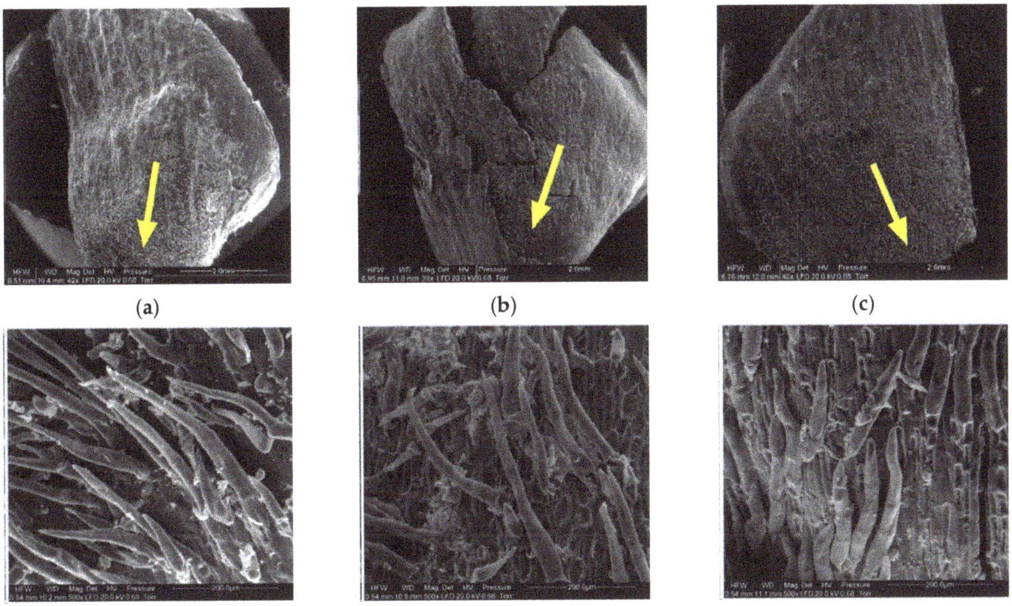

Figure 13. Outer side of larch scale. Proximal scale area: (**a**) MC = 20%, zoom 50×; (**b**) MC = 10%, zoom 50×; (**c**) MC = 5%, zoom 50×; (**d**) MC = 20%, zoom 500×; (**e**) MC = 10%, zoom 500×; and (**f**) MC = 5%, zoom 500×.

On the scale margin, outside of the wing area (1) there are elongated cells differing in their wall thickness. Table 3 shows means with standard deviations as well as minimum and maximum values for the studied cells.

Table 3. Cell sizes at the scale margin outside of the wing area (1) on the inner side of scales with a moisture content of 5%, 10%, and 20%.

Data	Moisture Content of Scales								
	20%			10%			5%		
	Mean ± SD	Min.	Max.	Mean ± SD	Min.	Max.	Mean ± SD	Min.	Max.
Length [µm]	187.57 [a] ± 47.00	105.70	274.40	119.45 [b] ± 37.14	95.40	217.80	119.35 [b] ± 28.18	92.10	208.60
Width [µm]	17.69 [a] ± 3.48	9.80	24.00	15.71 [a] ± 2.40	11.40	20.60	15.67 [a] ± 3.93	10.70	23.80
Wall thickness [µm]	7.82 [a] ± 1.74	5.20	12.10	5.45 [b] ± 0.92	4.10	7.00	4.71 [b] ± 1.22	2.90	7.00

Note: [a,b]—homogeneous groups.

The largest dimensions and wall thicknesses were found for elongated cells in scales with a moisture content of 20% (length of 185.57 µm, width of 17.69 µm, and wall thickness of 7.82 µm). Cells in scales with a moisture content of 10% were much smaller and had thinner walls due to loss of moisture (length of 119.45 µm, width of 15.71 µm, and wall thickness of 5.45 µm). Cells in scales with a moisture content of 5% were slightly smaller than those in scales with a moisture content of 10% (length of 119.35 µm, width of 15.67 µm, and wall thickness of 4.71 µm), with the difference not being statistically significant in the Duncan test. Significant differences were found for the length and width of cells in scales with a moisture content of 20% ($p < 0.001$).

Another region on the inner side of the scale, the wing area (2), featured elongated cells similar to those at the wing margin (1).

Cells in scales with a moisture content of 20% were convex, cylindrical, and overlapping, and had the thickest cell walls (from 4.90 μm to 13.50 μm, with a mean of 9.60 ± 1.44 μm). Their surface featured clusters of particles, probably consisting of resin. Cells in scales with a moisture content of 10% were not convex, having a concave interior, and a wall thickness ranging from 4.60 μm to 8.90 μm (on average 6.50 ± 1.31 μm). Cells in scales with a moisture content of 5% were characterized by thin, damaged walls with a thickness from 3.20 μm to 8.20 μm (on average 5.17 ± 2.09 μm); their surface layer revealed defragmentation.

On the inner side of the proximal part of scales, there were usually two seed nests, but they were not necessarily well-defined or developed on all scales. The seed depression on the inner side of the scale (3) consisted of irregularly-shaped cells. Following seed detachment, the cells were irregular in shape and frayed, as can be seen from the figures presenting this region of the scale (3).

The apical part of the outer side of scales (4) contained closely arranged elongated cells differing in cell wall thickness. The lower the moisture content of the cone, the thinner the cell wall. The wall thickness of cells ranged from 6.20 μm to 10.90 μm, with a mean of 8.02 ± 1.29 μm in scales with a moisture content of 20%; from 5.40 μm to 9.20 μm, with a mean of 6.57 ± 1.08 μm in scales with a moisture content of 10%; and from 4.50 μm to 7.60 μm, with a mean of 6.13 ± 0.81 μm in scales with a moisture content of 5%. Cell wall thickness in scales with a moisture content of 20% was significantly different from that in scales with a moisture content of 5% and 10% ($p < 0.001$ in the Duncan test).

In the case of cones with a moisture content of 20%, the middle part of scales, to which lower scales are adjacent (5), revealed elongated cells with projections in the form of hairs with a mean length of 108.12 ± 54.81 μm and a width at the base of 25.84 ± 3.08 μm. Scales with a moisture content of 10% featured projections with a mean length of 61.78 ± 24.42 μm and a length at the base of 23.20 ± 3.40 μm. No projections were found on scales with a moisture content of 5%; instead, they revealed pores of different diameters—on average 10.76 ± 2.04 μm.

Numerous projections were found on the outer side of the proximal part of scales (6) with all the studied moisture content values. The mean length and width of hair cells on scales with a moisture content of 20% was 265.88 ± 116.72 μm and 28.09 ± 3.39 μm, respectively. Hairs on scales with a moisture content of 10% were 219.99 ± 71.56 μm long and 28.42 ± 5.50 μm wide, while those on scales with a moisture content of 5% were 207.30 ± 48.74 μm long and 31.19 ± 2.82 μm wide. Projections on scales with a moisture content of 5% were the shortest and widest at the base.

4. Discussion

As reported by Lin et al. [41] for *Pinus pinaster* cones, the cone opening and closing mechanism can be attributed to the self-bending of their scales, which undergo three states of humidity-driven deformation in terms of Föppl–von Kármán plate theory [42]. Based on three other reports [29,33,35], it may be concluded that it is moisture and the shape and size of cells that trigger opening and closing of cones.

Loss of moisture in the course of drying causes changes in the shape of the treated material [43]. The process of seed extraction involves the contraction of cell walls into the space previously occupied by water and a decrease in the volume of the material [44]. In contrast to other conifer species (spruce or pine), larch cones do not open sufficiently to release seeds freely even in very dry air. The scale structure and opening mechanism for *Pinus radiata* cones were described in detail by Dawson et al. [35], who identified two types of scales growing from the main body of the cone, with the larger ones responding to changes in relative humidity. Therefore, the crucial issue in larch seed extraction is to stimulate cone opening by alternating seed drying and moistening [20], which leads to gradual seed displacement from between the scales.

Larch cones may open to a greater or lesser extent or close depending on air humidity [1], but in the literature there is a dearth of information about the opening of *Larix* cones.

In this study, the opening angle of larch scales, defined by the aforementioned three points, increased with decreasing moisture content in the cones.

In the course of four-day seed extraction, the greatest increment in the opening angle was observed on day 1 for scales in the middle cone segment (by approx. 39°), followed by those at the base (by approx. 33°) and at the apex (by approx. 30°). The largest mean increment was found for scales in the middle cone segment. It was calculated that at the end of the extraction process, the scale opening angle ranged from 140° to 145° at a moisture content of 5% to 10%. For larch, the maximum values of the scale opening angle were much higher than those obtained by Bae and Kim [29], who differentiated between the right (120.7°) and left (111.6°) bracts of *Pinus* cones. However, these figures cannot be directly compared due to differences in the angle measurement methodology. The values obtained for larch are close to the maximum angles (approx. 145°) reported by Reyssat and Mahadevan [36] for *Pinus coulteri*.

In view of the finding that neither the scale position in the cone nor the process duration (for various hours on subsequent days) had a significant impact on the scale opening angle, it seems reasonable to shorten the seed extraction process, for example, from four to three days. The yield should be monitored, and if it no longer increases, the extraction process can be terminated.

The initial moisture content in larch scales increased, while the final content decreased, with each day of the process. Cone moistening caused scale closure at the beginning of each day (Figure 3). In the literature there are insufficient data on the number and duration of cone drying and moistening steps needed to maximize seed yield. It is known that an hour-long cone immersion in water is inadvisable due to the swelling of seeds (which must be then promptly sown) [1]. In another study (forthcoming), the authors reported the effects of the number and duration of seed extraction and cone moistening steps on the yield of larch seeds of first class quality. It was found that three 8 h seed extraction steps with two 10 min water immersion treatments in between led to a 59% yield (seeds obtained as compared to the overall number of seeds in the cone).

As reported by, inter alia, Tyszkiewicz [1,20], Bae and Kim [29], Fahn and Werker [30], Bar-On et al. [31], and Reyssat and Mahadevan [36] for cones of *Pinus* or other trees, the structural tissue responds to moisture changes, which trigger the gradual opening of tightly closed cones. The mechanism of cone opening or closing associated with the loss or gain of moisture, respectively, is based on the two-layer structure of cells that transform with changes in moisture content. In the outer layer of the tissue, thick-walled cells respond by expanding in the longitudinal direction when exposed to moisture increase [35,45] and by shrinking in response to to drying, while the simultaneous reaction of cells in the inner layer is feebler.

The microscopic structure of scales revealed cells differing in terms of their wall thickness. Larch scales consist of three types of cells: small, medium-sized, and large. During seed extraction, the cells changed their dimensions, due to which they wrinkled and deflected from the cone rachis [4] to release seeds [37]. This is associated with the close adjoining of epidermal cells on the outer side (Figure 6a), where the cell lumen is much smaller than in the case of inner epidermal cells (Figure 6d). This may be related to the wall thickness of medium-sized cells, which was not affected by loss of moisture to the same extent as the wall thickness of large cells (a decrease of approx. 57%). Furthermore, as reported by Aniszewska [34], between three and five layers of cells may be identified on cross-sections of spruce cones, depending on the scale position. Near the stem, the small, medium-sized, and large cells have diameters of 56.7 µm, 32.3 µm, and 15.3 µm, respectively.

Cells in *Pinus* scales with a moisture content of 20% had thicker cell walls and smaller lumina than those in scales with a lower moisture content. A study on the thickness of cell walls in larch wood reported 20–23 µm for wood samples dried to a moisture content of 5%–15% [46], which is consistent with the results obtained in this paper.

The outer and inner structures of cones scales are different. In the middle segment of the outer surface (5), scales with the highest moisture content revealed short projections, which decreased in length and width with the degree of moisture loss from the cone. In this region (5), scales with the lowest moisture content featured pore-like structures, which probably enabled the elimination of excess water from the cones [4]. The projections were situated on the outer side of scales with all studied levels of moisture content. The proximal part of the scale (6) exhibited hairs, whose length increased with the moisture content of the cone. Hairs (6) on scales with a moisture content of 5% had the greatest width at the base and adhered to the scale surface, while hairs (6) on scales with a moisture content of 20% and 10% formed a bristle. *Pseudotsuga menziesii* and *Abies alba* have scales of a similar structure between seeds [47].

Resin particles were found on scales with the highest moisture content, while scales with lower moisture content values did not reveal such particles; in the latter case they tended to come off, which indicates that the first by-product of seed extraction from conifer cones is dried resin (colophony) [1].

Studies show that the seed extraction process does not have to be conducted over four days since three days with two moistening treatments in between is sufficient. On the last day, the change in the opening angle is lower than that on the preceding days, while moisture content does not decrease below the level obtained on day 3.

5. Conclusions

The opening angle of larch scales increased with decreasing moisture content in the cones. The greatest increment in the opening angle was observed on the first day of seed extraction (on average 34° for the three types of scales). The largest mean opening angle increment was found for scales in the middle segments of the studied cones; the largest mean opening angle was 145.99°.

The size and thickness of cell walls in scales is determined by the moisture content of the cones: the higher it is, the thicker the cell walls (up to the fiber saturation point of approx. 30%). Conversely, the lower the moisture content, the larger the cell lumen. The thickest walls were found in the inner epidermal cells (9.458 μm), and the thinnest walls in vascular bundle cells (3.313 μm). In turn, the greatest change in wall thickness was identified in the large cells found in the middle scale segment, with the mean difference between dry and moist states amounting to 4.708 μm. The mean wall thickness of large cells in scales with a 5% moisture content amounted to 42% of that in scales with a 20% moisture content.

The outer and inner scale structures differed depending on moisture content. The greatest differences in the surface structure of scales with 5%, 10%, and 20% moisture contents could be observed on the outer side. On scales with a 20% moisture content, the hair-like cells were elongated and strongly deflected outwards; in contrast, on scales with a 10% moisture content, there were fewer such cells, which exhibited constrictions and leant towards the cone rachis. On scales with a 5% moisture content, the hairs were short and adhered to the outer scale surface. The inner sides of scales with moisture contents of 5%, 10%, and 20% differed significantly at the scale margin outside of the wing area. Resin particles were found on both sides of scales with a 20% moisture content but not on scales with a 5% moisture content.

The results of our investigation of the scale opening kinematics and the cellular structure of larch cones depending on the cone moisture content and the duration and stage of seed extraction may contribute to determining the conditions for the automation of this process.

Author Contributions: Conceptualization, E.T. and M.A.; methodology, E.T. and M.A.; formal analysis, E.T.; writing—original draft preparation, E.T.; writing—review and editing, M.A. and A.G.; supervision M.A. and A.G. All authors have read and agreed to the published version of the manuscript.

Funding: This research did not receive any specific grant from funding agencies in the public, commercial, or not-for-profit sectors.

Institutional Review Board Statement: Not applicable.

Informed Consent Statement: Not applicable.

Data Availability Statement: Not applicable.

Acknowledgments: The authors would like to express their gratitude to the Department of Silviculture and Genetics of Forest Trees in Sękocin Stary for enabling cellular analysis of scales with a 5% moisture content, as well as to Marcin Klisz, for his valuable comments.

Conflicts of Interest: The authors declare no conflict of interest.

References

1. Tyszkiewicz, S. *Wyłuszczanie Nasion Leśnych [Forest Seed Extraction]*; Państwowe Wydawnictwo Rolnicze i Leśne: Warszawa, Poland, 1951.
2. Rohmeder, E. *Das Saatgut in der Forstwirtschaft*; 57 Übersichten, 142 Abb.; Verlag Paul Parey: Berlin, Germany, 1972; ISBN 978-3-490-06916-0.
3. Puritch, G.S. *Cone Production in Conifers. A Review of the Literature and Evaluation of Research Needs*; Pacific Forest Research Centre, Canadian Forestry Service: Victoria, BC, Canada, 1977.
4. Aniszewska, M. *Dynamics of the Process of Seed Extraction from Pine Cones Pinus Silvestris L. Performed in One Two-Steps*; Rozprawy Naukowe i Monografie, Wydawnictwo SGGW: Warszawa, Poland, 2012.
5. Calama, R.; Montero, G. Cone and Seed Production from Stone Pine (*Pinus pinea* L.) Stands in Central Range (Spain). *Eur. J. For. Res.* **2007**, *126*, 23–35. [CrossRef]
6. Owens, J.N.; Kittirat, T.; Mahalovich, M.F. Whitebark Pine (*Pinus albicaulis* Engelm.) Seed Production in Natural Stands. *For. Ecol. Manag.* **2008**, *255*, 803–809. [CrossRef]
7. Kantorowicz, W. Half a Century of Seed Years in Major Tree Species of Poland. *Silvae Genet.* **2000**, *49*, 245–249.
8. Broome, A.; Hendry, S.; Peace, A. Annual and Spatial Variation in Coning Shown by the Forest Condition Monitoring Programme Data for Norway Spruce, Sitka Spruce and Scots Pine in Britain. *For. Int. J. For. Res.* **2007**, *80*, 17–28. [CrossRef]
9. Chałupka, W.; Barzdajn, W.; Blonkowski, S.; Burczyk, J.; Fonder, W.; Grądzki, T.; Gryzło, Z.; Kacprzak, P.; Kowalczyk, J.; Kozioł, C.; et al. *Program of Conserving Forest Genetic Resources and Breeding of Trees in Poland for the Years 2011–2035*; The State Forests Information Centre: Warszawa, Poland, 2011; ISBN 978-83-61633-61-7.
10. Fonder, W.; Matras, J.; Załęski, A. *Leśna Baza Nasienna w Polsce [The Forest Seed Base in Poland]*; Centrum Informacyjne Lasów Państwowych: Warszawa, Poland, 2007; ISBN 978-83-89744-60-4.
11. Skrzypczyńska, M.; Furgała, E. The Qualitative Estimation of Insects Inhabiting European Larch *Larix decidua* Mill. Cones Originating from the Forest District Myślenice (the Kraków Forest Region). Acta Agr. Silv. 2015, 61–70. *Acta Agrar. Silvestria Ser. Silvestris* **2015**, *53*, 61–70.
12. Klisz, M.; Ukalska, J.; Koprowski, M.; Tereba, A.; Puchałka, R.; Przybylski, P.; Jastrzębowski, S.; Nabais, C. Effect of Provenance and Climate on Intra-Annual Density Fluctuations of Norway Spruce *Picea abies* (L.) Karst. in Poland. *Agric. For. Meteorol.* **2019**, *269–270*, 145–156. [CrossRef]
13. Saulnier, M.; Corona, C.; Stoffel, M.; Guibal, F.; Edouard, J.-L. Climate-Growth Relationships in a *Larix decidua* Mill. Network in the French Alps. *Sci. Total Environ.* **2019**, *664*, 554–566. [CrossRef]
14. Gendek, A.; Aniszewska, M.; Malaťák, J.; Velebil, J. Evaluation of Selected Physical and Mechanical Properties of Briquettes Produced from Cones of Three Coniferous Tree Species. *Biomass Bioenergy* **2018**, *117*, 173–179. [CrossRef]
15. Aniszewska, M.; Gendek, A.; Hýsek, Š.; Malaťák, J.; Velebil, J.; Tamelová, B. Changes in the Composition and Surface Properties of Torrefied Conifer Cones. *Materials* **2020**, *13*, 5660. [CrossRef] [PubMed]
16. Malaťák, J.; Gendek, A.; Aniszewska, M.; Velebil, J. Emissions from Combustion of Renewable Solid Biofuels from Coniferous Tree Cones. *Fuel* **2020**, *276*, 118001. [CrossRef]
17. Bożym, M.; Gendek, A.; Siemiątkowski, G.; Aniszewska, M.; Malaťák, J. Assessment of the Composition of Forest Waste in Terms of Its Further Use. *Materials* **2021**, *14*, 973. [CrossRef]
18. Aniszewska, M.; Gendek, A.; Drożdżek, M.; Bożym, M.; Wojdalski, J. Physicochemical Properties of Seed Extraction Residues and Their Potential Uses in Energy Production. *Rocz. Ochr. Srodowiska* **2017**, *19*, 302–334.
19. Załęski, A. Comparison of Methods of Seed Extraction by Drying Cones and Mechanical One of European Larch Seeds *Larix decidua* Mill. *Pr. Inst. Badaw. Lesn. Ser. A* **2002**, *3*, 5–17.
20. Tyszkiewicz, S. *Nasiennictwo Leśne [Forestry Seed]*; Instytut Badawczy Leśnictwa: Warszawa, Poland, 1949.
21. Suszka, B. *Nowe Technologie i Techniki w Nasiennictwie Leśnym [New Technologies and Techniques in Forest Seed]*; Bogucki Wydawnictwo Naukowe S.C.: Warszawa, Poland, 2000; ISBN 83-88163-14-0.
22. Aniszewska, M. Multiphase Process of European Larch *Larix decidua* Mill. Seed Extraction: A Case of Czarna Białostocka. *For. Res. Pap.* **2008**, *69*, 155–163.

23. Tylek, P.; Walczyk, J. Drum de-winger of the seeds by using dry and wet methods. *Zesz. Probl. Postępów Nauk Rol.* **2009**, *543*, 365–370.
24. Sarnowska, G.; Więsik, J. Wyłuszczarnia w Czarnej Białostockiej. Część II. Proces Wyłuszczania i Obróbki Nasion [Extraction Plant in Czarna Białostocka. Part II. The Process of Seed Extraction and Treatment]. *Przegląd Tech. Rol. Lesn.* **1997**, *12*, 8–10.
25. Załęski, A. Technologie i Sprzęt Stosowane w Wyłuszczaniu, Oczyszczaniu, Suszeniu i Pakowaniu Nasion [Technologies and Equipment Used in Seed Extraction, Cleaning, Drying and Packing]. *Postępy Tech. W Lesn.* **2002**, *82*, 21–33.
26. Tyszkiewicz, S.; Tomanek, J. *Tablice Psychrometryczne Do Użytku w Wyłuszczarniach Nasion Sosny i Świerka [Psychrometric Tables for Use in Cone Kilns for Pine and Spruce]*; Seria B, Nr 9.; Instytut Badawczy Leśnictwa; Wydawnictwa Pomocnicze i Techniczno Gospodarcze: Kraków, Poland, 1946.
27. Aniszewska, M.; Zychowicz, W.; Gendek, A. The Effectiveness of Short-Term Microwave Irradiation on the Process of Seed Extraction from Scots Pine Cones (*Pinus sylvestris* L.). *IForest-Biogeosci. For.* **2020**, *13*, 73. [CrossRef]
28. Aniszewska, M.; Słowiński, K. Effects of Microwave Irradiation by Means of a Horn Antenna in the Process of Seed Extraction on Scots Pine (*Pinus sylvestris* L.) Cone Moisture Content and Seed Germination Energy and Capacity. *Eur. J. For. Res.* **2016**, *135*, 633–642. [CrossRef]
29. Bae, H.; Kim, J. Functional Principles of Morphological and Anatomical Structures in Pinecones. *Plants* **2020**, *9*, 1343. [CrossRef] [PubMed]
30. Fahn, A.; Werker, E. Chapter 4-Anatomical Mechanisms of seed dispersal. In *Seed Biology*; Kozlowski, T.T., Ed.; Academic Press: New York, NY, USA; London, UK, 1972; pp. 151–221. ISBN 978-0-12-424301-9.
31. Bar-On, B.; Sui, X.; Livanov, K.; Achrai, B.; Kalfon-Cohen, E.; Wiesel, E.; Daniel Wagner, H. Structural Origins of Morphing in Plant Tissues. *Appl. Phys. Lett.* **2014**, *105*, 033703. [CrossRef]
32. Song, K.; Yeom, E.; Seo, S.-J.; Kim, K.; Kim, H.; Lim, J.-H.; Joon Lee, S. Journey of Water in Pine Cones. *Sci. Rep.* **2015**, *5*, 9963. [CrossRef] [PubMed]
33. Aniszewska, M. Analysis of Opening Cones of Selected Coniferous Trees. *Ann. Wars. Univ. Life Sci.-SGGW Agric.* **2010**, *55*, 57–64.
34. Aniszewska, M. Anatomical structure of spruce cones. *Sylwan* **2002**, *146*, 85–91.
35. Dawson, C.; Vincent, J.F.V.; Rocca, A.-M. How Pine Cones Open. *Nature* **1997**, *390*, 668. [CrossRef]
36. Reyssat, E.; Mahadevan, L. Hygromorphs: From Pine Cones to Biomimetic Bilayers. *J. R. Soc. Interface* **2009**, *6*, 951–957. [CrossRef] [PubMed]
37. Le Duigou, A.; Castro, M. Evaluation of Force Generation Mechanisms in Natural, Passive Hydraulic Actuators. *Sci. Rep.* **2016**, *6*, 18105. [CrossRef] [PubMed]
38. Aniszewska, M.; Olejnikowski, J. Noż Do Cięcia Szyszek, Podstawa Do Noża, Zestaw Oraz Sposób Cięcia Szyszek [A Knife for Cutting Cones, a Knife Base, a Set and a Method for Cutting Cones]. Patent Application: P.436963 [WIPO ST 10/C PL436966], 15 February 2020.
39. Aniszewska, M.; Stadnik, S.; Gendek, A. Variability of Cones and Scale Surface Area of European Larch (*Larix decidua* Mill.). *For. Res. Pap.* **2017**, *78*, 198–209. [CrossRef]
40. Barthlott, W.; Mail, M.; Bhushan, B.; Koch, K. Plant Surfaces: Structures and Functions for Biomimetic Innovations. *Nano-Micro Lett.* **2017**, *9*, 23. [CrossRef]
41. Lin, S.; Min, X.Y.; Li, Q.; Huang, X.; Zhou, S. On the Shape Transformation of Cone Scales. *Soft Matter.* **2016**, *12*, 9797–9802. [CrossRef]
42. Mansfield, E.H. *The Bending and Stretching of Plates*, 2nd ed.; Cambridge University Press: Cambridge, UK, 1989; ISBN 978-0-521-33304-7.
43. Górnicki, K.; Kaleta, A. Modelling Convection Drying of Blanched Parsley Root Slices. *Biosyst. Eng.* **2007**, *97*, 51–59. [CrossRef]
44. Wu, X.-F.; Zhang, M.; Li, Z. Dehydration Modeling of Cordyceps Militaris in Mid-Infrared-Assisted Convection Drying System: Using Low-Field Nuclear Magnetic Resonance with the Aid of ELM and PLSR. *Dry. Technol.* **2019**, *37*, 2072–2086. [CrossRef]
45. Harlow, W.M.; Côté, W.A.; Day, A.C. The Opening Mechanism of Pine Cone Scales. *J. For.* **1964**, *62*, 538–540. [CrossRef]
46. Klisz, M. Genetic Aspects of Wood Properties of European Larch (*Larix decidua* Mill.). Ph.D. Thesis, Instytut Badawczy Leśnictwa, Sękocin Stary, Poland, 2011.
47. Aniszewska, M.; Gendek, A.; Śliwińska, J. Variability of Silver Fir (*Abies alba* Mill.) Cones–Variability Structure of Scale Surface Area. *For. Res. Pap.* **2017**, *78*, 5–13. [CrossRef]

materials MDPI

Article

Numerical Reliability Study Based on Rheological Input for Bingham Paste Pumping Using a Finite Volume Approach in OpenFOAM

Robin De Schryver, Khadija El Cheikh, Karel Lesage, Mert Yücel Yardimci and Geert De Schutter *

Magnel-Vandepitte Laboratory, Department of Structural Engineering and Building Materials, Faculty of Engineering and Architecture, Ghent University, Technologiepark-Zwijnaarde 60, B-9052 Ghent, Belgium; Robin.DeSchryver@UGent.be (R.D.S.); Khadija.El.Cheikh@bbri.be (K.E.C.); Karel.Lesage@UGent.be (K.L.); MertYucel.Yardimci@UGent.be (M.Y.Y.)

* Correspondence: Geert.DeSchutter@UGent.be

Abstract: Rheological quantification is important in many industries, the concrete industry in particular, e.g., pumping, form filling, etc. Instead of performing expensive and time-consuming experiments, numerical simulations are a powerful means in view of rheological assessment. However, due to the unclear numerical reliability and the uncertainty of rheological input data, it is important for the construction industry to assess the numerical outcome. To reduce the numerical domain of cementitious suspensions, we assessed the numerical finite volume simulations of Bingham paste pumping flows in OpenFOAM. We analysed the numerical reliability, first, irrespective of its rheological input by comparison with the literature and theory, and second, dependent on a certain rheological quantification by comparison with pumping experiments. Irrespective of the rheological input, the numerical results were significantly accurate. Dependent on the rheological input, a numerical mismatch, however, existed. Errors below 1% can be expected for proposed numerical rules of thumb: a bi-viscous regularisation, with pressure numbers higher than 5/4. To improve bias due to uncertain rheology, a rheological configuration close to the engineer's aimed application should be used. However, important phenomena should not be overlooked. Further assessment for lubrication flows, in, e.g., concrete pumping, is still necessary to address concerns of reliability and stability.

Keywords: numerical simulation; computational fluid dynamics; OpenFOAM; cementitious materials; Bingham rheology; pumping; reliability

Citation: De Schryver, R.; El Cheikh, K.; Lesage, K.; Yardimci, M.Y.; De Schutter, G. Numerical Reliability Study Based on Rheological Input for Bingham Paste Pumping Using a Finite Volume Approach in OpenFOAM. *Materials* **2021**, *14*, 5011. https://doi.org/10.3390/ma14175011

Academic Editors: Michele Bacciocchi and Abbas S. Milani

Received: 6 July 2021
Accepted: 20 July 2021
Published: 2 September 2021

1. Introduction

Concrete structures and their finishing quality, durability or even structural integrity passively rely on the concrete's fresh state properties [1], more specifically, the fresh state flow characteristics, which are usually characterised by so-called rheological parameters. The rheology of concrete and more general cementitious-like suspensions, which are mostly governed by Bingham flow behaviour, therefore determines the outcome of its application [1]. For simple and, more generally, industrial processes, one could predict the outcome based on existing theories. For more complicated processes, this is not the case. As a first alternative, the outcome of an industrial process could then be assessed based on experiments. Experimental tests can, however, be inconvenient, time consuming and/or expensive. As a second alternative, one could therefore use numerical simulations to overcome this issue.

The numerical simulation of cementitious suspensions, Bingham flows in general, shows a huge potential for predicting the flow behaviour in several construction or industrial processes. Numerical simulations for predicting industrial (construction) processes could for instance provide the required pumping pressure, form filling ability, etc. One could even use numerical simulations to predict more complicated behavioural aspects such as particle migration or time-dependent behaviour such as thixotropy, which can be a major cause for blocking problems [2–8]. Moreover, one could use numerical simulations for industrial design.

One popular approach to numerically simulate cementitious suspensions or Bingham flows in general is computational fluid dynamics (CFD) [3,5,6,8–12]. CFD considers a homogenous fluid that can be justified to simulate flows of cementitious materials [3,6]. Since a continuous approach is involved in CFD, particle behaviour aspects may only be considered from a macroscopic point of view [3,5]. If one is interested in specific particle phenomena occurring in, e.g., concrete flows, other techniques such as discrete element method (DEM) and lattice Boltzmann method (LBM) are advised [5]. The advantage of CFD over other techniques is that it allows for faster computations and larger computation domains, which are suitable for practical simulations such as pumping, formwork filling, on-site rheological test cases, etc. [6,8,13]. The amount of numerical mesh cells is still considerable (ca. 10^4–10^6 cells [3,5,6,8,9,13]) without the requirement of special computational infrastructure. Therefore, CFD could serve as a useful tool within practical reach for design engineers. Many commercial CFD softwares on the market, however, are bounded by licences that can be expensive. Therefore, open-source CFD codes have been rising in popularity because they are accessible for anyone. Instead of a black box-embedded CFD code, the source code is accessible and modifiable for any prerequisite needed. One open-source software in particular, named OpenFOAM, has been rising in popularity over the past decades and is therefore convenient or suitable for use in applications of the modern industries, particularly among the concrete industry. Since numerical simulations use rheological data for their input values, the numerical simulations themselves are an immediate consequence of the rheological device or characterisation procedure. Several devices, known as rheometers, exist to quantify rheological data for cementitious suspensions together with their respective geometries, such as ConTech Viscometer, Tattersall MK-II, ICAR, Anton Paar MCR device, Sliper and more [14–17]. Despite the numerous existence of rheological devices, the rheological parameters obtained by these devices differ [14–19]. Worse is that the difference among devices can be significant [14,15,19–21].

Irrespective of the rheological input, some works assessing the numerical accuracy for Bingham flows exist [21–29]. However, these studies are in terms of error in numerical residuals or numerical stability criteria with regard to numerical regularisation techniques. Therefore, they may somewhat lack genuine interpretation of the actual numerical outcome or reliability. In the context of cementitious materials, some numerical studies have also been conducted to numerically simulate cementitious paste suspensions or even concrete suspensions [2,3,5–9,13,30–35]. However, in most of these works, it is unclear how reliable the numerical simulations actually are, independent of whether a mismatch or good match was found, let alone to know the actual influence of the rheological input with regard to the numerical outcome. Therefore, the question remains how reliable are numerical simulations based on their rheological input and how can more reliable results be obtained, especially because of the disparity in rheological data? Moreover, some devices are reliable for only certain rheological properties and/or types of concretes or suspensions, e.g., paste, traditional concrete and self-compacting concrete. Since numerical simulations rely on rheological input values and therefore on their uncertain character, it is important to assess how reliable the numerical simulations are based on their input. In that way, confidence can be built in the simulation of Bingham suspension flows and, by extension, for engineers practicing in the industry. One example, in particular, is to model fresh concrete construction processes in the concrete industry. The uncertainty problem is twofold: on the one hand, the uncertain character of the rheological input data and, on the other, the accuracy of the numerical outcome itself.

To simplify the vast numerical uncertainty problem in this study, we assessed the accuracy of the numerical simulations for the simple case of pumping of Bingham pastes in the context of cementitious paste suspensions. A potential follow-up in the context of the concrete industry is the more complicated concrete pumping flows, in which an additional pipe wall lubrication layer is expected to contribute to the flow. This follow-up has not been undertaken to not further complicate the conciseness and findings of this work. In that way, we provide a validation framework for engineers in practice to model laminar Bingham flows; simple cementitious suspensions; or perhaps, by extension, even projections for fresh concrete construction processes. To deal with the twofold character of the uncertainty problem, we decomposed it into a first part by assessing numerical simulations using the expected theory or literature and a second part by assessing the simulations based on experiments. The twofold problem decomposition is graphically depicted in Figure 1. Hence, the first decomposed part is irrespective of its rheological input, while the second decomposed part depends on a certain rheological input.

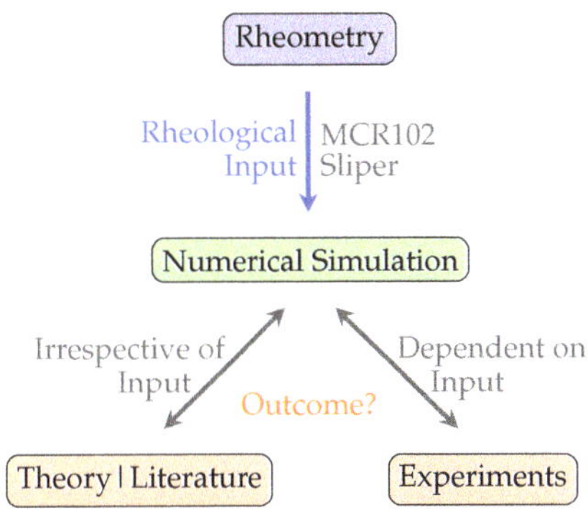

Figure 1. Reliability overview of the input for numerical simulations.

To do so, we first elaborate upon the numerical methodology used, which is a CFD finite volume approach in OpenFOAM to simulate horizontal, laminar pipe flow of cementitious-like Bingham pastes. The numerical framework is followed by the experimental framework for all performed tests. Then, the results for the first and respective second problem decomposition parts are outlined. In the first part, numerical simulations are compared to expected literature results and theories. In the second part, simulations are compared to experimental pumping observations based on a certain rheological input. In that way, we are able to assess the numerical outcome irrespective of the rheological input as well as the reliability based on certain rheological data. Finally, the results are discussed and concluded.

2. Numerical Framework

The popular open-source software OpenFOAM (v5, The OpenFOAM Foundation Ltd., London, UK) is used in this framework to assess the numerical adequacy of modelling Bingham cementitious-like suspension pipe flows. We therefore outline the numerical methodology, regularisation, control, case set-up and mesh independence.

2.1. Numerical Methodology

Proper to CFD, OpenFOAM models the fluid domain as a continuum with macroscopic properties [3,8]. Doing so, the numerical solution is in fact an approximation of reality constrained by its discretisation (mesh), solving algorithm (solver), solver settings and imposed assumptions. The numerical solution is obtained by solving the fluid continuum's conservation laws of mass and momentum. This results in solving the continuum Equation (1) and the Navier–Stokes Equation (2) because cementitious-like material flows are incrompressible with a constant density ρ [3,8,36].

$$\nabla \cdot U = 0 \tag{1}$$

$$\rho \frac{DU}{Dt} = \rho g - \nabla p + \nabla \cdot (\nu \nabla U) \tag{2}$$

In these equations, U is the velocity vector field and p the pressure scalar field. In the used solver, gravity g is omitted because gravity is invariant in the case of horizontal pipe flow, as considered in this paper. Therefore, the pressure p and viscosity ν are computed in kinematic form, which is relative to the density ρ. The material derivative is denoted by $\frac{DX}{Dt} = \frac{\partial X}{\partial t} + U \cdot \nabla X$, in which $\cdot$ is the inner (tensor) product.

Cement pastes exhibit non-Newtonian flow behaviour. By omitting time-dependent behaviour, they are usually modelled as a Bingham fluid with a yield stress τ_0 and a plastic viscosity μ [10,37–42]. The non-Newtonian character is therefore numerically coped with by using a Generalised Newtonian Fluid (GNF) approach in which the Newtonian viscosity ν is replaced by an apparent viscosity η computed for each mesh cell corresponding to the considered constitutive rheological model, i.e., the Bingham model. In doing so, a numerical singularity occurs for zones with zero or low shear rates (i.e., quiescent zones). Eventually, this can lead to numerical stability issues. This is the so-called unregularised viscoplastic problem [43]. In the literature, this problem is either coped with by making use of a regularisation approach or a different modelling technique (e.g., augmented Langrangian method) [25,43]. Other examples of numerical techniques where this problem is coped with were conducted by, e.g., Jahromi et al. [28], Bleyer et al. [44], Pimenta and Alves [45]. Using such advanced numerical techniques is however not straightforward. Therefore, the bi-linear regularisation approach is used in this work.

2.2. Numerical Regularisation

Several regularisation approaches have been investigated in the literature such as the bi-linear [23,24,46], Bercovier and Engelman [22], Papanastasiou [10,12,47–49] and other approaches [25,43,50]. These regularisation approaches restrain the infinite apparent viscosity η for zero shear rates, usually defined by a so-called epsilon ε environment. The smaller ε, the higher the apparent viscosity can reach, resulting in an overall more stiff or viscous flow behaviour in quiescent zones [43]. Higher apparent viscosities lead to better approximations of the non-Newtonian yield character; it may however also inflict numerical stability issues [43].

For ease of implementation and interpretation, the bi-linear regularisation approach [23] is used in this work by restraining the apparent viscosity to a maximum value, similar to [21]. A constant Generalised Newtonian Approach Ratio ($GNAR$) is therefore defined. This is the ratio of the maximum apparent viscosity η_{max} and plastic viscosity μ because it defines the order of magnitude and therefore the dominance of the numerical matrices in which the apparent viscosity η is used. This approach of limiting the ratio to 1000 (or an equivalent regularisation parameter) is in line with the philosophy by Schaer et al. [21], Bercovier and Engelman [22], Ahmadi and Karimfazli [29], Papanastasiou [47]. Moreover, this is exactly the same approach and recommendations as Beverly and Tanner [24], Bullough et al. [46] and Burgos et al. [51]. Thus, the higher the $GNAR$, the better the non-Newtonian character but the more the numerical simulation is subjected to stability issues, especially in quiescent zones. Quiescent zones are zones with very low or zero shear rates, such as the central plug zone in pipe flows. Based on a parametric sensitivity study, we used a $GNAR$ of 1000 to obtain better yield behaviour predictions.

2.3. Numerical Control

To additionally cope with numerical instability problems for simulations with big plug zones (quiescent zones), under-relaxation of the velocity field U was used to a limited extent of 0.95. This could be considered a similar but more convenient and easy to implement relaxation approach as used by Chupin and Dubois [52]. Since under-relaxation smears out the numerical solution over time or iterations, we advise limiting the velocity under-relaxation between $[0.95, 1.0]$. This is based on a time-dependent parameter study of the under-relaxation. This helped us to overcome all numerical GNF instability issues while still respecting the temporal flow behaviour to a reasonable extent.

In order to ascertain temporal numerical stability, the well-known CFL (Courant–Friedrichs–Lewy) number was controlled and limited to 0.5. We gradually increased the CFL number after flow initiation from 0.001 to 0.5 to allow for faster numerical progress because no temporal boundary conditions were considered and hence the initiation phase is an immediate flow jump, starting from a zero velocity to a velocity corresponding to the imposed discharge Q. Hence, the initiation may lead to initial numerical instability issues, which have been coped with by gradually increasing the CFL number.

The total simulation time was 60 s, performed by the transient (PISO: Pressure-Implicit with Splitting of Operators) OpenFOAM solver called nonNewtonianIcoFoam, adopted for a customly developed Bingham model. The numerical data were analysed using a Python script, considering steady-state only for the data analysis.

2.4. Numerical Cases

Several numerical pipe flow simulations were performed in order to assess their reliability irrespective of the rheological input. On the one hand, 10 pipe flow simulations were performed to compare the results with the literature results from Tichko [13] for a DN100 (106 mm) pipe. On the other hand, to compare simulation results with the Buckingham–Reiner theory, 213 simulations were additionally conducted, re-simulating small-scale pumping experiments that were performed with a DN25 (26.64 mm) pipeline. Simultaneously, these 213 pipe flow simulations served to assess the reliability dependent on the rheological input by comparing the numerical results with the experimental observations. Hence, a total of 223 numerical cases were considered, disregarding the 660 numerical cases conducted for sensitivity studies.

The boundary conditions (BC) are also important, since they determine how a simulation domain interacts with the physical outer world. Therefore, the pipe boundary conditions are depicted in Table 1.

Table 1. Pipe flow boundary conditions of the velocity field U and the pressure field p. Herein, Q is the imposed discharge, uniformly distributed over the inlet surface A.

Boundary	Field	Type	Definition	
Inlet	U	Dirichlet	uniform value	$U = Q/A$
	p	Neumann	zero gradient	$\nabla p = 0$
Wall	U	Dirichlet	noSlip	$U = 0$
	p	Neumann	zero gradient	$\nabla p = 0$
Outlet	U	Neumann	zero gradient	$\nabla U = 0$
	p	Dirichlet	zero value	$p = 0$

A default set of numerical solvers and schemes (mostly linear) are used for the considered simulations. Even though higher-order schemes and better-performing solvers would facilitate the numerical accuracy—though it has an additional computational cost—it is not the aim of this work to outline the most optimal settings. Our aim is rather to give a global idea on the accuracy, be it with rather default-like or CFD-advised settings. Further details on the numerical schemes and settings can be consulted in the respective Appendix A. In that way, a benchmark may be obtained for simulation engineers in practice or in related research domains.

Examples of the typical numerical outcome of the pipe flow simulations are illustrated in Figure 2. For every simulation, the 3D pressure and velocity field are simulated, after which the fully developed pressure loss is processed for further analysis in this work. For steady-state, these illustrations show a linear pressure loss as well as a quadratic velocity profile with central plug flow. This is expected from laminar Bingham simulations [53].

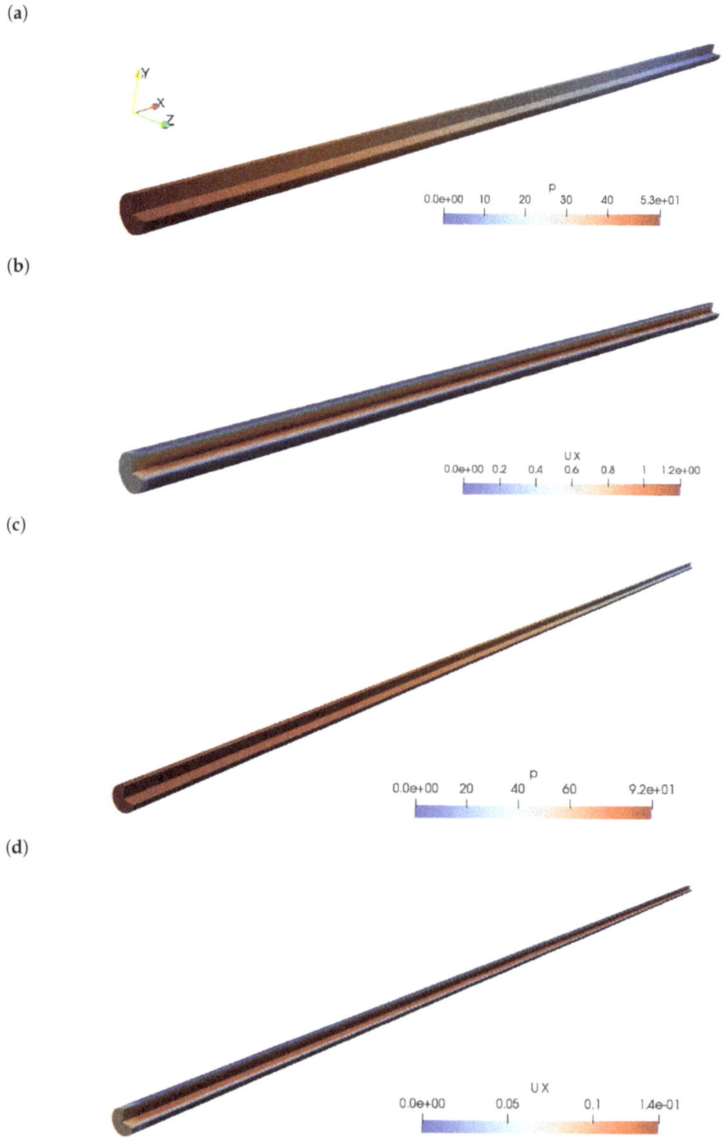

Figure 2. Illustrations of the numerical outcome. Examples of the kinematic pressure field p (m^2/s^2) and the x-component velocity field U_x (m/s) are shown for both the DN100 and DN25 pipe cases. (**a**) Pressure field p for a DN100 pipe. (**b**) Velocity field U_x for a DN100 pipe. (**c**) Pressure field p for a DN25 pipe. (**d**) Velocity field U_x for a DN25 pipe.

2.5. Mesh Independence

A last aspect that is to be considered for a simulation is the mesh quality. The mesh quality has a big influence on the obtained results in numerical modelling studies. Therefore, the simulations should be mesh independent. A mesh independence study was performed for pipe flow simulations. The acceptance criteria for mesh independence were based on mesh orthogonality; mesh angularity; a mesh verification check by OpenFOAM; and more importantly, convergence of simulated pressure loss results, indicating true independence. To respect orthogonality, a cylindrical hexagonal mesh configuration of the pipe was used as depicted in Figure 3. Based on the independence analysis, the final cross-sectional mesh configuration consisted of 9 × 9 inner square elements (to respect orthogonality) and compatible 9 × 9 outer circular section elements (on the four sides). The inner square diagonal to pipe diameter ratio was 0.5 (Figure 3a). The longitudinal mesh was uniformly subdivided into elements with a length of 2 cm.

(**a**) (**b**) (**c**)

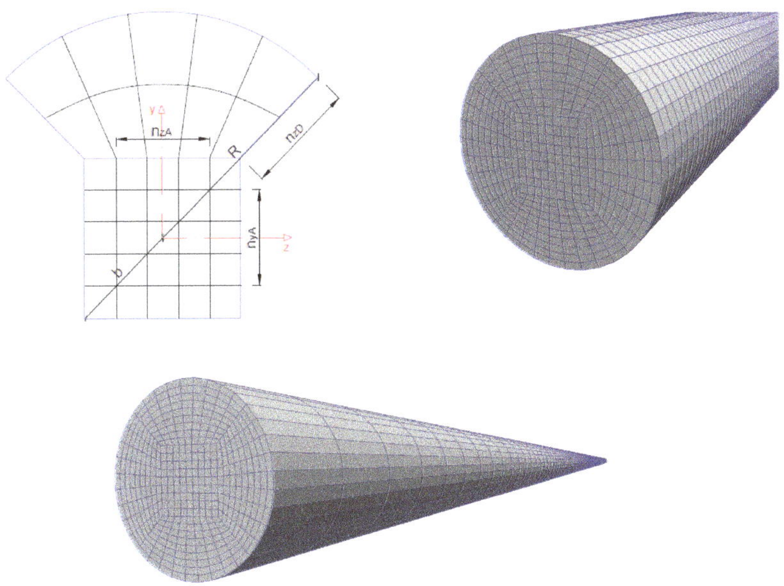

Figure 3. Hexagonal cross-sectional mesh configuration, with an inner square ($n_{yA} \times n_{zA}$) with diagonal $2b$ and outer circle segments ($n_{zA} \times n_{zD}$) to pipe radius R. (**a**) Cross-sectional view. (**b**) Three-dimensionally rendered mesh view of a DN100 pipe. (**c**) Rescaled mesh of experimental pumping circuit, with an equal radial mesh resolution for a DN25 pipe.

The mesh independence analysis was based on a pipe with a diameter of 106 mm (DN100), which was used to compare the simulations with literature. Since the pipe diameter of the conducted pumping experiments is smaller, the radial configuration is scaled down by maintaining the same mesh numbers (Figure 3c). Longitudinally, it is assumed to be sufficiently refined, and a mesh cell length of 2 cm was not further refined. This mesh scaling is justified since the flow can radially develop over the same amount of mesh cells. Longitudinally, it is also justified because the simulated pressure losses used in the analysis are verified fully developed pressure losses. Moreover, this re-scaling is even proven to be justified by the fact that the simulation results of DN100 pipes have the same relative theoretical error (ca. 1%) as DN25 pipes, as outlined in Sections 4.1 and 4.2 respectively.

3. Experimental Framework

The experimental work consisted of two stages. First, a rheological characterisation was performed for all considered samples by rheometer tests. Second, pumping experiments were performed to assess the reliability of the numerical simulations based on rheological input data. In total, four different paste mixture designs (A1 to A4 in Table 2) were tested at least three times to ensure the repeatability.

Table 2. Conceptual overview of rheological mixture designs of the considered four Bingham model pastes (A1 to A4), as a function of the water (W), powder (P) and superplasticiser (SP) content.

Mix Designs	Plastic Viscosity μ				
	Low		High		
Yield Stress τ_0	Mix	*W*/*P*	Mix	*W*/*P*	*SP*/*P*
Low	A1	0.40	A2	0.20	0.175%
High	A3	0.33	A4	0.25	0.100%

The pastes used in the experiments were in fact Bingham model fluid pastes, without thixotropic and hydration-related properties, to increase the experimental time window, instead of actual cement pastes. The model pastes consisted of limestone powder (P; with a median particle size of 3.7 μm), water (W) and/or a polycarboxylate ether-based superplasticiser (SP). The four different paste designs represent a rheological range of cement pastes with a low/high yield stress τ_0 and a low/high plastic viscosity μ. Hence, the terminology of cementitious-like pastes is preferred, since the model fluids represent typical cement pastes.

3.1. *Rheometry*

The Bingham behaviour of the model pastes were rheologically characterised by two approaches. The first approach is a characterisation based on a flow curve protocol in a rotational rheometer (MCR-102, Anton Paar Benelux BVBA, Gentbrugge, Belgium). The second rheological characterisation is based on a sliding pipe rheometer (Sliper, Schleibinger Geräte Teubert und Greim GmbH, Buchbach, Germany) [54].

3.1.1. Rotational Rheometry

For all paste samples, rheological tests were performed with a parallel plate configuration in a rotational rheometer (MCR-102, Anton Paar Benelux BVBA, Gentbrugge, Belgium). A constant temperature of ca. 20 °C temperature was maintained during the testing procedure by the rheometer's temperature control unit. The considered parallel plate geometry (1 mm gap) is shown in Figure 4.

The testing procedure consisted of a flow curve test anticipated by a pre-shear (20 s at 120 rad/s). The flow curve test consisted of a shear step-up and step-down protocol in which each shear step had a constant shear duration of 20s after an increment or decrement of 20 rad/s. Steady-state values of the step-down parts were used to compute the flow curves.

3.1.2. Sliding Pipe Rheometry

In addition to the MCR-102 rheometer tests, sliding pipe rheometer (Sliper, Schleibinger Geräte Teubert und Greim GmbH, Buchbach, Germany) tests were also performed for all samples. The Sliper test consists of a pipe (D = 125 mm) filled with paste that slides down a platform with a centrally positioned pressure sensor. During this downward motion over a pipe length of 0.5 m, a laser distance sensor captures the velocity of the sliding pipe and thus the discharge of the flow. The geometrical layout of the Sliper is depicted in Figure 4. This sliding procedure is repeated several times each with a different weight attached to the Sliper, allowing for variation in the imposed discharge. In that way, a pressure loss–discharge diagram can be recorded as depicted in Figure 5.

(a) 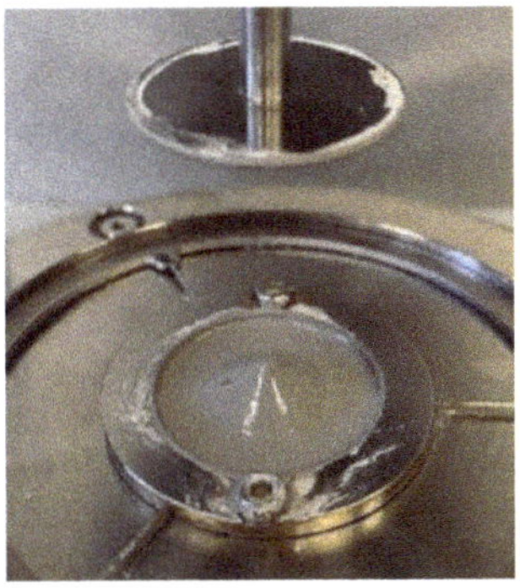 (b) 

Figure 4. Rheometer test set-up configurations: (**a**) rotational parallel plate rheometer configuration (Anton Paar MCR-102 with a 1 mm gap); (**b**) sliding pipe rheometer (Sliper).

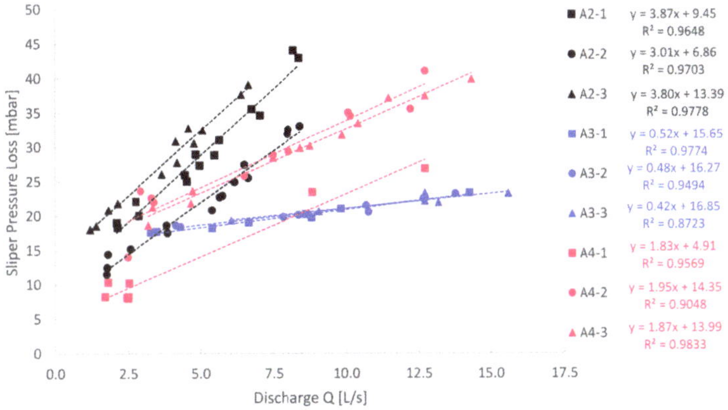

Figure 5. Sliper pressure loss recorded as a function of the discharge for all conducted experiments (mix design A2 to A4). No feasible results could be obtained for mix design A1. Rheological parameters were obtained from the linear regressions.

These pressure loss and discharge values were correlated by a linear regression. The linear regressions were then used to characterise the rheological parameters for each of the model pastes based on a reversed engineered Buckingham–Reiner approach. In that way, the Bingham yield stress and plastic viscosity could be obtained by a second rheometric approach. Some Sliper measurements were jeopardised due to bad sensor connectivity. Due to severe leakage, no Sliper measurements could be obtained for mix design A1.

An overview of rheological parameters (i.e., density ρ, Bingham yield stress τ_0 and plastic viscosity μ) obtained by both the MCR-102 rheometer and Sliper tests is depicted in Table 3.

Table 3. Rheological overview of several mixture design samples, depicting the paste density ρ, Bingham yield stress τ_0 and plastic viscosity μ obtained from different rheometers.

		MCR-102		Sliper	
Mixture	ρ	τ_0	μ	τ_0	μ
Sample	**[kg/L]**	**[Pa]**	**[Pa.s]**	**[Pa]**	**[Pa.s]**
A1-1	1.776	55.5	0.5	N/A	N/A
A1-2	1.786	69.7	0.7	N/A	N/A
A1-3	1.792	67.7	0.7	N/A	N/A
A1-4	1.794	94.1	0.9	N/A	N/A
A2-1	2.035	57.2	8.5	44.3	4.6
A2-2	2.075	65.7	9.1	32.2	3.6
A2-3	2.047	43.6	10.3	62.8	4.6
A3-1	1.903	150.5	2.6	73.4	0.6
A3-2	1.865	165.6	2.7	76.3	0.6
A3-3	1.896	181.1	2.8	79.0	0.5
A4-1	1.980	135.1	4.6	23.0	2.2
A4-2	1.979	129.3	4.8	67.3	2.3
A4-3	1.981	110.7	5.0	65.6	2.2

3.2. *Small-Scale Pumping*

To quantitatively assess the numerical performance, a validation experiment was set up based on a pumping experiment. A pumping circuit was designed for stainless steel pipes, with an internal diameter of $D = 26.64$ mm (DN25). Several pressure sensors were inserted in the pipeline using T-connections in which the pressure sensors were screwed as close as possible to the (virtual) pipe wall (see Figure 6). In that way, the pressure loss was recorded over respective distances (between pressure sensor 2 and 4 in Figure 7). To also capture the induced discharge, a container was supported from a load cell. The paste was injected into the pumping circuit by a pressure controlled barrel, allowing for a variation in driving pressure and discharge, which were recorded accordingly. The geometric configuration of the small scale pumping set-up (SPS) is depicted in Figure 7. In that way, an experimental validation data set was provided by a pumping experiment.

Figure 6. Example of a pressure sensor being screwed in a T-connector as close as possible to the virtual pipe wall.

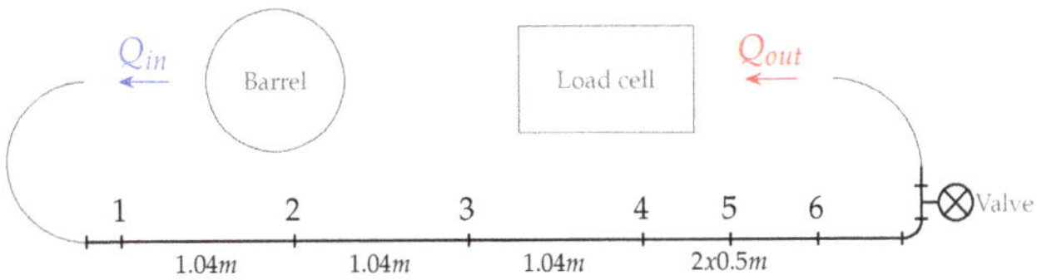

Figure 7. Layout of a small-scale pumping circuit. Each number corresponds to a pressure sensor position. At the inlet, the paste was injected via a pressure barrel, for which the outlet discharge Q_{out} was measured by a load cell.

4. Reliability Irrespective of Rheological Input

To assess the reliability of the numerical outcome irrespective of the rheological input, two comparisons were made. First, pipe flow simulations were compared to simulations performed in the literature. Second, simulations were compared with the expected Buckingham–Reiner theory of Bingham Poiseuille flows. This was performed for re-simulated cases of performed small-scale pumping experiments.

4.1. Comparison with the Literature

To compare the numerical pipe flow simulations with the literature, pumping simulations of concrete suspensions are compared in particular. This is because cementitious paste suspensions are the main scope of this work. More specifically, we compared the simulations performed by Tichko [13]. Tichko [13] simulated the flow of concrete pumping experiments using a homogeneous suspension approach. Two pumping series A and B were considered in a DN100 pipeline. The lack of wall lubrication layer effect in these simulations is justified from a pure comparison point of view. Figure 8 outlines the comparison of concrete pipe flow simulations by Tichko [13] performed in ANSYS FLUENT®, with the re-simulated ones performed in this work by OpenFOAM. No under-relaxation was considered, and a more strict regularisation was applied ($GNAR \approx 1500$). The error is presented relative to the expected Buckingham–Reiner pressure loss.

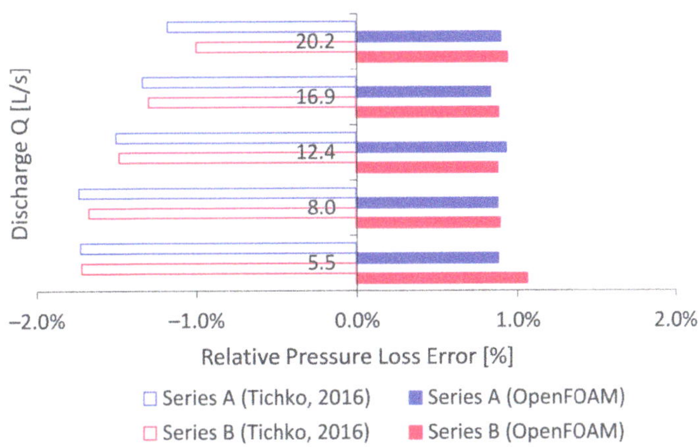

Figure 8. Literature comparison of the numerical concrete pipe flow simulations performed by Tichko [13], relative to the expected Buckingham–Reiner theory. The results indicate that OpenFOAM performs at least as good as, if not better than, commercial CFD software ANSYS FLUENT®.

Comparing several discharges, it can be seen that the results in this framework are more accurate with a relative pressure loss error below 1%. Hence, it is concluded that homogeneous Bingham pipe flow simulations performed by OpenFOAM are at least as accurate as, if not better than, simulations by commercial software ANSYS FLUENT®.

4.2. Comparison with the Theory

To assess the numerical simulations irrespective of their rheological input, they are compared with the expected theory of Buckingham–Reiner. This is based on re-simulated cases of all performed small-scale experiments, as depicted in Table 3.

All small-scale pumping experiments were simulated based on their respective rheological input values. To assess these simulations irrespective of their input, they were analysed in comparison with the expected theory of Buckingham–Reiner. This totals to 213 simulation cases. An overview of the simulation outcome is depicted in Figures 9 and 10. Here, the simulated pressure loss is plotted as a function of the discharge. Additionally, the pressure loss as well as the discharge are transformed into dimensionless form via the pressure number $Pn = \Delta pR/(2L\tau_0)$ and dimensionless discharge $\widehat{Q} = Q/(R^3\tau_0/\mu)$ after De Schryver and De Schutter [55].

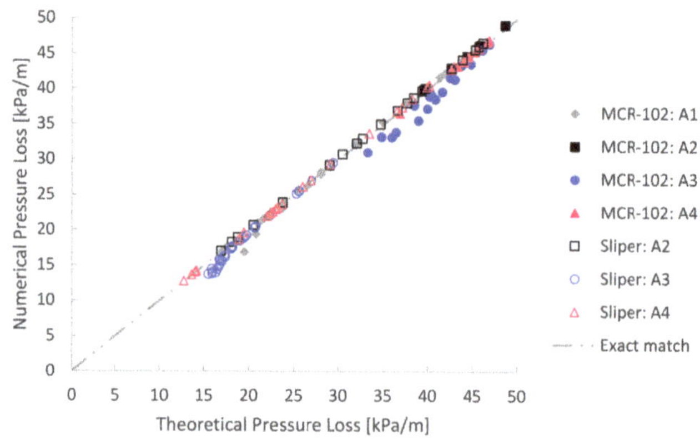

Figure 9. A good agreement is obtained between the simulated pressure loss and the expected theoretical Buckingham–Reiner theory.

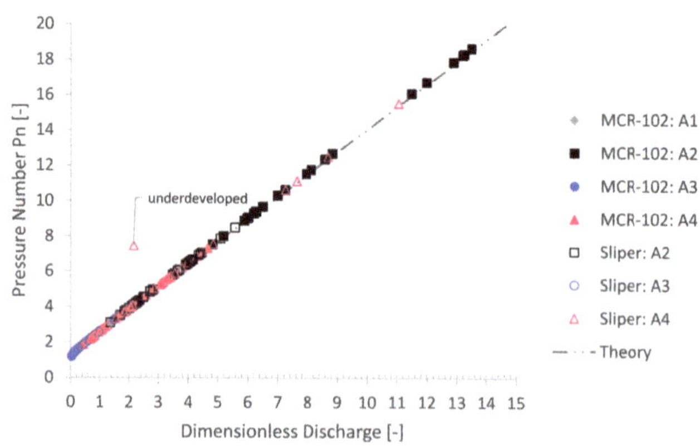

Figure 10. Using a dimensionless form of the Bingham Poiseuille flow, again, a good agreement is achieved between the simulated pressure loss and the expected Buckingham–Reiner theory. Indeed, the simulations map onto a single Bingham discharge curve.

It is clear that the numerical simulations, irrespective of their rheological input, are in good agreement with the Buckingham–Reiner theory (cf. Figures 9 and 10). Indeed, apart from one underdeveloped case, all simulation results map onto a single discharge diagram curve, as expected for pumping of a Bingham fluid [55,56].

To quantitatively assess the agreement, a relative pressure loss error is defined in Figure 11 by comparing the pressure loss from the simulation, with the pressure loss expected by Buckingham–Reiner. The results in Figure 11 show that, for higher discharges Q, the simulations are significantly accurate with a relative error below 1%. The convergence to a relative error of 1% for higher discharges may be restrained by the mesh resolution and default-like solving settings. Therefore, it would be expected that, for a proportionally finer mesh resolution as well as higher-order numerical schemes and more restricted solver tolerances, the relative error can be reduced even more. Hence, it is concluded that, if one considers pipe flow with a mesh resolution and default-like solver settings, as considered in this work, no error higher than 1% would be expected. Given this result, one may have confidence in the numerical simulation.

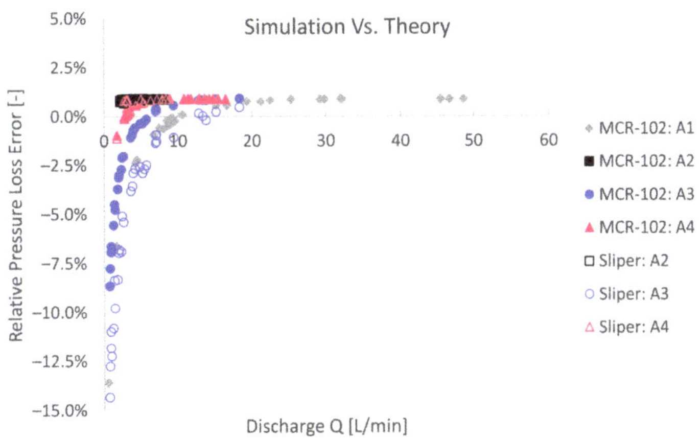

Figure 11. Significantly accurate simulations can be obtained in comparison with the Buckingham–Reiner theory. For higher discharges, a relative pressure loss error below 1% can be obtained. Simulations with lower discharges or higher yield stresses can be more influenced by the unregularised viscoplastic problem [25,43].

For lower discharges on the other hand, the relative error increases. Since it concerns a relative error, this is partially explained by the lower reference pressure in the division of the relative error. The relative error is namely the absolute error divided by the expected pressure loss, which in its turn decreases for lower discharges.

A second explanation for higher relative theoretical errors is related to the numerical methodology. The Bingham model is approximated by a bi-viscous regularisation approach, in which an apparent viscosity is used, restrained by a pre-defined maximum. On the one hand, this has to be performed to avoid numerical instability. On the other hand, this impedes numerical accuracy. To depict the numerical inaccuracy caused by the modelling approach, again dimensionless Bingham formulations are used in Figures 12 and 13 after De Schryver and De Schutter [55]. From these results, it is clear that the lower the pressure number Pn or the lower the discharge number $\widehat{Q}$, the relatively less accurate the simulation becomes compared to the Buckingham–Reiner theory.

Reminiscent to the numerical Bingham stability [25,57,58], one could similarly define a certain flow condition boundary below which a lower relative accuracy could be expected. Although former studies were conducted in the framework of turbulent transition instability, one may also define stability towards the lower Reynolds number end, i.e., laminar unyielded flow, potentially leading to numerical instability. Doing so, based on the considered bi-viscous approach ($GNAR = 1000$), one could state that below a dimensionless discharge $\widehat{Q}$ of ca. 0.069 or a pressure number Pn below ca. 1.25 (5/4), less numerically accurate results could be expected. This means that, when the yield stress contribution becomes more than 4/5 ($PN \leq 1.25$) of the pressure loss or flow regime, less accurate simulations are expected. Numerical inaccuracy or even instability occur for quiescent flow conditions (i.e., low or zero shear zones) [25,28,29]. When the unyielded region becomes relatively large, less accurate results are obtained due to the (un)regularised viscoplastic problem [25,28,29]. A pressure loss number of below 1.25 ($Pn \leq 5/4$) means a plug radius of at least 80% of the pipe radius (i.e., $R_p/R = 1/Pn \geq 4/5$) in accordance with Figure 14 or 64% in terms of plug area (i.e., $A_p/A = 1/Pn^2 \geq 16/25$). This of course is only valid for the considered bi-viscous regularisation approach ($GNAR = 1000$) and pipe flow simulations.

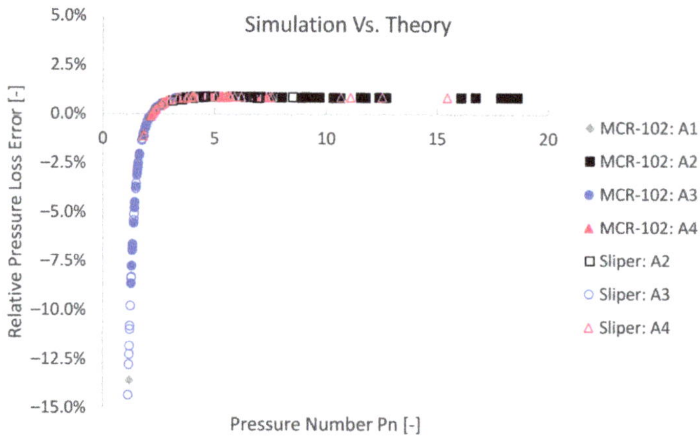

Figure 12. The relative simulation pressure loss error (compared to Buckingham–Reiner) as a function of the pressure number Pn reveals the influence of the flow regime or unregularised viscoplastic problem. The accuracy decreases for lower pressure numbers or for more yield stress dominant flow regimes.

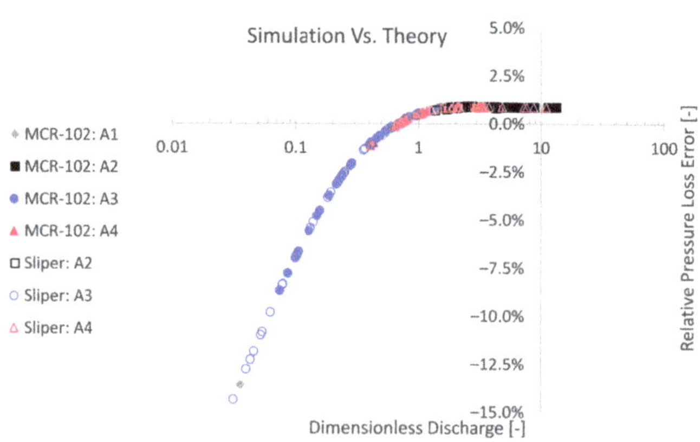

Figure 13. Plotting the relative simulation error as a function of the dimensionless discharge $\widehat{Q}$ also shows that low discharges or more yield stress dominated pipe flows impede numerical accuracy.

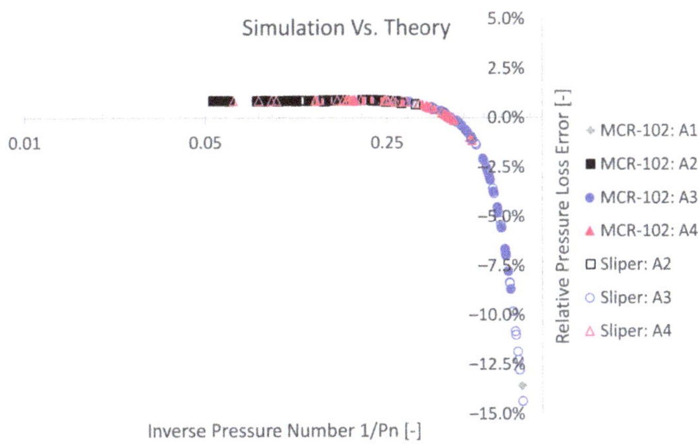

Figure 14. The relative pressure loss error as a function of the inverse pressure number is equivalent to the degree of plug formation. The ratio of plug radius R_p to pipe radius R equals the inverse pressure number $1/Pn$.

This conclusion is in line with Frigaard and Nouar [25] and Jahromi et al. [28], where the closer the applied stress comes to a fully unyielded flow, the higher the error becomes. These errors can become high, even up to 100%, and hence flow start-up problems using a regularised approach are of questionable nature with regard to accuracy. Therefore, the yielded region and regularisation approach should always be verified in the numerical approach to ascertain proficient numerical outcome.

Nevertheless, dimensionless numbers, such as the Bingham number Bn or more specifically the discharge and pressure number ($\widehat{Q}$ and Pn), could serve as an indication of whether numerical inaccuracies or even numerical stability problems could be expected due to the so-called unregularised viscoplastic problem. Concluding the numerical reliability irrespective of its rheological input, based on literature and the Buckingham–Reiner theory, it can be stated that the OpenFOAM simulations are significantly accurate and in significant agreement with the theory. Some rules of thumb have been outlined to ensure adequate numerical accuracy related to the considered regularisation.

To further assess the reliability of the numerical outcome, an assessment was also performed with respect to its rheological input.

5. Reliability Dependent on Rheological Input

To assess the numerical reliability of pipe flow simulations (using OpenFOAM) dependent on its rheological input, a comparison was made with the experimental results obtained from a small-scale pumping experiment. As described in the experimental framework section, four mix designs (A1 to A4) were rheologically characterised based on two different devices (MCR-102 and Sliper, apart from A1) and tested in a small-scale pumping circuit. All experimental pumping series were simulated, each based on two different rheological input values. The first rheological input is the characterisation by the MCR-102 rheometer, the second by Sliper. A single discharge value translates to a single simulation. For all rheological values (Table 3) and discharges combined, this resulted in 213 simulations in total.

Comparison with Experiments

For the analysis, the relative pressure losses obtained by the simulations were compared with the experimental ones. Comparing the numerical prediction with the experiment on a one-to-one basis in Figure 15, it is clear that an exact match was rarely the case. The numerical simulations overpredicted the pumping experiment, whether the rheological input was based on the MCR-102 rheometer or the Sliper. It is however remarkable that the rheological input based on Sliper performs better than the results based on the MCR-102. Some Sliper results even lie in the near vicinity of an exact match. The origin and potential influences of this mismatch are discussed in Section 6.

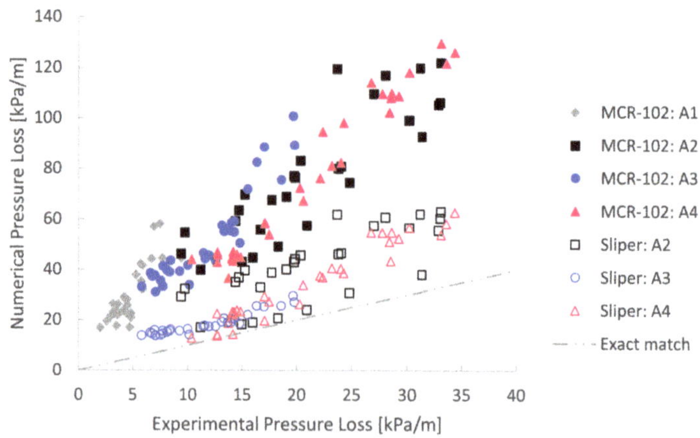

Figure 15. Comparison of numerically simulated pressure loss with experimentally obtained pumping results, for four different mix designs (A1 to A4) and based on rheological input from the MCR-102 rheometer and the Sliper.

For a quantitative reliability assessment of the pipe flow simulations in comparison with the conducted pumping experiments, again, a relative pressure loss error is computed. The relative error depicted in Figures 16 and 17 is, this time, the absolute difference between the simulation and the pumping experiment and relative to the experiment.

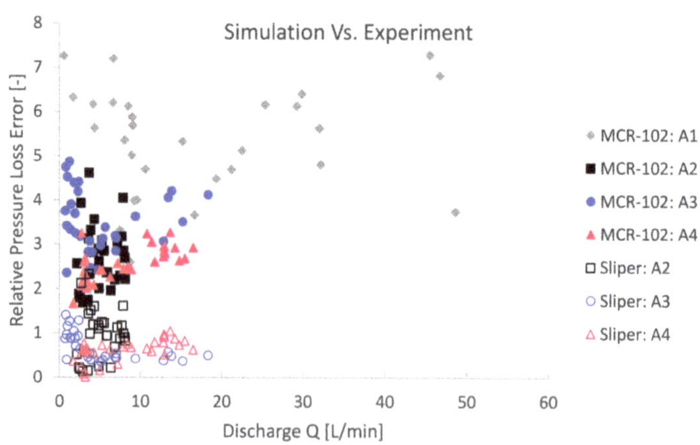

Figure 16. The relative simulation pressure loss error compared to the pumping experiment series indicates that the rheolocigal input from the Sliper is in better agreement than that based on MCR-102.

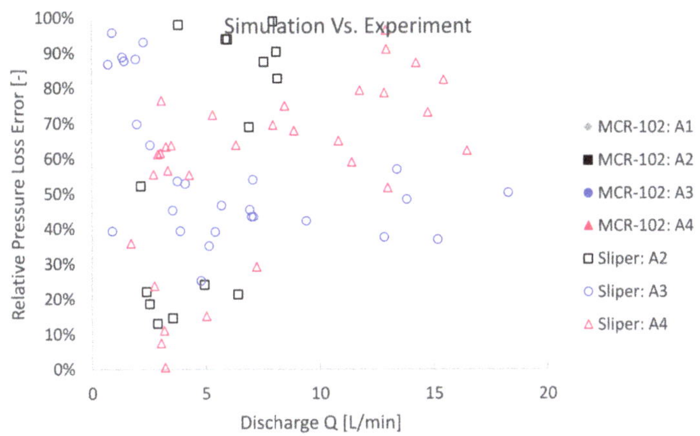

Figure 17. A close-up of the comparison between numerical simulations and pumping experiments shows that the rheological input based on Sliper is significantly better than that for MCR-102, with relative errors ranging from 0% to ca. 100%.

High experimental relative errors were observed. The simulation errors based on MCR-102 were 2 to even 7 times higher than the experiment (Figure 16). For rheological input from the Sliper, the relative error is more limited from 0 to ca. 100% error (Figure 17). This could mean several things. First, this could mean that the rheological characterisation is biased or completely unreliable in a different application or context. Second, this could mean that the theory is biased and that a dominant physical phenomenon was overlooked and, therefore, not simulated. The third explanation is a combination of the former two to some degree: an unreliable rheological bias and physical phenomenon bias. Last, there might be an experimental source of error to an unknown extent.

6. Discussion

The general rules of conduct for appropriate regularisation were not explored in this work. As for most works [21,25,28,29], the regularisation itself is ad hoc and applied for a specific application or problem [25]. Restrained to the context of this work, some general rules of thumbs may be defined, for which less accurate numerical results can be expected. These are only valid for an apparent viscosity ratio $GNAR = 1000$, as advised by Bercovier and Engelman [22], and for the specific case of pipe flow of cementitious paste suspensions. A high maximum apparent viscosity ratio up to 1000 can be at the cost of numerical stability. In the work of Syrakos et al. [26], the solver could only cope with simulations with ratios up to 400. Therefore, it is noteworthy to use under-relaxation to a limited extent, as in this work. This is highly beneficial for numerical stability while still allowing for a very strict regularisation parameter. Even though relaxation smears out the simulation to a limited extent, it allows for simulations that were otherwise infeasible.

In line with O'Donovan and Tanner [23], Jahromi et al. [28], Ahmadi and Karimfazli [29], it is concluded that, for flows where the unyielded region is large—or equivalently, a low pressure number Pn and dimensionless discharge $\widehat{Q}$—less accurate results are expected. Therefore, similar to that stated by Frigaard and Nouar [25], lubrication layer flows are more of a concern because they convey a very large plug zone. This is especially the case for concrete flows. Even though higher shear rates are expected near the pipe wall and therefore higher local related dimensionless discharges (based on the lubrication layer properties) are expected, the numerical outcome and especially the stability are questionable. To ascertain accuracy for these kind of flows, a regularisation parameter as close as possible to zero (or an apparent viscosity cut-off close to infinity: $GNAR \to +\infty$) would be necessary. However, as the regularisation parameter tends towards zero, the simulation tends to blow up [27].

Therefore, an extensive regularisation study on lubrication pipe flows is an important study that could be conducted in the future, especially for the practice of concrete flow simulations.

Despite very accurate results compared to the theory, a high relative experimental error exists, and rarely, an exact match was found between the numerical simulation and experiments. This could be explained by a bias in rheological input, a bias in overlooked underlying physical phenomenon or a combination of both. A fourth error source could be the experimental accuracy or set-up. The pressure sensors have a finite accuracy. Additioanlly, a residual paste membrane found in front of the pressure sensor—but limited to only a few millimetres thickness—could have reduced recorded pressure losses to an unknown extent. Lastly, the accuracy of the density measurement may have had an influence on the discharge computed from the load cell and therefore an indirect influence on the results.

A rheological bias may very well be the case, since significant differences exist in rheological characterisations of cementitious suspensions [14,15]. Hence, the rheological input is certainly a source of uncertainty, as also reported by Schaer et al. [21]. In order to cope with the potentially overlooked rheological bias, it would be more appropriate to obtain rheological input from an application that is closely related to the nature of the aimed engineering application. However, one should be cautious not to overlook any other physical phenomena biases. Doing as such, more reliable numerical predications can be achieved. In fact, one could even consider this to be a model calibration. If for instance rheological data are obtained from a pumping experiment itself, which is then used as the rheological input, it is by definition calibration. This calibration could then be used to predict a new pumping experiment, be it with the same mix design or perhaps even for a different application of the same mix sample.

Despite the uncertain character of rheometers, it does not mean that they cannot be used to assess rheological properties. They still serve as a reference framework within which rheological differences can be addressed. The difficulty lies in a change of context.

A theoretical bias could also lie at the origin of the simulation mismatch. In order not to be biased, it is the researcher's or engineer's task to consider the adequacy of the considered numerical or, by extension, the theoretical modelling approach. In the case of simple cementitious paste pipe flows, the behaviour is well established by a Bingham approach and a behavioural bias by, for instance slippage, or a wall lubrication layer is therefore not expected [38,59]. Even though the rheological input based on the Sliper tests was significantly better than for the MCR-102, the results mostly overestimated the pumping experiment. Since the flow inside the Sliper device is not fully developed, the rheological parameters are expected to be at least overestimated, and therefore, it could partially explain the overprediction.

Although no wall lubrication layer is expected in the case of paste, there may have been some influence by the pre-usage water cleaning procedure. Even though all residual water was pumped out, limited intermixed water near the pipe wall surface, slightly altering the mixture composition, may have had a contribution in reduced flow resistance. From a future perspective, it may therefore be important to experimentally investigate the influence of a pre-wetting procedure of pumping pipelines, whether that is using water or cement-paste itself before the application of concrete pumping.

In the case of more complicated pipe flows of fresh concrete, a bias may occur due to not considering the so-called lubrication layer approach. Although no concrete and only cementitious pastes and their modelling were considered in this work, knowing whether such a layer could adequately be simulated is another important interpretation of this work since the lubrication layer is assigned to the formation of a cementitious-like layer of paste attached to the vicinity of the pipeline wall. Over the past decade, this phenomenon has been investigated and even though some promising approaches were established, it is still an ongoing research subject on how to adequately model this so-called lubrication layer phenomenon [34,60–62].

From a preliminary projective point of view, one could perhaps extend the reliability interpretation of this work to more complicated flows such as concrete. By making use of rheological input closely related to the pumping application, it could be expected that modelling an additional wall lubrication layer effect would be in significant agreement with related theoretical models, such as the model of Kaplan [63] or modified equations of Kwon et al. [16]. Indeed, in the work of Secrieru et al. [34], such numerical simulations with a lubrication layer have been performed and a good agreement can be observed between the numerical simulation and the analytical equations. Despite a good agreement between their successful simulations, lubricational flows in general, however, form a concern from the viewpoint of numerical accuracy and stability, similar to that in Frigaard and Nouar [25]. As mentioned before, an elaborate regularisation accuracy study for lubrication layer pipe flows would especially facilitate engineering practices to simulate concrete pumping. Aside from that, the fact that their numerical model was calibrated for the lubrication layer thickness based on the Sliper rheometry, which lies more close to the nature of concrete pumping itself, forms additional evidence that more reliable simulation results can be obtained by rheological parameters obtained from rheometry configurations as close as possible to the aimed application.

Despite such a preliminary interpretation and from a future perspective, it would be relevant to assess the numerical reliability including a lubrication layer approach. Since not only cementitious paste but also concrete flow simulations are important in construction processes, where the wall lubrication layer effect may not be overlooked. Again, the reliability could be assessed from both a theoretical point of view as an experimental viewpoint. Since no closed analytical expressions are available for a modelling approach by particle migration, the reliability could be assessed for a dual Bingham fluid approach. Closed analytical Poiseuille flow extensions are namely available, such as the model of Kaplan [63] or modified equations of Kwon et al. [16]. In the meantime, the influence of numerical regularisation could be investigated for such flows.

7. Conclusions

To build confidence in and to answer the question on numerical reliability, simplified numerical simulations of horizontal, laminar Bingham pipe flow were assessed for their outcome. Specific to flow simulations of cementitious Bingham paste suspensions, and by extension to model fresh concrete construction processes for engineers in practice, the assessment was performed irrespective of and dependent on their rheological input.

The reliability assessment irrespective of the rheological input revealed that numerical finite volume simulations by OpenFOAM are very accurate compared to the theoretical equations. From a practical, construction process engineering viewpoint, an accuracy of 1% is reasonable. At least if the simulations are performed by meeting the defined rules of thumb: a proper regularisation approach (e.g., a bi-viscous approach with $GNAR = \mu_{max}/\mu \approx 1000$) and for higher discharge regimes defined by the dimensionless pressure number ($Pn > 1.25$) and/or discharge ($\hat{Q} > 0.069$). More accurate results can be obtained if a higher mesh resolution is considered and if higher order numerical schemes are considered. Although it should be limited, artificial under-relaxation appeared to be a promising technique to cope with potentially occurring numerical instabilities due to the (un)regularised viscoplastic problem [25,43]. This is especially important for lubrication flows, such as concrete pumping, although accuracy and stability are still of concern for lubrication type flows.

Based on the reliability dependent on the rheological input, it was shown that, due to the uncertain character of rheological input data, more reliable numerical predictions can be obtained by considering rheometry that lies more close to the configuration of the engineer's aimed application. In that way, one is able to cope with a possible bias due to rheological uncertainty. One should however be cautious not to overlook a physical phenomenon. All significant physical phenomena should be modelled as well. Unlike cementitious pastes, this is, for instance, the case for concrete pumping where the wall lubrication layer effect cannot be overlooked.

Even though the numerical simulations used in this work may not cover the full spectrum of numerical simulations for the behaviour of cementitious paste suspensions, it still forms a certain benchmark for a CFD finite volume modelling approach in OpenFOAM to model cementitious paste suspensions as a Bingham fluid important for the construction industry.

From a future perspective, it would be especially of interest for engineering practice to build confidence in CFD simulations by performing a similar reliability assessment considering an additional lubrication layer effect at the pipe wall. One could similarly compare the numerical outcome irrespective of the rheological input with related theories as well as make an assessment based on rheological input by comparing it with experiments. Simultaneously, with regard to accuracy, one may also address the concern of regularisation and stability for lubrication layer flows. Another important study that could be of experimental interest is the influence of a pre-wetting or pre-lubricating procedure for concrete pumping.

Author Contributions: Conceptualisation, R.D.S.; methodology, R.D.S. and K.E.C.; software, R.D.S.; validation, R.D.S.; formal analysis, R.D.S.; investigation, R.D.S., K.E.C. and M.Y.Y.; resources, R.D.S. and K.L.; data curation, R.D.S.; writing—original draft preparation, R.D.S.; writing—review and editing, R.D.S., K.E.C, K.L., M.Y.Y. and G.D.S.; visualisation, R.D.S.; supervision, G.D.S.; project administration, K.L. and G.D.S.; funding acquisition, K.L. and G.D.S. All authors have read and agreed to the published version of the manuscript.

Funding: This research was funded by Horizon 2020 European Research Council Advanced Grant project "SmartCast", grant number 693755.

Institutional Review Board Statement: Not applicable.

Informed Consent Statement: Not applicable.

Data Availability Statement: The data presented in this study are available on reasonable request from the corresponding author.

Conflicts of Interest: The authors declare no conflict of interest. The funders had no role in the design of the study; in the collection, analyses or interpretation of data; in the writing of the manuscript; or in the decision to publish the results.

Appendix A. Simulation Settings

An overview of the numerical discretisation schemes used is depicted in Table A1. The numerical solver and respective settings are summarised in Table A2.

Table A1. Discretisation scheme settings in OpenFOAM.

Discretisation	Field	Scheme
ddtSchemes	default	Gauss linear
gradSchemes	default	Gauss linear
	grad(p)	leastSquares
divSchemes	default	none
	div(phi,U)	Gauss linear corrected
laplacianSchemes	default	none
	laplacian(nu,U)	Gauss linear corrected
	laplacian(1 ǀ A(U),p)	Gauss linear corrected
interpolationSchemes	default	linear
snGradSchemes	default	corrected

Table A2. Solver solution settings in OpenFOAM.

Application	nonNewtonianIcoFoam	
Pressure-Velocity Coupling	PISO	
nCorrectors	5	
nNonOrthogonalCorrectors	0	
	Fields	
Solver Settings	*p*	*U*
Solver	GAMG	smoothSolver
Smoother	GaussSeidel	symGaussSeidel
Tolerance	1×10^{-6}	1×10^{-5}
Relative Tolerance	0.1	0
Final Relative Tolerance	0	
Relaxation Factor	1.0	0.95

References

1. De Schutter, G.; Lesage, K. Active control of properties of concrete: A (p)review. *Mater. Struct. Constr.* **2018**, *51*, 1–16. [CrossRef]
2. Roussel, N.; Geiker, M.R.; Dufour, F.; Thrane, L.N.; Szabo, P. Computational modeling of concrete flow: General overview. *Cem. Concr. Res.* **2007**, *37*, 1298–1307. [CrossRef]
3. Gram, A.; Silfwerbrand, J. Numerical simulation of fresh SCC flow: applications. *Mater. Struct.* **2011**, *44*, 805–813. [CrossRef]
4. Thrane, L.N. 10—Modelling the flow of self-compacting concrete. In *Understanding the Rheology of Concrete*; Roussel, N., Ed.; Woodhead Publishing Series in Civil and Structural Engineering, Woodhead Publishing Limited: Sawston, UK 2012; pp. 259–285; [CrossRef]
5. Roussel, N.; Gram, A. *Simulation of Fresh Concrete Flow*; RILEM State-of-the-Art Reports; Springer: Dordrecht, The Netherlands, 2014; Volume 15. [CrossRef]
6. Tichko, S.; Van De Maele, J.; Vanmassenhove, N.; De Schutter, G.; Vierendeels, J.; Verhoeven, R.; Troch, P. Numerical simulation of formwork pressure while pumping self-compacting concrete bottom-up. *Eng. Struct.* **2014**, *70*, 218–233. [CrossRef]
7. Roussel, N.; Gram, A.; Cremonesi, M.; Ferrara, L.; Krenzer, K.; Mechtcherine, V.; Shyshko, S.; Skocec, J.; Spangenberg, J.; Svec, O.; et al. Numerical simulations of concrete flow: A benchmark comparison. *Cem. Concr. Res.* **2016**, *79*, 265–271. [CrossRef]
8. Wallevik, J.E.; Wallevik, O.H. Analysis of shear rate inside a concrete truck mixer. *Cem. Concr. Res.* **2017**, *95*, 9–17. [CrossRef]
9. Wallevik, J.E. Rheological properties of cement paste: Thixotropic behavior and structural breakdown. *Cem. Concr. Res.* **2009**, *39*, 14–29. [CrossRef]
10. Patzák, B.; Bittnar, Z. Modeling of fresh concrete flow. *Comput. Struct.* **2009**, *87*, 962–969. [CrossRef]
11. Kolařík, F.; Patzák, B.; Thrane, L. Modeling of fiber orientation in viscous fluid flow with application to self-compacting concrete. *Comput. Struct.* **2015**, *154*, 91–100. [CrossRef]
12. Kolařík, F.; Patzák, B.; Zeman, J. Computational homogenization of fresh concrete flow around reinforcing bars. *Comput. Struct.* **2018**, *207*, 37–49. [CrossRef]
13. Tichko, S. Hydrodynamic Modelling of the Flow of Self-Compacting Concrete in Formworks. Ph.D. Dissertation, Ghent University, Ghent, Belgium, 2016.
14. Hočevar, A.; Kavčič, F.; Bokan-Bosiljkov, V. Rheological parameters of fresh concrete-comparison of rheometers. *Gradevinar* **2013**, *65*, 99–109. [CrossRef]
15. Wallevik, O.H.; Feys, D.; Wallevik, J.E.; Khayat, K.H. Avoiding inaccurate interpretations of rheological measurements for cement-based materials. *Cem. Concr. Res.* **2015**, *78*, 100–109. [CrossRef]
16. Kwon, S.H.; Jang, K.P.; Kim, J.H.; Shah, S.P. State of the Art on Prediction of Concrete Pumping. *Int. J. Concr. Struct. Mater.* **2016**, *10*, 75–85. [CrossRef]
17. Kim, J.S.; Kwon, S.H.; Jang, K.P.; Choi, M.S. Concrete pumping prediction considering different measurement of the rheological properties. *Constr. Build. Mater.* **2018**, *171*, 493–503. [CrossRef]
18. Ley-Hernández, A.M.; Feys, D.; Kumar, A. How do different testing procedures affect the rheological properties of cement paste? *Cem. Concr. Res.* **2020**, *137*, 106189. [CrossRef]
19. Haist, M.; Link, J.; Nicia, D.; Leinitz, S.; Baumert, C.; von Bronk, T.; Cotardo, D.; Eslami Pirharati, M.; Fataei, S.; Garrecht, H.; et al. Interlaboratory study on rheological properties of cement pastes and reference substances: comparability of measurements performed with different rheometers and measurement geometries. *Mater. Structs.* **2020**, *53*, 92. [CrossRef]
20. Wallevik, O.H.; Wallevik, J.E. Rheology as a tool in concrete science: The use of rheographs and workability boxes. *Cem. Concr. Res.* **2011**, *41*, 1279–1288. [CrossRef]
21. Schaer, N.; Vazquez, J.; Dufresne, M.; Isenmann, G.; Wertel, J. On the determination of the yield surface within the flow of yield stress fluids using computational fluid dynamics. *J. Appl. Fluid Mech.* **2018**, *11*, 971–982. [CrossRef]

22. Bercovier, M.; Engelman, M. A finite-element method for incompressible non-Newtonian flows. *J. Comput. Phys.* **1980**, *36*, 313–326. [CrossRef]
23. O'Donovan, E.J.; Tanner, R.I. Numerical study of the Bingham squeeze film problem. *J. Non-Newton. Fluid Mech.* **1984**, *15*, 75–83. [CrossRef]
24. Beverly, C.R.; Tanner, R.I. Numerical Analysis of Extrudate Swell in Viscoelastic Materials with Yield Stress. *J. Rheol.* **1989**, *33*, 989–1009. [CrossRef]
25. Frigaard, I.A.; Nouar, C. On the usage of viscosity regularisation methods for visco-plastic fluid flow computation. *J. Non-Newton. Fluid Mech.* **2005**, *127*, 1–26. [CrossRef]
26. Syrakos, A.; Georgiou, G.C.; Alexandrou, A.N. Performance of the finite volume method in solving regularised Bingham flows: Inertia effects in the lid-driven cavity flow. *J. Non-Newton. Fluid Mech.* **2014**, *208-209*, 88–107. [CrossRef]
27. El Khouja, N.; Roquet, N.; Cazacliu, B. Analysis of a Regularized Bingham Model with Pressure-Dependent Yield Stress. *J. Math. Fluid Mech.* **2015**, *17*, 723–739. [CrossRef]
28. Jahromi, H.R.; Raisi, A.; Ghasemi, B.; Nadooshan, A.A. Numerical study of the Bingham fluid flow in a cylindrical enclosure with exact Bingham model. *J. Braz. Soc. Mech. Sci. Eng.* **2020**, *42*, 1–15. [CrossRef]
29. Ahmadi, A.; Karimfazli, I. A quantitative evaluation of viscosity regularization in predicting transient flows of viscoplastic fluids. *J. Non-Newton. Fluid Mech.* **2021**, *287*, 104429. [CrossRef]
30. Gram, A.; Lagerblad, B. Obtaining Rheological Parameters from Slump Flow Test for Self-Compacting Concrete. In Proceedings of the International Conference on Sustainable Construction Materials & Technologies, SCMT3, Kyoto, Japan, 18–21 August 2013.
31. Choi, M.S.; Kim, Y.J.; Kwon, S.H. Prediction on pipe flow of pumped concrete based on shear-induced particle migration. *Cem. Concr. Res.* **2013**, *52*, 216–224. [CrossRef]
32. Roussel, N.; Spangenberg, J.; Wallevik, J.; Wolfs, R. Numerical simulations of concrete processing: From standard formative casting to additive manufacturing. *Cem. Concr. Res.* **2020**, *135*, 106075. [CrossRef]
33. Wallevik, J.E. Thixotropic investigation on cement paste: Experimental and numerical approach. *J. Non-Newton. Fluid Mech.* **2005**, *132*, 86–99. [CrossRef]
34. Secrieru, E.; Mohamed, W.; Fataei, S.; Mechtcherine, V. Assessment and prediction of concrete flow and pumping pressure in pipeline. *Cem. Concr. Compos.* **2020**, *107*, 103495. [CrossRef]
35. Perrot, A.; Al, E. From analytical methods to numerical simulations: a process engineering toolbox for 3D-concrete printing. *Cem. Concr. Compos.* **2021**, *122*, 104164. [CrossRef]
36. Welty, J.R.; Wicks, C.E.; Wilson, R.E.; Rorrer, G.L. *Fundamentals of Momentum, Heat, and Mass Transfer*; John Wiley & Sons, Inc.: Hoboken, NJ, USA, 2008; p. 201,
37. Wallevik, J.E. Rheology of Particle Suspension: Fresh Concrete, Mortar and Cement Paste with Various Types of Lignosulfates. Ph.D. Thesis, The Norwegian University of Science and Technology, Trondheim, Norway, 2003.
38. Banfill, P.F.G. Rheology of Fresh Cement and Concrete. *Rheol. Rev.* **2006**, *2006*, 61–130. [CrossRef]
39. Feys, D.; Verhoeven, R.; De Schutter, G. Evaluation of time independent rheological models applicable to fresh self-compacting concrete. *Appl. Rheol.* **2007**, *17*, 1–10. [CrossRef]
40. Ferraris, C.F.; Martys, N.S. 3—Concrete Rheometers. In *Understanding the Rheology of Concrete*; Roussel, N., Ed.; Woodhead Publishing Series in Civil and Structural Engineering, Woodhead Publishing Limited: Sawston, UK 2012; pp. 63–82. [CrossRef]
41. Feys, D.; Wallevik, J.E.; Yahia, A.; Khayat, K.H.; Wallevik, O.H. Extension of the Reiner-Riwlin equation to determine modified Bingham parameters measured in coaxial cylinders rheometers. *Mater. Struct.* **2013**, *46*, 289–311. [CrossRef]
42. Yim, H.; Kim, J.; Kwon, S. Effect of Admixtures on the Yield Stresses of Cement Pastes under High Hydrostatic Pressures. *Materials* **2016**, *9*, 147. [CrossRef] [PubMed]
43. Saramito, P.; Wachs, A. Progress in numerical simulation of yield stress fluid flows. *Rheol. Acta* **2017**, *56*, 211–230. [CrossRef]
44. Bleyer, J.; Maillard, M.; De Buhan, P.; Coussot, P. Efficient numerical computations of yield stress fluid flows using second-order cone programming. *Comput. Methods Appl. Mech. Eng.* **2015**, *283*, 599–614. [CrossRef]
45. Pimenta, F.; Alves, M.A. Stabilization of an open-source finite-volume solver for viscoelastic fluid flows. *J. Non-Newton. Fluid Mech.* **2017**, *239*, 85–104. [CrossRef]
46. Bullough, W.; Ellam, D.; Wong, A.; Tozer, R. Computational fluid dynamics in the flow of ERF/MRF in control devices and of oil through piezo-hydraulic valves. *Comput. Struct.* **2008**, *86*, 266–280. [CrossRef]
47. Papanastasiou, T.C. Flows of Materials with Yield. *J. Rheol.* **1987**, *31*, 385–404. [CrossRef]
48. Papanastasiou, T.C.; Boudouvis, A.G. Flows of viscoplastic materials: Models and computations. *Comput. Struct.* **1997**, *64*, 677–694. [CrossRef]
49. Cremonesi, M.; Frangi, A.; Perego, U. A Lagrangian finite element approach for the simulation of water-waves induced by landslides. *Comput. Struct.* **2011**, *89*, 1086–1093. [CrossRef]
50. Saramito, P.; Roquet, N. An adaptive finite element method for viscoplastic fluid flows in pipes. *Comput. Methods Appl. Mech. Eng.* **2001**, *190*, 5391–5412. [CrossRef]
51. Burgos, G.R.; Alexandrou, A.N.; Entov, V. On the determination of yield surfaces in Herschel-Bulkley fluids. *J. Rheol.* **1999**, *43*, 463–483. [CrossRef]
52. Chupin, L.; Dubois, T. A bi-projection method for Bingham type flows. *Comput. Math. Appl.* **2016**, *72*, 1263–1286. [CrossRef]
53. De Schutter, G.; Feys, D. Pumping of Fresh Concrete: Insights and Challenges. *RILEM Tech. Lett.* **2016**, *1*, 76. [CrossRef]

54. Mechtcherine, V.; Nerella, V.N.; Kasten, K. Testing pumpability of concrete using Sliding Pipe Rheometer. *Constr. Build. Mater.* **2014**, *53*, 312–323. [CrossRef]
55. De Schryver, R.; De Schutter, G. Insights in thixotropic concrete pumping by a Poiseuille flow extension. *Appl. Rheol.* **2020**, *30*, 77–101. [CrossRef]
56. Billingham, J.; Ferguson, J.W. Laminar, unidirectional flow of a thixotropic fluid in a circular pipe. *J. Non-Newton. Fluid Mech.* **1993**, *47*, 21–55. [CrossRef]
57. Frigaard, I.A.; Howison, S.D.; Sobey, I.J. On the Stability of Poiseuille Flow of a Bingham Fluid. *J. Fluid Mech.* **1994**, *263*, 133–150. [CrossRef]
58. Frigaard, I.; Nouar, C. On three-dimensional linear stability of Poiseuille flow of Bingham fluids. *Phys. Fluids* **2003**, *15*, 2843–2851. [CrossRef]
59. Feys, D.; Verhoeven, R.; Schutter, G.D. Pipe flow velocity profiles of complex suspensions, like concrete. In Proceedings of the National Congress on Theoretical and Applied Mechanics, 8th, Proceedings, Brussels, Belgium, 28–29 May 2009; pp. 66–73.
60. Haustein, M.A.; Kluwe, M.N.; Schwarze, R. Experimental investigation of the pumping of a model-concrete through pipes. *Materials* **2020**, *13*, 1161. [CrossRef] [PubMed]
61. Fataei, S.; Secrieru, E.; Mechtcherine, V. Experimental insights into concrete flow-regimes subject to shear-induced particle migration (SIPM) during pumping. *Materials* **2020**, *13*, 1233. [CrossRef] [PubMed]
62. Salinas, A.; Feys, D. Estimation of Lubrication Layer Thickness and Composition through Reverse Engineering of Interface Rheometry Tests. *Materials* **2020**, *13*, 1799. [CrossRef]
63. Kaplan, D. Pompage Des Bétons. Ph.D. Thesis, École Nationale des Ponts et Chaussées, Marne-la-Vallée, France, Laboratoire Central des Ponts et Chaussées, Paris, France, 2001.

materials

MDPI

Article

Phonon Transport and Thermoelectric Properties of Imidazole-Graphyne

Yanyan Chen [1,2], Jie Sun [1,2], Wei Kang [2] and Qian Wang [1,2,*]

1 School of Materials Science and Engineering, Peking University, Beijing 100871, China; yanyanchen@pku.edu.cn (Y.C.); sunjie6@pku.edu.cn (J.S.)
2 Center for Applied Physics and Technology, HEPDS, College of Engineering, Peking University, Beijing 100871, China; weikang@pku.edu.cn
* Correspondence: qianwang2@pku.edu.cn

Abstract: The pentagon has been proven to be an important structural unit for carbon materials, leading to different physical and chemical properties from those of hexagon-based allotropes. Following the development from graphene to penta-graphene, a breakthrough has very recently been made for graphyne—for example, imidazole-graphyne (ID-GY) was formed by assembling experimentally synthesized pentagonal imidazole molecules and acetylenic linkers. In this work, we study the thermal properties and thermoelectric performance of ID-GY by combining first principle calculations with the Boltzmann transport theory. The calculated lattice thermal conductivity of ID-GY is 10.76 W/mK at 300 K, which is only one tenth of that of γ-graphyne (106.24 W/mK). A detailed analysis of the harmonic and anharmonic properties, including the phonon group velocity, phonon lifetime, atomic displacement parameter, and bond energy curves, reveals that the low lattice thermal conductivity can be attributed to the low Young's modulus, low Debye temperature, and high Grüneisen parameter. Furthermore, at room temperature, ID-GY can reach a high ZT value of 0.46 with a 5.8×10^{12} cm^{-2} hole concentration, which is much higher than the value for many other carbon-based materials. This work demonstrates that changing structural units from hexagonal to pentagonal can significantly reduce the lattice thermal conductivity and enhance the thermoelectric performance of carbon-based materials.

Keywords: pentagon-based 2D material; thermal conductivity; thermoelectric properties; anharmonicity

Citation: Chen, Y.; Sun, J.; Kang, W.; Wang, Q. Phonon Transport and Thermoelectric Properties of Imidazole-Graphyne. *Materials* **2021**, *14*, 5604. https://doi.org/10.3390/ma14195604

Academic Editors: Michele Bacciocchi and Abbas S. Milani

Received: 5 September 2021
Accepted: 22 September 2021
Published: 27 September 2021

1. Introduction

Thermoelectric materials that can convert waste heat to electricity based on the Seebeck effect have aroused great attention in the energy field. The conversion efficiency of thermoelectric materials is evaluated by a dimensionless figure of merit (ZT), $ZT = S^2\sigma T/(k_e + k_l)$, which depends on the synergetic effect of the Seebeck coefficient (S), electrical conductivity (σ), absolute temperature (T), electronic thermal conductivity (k_e), and lattice thermal conductivity (k_l). However, most commercial thermoelectric materials are based on elements that are relatively scarce and/or toxic, such as Bi_2Te_3 [1], PbTe [2], and Sb_2Te_3 [3]. Therefore, there is a need to find other earth-abundant and environmentally friendly materials with a good thermoelectric performance.

For this, carbon-based materials can be candidates because of their nontoxicity, light weight, low cost, and high compatibility. More importantly, the lattice thermal conductivity of carbon materials can vary within a huge range of five orders of magnitude depending on the atomic configuration [4]. Usually, materials with a low lattice thermal conductivity are desirable in thermoelectric applications for energy conversion. It has been found that carbon materials can reach very low lattice thermal conductivities and exhibit a good thermoelectric performance [5–7]. For instance, Yan's group reported that the thermoelectric properties of carbon nanotubes (CNTs) can be significantly enhanced by changing their morphology to CNT bulky papers with Ar plasma treatment. The ZT value of

CNT bulky papers is increased from 0.01 for pristine CNTs to 0.4 for Ar plasma-treated CNTs [8]. Chen et al. found that the thermal conductivity of single-walled carbon nanotube (SWNT)/polyaniline (PANI) hybrid film is only 0.43 W/mK and that the *ZT* value reaches 0.12 at room temperature, remarkably higher than that of either of the individual components of the composite [9]. Meanwhile, for two-dimensional (2D) carbon materials, graphene is a typical representative and exhibits a high electrical conductivity, which is one of the essential requirements for thermoelectric materials. However, graphene possesses an ultra-high lattice thermal conductivity (3151.53 W/mK at 300 K) [10] with a low Seebeck coefficient (about 100 μV/K at 300 K) [11] because of its gapless band structure and strong sp^2 covalent bonds, hindering its application in the thermoelectric field. As an allotrope of graphene, the recently synthesized γ-graphyne [12] has provided a new possibility for the application of carbon-based materials in the thermoelectric field due to its high Seebeck coefficient of 690 μV/K [13] and low k_l of 106.24 W/mK [14], which are superior to the corresponding values of graphene.

On the other hand, it has been found that changing structural units can not only change the geometrical structures of materials but also significantly change the values of the lattice thermal conductivity. For instance, the lattice thermal conductivity of penta-graphene is found to be 645 W/mK at room temperature [15], much lower than the 3151.53 W/mK of graphene [10]. When going from 2D carbon sheets to one-dimensional (1D) and three-dimensional (3D) carbon structures, a similar trend also exists. For 1D carbon, the lattice thermal conductivity of a pentagon-based nanotube is only 95.87 W/mK, which is less than one tenth of the value of (6,6) a carbon nanotube with a similar tube radius [16]. For penta-diamond, which is a new 3D carbon allotrope consisting of five-membered rings, the lattice thermal conductivity is 490.88 W/mK at room temperature [17], much lower than that of diamond (2664.93 W/mK). These results clearly show that the pentagonal unit can effectively modulate the thermal transport of carbon materials. Very recently, we proposed a 2D pentagon-based derivative of graphyne, imidazole-graphyne [18], named ID-GY, which has a direct band gap of 1.10 eV, a low Young's modulus, and strong refraction near infrared (IR), with potential applications in nanoelectronics and optical devices. ID-GY could be formed by assembling experimentally synthesized five-membered imidazole molecules with acetylenic linkers, as exhibited in our previous work [18]. In this work, we further study the thermal transport and thermoelectric properties of ID-GY.

2. Computational Methods

Geometry optimization and electronic band structure calculation are carried out using density functional theory (DFT), as implemented in the Vienna ab initio simulation package (VASP) [19], using the projector augmented wave (PAW) method [20,21]. The Perdew–Burke–Ernzerhof (PBE) functional [22] within the generalized gradient approximation (GGA) [23] is used to treat the exchange–correlation interaction of electrons, while the Heyd–Scuseria–Ernzerhof hybrid functional (HSE06) [24] is used for more accurate band-structure calculations. The kinetic energy cutoff of wave function is set to 520 eV, and the Monkhorst–Pack [25] *k*-point, with a grid density of $2\pi \times 0.02$ Å^{-1}, is used to sample the Brillouin zone for integration in the reciprocal space. All atomic positions are fully optimized with convergence thresholds of 10^{-8} eV and 10^{-6} eV/Å for the total energy and force component, respectively. During the calculations of geometry optimization and band structure, 2D periodic boundary conditions along the x and y directions are applied to ID-GY, while a vacuum region of 16.59 Å is set along the z direction to exclude the mirror interactions between adjacent images.

The electrical transport properties, including the Seebeck coefficient, electrical conductivity, and electronic thermal conductivity are calculated using BoltzTraP2 software [26]. The lattice thermal conductivity is calculated using the ShengBTE package [27]. The second and third interatomic force constants (IFCs) are obtained based on a 2×2 supercell using phonopy software [28] and thirdorder.py code [27], respectively. To calculate the anharmonic IFCs, we include the interactions up to the ninth-nearest neighbor atoms with the cutoff radius of 4.72 Å. The q-grids of 40 × 40 and the thickness of 3.4 Å are chosen to solve the phonon Boltzmann transport equation.

3. Results and Discussion

3.1. Phonon Spectrum and Band Structure

ID-GY crystals in a tetragonal unit cell with the lattice constants of $a = b = 12.14$ Å containing 32 carbon and 8 nitrogen atoms in the unit cell with the space group symmetry of P4/mbm (No. 127). Chemically nonequivalent atoms are marked in Figure 1a, where C_1, C_2, and N atoms are in sp^2 hybridization while C_3, C_4, and C_5 atoms are in sp hybridization. Unlike γ-graphyne, composed of hexagonal units, ID-GY is composed of pentagonal units connected by acetylenic linkers. Compared with the highly symmetric γ-graphyne, the complex geometric structure and hybridized bonding in ID-GY make it a promising material with a low lattice thermal conductivity, just like graphyne and graphdiyne [29,30].

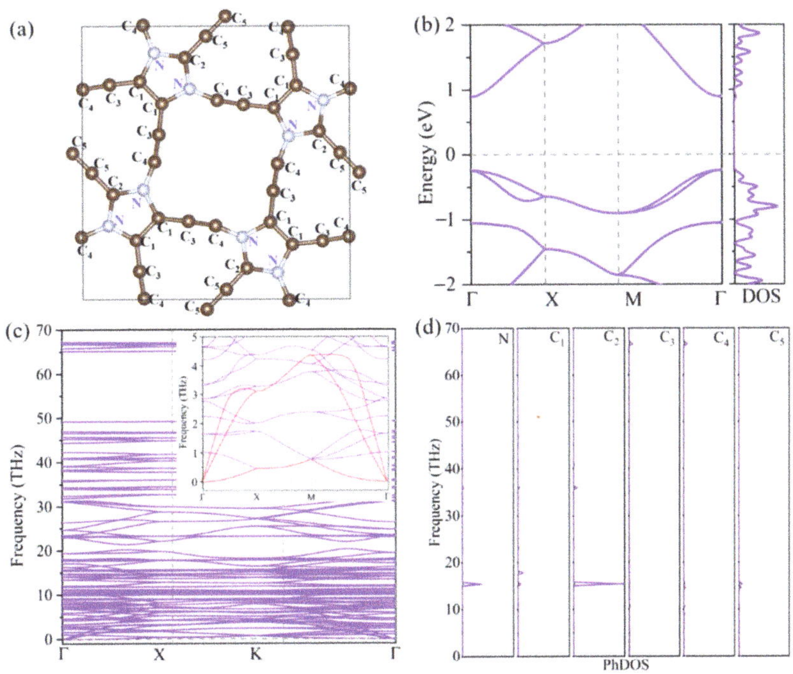

Figure 1. (**a**) Geometric structure, (**b**) electronic band structure, (**c**) phonon spectrum, and (**d**) partial phonon density of states (PhDOS) of ID-GY.

The phonon dispersion spectrum of ID-GY along the high symmetry k-point path (Γ-X-M-Γ) in the first Brillouin zone is shown in Figure 1c. All the vibrational modes are real in the entire Brillouin zone, confirming that ID-GY is dynamically stable. Since both carbon and nitrogen are light atoms, the highest frequency in the phonon spectrum reaches 67 THz. In addition, one can see that there is a large phonon band gap (about 15 THz) in the high-frequency region, and the corresponding phonon density of states (PhDOS) (Figure 1d), indicate that the high frequency can be attributed to the C_3 and C_4 atoms. The bond length of C_3-C_4 is 1.23 Å, showing the characteristics of alkyne bonds. Because the large portion of heat is carried by low-frequency phonons, especially the acoustic phonons, the low-frequency region of the phonon spectrum is magnified, and the acoustic phonon branches are highlighted in red. The longitudinal acoustic (LA) and transverse acoustic (TA) branches of ID-GY are linear when the wave vector q is close to the Γ point, while the out-of-plane acoustic (ZA) branch exhibits parabolic dispersion, which is a characteristic of monolayer 2D materials [31]. The highest frequency of the acoustic phonon is relatively low (<5 THz), lower than that of γ-graphyne (about 8 THz) [14]. The low frequency of the acoustic phonons is associated with a low acoustic Debye temperature, as discussed in the following paragraph. Moreover, there is a strong overlap between the acoustic and low-frequency optical branches. These characteristics indicate that the lattice thermal conductivity of ID-GY might be low.

To examine the mechanical stability of ID-GY, the elastic constants were calculated and are listed in Table 1. It is obvious that ID-GY satisfies the Born–Huang criteria [18,32] for 2D tetragonal materials—namely, $C_{11} > 0$, $C_{66} > 0$ and $C_{11} > C_{12}$. The Young's modulus Y, Poisson's ratio v, bulk modulus B, and shear modulus G were also calculated and are presented in Table 1. It was found that the stiffness (122.20 N/m) of ID-GY is only half that of graphene (342 N/m) [33], owing to weak in-plane bonds. Moreover, the sound velocity, which is usually used to measure the speed of phonons propagating through the lattice, can be determined from bulk modulus B and shear modulus G by the following formulas [34]: longitudinal sound velocity $v_l = \sqrt{\frac{B+G}{\rho}}$, transverse sound velocity $v_t = \sqrt{\frac{G}{\rho}}$, and average sound velocity vs. $= 1/\sqrt[3]{\frac{1}{3}\left(\frac{1}{v_l^2} + \frac{2}{v_t^2}\right)}$, where ρ is the mass density. Based on the sound velocity, we obtained the Debye temperature using $\theta_D = \frac{\hbar v_s}{k_B}\left(\frac{4\pi N}{S}\right)^{1/2}$, where N is the number of atoms in the cell and S is the area of the unit cell. Debye temperature measures the temperature above which all modes begin to be excited; therefore, a high θ_D indicates weak three-phonon scattering and hence a high k_l. The calculated Debye temperature of ID-GY is 647 K, which is much lower than the corresponding value of 805 K of γ-graphyne. Consequently, it is natural to expect that ID-GY possesses a lower lattice thermal conductivity than γ-graphyne.

To study the electrical transport property, we calculated the band structure of ID-GY. As shown in Figure 1b, ID-GY exhibits semiconducting electronic features with a direct bandgap value of 1.10 eV. Compared with the bandgap (0.47 eV) of γ-graphyne [14], the larger bandgap can effectively overcome the high-temperature bipolar conduction problem, benefitting thermoelectric performance [35]. Moreover, the sharp conduction band and valence band around the Γ point suggest low carrier effective masses and a possible large carrier mobility. It is worth noting that the valence band maximum (VBM) is doubly degenerated, leading to a sharp density of states (DOS). The high degeneracy of the valence band and the sharp DOS would enhance the Seebeck coefficient of p-type ID-GY, as is the case with bilayer MoS_2 [36].

Table 1. Calculated elastic coefficients C_{ij} (in N/m), Young's modulus Y (in N/m), Poisson's ratio v, bulk modulus B (in N/m), shear modulus G (in N/m), longitudinal sound velocity v_l (in km/s), transverse sound velocity v_t (in km/s), average sound velocity vs. (in km/s), and Debye temperature θ_D (in K) for ID-GY. For comparison, the corresponding values for γ-graphyne (γ-GY) are also listed here.

	C_{11}	C_{12}	C_{66}	Y	v	B	G	v_l	v_t	v_s	θ_D
ID-GY	164.26	83.12	12.79	122.20	0.51	124.69	40.46	17.17	8.5	4.59	647
γ-GY	-	-	-	-	0.41 *	122.73 *	77.04 *	18.53	11.51	5.50	805

* Data from ref. [37].

3.2. Thermal Transport Properties

The lattice thermal conductivity (k_l) of ID-GY was calculated for different temperatures. As shown in Figure 2a, the lattice thermal conductivity of ID-GY is 10.76 W/mK at 300 K, which is two orders of magnitude lower than that of graphene (3151.53 W/mK) [10] and much lower than that of many other 2D carbon hexagonal structures, including α-graphyne (21.11 W/mK) [14], β-graphyne (22.3 W/mK) [14], γ-graphyne (106.24 W/mK) [14], and graphdiyne (22.3 W/mK) [29], at the same temperature. This shows the importance of structural units in affecting the thermal conductivity of a material. We fitted the relationship of k_l with temperature and found that k_l is proportional to $1/T^{1.05}$, indicating that the three-phonon scattering is dominant in ID-GY, as is the case with graphene [38] and penta-graphene [15]. This was further confirmed by comparing the scattering rates of three-phonon scattering with those of the isotopic scattering process. The calculated results for these two scattering processes are plotted in Figure S1 in the Supplementary Materials. We found that the three-phonon scattering rates are nearly 100 times larger than the isotopic scattering ones. The calculated cumulative k_l as a function of frequency is plotted in Figure S2, which shows that the phonons with frequencies lower than 20 THz contribute about 85% to the lattice thermal conductivity. Therefore, we focus on the low-frequency phonon branches (<20 THz) in the following discussion.

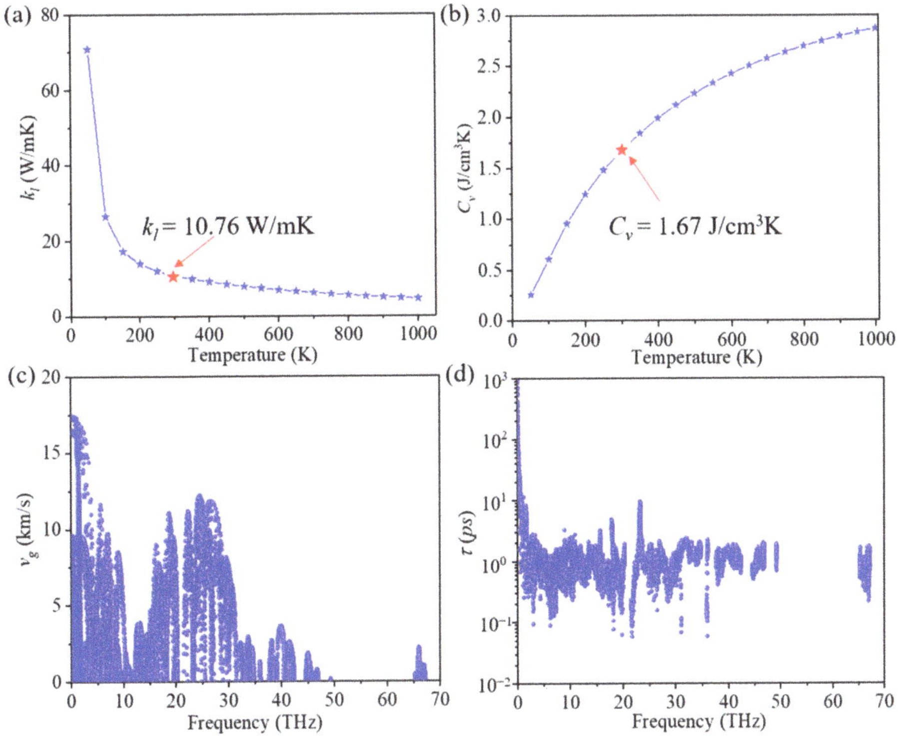

Figure 2. (**a**) Lattice thermal conductivity (k_l), (**b**) phonon volumetric-specific heat (C_v), (**c**) group velocity (v_g), and (**d**) phonon lifetime (τ) of ID-GY.

It is important to understand the reasons for the low k_l of ID-GY. k_l can be expressed in the following form through the summation of the contribution of all of the phonon modes $\lambda(q, j)$ with the wave vector q and branch index j:

$$k_l = \frac{1}{N}\sum_{\lambda} C_\lambda \nu_{\alpha,\lambda} \nu_{\beta,\lambda} \tau_\lambda, \tag{1}$$

where α and β denote the three directions (x, y, or z) and λ is the phonon mode consisting of both wave vector q and branch index j. C_λ, $\nu_{\alpha,\lambda}$, and τ_λ represent the phonon volumetric-specific heat, group velocity, and phonon lifetime, respectively.

Our calculated phonon specific heat value of ID-GY is 1.67 J/cm^3K at 300 K. The variation in the phonon volumetric-specific heat with temperature is plotted in Figure 2b.

It is worth mentioning that the phonon-specific heat value usually is not different from one material to another [39]. For instance, the phonon-specific heat value of γ-GY is 1.68 J/cm^{3}K [29], which is almost same as that of ID-GY. The change in group velocity of ID-GY with frequency is given in Figure 2c, which shows that at the long-wavelength limit, the group velocity reaches the highest value of 17.5 km/s, close to that of γ-GY (~17.9 km/s) [14]. In the low-frequency region (below 20 THz), the overall group velocity is only slightly lower than that of γ-GY. The variation in phonon lifetime with frequency at room temperature is plotted in Figure 2d. The lifetime for most low-frequency phonon modes (0~20 THz) is about 2 *ps*, while that for γ-GY is larger than 10 *ps* [14]. Therefore, the short phonon lifetime in ID-GY is the main reason for the low lattice thermal conductivity.

A short phonon lifetime is usually associated with strong phonon scattering. Therefore, we studied the phonon scatterings by carrying out additional calculations based on the scattering mechanism of phonon modes. Three-phonon scattering is the dominant process in the heat transport process and usually depends on two factors: the number of existing scattering channels and the strength of each scattering channel, which can be described by weighted phase space (*WP3*) and mode Grüneisen parameter (γ). The weighted phase space of each phonon mode with respect to phonon frequency is plotted in Figure 3a. We found that the value of the *WP3* is relatively large, which results from the large number of atoms per unit cell and the complex geometric structure of ID-GY. Moreover, the Grüneisen parameter, which quantifies the intensity of anharmonic interactions between the phonon branches, is another key factor in determining the phonon lifetime of a system. The average Grüneisen parameter of ID-GY was calculated to be 1.51, which is comparable to that of traditional thermoelectric material PbTe (1.65) [40], suggesting the existence of a strong anharmonicity in ID-GY.

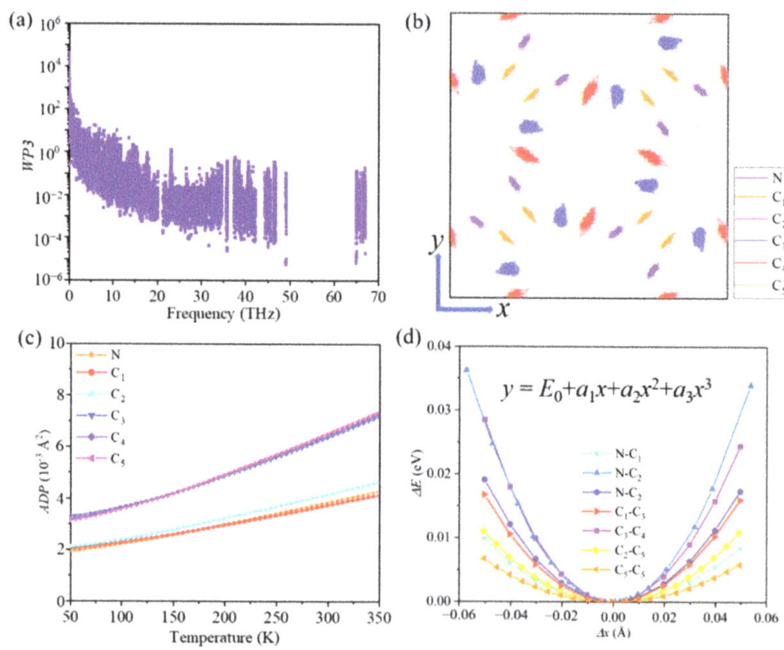

Figure 3. (**a**) Weighted phase space (*WP3*), (**b**) trajectory of the atoms in the *xy*-plane from ab initio molecular dynamics simulations at 800 K, (**c**) atomic displacement parameter (ADP), and (**d**) bond energy curves of ID-GY.

To find the origin of the strong anharmonicity from the point of view of geometric structure, we plotted the simulated trajectory of atoms in the *xy*-plane at 800 K during an ab initio molecular dynamics (AIMD) simulation. As shown in Figure 3b, the C_3 and C_4 atoms vibrate strongly around their equilibrium positions. The atomic trajectory shows that -$C_3{\equiv}C_4$- pairs in acetylenic linkers are weakly bonded to the pentagonal rings. These conclusions are further confirmed by the atomic displacement parameter (ADP) and the bond energy curve, which can provide a visualization of the anharmonicity. As shown in Figure 3c, the ADPs of the C_3 and C_4 atoms are much larger. The bond energy curve shows the vibration of a relative energy change ΔE (in eV/per atom) when the bond length changes and reveals the phonon anharmonicity. By fitting the bond energy curve (Figure 3d), we obtain the anharmonic parameters (a_3) for different type of bonds in ID-GY. The data in Table 2 show that the strong anharmonicity in ID-GY mainly originates from the -$C_3{\equiv}C_4$- pairs, which are weakly bonded to the pentagonal rings. The single *N*-C_4 and C_1-C_3 bonds are too weak to yield an inefficient thermal transport by lattice vibration. The inhomogeneous bond environment and large lattice vibrational mismatch between the pentagonal rings and the acetylenic linkers hinder the transport of heat.

Table 2. Fitted anharmonic parameters (a_3) of the bond energy curves for ID-GY.

	C_3-C_4	N-C_4	N-C_1	C_1-C_3	C_5-C_5	N-C_2	C_2-C_5
a_3	15.41	7.13	5.90	4.28	3.28	0.68	0.60

3.3. Electrical Transport Properties

ID-GY possesses an appropriate band gap and a low lattice thermal conductivity; therefore, it could be a high-performance thermoelectric material. To study the thermoelectric performance of ID-GY, we calculated its electrical transport properties. To this end, the Seebeck coefficient (*S*) was calculated using the BoltzTraP2 software [26]. As shown in Figure 4a, the *S* can reach a peak of 1150 μV/K at 300 K, which is much larger than that of γ-graphyne (690 μV/K) [13] owing to the larger bandgap and doubly degenerate valence bands of ID-GY. To obtain the electronic conductivity, the carrier relaxation time (τ) is necessary, which can be expressed as:

$$\tau = \frac{\mu|m^*|}{e} = \frac{2\hbar^3 C}{3k_B T|m^*|E_1^2}, \tag{2}$$

where μ is the carrier mobility, *C* is the in-plane elastic constant, and E_1 is the deformation-potential constant. The effective mass of carrier m^* was calculated from the curvature of the conduction band minimum or valence band maximum by the parabolic fitting of the band edge using the formula $m^* = \hbar\left[\partial^2 E/\partial k^2\right]^{-1}$. These calculated results are summarized in Table 3.

Table 3. In-plane elastic constant *C* (in N/m), deformation-potential constant E_1 (in eV), effective mass of carrier m^* (in m_e), carrier mobility μ (in cm^2/Vs), and carrier relaxation time τ (in 10^{-14} s) of ID-GY at 300 K.

Carrier Type	C	E_1	m^*	μ	τ
electron	164.26	4.94	0.11	7932.54	49.56
hole	164.26	4.53	0.25	1826.31	25.93

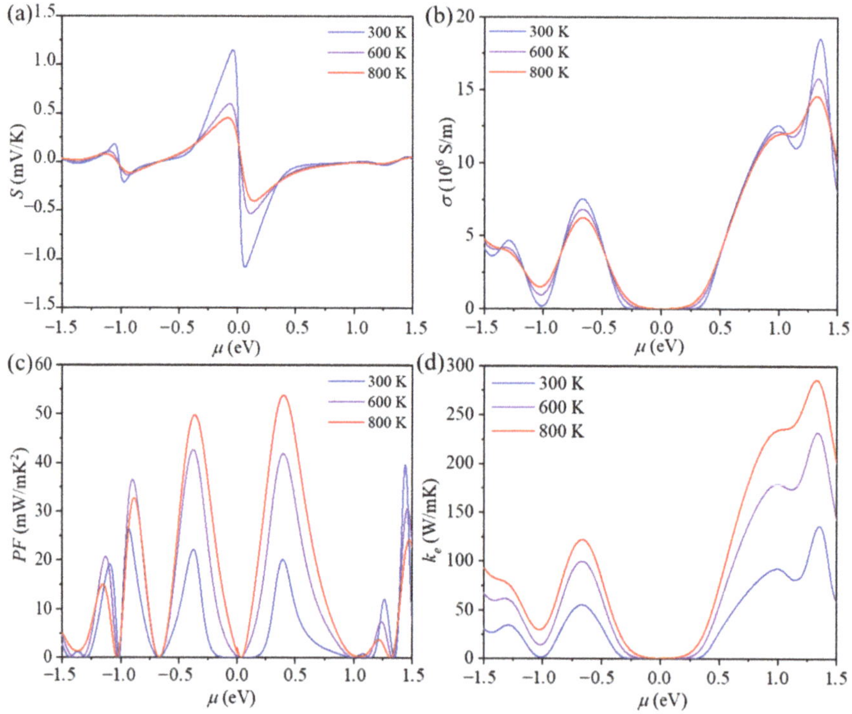

Figure 4. (**a**) Seebeck coefficient (S), (**b**) electrical conductivity (σ), (**c**) power factor (PF), and (**d**) electronic thermal conductivity (k_e) of ID-GY as a function of chemical potential μ, respectively.

As expected from the band structure, the effective mass of the carrier would be very low. The calculated value is indeed only 0.11~0.25 m_e, indicating that ID-GY could have a considerably high carrier mobility. The obtained carrier mobilities of ID-GY are 7933 and 1826 cm^2/Vs for electrons and holes, respectively; accordingly, the electron relaxation time (τ_e) is much longer than the hole relaxation time (τ_h). It is worth noting that although the deformation potential approximation [41] has been widely used for predicting the carrier mobility of new thermoelectric materials [42,43], the carrier mobility is usually overestimated as compared to the experimental result due to the neglect of scattering between the carrier and either the defect or the substrate [44]. Conversely, these scattering processes can also reduce the thermal conductivity. Thus, the overestimated electrical conductivity and overestimated thermal conductivity may cancel each other out to some extent, resulting in a more reliable prediction.

Based on the carrier relaxation time, the electrical conductivity (σ) is obtained and presented in Figure 4b. The electronic conductivity of n-type ID-GY is higher than that of p-type ID-GY due to the longer electron relaxation time. We note that the absolute values of the Seebeck coefficient and electrical conductivity show opposite trends when the chemical potential changes. Thus, to obtain a good power factor ($PF = S^2\sigma$), an optimum chemical potential is needed. The maximum PF value for p-type ID-GY is 22.14 mW/mK^2 at 300 K, while that for p-type ID-GY is 20.18 mW/mK^2.

Thermal conductivity is the sum of the lattice thermal conductivity and electronic thermal conductivity. The former is described in Equation (1) and the latter follows the Wiedemann–Frranz law $k_e = L\sigma T$, where the Lorenz number L is equal to 2.44×10^{-8} WΩ/K^2 [45]. As shown in Figure 4d, electronic thermal conductivity has a similar tendency to electrical conductivity.

Finally, the thermoelectric performance of ID-GY was evaluated using the ZT value. The variations in our calculated ZT values with chemical potential μ at different temperatures are plotted in Figure 5. At 300 K, the optimized ZT values of ID-GY can reach 0.46 and 0.38 for p-type and n-type with the hole- and electron-doping concentrations of 5.8×10^{12} and 1.75×10^{12} cm^{-2}, respectively, which are higher than the ZT values of many other 2D carbon materials [30], including graphene (0.01), α-graphyne (0.03), β-graphyne (0.12), γ-graphyne (0.17), and 6,6,12-graphyne (0.05). While at 800 K, the ZT values are 2.20 and 2.21 for p-type and n-type ID-GY.

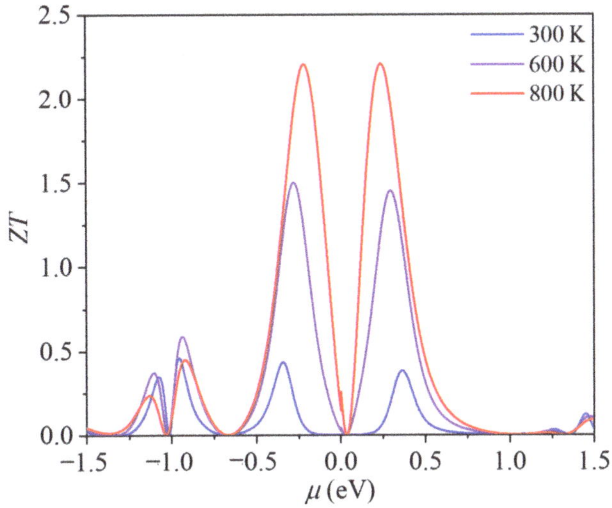

Figure 5. Variation in the thermoelectric figure of merit (ZT) of ID-GY with the chemical potential (μ).

4. Conclusions

In this work, based on first principle calculations, we investigated the thermal transport and thermoelectric properties of ID-GY, a new pentagon-based 2D material that is constructed by assembling an experimentally synthesized pentagonal imidazole molecule and acetylenic linkers. We showed that the thermoelectric properties of graphyne are significantly improved by changing its structural unit to a pentagonal imidazole molecule. The lattice thermal conductivity was decreased from 106.24 W/mK for γ-graphyne to 10.76 W/mK for ID-GY at 300 K, while the Seebeck coefficient was increased to 1150 V/K for ID-GY from 690 μV/K for γ-graphyne at room temperature. The mechanism of the low lattice thermal conductivity was further studied by analyzing the group velocity, scattering rate, weighted phase space, and bond energy curve. The mismatched lattice vibration between the pentagonal rings and the linkers in ID-GY resulted in a strong anharmonicity. Moreover, as compared to γ-graphyne, the doubly degenerated VBM and larger band gap resulted in a higher Seebeck coefficient in ID-GY. The calculated thermoelectric figure of merit at 300 K was 0.46, suggesting the effectiveness of using the pentagonal structural unit for enhancing the thermoelectric performance of carbon-based materials.

Supplementary Materials: The following are available online at https://www.mdpi.com/article/10.3390/ma14195604/s1, Figure S1: the isotopic scattering rates of ID-GY, Figure S2: viaration of the normalized cumulative lattice thermal conductivity with frequency for ID-GY.

Author Contributions: Calculation, Y.C.; formal analysis, Y.C., J.S. and W.K.; writing—original draft preparation, Y.C.; writing—review and editing, Q.W. and J.S.; supervision, Q.W.; project administration, Q.W.; funding acquisition, Q.W. and W.K. All authors have read and agreed to the published version of the manuscript.

Funding: This research was funded by grants from the National Key Research and Development Program of the Ministry of Science and technology of China (Grant No. 2017YFA0205003) and the National Natural Science Foundation of China (Grants Nos. NSFC-11974028 and NSFC-21773004). It is also supported by the high-performance computing platform of Peking University, China.

Institutional Review Board Statement: Not applicable.

Informed Consent Statement: Not applicable.

Data Availability Statement: The data presented in this study are available on request from the corresponding author.

Acknowledgments: This work is partially supported by grants from the National Key Research and Development Program of the Ministry of Science and Technology of China (Grant No. 2017YFA0205003) and the National Natural Science Foundation of China (Grants Nos. NSFC-11974028 and NSFC-21773004). It is also supported by the high-performance computing platform of Peking University, China.

Conflicts of Interest: There are no conflicts to declare.

References

1. Poudel, B.; Hao, Q.; Ma, Y.; Lan, Y.; Minnich, A.; Yu, B.; Yan, X.; Wang, D.; Muto, A.; Vashaee, D.; et al. High-thermoelectric performance of nanostructured bismuth antimony telluride bulk alloys. *Science* **2008**, *320*, 634. [CrossRef] [PubMed]
2. Heremans, J.P.; Jovovic, V.; Toberer, E.S.; Saramat, A.; Kurosaki, K.; Charoenphakdee, A.; Yamanaka, S.; Snyder, G.J. Enhancement of thermoelectric efficiency in PbTe by distortion of the electronic density of states. *Science* **2008**, *321*, 554. [CrossRef] [PubMed]
3. Kudryashov, A.A.; Kytin, V.G.; Lunin, R.A.; Kulbachinskii, V.A.; Banerjee, A. Effect of thallium doping on the mobility of electrons in Bi_2Se_3 and holes in Sb_2Te_3. *Semic* **2016**, *50*, 869–875. [CrossRef]
4. Balandin, A.A. Thermal properties of graphene and nanostructured carbon materials. *Nat. Mater.* **2011**, *10*, 569–581. [CrossRef]
5. Zhang, Y.; Zhang, Q.; Chen, G. Carbon and carbon composites for thermoelectric applications. *Carbon Energy* **2020**, *2*, 408–436. [CrossRef]
6. Cho, C.; Wallace, K.L.; Tzeng, P.; Hsu, J.; Yu, C.; Grunlan, J.C. Outstanding low temperature thermoelectric power factor from completely organic thin films enabled by multidimensional conjugated nanomaterials. *Adv. Energy Mater.* **2016**, *6*, 1502168. [CrossRef]
7. Blackburn, J.L.; Ferguson, A.J.; Cho, C.; Grunlan, J.C. Carbon-nanotube-based thermoelectric materials and devices. *Adv. Mater.* **2018**, *30*, 1704386. [CrossRef]
8. Zhao, W.; Fan, S.; Xiao, N.; Liu, D.; Tay, Y.Y.; Yu, C.; Sim, D.; Hng, H.H.; Zhang, Q.; Boey, F.; et al. Flexible carbon nanotube papers with improved thermoelectric properties. *Energy Environ. Sci.* **2012**, *5*, 5364–5369. [CrossRef]
9. Yao, Q.; Wang, Q.; Wang, L.; Chen, L. Abnormally enhanced thermoelectric transport properties of SWNT/PANI hybrid films by the strengthened PANI molecular ordering. *Energy Environ. Sci.* **2014**, *7*, 3801–3807. [CrossRef]
10. Yue, S.; Qin, G.; Zhang, X.; Sheng, X.; Su, G.; Hu, M. Thermal transport in novel carbon allotropes with sp^2 or sp^3 hybridization: An ab initio study. *Phys. Rev. B* **2017**, *95*, 085207. [CrossRef]
11. Zuev, Y.M.; Chang, W.; Kim, P. Thermoelectric and magnetothermoelectric transport measurements of graphene. *Phys. Rev. Lett.* **2009**, *102*, 096807. [CrossRef] [PubMed]
12. Li, Q.; Li, Y.; Chen, Y.; Wu, L.; Yang, C.; Cui, X. Synthesis of γ-graphyne by mechanochemistry and its electronic structure. *Carbon* **2018**, *136*, 248–254. [CrossRef]
13. Jiang, P.H.; Liu, H.J.; Cheng, L.; Fan, D.D.; Zhang, J.; Wei, J.; Liang, J.H.; Shi, J. Thermoelectric properties of γ-graphyne from first-principles calculations. *Carbon* **2017**, *113*, 108–113. [CrossRef]
14. Yang, X.; Dai, Z.; Zhao, Y.; Meng, S. Phonon thermal transport in a class of graphene allotropes from first principles. *Phys. Chem. Chem. Phys.* **2018**, *20*, 15980–15985. [CrossRef]
15. Wang, F.Q.; Yu, J.; Wang, Q.; Kawazoe, Y.; Jena, P. Lattice thermal conductivity of penta-graphene. *Carbon* **2016**, *105*, 424–429. [CrossRef]
16. Sun, J.; Chen, Y.; Wang, Q. Low lattice thermal conductivity of a 5-8-peanut-shaped carbon nanotube. *Phys. Chem. Chem. Phys.* **2021**, *23*, 5460–5466. [CrossRef]
17. Chen, Y.; Sun, J.; Li, T.; Wang, Q. Low lattice thermal conductivity of pentadiamond. *J. Appl. Phys.* **2021**, *129*, 215107. [CrossRef]
18. Zhou, W.; Guo, Y.; Shen, Y.; Wang, Q.; Jena, P. Imidazole-graphyne: A new 2D carbon nitride with a direct bandgap and strong IR refraction. *Phys. Chem. Chem. Phys.* **2021**, *23*, 10274–10280. [CrossRef] [PubMed]
19. Kresse, G.; Furthmüller, J. Efficient iterative schemes for *ab initio* total-energy calculations using a plane-wave basis set. *Phys. Rev. B* **1996**, *54*, 11169–11186. [CrossRef] [PubMed]
20. Kresse, G.; Joubert, D. From ultrasoft pseudopotentials to the projector augmented-wave method. *Phys. Rev. B* **1999**, *59*, 1758–1775. [CrossRef]
21. Blöchl, P.E. Projector augmented-wave method. *Phys. Rev. B* **1994**, *50*, 17953–17979. [CrossRef]

22. Perdew, J.P.; Burke, K.; Ernzerhof, M. Generalized gradient approximation made simple. *Phys. Rev. Lett.* **1996**, *77*, 3865–3868. [CrossRef]
23. Hu, Y.F.; Storey, C. Efficient generalized conjugate gradient algorithms, part 2: Implementation. *J. Optim. Theory Appl.* **1991**, *69*, 139–152. [CrossRef]
24. Heyd, J.; Scuseria, G.E.; Ernzerhof, M. Hybrid functionals based on a screened Coulomb potential. *J. Chem. Phys.* **2003**, *118*, 8207. [CrossRef]
25. Monkhorst, H.J.; Pack, J.D. Special points for Brillouin-zone integrations. *Phys. Rev. B* **1976**, *13*, 5188–5192. [CrossRef]
26. Madsen, G.K.H.; Carrete, J.; Verstraete, M.J. BoltzTraP2, a program for interpolating band structures and calculating semi-classical transport coefficients. *Comput. Phys. Commun.* **2018**, *231*, 140–145. [CrossRef]
27. Li, W.; Carrete, J.; AKatcho, N.; Mingo, N. ShengBTE: A solver of the Boltzmann transport equation for phonons. *Comput. Phys. Commun.* **2014**, *185*, 1747–1758. [CrossRef]
28. Togo, A.; Tanaka, I. First principles phonon calculations in materials science. *Scr. Mater.* **2015**, *108*, 1–5. [CrossRef]
29. Tan, X.; Shao, H.; Hu, T.; Liu, G.; Jiang, J.; Jiang, H. High thermoelectric performance in two-dimensional graphyne sheets predicted by first-principles calculations. *Phys. Chem. Chem. Phys.* **2015**, *17*, 22872–22881. [CrossRef]
30. Sevinçli, H.; Sevik, C. Electronic, phononic, and thermoelectric properties of graphyne sheets. *Appl. Phys. Lett.* **2014**, *105*, 223108. [CrossRef]
31. Şahin, H.; Cahangirov, S.; Topsakal, M.; Bekaroglu, E.; Akturk, E.; Senger, R.T.; Ciraci, S. Monolayer honeycomb structures of group-IV elements and III-V binary compounds: First-principles calculations. *Phys. Rev. B* **2009**, *80*, 155453. [CrossRef]
32. Born, M.; Huang, K. *Dynamical Theory of Crystal Lattices*; Oxford University Press: Oxford, UK, 1954.
33. Peng, B.; Zhang, H.; Shao, H.; Xu, Y.; Ni, G.; Zhang, R.; Zhu, H. Phonon transport properties of two-dimensional group-IV materials from ab initio calculations. *Phys. Rev. B* **2016**, *94*, 245420. [CrossRef]
34. Zhuo, Y.; Mansouri Tehrani, A.; Oliynyk, A.O.; Duke, A.C.; Brgoch, J. Identifying an efficient, thermally robust inorganic phosphor host via machine learning. *Nat. Commun.* **2018**, *9*, 4377. [CrossRef]
35. Wang, F.Q.; Zhang, S.; Yu, J.; Wang, Q. Thermoelectric properties of single-layered SnSe sheet. *Nanoscale* **2015**, *7*, 15962–15970. [CrossRef]
36. Hong, J.; Lee, C.; Park, J.-S.; Shim, J.H. Control of valley degeneracy in MoS_2 by layer thickness and electric field and its effect on thermoelectric properties. *Phys. Rev. B* **2016**, *93*, 035445. [CrossRef]
37. Asadpour, M.; Malakpour, S.; Faghihnasiri, M.; Taghipour, B. Mechanical properties of two-dimensional graphyne sheet, analogous system of BN sheet and graphyne-like BN sheet. *Solid State Commun.* **2015**, *212*, 46–52. [CrossRef]
38. Choudhry, U.; Yue, S.; Liao, B. Origins of significant reduction of lattice thermal conductivity in graphene allotropes. *Phys. Rev. B* **2019**, *100*, 165401. [CrossRef]
39. Wu, X.; Varshney, V.; Lee, J.; Zhang, T.; Wohlwend, J.L.; Roy, A.K.; Luo, T. Hydrogenation of penta-graphene leads to unexpected large improvement in thermal conductivity. *Nano Lett.* **2016**, *16*, 3925–3935. [CrossRef] [PubMed]
40. Xiao, Y.; Chang, C.; Pei, Y.; Wu, D.; Peng, K.; Zhou, X.; Gong, S.; He, J.; Zhang, Y.; Zeng, Z.; et al. Origin of low thermal conductivity in SnSe. *Phys. Rev. B* **2016**, *94*, 125203. [CrossRef]
41. Bardeen, J.; Shockley, W. Deformation potentials and mobilities in non-polar crystals. *Phys. Rev.* **1950**, *80*, 72–80. [CrossRef]
42. Gao, Z.; Wang, J. Thermoelectric penta-silicene with a high room-temperature figure of merit. *ACS Appl. Mater. Interfaces* **2020**, *12*, 14298–14307. [CrossRef] [PubMed]
43. Lan, Y.; Chen, X.; Hu, C.; Cheng, Y.; Chen, Q. Penta-PdX_2 (X = S, Se, Te) monolayers: Promising anisotropic thermoelectric materials. *J. Mater. Chem. A* **2019**, *7*, 11134–11142. [CrossRef]
44. Poljak, M.; Suligoj, T.; Wang, K.L. Influence of substrate type and quality on carrier mobility in graphene nanoribbons. *J. Appl. Phys.* **2013**, *114*, 053701. [CrossRef]
45. Xu, Y.; Li, Z.; Duan, W. Thermal and thermoelectric properties of graphene. *Small* **2014**, *10*, 2182–2199. [CrossRef] [PubMed]

materials

MDPI

Article

FDM Layering Deposition Effects on Mechanical Response of TPU Lattice Structures

Chiara Ursini * and Luca Collini

Department of Engineering and Architecture, University of Parma, Viale delle Scienze 181/A, 43124 Parma, Italy; luca.collini@unipr.it
* Correspondence: chiara.ursini@unipr.it

Abstract: Nowadays, fused deposition modeling additive technology is becoming more and more popular in parts manufacturing due to its ability to reproduce complex geometries with many different thermoplastic materials, such as the TPU. On the other hand, objects obtained through this technology are mainly used for prototyping activities. For this reason, analyzing the functional behavior of FDM parts is still a topic of great interest. Many studies are conducted to broaden the spectrum of materials used to ensure an ever-increasing use of FDM in various production scenarios. In this study, the effects of several phenomena that influence the mechanical properties of printed lattice structures additively obtained by FDM are evaluated. Three different configurations of lattice structures with designs developed from unit cells were analyzed both experimentally and numerically. As the main result of the study, several parameters of the FDM process and their correlation were identified as possible detrimental factors of the mechanical properties by about 50% of the same parts used as isotropic cell solids. The best parameter configurations in terms of mechanical response were then highlighted by numerical analysis.

Keywords: additive manufacturing; fused deposition modeling; lattice structures; TPU; layering

Citation: Ursini, C.; Collini, L. FDM Layering Deposition Effects on Mechanical Response of TPU Lattice Structures. *Materials* **2021**, *14*, 5645. https://doi.org/10.3390/ma14195645

Academic Editors: Michele Bacciocchi and Abbas S. Milani

Received: 7 September 2021
Accepted: 23 September 2021
Published: 28 September 2021

1. Introduction

As opposed to traditional production, which is mainly subtractive in nature, additive manufacturing represents the idea of future production, thanks to its considerable savings in materials and the extended possibilities in producing complex geometries. The major advantage of the fused deposition modeling (FDM) additive process lies in the customized production of printed objects. At the same time, various limitations prevent their dominance in the production of fully functional mechanical components, for example, the limited size of produced parts, but the cost should not be underestimated as well [1]. Furthermore, many printing process parameters have an influence on the resulting microstructure of printed objects, as observed by Ziemian et al. and Durgun et al. and discussed below [2,3].

FDM additive process is based on the extrusion of material and on the principle of stacking layer by layer, in order to create parts that can have sophisticated 3D geometries. Due to this principle of deposition of the material, this technology presents some pitfalls. In fact, by building FDM parts from bottom to top, the material in the current layer solidifies before the next one is placed on it [4], causing incomplete interlayer adhesion and often leaving voids in solidified structure, resulting in a decrease in mechanical performance [5]. In addition to this, a combination of several parameters such as raster orientation, air gap, bead width, color, model temperature, infill, etc. causes a decrease in the compressive strength of FDM products [6,7]. The layering effect represents the main challenge in additive manufacturing with fused deposition technology and will be analyzed in depth in the discussion below.

In this study, the printed parts under examination were lattice structures made of the repetition of three different unit cells with the same relative density—namely, open cell, closed thin-walled cell, and closed thick-walled cell. The unit cells repeated with the principle of tessellation along the three principal directions constitute the resulting lattice structures [8] (Figure 1). Fabrication of these structures by conventional manufacturing is not trivial, but Kumar et al. have found that with additive technologies, it is possible to produce them without any support structure [9]. An undoubted advantage of lattice structures, which are basically metamaterials since they are specifically designed to impart special physical properties normally absent in the constituent materials [10], is lightness, often a valuable attribute in engineering applications [11]. The material here used for printing is the thermoplastic polyurethane (TPU) 90A, which has hyperelastic and viscoelastic properties [12]. Three specimens of each configuration were additively realized and were tested under compression loads. Comparative analysis among the three configurations was carried out based on the mechanical property of stiffness. At present, the mechanical properties of most 3D-FDM printed polymeric parts often do not meet the requirements of industrial applications [13].

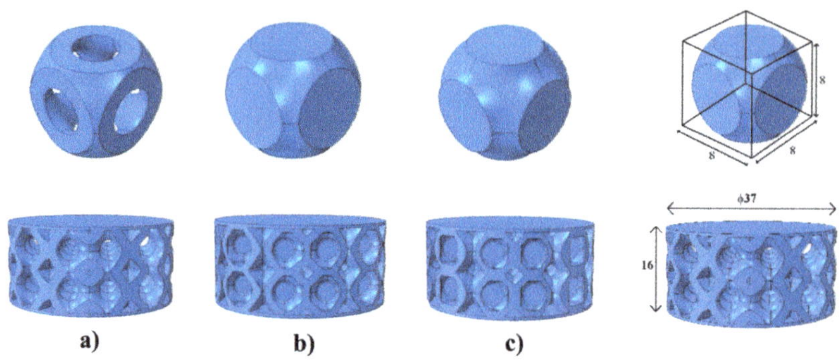

Figure 1. Unit cells and, respectively, lattice structures with their appropriate geometries and sizes [mm]: (**a**) open cell; (**b**) closed thin-walled cell; (**c**) closed thick-walled cell.

Hence, this paper aimed at evaluating the stiffness of produced lattice structures under compression, both numerically and experimentally, to make a comparison between the consistency assumed by finite element software and layering effects of experimental samples.

2. Fused Deposition Modeling on Lattice Structures

2.1. Design

Lattice structures were made by means of tessellation of a unit cell repeating it along the three main directions, essentially creating a honeycomb structure; see Figure 1. As known, honeycomb structures offer great advantages, especially in terms of minimum weight and great resistance to high stresses [14].

Three configurations of lattice structures were designed and analyzed, starting from three different unit cells with the same relative density, respectively, open cell, closed thin-walled cell, and closed thick-walled cell. These configurations of cells were designed based on the design for additive manufacturing. In detail, in accordance with the printing parameters used in the laboratory shown in Table 1, in particular the minimum thickness, the unit cells were produced, as illustrated in Figure 2.

It is immediately evident that the number of adjacent contours, N, in the walls of cells varies from 0 for open cell, is equal to 1 for thin-walled type, and is equal to 2 for thick-walled type. This choice of wall sizes will be the determining feature of the best geometry in terms of specific stiffness since layer thickness is one of the various parameters influencing the 3D-FDM printed objects [15,16].

The building orientation for all the configurations of lattice structures was the horizontal building direction, shown in Figure 3. This is another important printing process parameter in the mechanical response of the printed samples since it is found that building orientation plays an important role in the compressive strength of FDM printed parts [17,18].

Table 1. Printing process parameters.

Printing Phase: FDM Parameters	
Nozzle diameter [mm]	0.4
Layer height [mm]	0.2
Printing speed [mm/min]	1100
Print infill [%]	100
Printing temperature [°C]	230
Bed temperature [°C]	70
FDM Machine Parameters	
Minimum thickness [mm]	0.6
Maximum overhang angle [°]	50

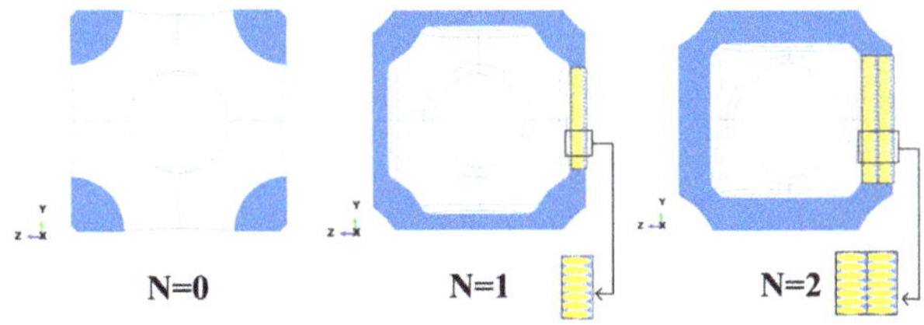

Figure 2. Section of unit cells: wall's contours.

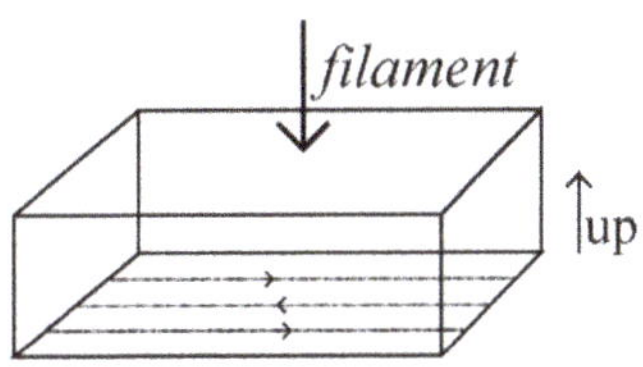

Figure 3. Building orientation of lattice structures.

Hence, the 3D printing of self-supported structures is definitely convenient for saving material, printing time, and post-printing processing. There is also no risk of damaging the printed object when removing the support structure.

2.2. Experimental Tests

Three specimens of each configuration of lattice structure were printed by MEX 3D printer Flashforge dreamer© with TPU 90A filament. Monotonic compression tests were then performed on lattice structures under displacement control of 5 mm/min for three deformation levels, respectively, 10, 20, and 30% of the specimens' height. For each configuration, load-displacement data were plotted in one graph to understand the geometry effect on compression behavior, in particular the number of contours with specific stiffness.

The greatest advantage of these designed lattice structures, in addition to lightness, is that no support structures are required during the printing phase. This is due to the maximum overhang angle of 50 degrees of the used MEX 3D printer machine (Figure 4) (High speed 3D printing research center, National Taiwan University of Science and Technology, Taipei, Taiwan). In fact, 3D printers use a barely appreciable horizontal offset between consecutive layers. In this way, the upper layer does not perfectly overlap the underlying layer but stacks with this small offset, allowing to print overhangs with respect to the vertical of an overhang angle that varies from machine to machine, in a range from 20 to 70 degrees. In any 3D printer, the test for identifying the maximum overhang angle that allows printing without support is called the "massive overhang test".

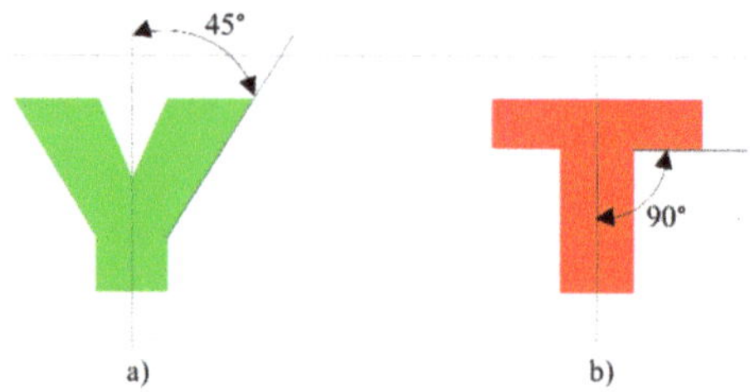

Figure 4. Overhang angle: (**a**) does not need support structures; (**b**) needs support structure.

This was performed while bearing in mind that geometry in FDM parts is an important factor that influences the stress distribution and mechanical properties of samples. Preliminary tests were carried out on cubic and cylindrical TPU 90A FDM samples for different deformation levels at the same strain rate. In order to highlight differences in the mechanical behavior of the two geometric configurations, the nominal stress-strain curves corresponding to 36% of deformation, shown in Figure 5, were obtained using the classical Equations (1) and (2) as follows:

$$\sigma = \frac{L}{A_0} \tag{1}$$

$$\varepsilon = \frac{d}{h_0} \tag{2}$$

where L is the compression applied load, A_0 is the cross-sectional area of the samples, d is the displacement, and h_0 is the initial height of the samples.

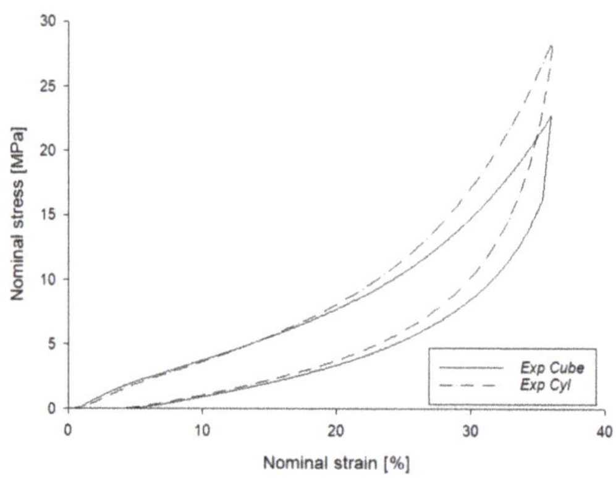

Figure 5. Compression-release tests on cubic and cylindrical TPU FDM samples.

The geometric configuration's significant effect on the mechanical behavior of FDM printed parts became suddenly clear. In particular, at the beginning of the test, the cubic sample shows a higher Young's modulus with a consequently lower plateau regime than the cylindric one. Experimental tests were performed by a servo-hydraulic MTS 810 with a 100 kN load cell. The load was applied perpendicularly to the building orientation of lattice structures so that the layers tend to crush each other, not stressing the weaker adhesion layer. In fact, in FDM printing, an anisotropic layering effect occurs, i.e., the printed parts are stronger in one direction than in the other [19].

A compression test is useful for unveiling the isotropic or anisotropic behavior of printed objects [20]. Given its layer-by-layer nature, the FDM process and many other 3D printing technologies, by definition, give rise to anisotropic parts in their structure [21,22]. This is due to the printing technique: when placing one layer on top of another, the underlying layer begins to partially solidify, causing incorrect and incomplete adhesion between the layers. This tends to create voids between the two adjacent layers, giving the printed object a consequent stiffness depending on the load application direction and also leads to its anisotropic behavior [23]. Obviously, this is one of the main difficulties in using and studying objects obtained by additive processes. Furthermore, the anisotropy present in FDM printed parts have different properties depending on the process parameters used in the printing phase [24]. This can have major consequences on the mechanical and functional properties of the printed parts in industrial applications.

2.3. FDM Process in Conjunction with TPU

As previously mentioned, the FDM additive process does not ensure mechanical properties equal to those of traditionally manufactured objects. This is certainly due to the considerable variability of the numerous process parameters of this type of 3D printing. Each of these parameters, as well as their combination, has different impacts on the mechanical properties of printed parts [25–27]. Through an extensive literature review, many FDM process parameters, such as layer thickness, air gap, bed temperature, raster orientation, model temperature, building orientation, etc. [28], have been found to influence the compressive strength of the samples obtained with the same technology. It must be noted that in all these works, the material used for printing is mostly ABS, while TPU has never been studied. This is of interest in our research, as TPU is increasingly used in conjunction with 3D printing in the manufacturing of thermoplastic printed parts since it offers a wide range of applications. This is due to the capability of TPU to combine mechanical performance characteristics of rubber with the possibility of being processed as a thermoplastic material.

At the same time, TPU is also a very complex material due to its hyperelastic and viscoelastic properties, hygroscopic nature of its filament, and wide range of values that its Young's modulus can assume (from 10 to 2000 MPa).

3. Finite Element Method on Lattice Structures

3.1. Material: Models

To simulate the behavior of lattice structures in TPU, an advanced model of hyperelastic material with hysteretic capability was defined in the Abaqus/CAE software (2020, Simulia Dassault Systèmes, Vélizy-Villacoublay, France). Uniaxial tensile test data from a TPU 90A dog bone sample were included in the Abaqus material model to identify the best match between experimental behavior and the different strain energy potential models available in the software, in the strain range of interest. It was found that in the strain energy potential model, described by means of Equation (3), the best approximate value of the experimental nominal stress-strain trend is the second-order Ogden model (Figure 6).

$$U^{def} = \sum_{i=1}^{N} \frac{2\mu_i}{\alpha_i^2}\left(\overline{\lambda}_1^{\alpha_i} + \overline{\lambda}_2^{\alpha_i} + \overline{\lambda}_3^{\alpha_i} - 3\right) + \sum_{i=1}^{N} \frac{1}{D_i}(J_{el} - 1)^{2i} \quad (3)$$

The strain energy potential U^{def} of Ogden form is expressed through the parameters J_{el}, $\overline{\lambda}_i$, μ_i, D_i, α_i which represent, respectively, the elastic volume ratio, the deviatoric principal stretches, and the temperature-dependent material parameters, illustrated in Table 2. Using the second-order Ogden model and a compression-release test performed on a cubic FDM TPU sample, the best hysteresis loop was reproduced by modifying the Abaqus hysteresis parameters until the loop that best approximates the real behavior was reached (Figure 7). Numerically, in fact, the hysteretic behavior is governed by the following formulation:

$$\dot{\epsilon}_B^{cr} = A[\lambda_B^{cr} - 1 + E]^C (\sigma_B)^m \tag{4}$$

where the effective creep strain rate, and also the mechanical response, is described by means of two networks, A and B—network A identifies the equilibrium-relaxation part and network B, the non-linear one. In particular, for the term definitions of network B, $\lambda_B^{cr} - 1$ represents the nominal creep strain, and σ_B is the effective stress. Table 3 shows the editable parameters in the Abaqus formulation for Equation (4), identified in the literature [29,30], and after an optimization procedure on cubic and cylindrical numerical models, S is the stress scaling factor, m is an exponent usually bigger than 1, C is an exponent that can assume values from −1 to 0, and A and E are constants.

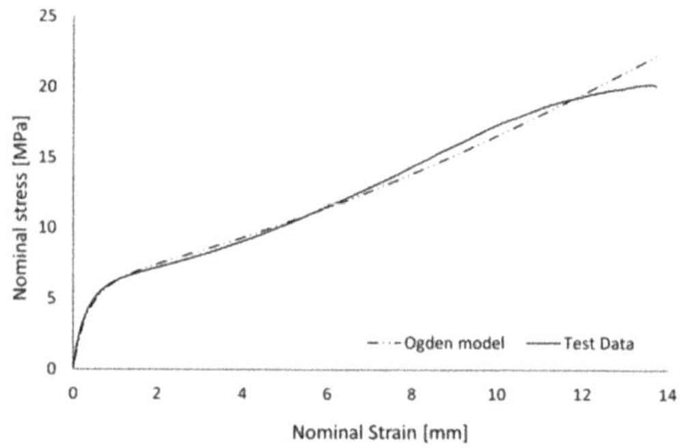

Figure 6. The test data for the second-order Ogden model vs. dog bone sample.

Table 2. The parameters of the second-order Ogden model.

Material Parameter	μ_i	D_i	α_i
i = 2, order	6.1298	0.0000	−1.9004

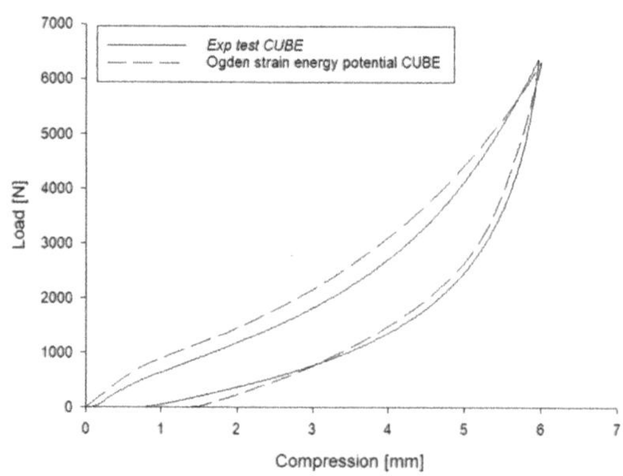

Figure 7. Determination of the hysteresis cycle.

Table 3. Abaqus hysteresis parameters.

Parameter	S	m	C	A	E
value	2.2	4	0	12×10^{-3}	0.01

Voids left in the structure by additive technique and layering effects also need to be considered and analyzed in depth in FE analysis. FDM, as previously mentioned, generates an anisotropic, layered structure. This was considered by defining a local orientation of the lattice structure. Material structure orientation was used in conjunction with a new material model, i.e., an anisotropic elastic material model with an "engineering constants" option, to compare the isotropic and anisotropic behaviors.

By means of the second-order Ogden model, the isotropic behavior was analyzed since the FE software considers structures as isotropic solid. Simulations were also performed for a linear elastic material, both for anisotropic and isotropic materials, to show the possible effects of anisotropy on the mechanical response of the samples.

A linear elastic anisotropic material model was then adopted, in order to tune and reproduce the experimental results. The engineering constants that define this model—namely, elastic modulus, Poisson ratio, and shear modulus, in three principal directions (Table 4) were arbitrarily chosen on the basis of literature data [31,32], imposing a level of anisotropy of 50%. Here, the direction of the applied load was set as the second direction, while the first and third directions were the axes that defined the layer plane. In this way, the intra-filament elasticity was halved, simulating the FDM layer deposition effect.

Table 4. Engineering constants: Young's modulus, Poisson ratio, and shear modulus in the three principal directions.

Material	E_{1*}	E_{2*}	E_{3*}	v_{12}	v_{13}	v_{23}	G_{12}	G_{13}	G_{23}
Linear elastic	13	26	13	0.49	0.39	0.49	4.36	9.35	4.36

* 1: x direction; 2: y direction; 3: z direction.

3.2. Simulations and Overviews

To simulate these tests in the Abaqus/CAE environment, the lattice structures were placed between two rigid analytical plates. The load was applied to the reference point of the upper plate by means of an imposed displacement (Figure 8). Three deformation levels were simulated for each geometrical configuration, which were, respectively, 10, 20, and 30% of specimens' height, at the same strain rate of 5 mm/min.

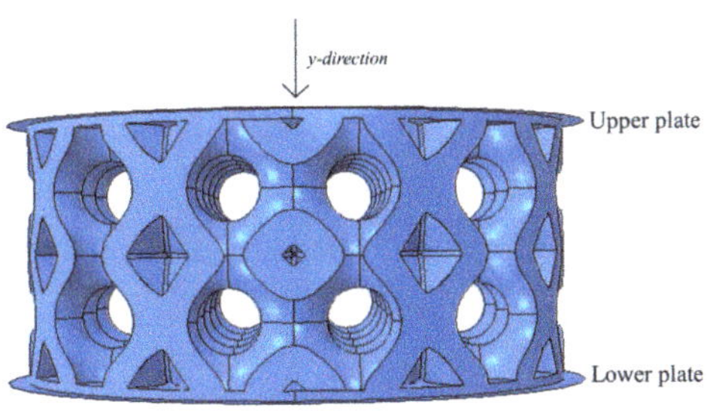

Figure 8. Analytical rigid plates and preferential direction y.

After a variability analysis that did not show major effects in a range from 0.025 to 0.3, a friction coefficient equal to $f = 0.1$ was defined between the contact surfaces. Finally, the lattice structure's typologies were formed as meshed units with linear solid tetrahedral elements, using a hybrid formulation that governs incompressible behaviors. The average mesh size of lattice structure elements was chosen as 1.6 mm, to have a good compromise between computational times and accuracy of the mechanical response in the simulations. This was carried out after a mesh independency analysis on varying the average mesh size from 0.8 mm to 2.4 mm, which showed no noticeably major changes in the mechanical response of the structures (Figure 9). Table 5 shows the resulting numbers of nodes and elements for the three geometrical configurations of lattice structures.

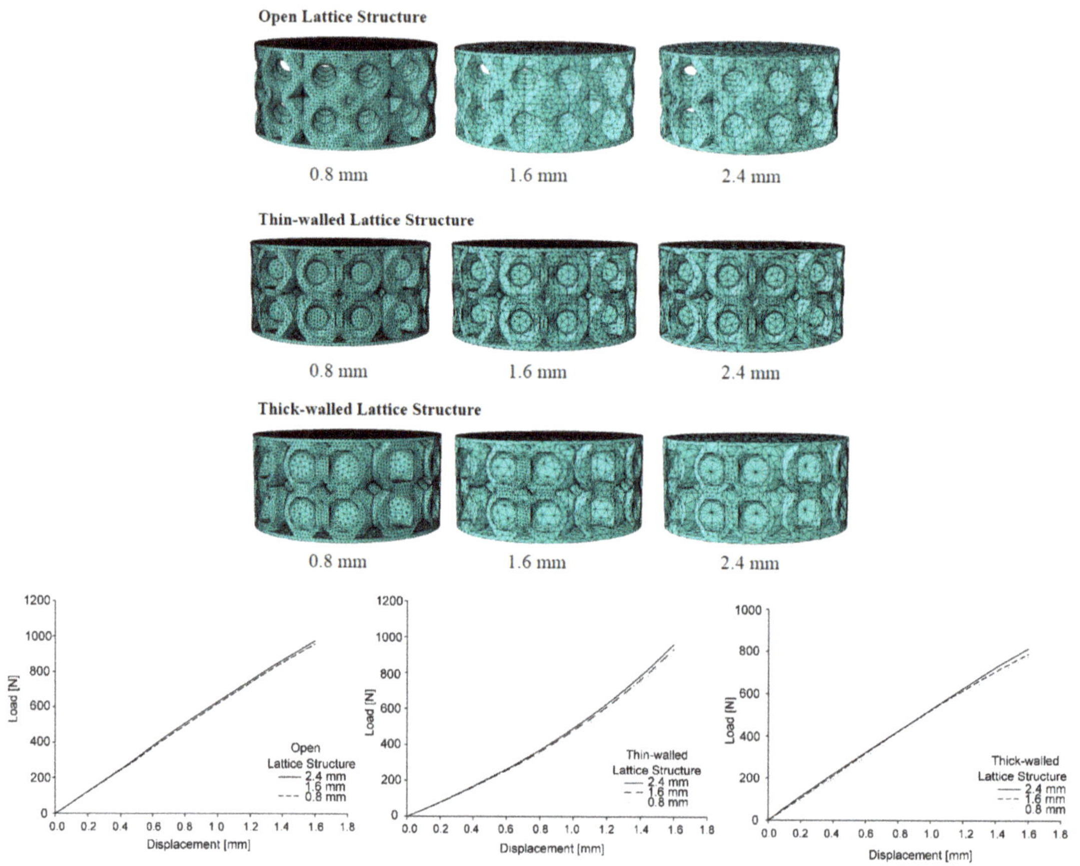

Figure 9. Mesh independency analysis of lattice structures and convergence analysis.

Table 5. Mesh of lattice structures: nodes and elements.

Topology	Nodes	Elements
Open cell	24,779	104,274
Closed thin-walled cell	33,735	145,385
Closed thick-walled cell	37,790	157,045

Load-displacement curves were obtained, as the reference point is affected by the sum of every single nodal response in the contact surface. From these curves, stiffness was calculated as the slope of the line that best fits the loading curve [33].

4. Results

All the obtained results, both experimental and numerical, are reported in Table 6. The nominal stress-strain curves in the range of interest, shown in Figure 10, were plotted through Equations (1) and (2), where the cross-sectional area of each typology of structures, $A_{0,\,eq}$, was calculated as follows:

$$A_{0,\mathrm{eq}} = \frac{V_L}{h_0} \tag{5}$$

where V_L is the volume of lattice, and h_0 is the height of the specimen.

Table 6. Specific stiffness of lattice structures.

10%	Stiffness K_0 [N/mm]			
	EXP	FEM		
		Ogden	Isotropic LE	Anisotropic LE
Open cell	331	534.6	480.8	256.4
Closed thin walled	336	588.8	508.7	260.7
Closed thick walled	278.9	531.2	500.8	256.3
20%	**Stiffness K_0 [N/mm]**			
	EXP	FEM		
		Ogden	Isotropic LE	Anisotropic LE
Open	220.3	421.7	548.1	290.2
Closed thin walled	219.2	432.2	592	294.8
Closed thick walled	181.9	403.8	524.6	266.2
30%	**Stiffness K_0 [N/mm]**			
	EXP	FEM		
		Ogden	Isotropic LE	Anisotropic LE
Open	165.5	389.6	568.4	307.7
Closed thin walled	161.7	380.1	609.5	306.6
Closed thick walled	137.5	366.7	521.6	253.3

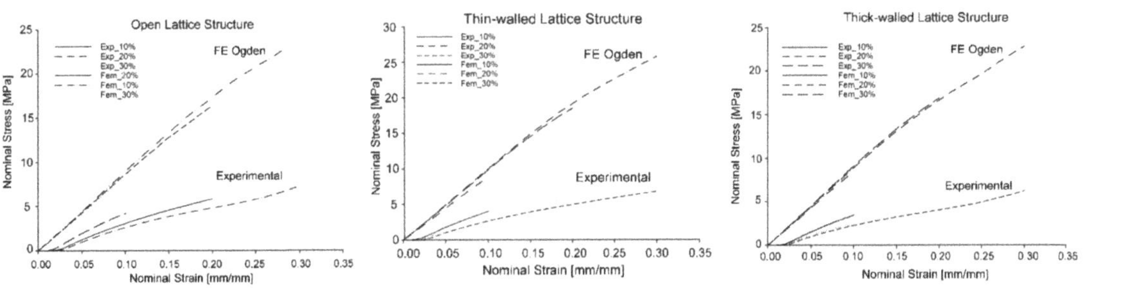

Figure 10. Nominal stress-strain curves of lattice structures, experimental and numerical.

In accordance with the results shown in Table 6, the comparison between experimental and numerical nominal stress-strain curves highlighted a good reproduction of the real experimental trend by Ogden model simulations, even if a visible gap was apparent between them.

By plotting the stiffness results obtained for the hyperelastic model and experimental tests in a distinctive graph (Figure 11), it was revealed that the Ogden model is able to reproduce the exact trend of the experimental response of the lattice structures, but with a discrete gap. The same gap was also found between the isotropic and anisotropic linear elastic results (Figure 12). The linear elastic material model proved useful for identifying and studying the effects of 3D printing on the mechanical response of FDM printed parts.

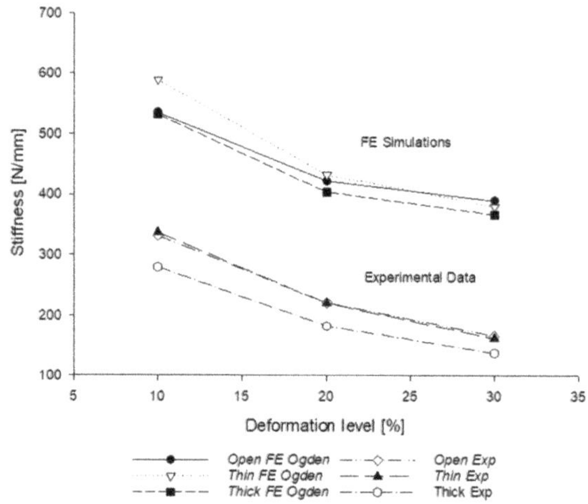

Figure 11. FEM Ogden model vs. experimental results.

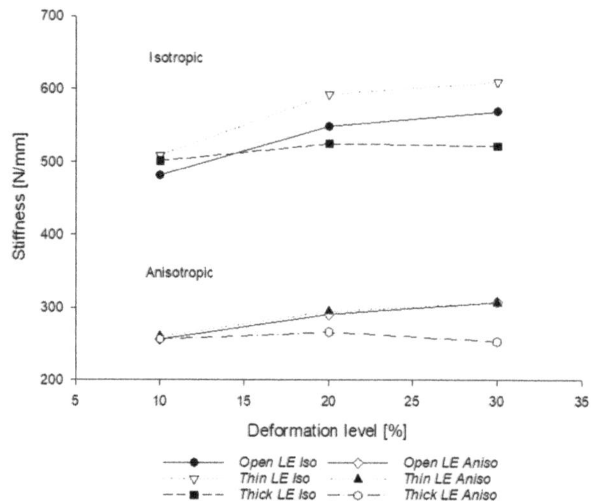

Figure 12. FEM simulations with linear elastic material: isotropic vs. anisotropic.

The two graphs in Figures 11 and 12 are similar; in fact, the simulation of the structures as isotropic cell solids showed a similar stiffness trend, but in the numerical results, it was double, compared to the anisotropic one. Therefore, the discrepancies between the experimental and numerical (by Ogden formulation) results are attributable to an anisotropic factor due to the printing process, i.e., to the various effects that occur in the printing phase. On the other hand, the simulations reflected the different mechanical responses of the three different geometrical configurations. For this reason, the aforementioned anisotropic layering factor is studied in depth in this discussion.

Geometrically, the graph in Figure 11 is able to highlight the lattice structure with the best mechanical response. In fact, in terms of stiffness, the thin-walled closed lattice structure showed the best performance, followed by the open typology, which showed a very similar mechanical response, especially in experimental tests. Finally, the thick-walled lattice structures showed a performance level always lower than the other two types, both numerically and experimentally.

Discrepancies in geometrical configurations response are certainly due to the effects of the FDM process [34]. In fact, as evident in Figure 2, the thick-walled structure had more deposited material in the walls, so it was more likely to have a high percentage of voids and layering effects inside. For this reason, it was desirable that these structures were weaker than the others, due to the greater and consistent presence of defects inside them.

5. Discussion

To understand the causes of discrepancies between experimental and numerical tests, all phenomena occurring in the printing process that influence the mechanical property of the FDM printed lattice structures were studied in depth. Several previous papers have found how FDM additive manufacturing technology widely affects the mechanical properties of printed parts [35,36]. FDM process is almost demanding due to the variability of its parameters and the uncontrollable printing effects such as porosity and layering.

Firstly, the porosity was studied both with SEM analysis and weight analysis. In their work, Abbot et al found that the simulated printed parts were at least 50% more solid than experimental printed samples [37]. This statement is in line with what was found in the present work, precisely shown in Figure 11, but it was found that this outcome is not due to the porosity of the FDM printed parts but to the layering effect that knocks down the mechanical properties of the additively manufacturing samples. In fact, SEM analysis on lattice structures conducted by Kumar et al. [38] revealed how the voids left by the FDM process have no major effects on the mechanical and functional properties of these structures. Moreover, through a weight analysis between real and virtual samples, no considerable differences in weight were found.

Finally, in the printing phase, there is another effect that is generated in the printed parts—the layering effect. It is extremely complex to simulate this phenomenon through a numerical model. For this reason, an *anisotropic layering factor* was identified, φ_l, which is able to describe how this effect acts on stiffness. Considering the similar precautions in the pre-printing phase (for example, care for the hygroscopicity of the TPU) and the same printing conditions for all the samples (i.e., the same process parameters on a unique batch and controllable boundary conditions), the focus was on the post-printing effects. In this way, all printed lattice structures were produced at the same time; hence, they had the same aging time, always showing comparable characteristics. Following a phenomenological approach, by defining φ_l as a function of geometry (i.e., number of wall's contours, N) and imposed deformation ε, Equation (6) can be written as

$$\varphi_l = \alpha(N) \cdot e^{\beta \cdot \varepsilon} \tag{6}$$

where $\alpha(N)$ is a function of the number of contours N in a cell's wall, equal to $\alpha = 1.3 + (0.15N)$, β is a constant equal to 0.02, and ε is the strain level in percentage (i.e., 10, 20 and 30%).

Firstly, the real trend of experimental curves was reproduced by means of a regression model with the R-squared coefficient of 0.98. Consequently, the parameters of Equation (6), α and β, were identified through an iterative process on varying the critical parameters that adversely affect stiffness, N and ε. Finally, when the optimal match between curves was reached, φ_l was described as used in Equation (6).

The plot of Equation (6) reveals an increasing effect of the number of layers N on the deformation level (Figure 13). This is due to the fact that with the increase of deformation, the stresses acting on layers and printing defects increase, determining a drastic decrease in the stiffness.

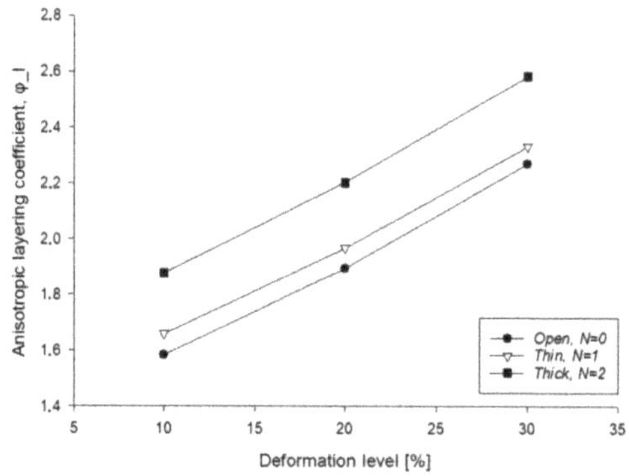

Figure 13. Anisotropic layering factor trend for each configuration.

Now, after recalibration of the simulated curves obtained by the Ogden hyperelastic model by dividing the stiffness values by the φ_l factor, a very good agreement can be found with the experimental results, as presented in the plot of Figure 14 and more precisely in the bar plot of Figure 15.

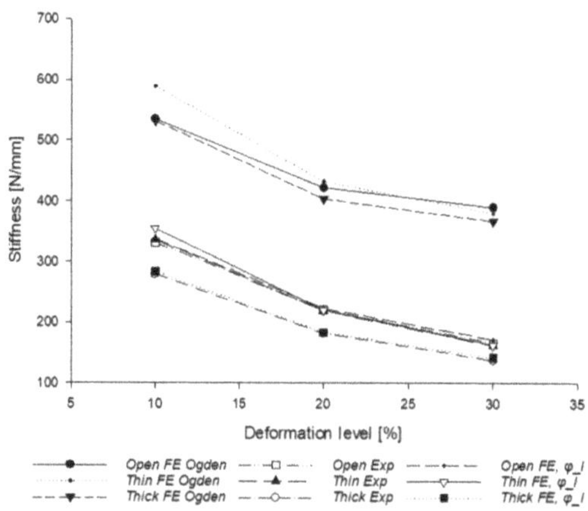

Figure 14. Experimental results and FE simulation results with the Ogden model.

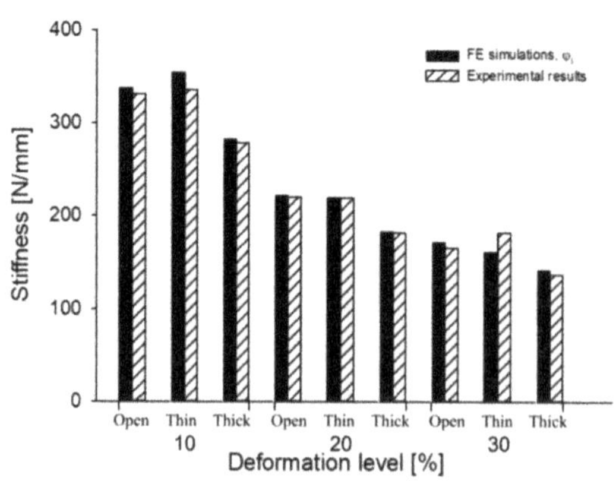

Figure 15. Experimental results and FE simulations corrected with φ_l.

6. Conclusions

This study was conducted in order to analyze and understand the mechanical behavior of specimens of cellular structures obtained by additive deposition process fused in thermoplastic polyurethane. Particular attention was paid to the layer-by-layer deposition effect on the resulting stiffness, showing that the mechanical behavior of printed samples with this technology is not easily predictable. Major findings can be summarized as follows:

- By the traditional FE analysis, an anisotropic behavior of such structures was proven;
- Anisotropy was ascribed to the layering process of filament, not always quantifiable a priori;
- A phenomenological layering factor φ_l was defined that tries to correlate the number of FDM contours, the deformation level, with the anisotropy degree;
- On the basis of the layering factor, thin-walled cell structures were confirmed to be the less affected, whereas larger walled structures were negatively affected;
- The mechanical and functional behaviors of this kind of structure were confirmed to be influenced by many parameters, related to material and process, as well asa specific geometry.

Author Contributions: Conceptualization and methodology, writing—review and editing, L.C.; running of simulations, formal analysis, and data curation, writing—original draft preparation, C.U. All authors have read and agreed to the published version of the manuscript.

Funding: This research has been possible thanks to the Overworld Mobility & Exchange Programme of the University of Parma.

Institutional Review Board Statement: Not applicable.

Informed Consent Statement: Not applicable.

Data Availability Statement: Data are available on request due to restrictions eg privacy or ethical.

Acknowledgments: Authors want to acknowledge Jeng and Kumar from the Department of Mechanical Engineering, National Taiwan University of Science and Technology, Taiwan, for providing experimental data and technical support.

Conflicts of Interest: The authors declare no conflict of interest.

References

1. Wang, P.; Zou, B.; Ding, S.; Li, L.; Huang, C. Effects of FDM-3D printing parameters on mechanical properties and microstructure of CF/PEEK and GF/PEEK. *Chin. J. Aeronaut.* **2021**, *34*, 236–246. [CrossRef]
2. Ziemian, C.; Sharma, M.; Ziemian, S. Anisotropic mechanical properties of ABS parts fabricated by fused deposition modelling. In *Mechanical Engineer*; Gokcek, M., Ed.; InTech: London, UK, 2017; pp. 159–180.
3. Durgun, I.; Ertan, R. Experimental investigation of FDM process for improvement of mechanical properties and production cost. *Rapid Prot. J.* **2014**, *20*, 228–235. [CrossRef]
4. Pagac, M.; Schwarz, D.; Petru, J.; Polzer, S. 3D printed polyurethane exhibits isotropic elastic behavior despite its anisotropic surface. *Rapid Prot. J.* **2020**, *26*. [CrossRef]
5. Garzon-Hernandez, S.; Garcia-Gonzalez, D.; Jérusalem, A.; Arias, A. Design of FDM 3D printed polymers: An experimental-modelling methodology for the prediction of mechanical properties. *Mater. Des.* **2020**, *188*, 108414. [CrossRef]
6. Ahn, S.H.; Montero, M.; Odell, D.; Roundy, S.; Wright, P.K. Anisotropic material properties of fused deposition modeling ABS. *Rapid Prot. J.* **2002**, *8*, 248–257. [CrossRef]
7. Baich, L.; Manogharan, G.; Marie, H. Study of infill print parameters on mechanical strength and production cost-time of 3D printed ABS parts. *Int. J. Rapid Manuf.* **2015**, *5*, 308–319. [CrossRef]
8. Bhate, D.; Penick, C.A.; Ferry, L.A.; Lee, C. Classification and selection of cellular materials in mechanical design: Engineering and biomimetic approaches. *Designs* **2019**, *3*, 19. [CrossRef]
9. Kumar, A.; Verma, S.; Jeng, J.-Y. Supportless Lattice structures for energy absorption fabricated by fused deposition modeling. *3D Print. Add. Manuf.* **2020**, *7*, 85–96. [CrossRef]
10. Askari, M.; Hutchins, D.A.; Thomas, P.J.L.; Astolfi, R.L.; Watson, M.; Abdi, M.; Ricci, S.; Laureti, L.; Nie, S.; Freear, R.; et al. Additive manufacturing of metamaterials: A. review. *Add. Manuf.* **2020**, *36*, 101562. [CrossRef]
11. Nazir, A.; Abate, K.M.; Kumar, A.; Jeng, J.-Y. A state-of-the-art review on types, design, optimization, and additive manufacturing of cellular structures. *Int. J. Adv. Manuf. Technol.* **2019**, *104*, 3489–3510. [CrossRef]

12. Bartolomé, L.; Aginagalde, A.; Martínez, A.B.; Urchegui, M.A.; Tato, W. Experimental characterization and modelling of large-strain viscoelastic behavior of a thermoplastic polyurethane elastomer. *Rubber Chem. Technol.* **2013**, *86*, 146–164. [CrossRef]
13. Solomon, I.J.; Sevvel, P.; Gunasekaran, J. A review on the various processing parameters in FDM. *Mater. Today Proc.* **2021**, *37*, 509–514. [CrossRef]
14. Panda, B.; Leite, M.; Biswal, B.B.; Niu, X.; Garg, A. Experimental and numerical modelling of mechanical properties of 3D printed honeycomb structures. *Measurement* **2018**, *116*, 495–506. [CrossRef]
15. Sood, A.K.; Ohdar, R.K.; Mahapatra, S.S. Experimental investigation and empirical modeling of FDM process for compressive strength improvement. *J. Adv. Res.* **2012**, *3*, 81–90. [CrossRef]
16. Wu, W.; Geng, P.; Li, G.; Zhao, D.; Zhang, H.; Zhao, J. Influence of layer thickness and raster angle on the mechanical study between PEEK and ABS. *Materials* **2015**, *8*, 5834–5846. [CrossRef]
17. Hernandez, R.; Slaughter, D.; Whaley, D.; Tate, J.; Asiabanpuor, B. Analyzing the tensile, compressive, and flexural properties of 3D printed ABS parts. In Proceedings of the 27th Annual International Solid Freeform Fabrication Symposium, San Marcos, TX, USA, 8–10 August 2016; pp. 939–950.
18. Motaparti, K.P. Effect of Build Parameters on Mechanical Properties of Ultem 9085 Parts by Fused Deposition Modeling. Masters' Thesis, Missouri University of Science and Technology, Rolla, MO, USA, 2016.
19. Vega, V.; Clements, J.; Lam, T.; Abad, A.; Fritz, B.; Ula, N.; Es-Said, O.S. The effect of layer orientation on the mechanical properties and microstructure of a polymer. *J. Mater. Eng. Perform.* **2011**, *20*, 978–988. [CrossRef]
20. Mohaed, O.A.; Masood, S.H.; Bhowmik, J.L. Optimization of FDM process parameters: A review of current research and future prospects. *Adv. Manuf.* **2015**, *3*, 42–52. [CrossRef]
21. Torrado, A.R.; Shemelya, C.M.; English, J.D.; Lin, Y.; Wicker, R.B.; Roberson, D.A. Characterizing the effect of additives to ABS on the echanical property anisotropy of specimens fabricated by material extrusion 3D printing. *Add. Manuf.* **2015**, *6*, 16–29.
22. Dawoud, M.; Taha, I.; Ebeid, S.J. Mechanical behaviour of ABS: An experimental study using FDM and injection moulding techniques. *J. Manuf. Process.* **2016**, *21*, 39–45. [CrossRef]
23. Hmeidat, N.S.; Pack, R.C.; Talley, S.J.; Moore, R.B.; Compton, B.G. Mechanical anisotropy in polymer composites produced by material extrusion additive manufacturing. *Add. Manuf.* **2020**, *34*, 101385. [CrossRef]
24. Rybachuk, M.; Mauger, C.A.; Fiedler, T.; Öchsner, A. Ochsner, Anisotropic mechanical properties of fused deposition modeled parts fabricated by using acrylonitrile butadiene styrene polymer. *J. Polym. Eng.* **2017**, *37*, 699–706. [CrossRef]
25. Anitha, R.; Arunachalam, S.; Radhakrishnan, P. Critical parameters influencing the quality of prototypes in fused deposition modelling. *J. Mater. Proc. Technol.* **2001**, *118*, 385–388. [CrossRef]
26. Rodríguez-Panes, A.; Claver, J.; Camacho, A.M. The influence of Manufacturing parameters on the mechanical behaviour of PLA and ABS pieces manufactured by FDM: A comparative analysis. *Materials* **2018**, *11*, 1333. [CrossRef]
27. Lee, B.; Abdullah, J.; Khan, Z. Optimization of rapid prototyping parameters for production of flexible ABS object. *J. Mater. Proc. Technol.* **2005**, *169*, 54–61. [CrossRef]
28. Popescu, D.; Zapciu, A.; Amza, C.; Baciu, F.; Marinescu, R. FDM process parameters influence over the mechanical properties of polymer specimens: A. review. *Polym. Test.* **2018**, *69*, 157–166. [CrossRef]
29. Qi, H.; Boyce, M. Stress-strain behavior of thermoplastic polyurethanes. *Mech. Mater.* **2005**, *37*, 817–839. [CrossRef]
30. Bergström, J.; Boyce, M. Constitutive modeling of the time-dependent and cyclic loading of elastomers and application to soft biological tissues. *Mech. Mater.* **2001**, *33*, 523–530. [CrossRef]
31. Hohimer, C.; Christ, J.; Aliheidari, N.; Mo, C.; Ameli, A. 3D printed thermoplastic polyurethane with isotropic material properties. In *Behavior and Mechanics of Multifunctional Materials and Composites*; SPIE: Bellingham, WA, USA, 2017; Article Number 1016511.
32. Elmrabet, N.; Siegkas, P. Dimensional consideration on the mechanical properties of 3D printed polymer parts. *Polym. Test.* **2020**, *90*, 106656. [CrossRef]
33. Kumar, A.; Collini, L.; Daurel, A.; Jeng, J.-Y. Design and additive manufacturing of closed cells from supportless lattice structure. *Add. Manuf.* **2020**, *33*, 101168. [CrossRef]
34. Collini, L.; Ursini, C.; Kumar, A. Design and optimization of 3D fast printed cellular structures. *Mater. Des. Proc. Commun.* **2021**, e227. [CrossRef]
35. Cantrell, J.T.; Rohde, S.; Damiani, D.; Gurmani, R.; DiSandro, L.; Anton, J.; Young, A.; Jerez, A.; Steinbach, D.; Kroese, C.; et al. Experimental characterization of the mechanical properties of 3D-printed ABS and polycarbonate parts. In *Advancement of Optical Methods in Experimental Mechanics*; Yoshida, S., Lamberti, L., Sciammarella, C., Eds.; Conference Proceedings of the Society for Experimental Mechanics Series; Springer: Cham, Switzerland, 2016; Volume 3.
36. Zou, R.; Xia, Y.; Liu, S.; Hu, P.; Hou, W.; Hu, Q.; Shan, C. Isotropic and anisotropic elasticity and yielding of 3D printed material. *Comp. Part B Eng.* **2016**, *99*, 506–513. [CrossRef]
37. Abbot, D.; Kallon, D.; Anghel, C.; Dube, P. Finite Element analysis of 3D printed model via compression tests. *Procedia Manuf.* **2019**, *35*, 164–173. [CrossRef]
38. Kumar, A.; Collini, L.; Ursini, C.; Jeng, J.-Y. Analyzing the functional properties of closed cell cellular lattice structure designed with thin and thick wall for additive manufacturing. *Mater. Des.* **2021**, submitted.

materials

MDPI

Article

The Influence of Specimen Geometry and Loading Conditions on the Mechanical Properties of Porous Brittle Media

Anatoly M. Bragov [1,*], Andrey K. Lomunov [1], Leonid A. Igumnov [1], Aleksandr A. Belov [1] and Victor A. Eremeyev [2,3]

1 Research Institute for Mechanics, National Research Lobachevsky State University of Nizhny Novgorod, 603950 Nizhny Novgorod, Russia; lomunov@mech.unn.ru (A.K.L.); igumnov@mech.unn.ru (L.A.I.); belov_a2@mech.unn.ru (A.A.B.)

2 Department of Mechanics of Materials and Structures, Faculty of Civil and Environmental Engineering, Gdansk University of Technology, 11/12 Gabriela Narutowicza Street, 80-233 Gdańsk, Poland; eremeyev.victor@gmail.com

3 Department of Civil and Environmental Engineering and Architecture (DICAAR), University of Cagliari, Via Marengo 2, 09123 Cagliari, Italy

* Correspondence: bragov@mech.unn.ru; Tel.: +7-831-465-1622

Abstract: Dynamic tests of fine-grained fired dioxide-zirconia ceramics under compression under uniaxial stress conditions were carried out. The influence of the specimen length on the obtained strength and deformation properties of ceramics is investigated. The thickness of the specimen has a significant impact on the course of the obtained dynamic stress–strain diagrams: short specimens have a much more sloping area of active loading branch. The main contribution to the modulus of the load branch resulting from tests of brittle porous media is made by the geometry of the specimens and the porosity of the material. When choosing the length of specimens for dynamic tests, the optimal geometry of the tested specimens is preferable in accordance with the Davies–Hunter criterion, when the contributions of axial and radial inertia are mutually compensated, and the contribution of the effects of friction in the resulting diagram is minimal. When choosing the geometry of specimens of brittle porous media, the structure of the material should be taken into account so that the size of the specimen (both length and diameter) exceeds the size of the internal fractions of the material by at least five times.

Keywords: zirconia ceramics; brittle medium; porosity; Kolsky method; compressibility; strength; fracture; stress growth rate

Citation: Bragov, A.M.; Lomunov, A.K.; Igumnov, L.A.; Belov, A.A.; Eremeyev, V.A. The Influence of Specimen Geometry and Loading Conditions on the Mechanical Properties of Porous Brittle Media. *Materials* **2021**, *14*, 7144. https://doi.org/10.3390/ma14237144

Academic Editors: Michele Bacciocchi and Abbas S. Milani

Received: 15 October 2021
Accepted: 19 November 2021
Published: 24 November 2021

1. Introduction

For many years, intensive work has been carried out all over the world to study the dynamic properties of structural materials. Brittle porous materials have a good ability to damp shock and explosive effects. So, they are widely used in various designs of new technology. Porous materials play an important role as shock-damping material in containers for air, automobile, and other transportation of radioactive or highly toxic materials. Porous materials significantly reduce the load on the main protective structural elements in the event of emergency situations or terrorist attacks, accompanied by shock or explosive influences and high temperatures. To reliably determine the behavior of containers with concordant damping materials under impact data on their properties are required, mainly, dynamic stress–strain curves. In addition, layered structures for protection against damaging effects by bullet and fragmentation elements are a possible field of application for porous low-density materials. In these designs, the gap between the metal layers is filled with a light material with good damping, dissipative and thermophysical properties.

The compressibility of metals under impact have been studied at present in more detail than the properties of brittle materials, such as concretes, ceramics, and refractories, which are associated with additional requirements imposed on test installations when studying this class of materials. Currently, there are no generally accepted standards for installations for the study of brittle and especially structurally inhomogeneous materials. A large number of types of various materials fall under the name of brittle and structurally heterogeneous, these are concretes, rocks, ceramics, various types of frozen materials (ice, bitumen, frozen soils, etc.). In their chemical and structural composition, all these materials in the field of dynamic loading are still insufficiently studied, so obtaining new data on their properties remains relevant.

The study of the mechanical properties of ceramic materials even under quasistatic effects is a very difficult and time-consuming task, since their properties depend on a large number of factors (multicomponent, porosity, humidity, etc.). Methodological difficulties in the study of the mechanical properties of ceramics and concrete repeatedly increase under dynamic loads, characterized by high intensities and short exposure times.

Brittle materials, in contrast to ductile ones, have a small deformation of destruction (often no more than 1%), and therefore, if the loading takes place too quickly, as in the usual SHPB test [1], the specimen can begin to break unevenly, i.e., the front of the specimen may be destroyed while the back remains intact. The conditions for the constancy of the strain rate and the equilibrium of stresses at the ends of the specimen should be satisfied during most of the test. A non-dispersive smoothly increasing pulse in the loading bar is required to test brittle materials (ceramics, rocks), which have an almost linear dependence of the stress–strain curve up to fracture. If the incident pulse is formed with a steep leading edge, then it is not possible to achieve dynamic stress equilibrium in the sample of a brittle material, and the specimen can immediately collapse at its end in contact with the loading bar after the arrival of the incident wave [2,3].

For SHPB test measurements to be valid, the dynamic load must increase slowly enough that the specimen is subjected to an almost quasi-static load; in this case, the deformation of the specimen will be uniform. The possibility of using the incident pulse shaper in the SHPB system was considered in [4]. It was noted that a smoothly rising incident pulse is preferred to minimize the dispersion and inertia effects and so to contribute to the dynamic equilibrium of the specimen stress state.

The easiest and most convenient way to create an incident pulse with an inclined front is to attach a small thin disk of soft material to the impact end of the loading pressure bar [4,5]. Such a disk, called a pulse shaper, can be made of copper, aluminum, brass or rubber with a thickness of 0.1–2.0 mm.

When studying the behavior of several ceramics under pulsed loading, it was noted [6] that the thickness of the specimen significantly affects the course of the obtained dynamic stress–strain curves. Specimens with a diameter of 20 mm and with a thickness from 2 mm to 10 mm were studied. It was determined that thicker specimens have a greater value of the modulus of the load branch and less energy absorption. Strength properties depend weakly on the length of the specimen. A similar trend was also noted in [7] when studying specimens of fine-grained concrete of 10 mm and 20 mm length: thicker specimens have a larger modulus of the load branch, while the breaking stress is practically independent of the length of the specimen.

The aim of this work was to evaluate the influence of the geometry of specimens of a brittle porous medium using the example of fine-grained firing ceramics based on zirconia on the deformation characteristics under compression at uniaxial stress condition.

2. Experimental Method

We used the original setup [5] that implements the Kolsky method [1] with a split Hopkinson pressure bar (SHPB) for investigation the dynamic properties of ceramics under compression. The experimental stand consists of a gas gun with a control system and a complex of measuring and recording equipment. Pressure bars with a diameter of 20 mm from D16T alloy are equipped with low-base strain gauges (Figure 1). The amplitude of the incident pulse, proportional to the striker velocity, varied from 60 MPa to 180 MPa, respectively, the strain rate (taking into account the different lengths of the specimens) was from 350 s^{-1} to 6000 s^{-1}.

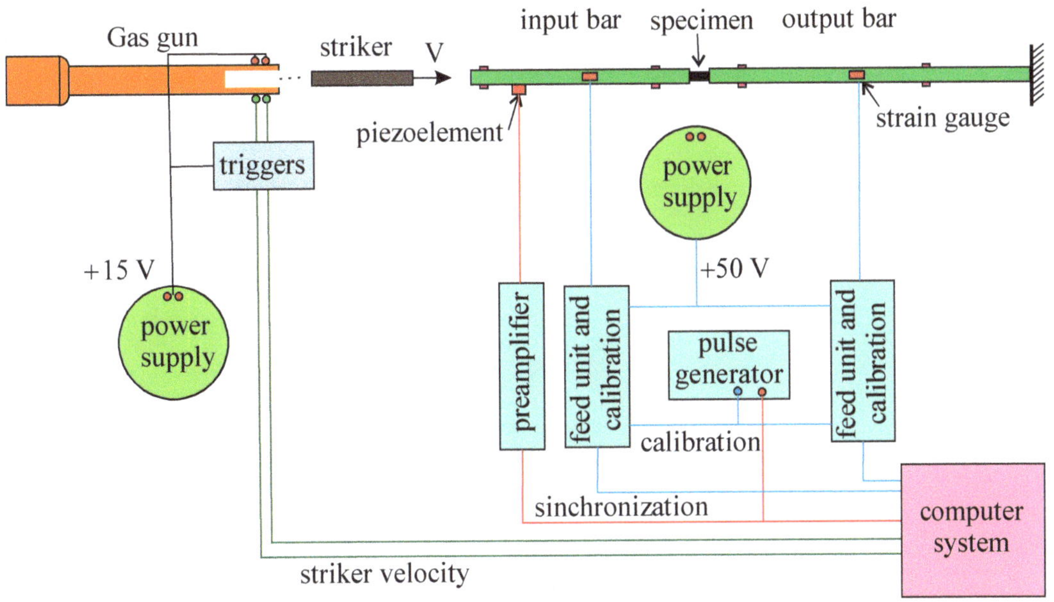

Figure 1. Experimental setup scheme.

In the test, a one-dimensional elastic loading (incident) impulse $\varepsilon^i(t)$ of the required amplitude and duration (determined by the speed and length of the striker), propagating along the incident bar with the speed of sound C, reaches the specimen and loads it; while part of the wave is reflected back by the reflected pulse $\varepsilon^r(t)$, and part passes into the supported bar by the transmitted pulse $\varepsilon^t(t)$. Based on these strain pulses recorded by strain gauges in measuring bars, the parametric dependences of the development of the axial stress $\sigma_s(t)$, strain $\varepsilon_s(t)$ and strain rate $\dot{\varepsilon}_s(t)$ components of the specimen over time by the Kolsky method formulas [8,9] were determined:

$$\sigma_s(t) = \frac{EA}{A_S^0}\varepsilon^T(t) \quad (1)$$

$$\varepsilon_s(t) = -\frac{2C}{L_0}\int_0^t \varepsilon^R(t)\cdot dt \quad (2)$$

$$\dot{\varepsilon}_s(t) = -\frac{2C}{L_0}\cdot\varepsilon^R(t) \quad (3)$$

where E and A are Young's modulus and cross-sectional area of the output bar, C is the speed of elastic waves in the input bar, L_0 is the initial length of the specimen, A_s^0 is the specimen initial cross-sectional area.

Then, after synchronizing the initial pulses, we can construct a dynamic diagram $\sigma_s(\varepsilon_s)$ with a dependence $\dot{\varepsilon}_s(\varepsilon_s)$.

When constructing dynamic stress–strain curves by the Kolsky method, sufficiently accurate matching in time of strain pulses is required, including expert selection of the starting point of reference for each pulse. A characteristic feature of the tests of porous brittle media by the Kolsky method is the presence of a significant (up to 20 μs) delay in the transmitted pulse relative to the reflected one, even in the case of gluing of both recording strain gauges at the same distance from the specimen. This is due to several reasons:

- Considerable difference in the acoustic impedances of the sample material and the material of the pressure bars;
- Low speed of wave propagation in a porous medium;
- The quality of the processing of the end surfaces of the specimen;
- Significant porosity of the material.

In addition, the registration of weak signals from strain gauges is sometimes accompanied by electromagnetic interference superimposed on the zero line. These factors lead to the fact that there is a probability of error when choosing the start points of these pulses.

The procedure of pulse synchronization is as follows. The beginning point of the incident pulse is the time point after which the first deflection of the recorded ray from the zero line is observed. Further, the positions of the starting points of the reflected and transmitted pulses are determined using the known elastic wave speed in the measuring bars and the known distance of the strain gauges from the specimen. When the sensors are glued at the same distance from the specimen, these points are selected synchronously.

At given starting points, it is checked the execution of the main condition of the Kolsky method, namely, the strains uniformity in the specimen body by checking the equality of deformations at the ends of the specimen during the test:

$$\varepsilon^{i}(t) + \varepsilon^{r}(t) = \varepsilon^{t}(t) \tag{4}$$

The original synchronization program allows one to automatically select and manually adjust the relative position of these pulses, but for most tests, this adjustment is not required. In this procedure, the main attention is paid to the fulfillment of condition (4) with the smallest error over the entire pulse duration, with the exception of the initial section (several microseconds), during which the stress–strain state of the specimen cannot be considered as uniform. Studies have shown that when testing concretes, ceramics, and some other porous brittle media, the assumption of equal forces at the ends of the specimen is quite good.

Elastic strain impulses in the pressure bars are measured using small-base foil strain gauges, then stored by a multi-channel digital oscilloscope and transmitted to a personal computer for processing and analysis. The original processing program allows one to synchronize the selected pulses and build true stress–strain diagrams. If necessary, a controlled smoothing of recorded pulses using integral splines is possible. It is also possible to conduct statistical and regression analyses of the results.

Because of the large contrast in the acoustic impedances ρC of the input pressure bar and the specimen of the porous material, the reflected pulse amplitude can reach 80–90% of the amplitude of the incident wave. So, the specimen will be exposed to several loading cycles. In order to authentically register a repeated loading cycles during one experiment, it is necessary to exclude the influence on the loading process in the second and subsequent cycles of the transmitted pulse reflected from the rear end of the transmitting bar. For that, the length of the transmitting bar should be increased in comparison with the length of the incident bar [10,11]. In this series of experiments, the length of incident and transmitting bars were 1.5 m and 4.5 m, respectively. As a result, it possible to register the main and two additional loading cycles.

3. Tested Specimens

There were tested specimens of zirconia ceramics, a promising material for use in the nuclear industry. From a chemical point of view, this material is relatively inert, does not form fusible compounds with uranium dioxide, and its melting point is about 3000 K. Materials based on zirconium dioxide have a thermal conductivity 1.5–2 times lower than materials based on other highly refractory oxides.

The raw material for the manufacture of ceramic specimens was a material obtained from cubic zirconia stabilized with yttrium oxide Y_2O_3 in a molar fraction of 11–12%. Manufacturing technology was close to factory. To obtain the required fractional composition of ceramics (Table 1), the technology included grinding, chemical and magnetic cleaning, the addition of a temporary plasticizing binder, pressing (pressure 100 MPa) and annealing (2000 K for 13 h).

Table 1. Granulometric composition of ceramics (wt.%).

Sizes of Fractions, mm					
<0.05	<0.2	0.2–0.315	0.315–0.4	0.63–1	1–2
40	-	20	-	40	-

These ceramics are characterized by the following physicomechanical properties under normal conditions: density 4.7–4.9 g/cm^3, porosity 20%, static compressive strength 39 MPa.

Specimens for testing were made in the form of a cylinder with a diameter of ~20 mm and a thickness of ~10 mm. This geometry of the specimens was chosen in accordance with the Davies—Hunter recommendation [12] to minimize the effects of inertia and friction. The ideal specimen slenderness ratio (that is, the ratio of its length to diameter) has been studied for a long time, as it plays an important role in inertial effects during dynamic SHPB testing. Based on a joint analysis of the effects of axial and radial inertia, Davies and Hunter [12] proposed an optimal ratio of sample slenderness $L/D = \sqrt{3} \cdot \nu_s /2$; where L and D are the length and diameter of the cylindrical specimen, respectively, and v_s is the Poisson's ratio of the material under test. With this L/D ratio, the components of axial and radial inertia are mutually compensated, therefore, the calculated stress in the sample is considered reliable. This ratio is valid for a variety of materials, including brittle media.

In addition, to assess the influence of the geometry of the specimens on the resulting stress–strain diagrams, some of the specimens were made with thicknesses of 5 mm and 2 mm.

One of the characteristic features of the deformation and fracture of brittle materials (and ceramics just refer to such materials) is a significant effect on the deformation and strength characteristics of the state of the surface of the specimens. The presence of micro- or macrocracks on the surface, barbs and cavities leads to a significant decrease in the strength properties of brittle materials. Therefore, the specimens before testing were subjected to manual grinding on sandpaper with a grain size of 0.01 mm. When establishing the specimen in the working position, to reduce friction and improve the acoustic contact between the ends of the measuring bars and the specimen, there were 2–3 layers of a thin (10 μm) fluoroplastic film.

Ceramic specimens were tested under compression conditions of a one-dimensional stress state. The specimen temperature in all experiments was 20 ± 2 °C.

4. Results and Discussion

Ceramic specimens were tested for compression using a device that implements the Kolsky method. During testing, by varying the striker velocity (i.e., the amplitude of the incident pulse), loading modes were selected in which the specimen after test either retained its apparent integrity and strength, or collapsed. Visual control of the samples after the experiments made it possible to assess the sample destruction degree. However, such an examination does not give an unambiguous answer to the question at what point the specimen collapsed. It is known that for tested materials with a low acoustic impedance ρC, the reflected pulse can have significant amplitude. This impulse, after reaching the impacted end-face of the incident bar, reflects from it as a compression wave and, after reaching the specimen, reloads it, then again partially reflects and so on. Such a process is repeated many times until this pulse is completely faded away. As a result of repeated loading cycles, the resulting microfractures have the ability to develop and enlarge. In addition, unloading waves from the free side surface of the specimen can affect the fracture process. However, the choice of the optimal specimen slenderness due to the Davies–Hunter recommendation [12], as well as the analysis of pulses in the bars, give reason to believe that there is no such effect.

A more accurate answer about the destruction of specimen in the first loading cycle may be given after the analysis of deformation pulses recorded in the pressure bars. As an example, Figures 2 and 3 show the initial pulses in the measuring bars, both during registration (a) and in the process of pulses synchronization (b). Moreover, the pulses in Figure 2 correspond to the case of maintaining the apparent integrity of the specimen, and the pulses in Figure 3 correspond to the case of complete destruction of the specimen (into powder). In these figures, the initial pulses recorded in the pressure bars are shown: 1—incident pulse, 2—reflected pulse and 3—transmitted pulse.

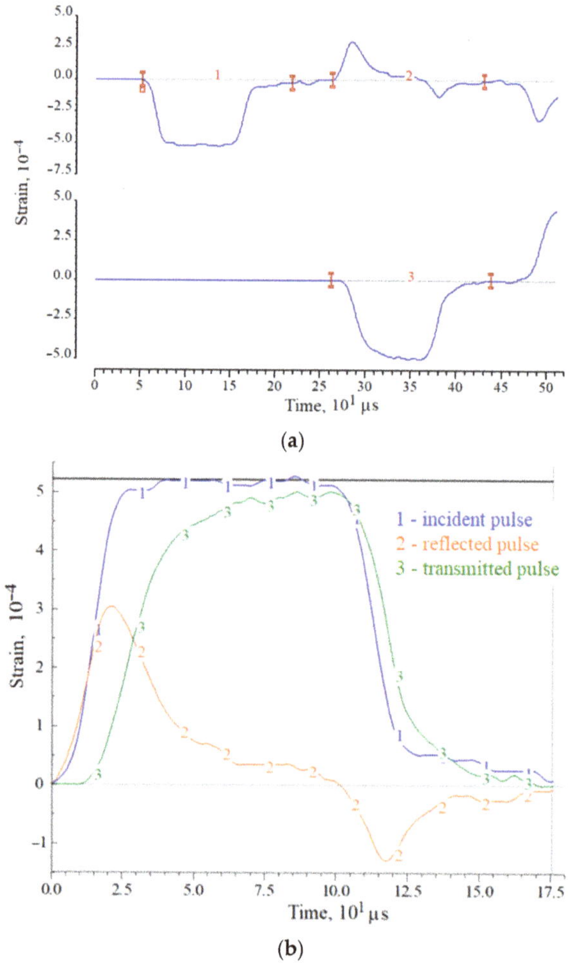

Figure 2. An example of strain pulses in measuring bars during registration (**a**) and during synchronization (**b**) when testing a ceramic specimen maintaining its visible integrity.

In the case of absence of the sample destruction under loading by a trapezoidal incident pulse $\varepsilon_i(t)$ with a flat top, the reflected pulse $\varepsilon_r(t)$ (the sample strain rate) first increases and then decreases due to an increase in the resistance of the sample during its deformation. After the end of affecting of the incident pulse on the sample, the transmitted pulse $\varepsilon_t(t)$ also begins to decrease, while the strain rate determined by the reflected pulse $\varepsilon_r(t)$ becomes negative. Thus, in the section of the sample active loading, both stress and deformation increase, then, with the beginning of the incident pulse decrease, the stress in the specimen decreases to almost zero, while the achieved deformation decreases by a certain amount, determined by the unloading capacity of the material and it is calculated by using the negative part of the reflected pulse.

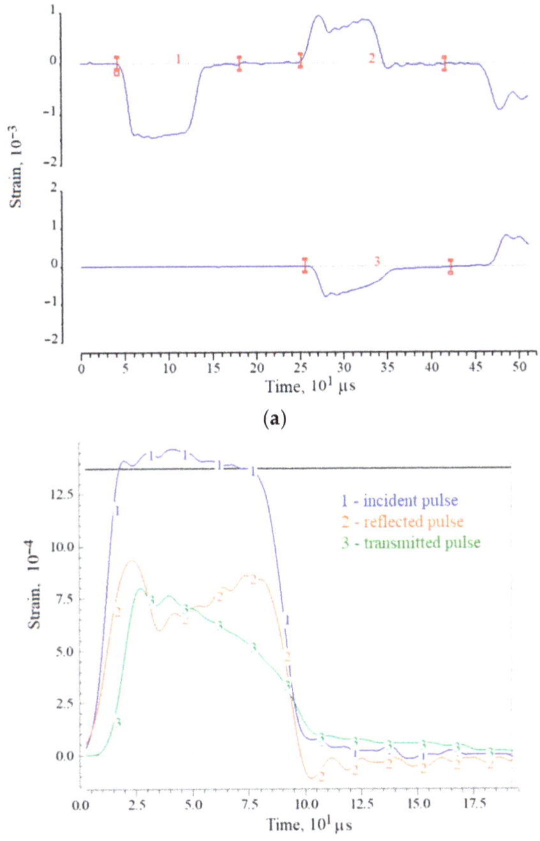

(**a**)

(**b**)

Figure 3. An example of strain pulses in measuring bars during registration (**a**) and during synchronization (**b**) when testing a ceramic specimen with its complete destruction.

A different picture occurs in the case of destruction of the specimen. A simultaneous increase in stress (pulse $\varepsilon_t(t)$) and strain rate (pulse $\varepsilon_r(t)$) takes place at the initial stage of specimen loading, however, after the point of maximum stress, the avalanche-like fracture process begins in the specimen. The stress after this point decreases, whereas the strain rate increases. So, although the amplitude of the incident pulse remains almost constant, the collapsing specimen does not completely pass the compression wave, its resistance to deformation is steadily decreasing.

Examples of resulting charts corresponding to these two types of tests are presented in Figure 4 in the form of parametric dependences $\sigma_s(t)$ and $\dot{\varepsilon}_s(t)$, as well as the diagrams $\sigma_s(\varepsilon_s)$ and $\dot{\varepsilon}_s(\varepsilon_s)$ themselves. The functions $\dot{\varepsilon}_s(t)$ and $\dot{\varepsilon}_s(\varepsilon_s)$ are represented on the graphs by dotted lines in the lower half of the figure field. The corresponding axis is located on the right side of the graphs. Digital markers on the lines are used to identify the curves and their mutual reference.

For both cases, one can note the nonlinearity of the initial section of the load branch and the significant difference between the load and unload branches. In the case of fracture of the specimen, the stress after reaching a maximum value begins to decrease with a constant increase in strain. Due to the short duration of the deformation process, the destroyed specimen particles that are not connected to each other during the test remain between the ends of the measuring rods, and their partial compaction occurs. This process is similar to high-speed deformation of non-cohesive soils. After the end of the incident pulse, the compacted ceramic particles have a small unloading capacity, which reveals in the form of a section of the diagram with partial restoration of the sample shape (negative portion of the function $\dot{\varepsilon}_s(\varepsilon_s)$).

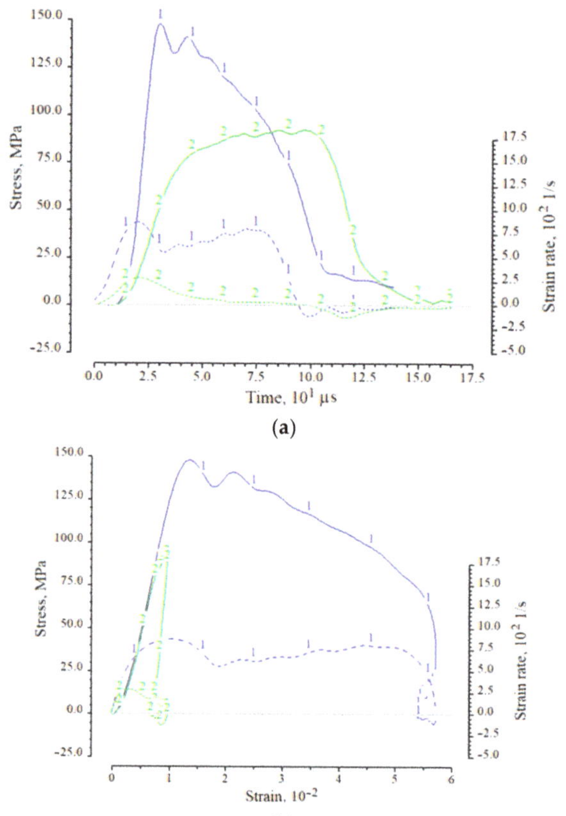

Figure 4. Examples of functions $\sigma_s(t)$ and $\dot{\varepsilon}_s(t)$ (**a**) and $\sigma_s(\varepsilon_s)$ and $\dot{\varepsilon}_s(\varepsilon_s)$ (**b**) for cases of maintaining integrity (curves 2) and complete destruction of a specimen (curves 1).

When the specimen is loaded with a high-intensity pulse, the destruction of the specimen occurs directly in the load wave; the proof of that can be illustrated in Figure 5, where the parametric processes $\sigma_i(t)$ and $\sigma_s(t)$ are shown.

The results of compression tests under uniaxial stress state of ceramic specimens of optimal geometry (diameter 20 mm, length 10 mm) are presented in Figure 6.

The solid lines in the figure show the characteristic diagrams of dynamic deformation, which are the result of averaging several experiments conducted under nominally identical conditions. The dashed lines of the corresponding colors at the bottom of the figure show the history of changes in the strain rate of the specimen. Curve 1 corresponds to the conditions under which, after the experiment, the specimen retained its visible integrity. Curve 3 was obtained under conditions of complete destruction of the specimen. Curve 2 was obtained under conditions when the specimens either had small (mainly peripheral) fractures, or, retaining the whole appearance, lost their structural connectivity.

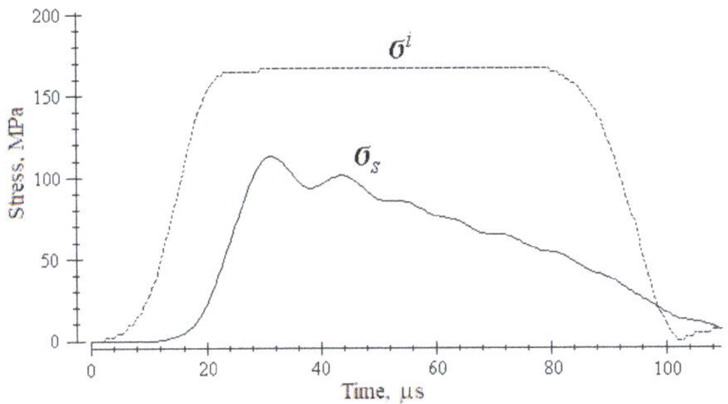

Figure 5. The parametric process of stress development in a specimen σ_s in case of its destruction in comparison with a loading pulse σ^i.

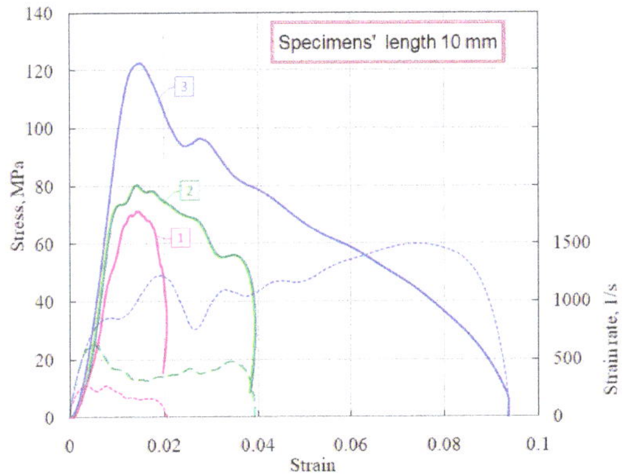

Figure 6. Deformation diagrams of ceramic specimens of optimal geometry under compression.

It is clearly seen that the structural strength of ceramics is about 70 MPa, however, during the dynamic loading, higher stress values were obtained, as evidenced by the upper curves 2 and 3. This behavior of the materials is due to the dynamic nature of the load and is determined by two competing processes occurring in specimen: the process of formation, growth and fusion of microcracks and micropores into macropores and cracks, on the one hand, and the wave nature of the increase of load in the material, on the other. It should be noted that the sample clamped between the ends of the measuring bars, due to the inertia of the process of its deformation, even when the process of internal damage has begun, remains in place and continues to transmit through itself a compressive stress wave of increasing amplitude. So, if the rate of increase of stress exceeds the velocity of the fracture process, then the specimen with already formed and developing fracture centers can be overloaded, i.e., it can withstand for some time ever-increasing loads. Similar phenomena were discussed in the analysis of the temporal dependence of compressive strength in [13,14].

Using the obtained stress–strain charts, the average values of the modules of the load branches ($d\sigma/d\varepsilon$) were measured (Figure 7), as well as there were determined the maximum stresses that the specimen withstood before failure began at various levels of the strain rate and the corresponding times at the onset of fracture (the beginning of the decay of the $\sigma_s(t)$ curve. These parameters for each curve are given in Table 2. When processing the experimental information for each diagram, in addition to the average strain rate $\dot{\varepsilon}_s$ of the sample, we determined the maximum values of the stress growth rate $\dot{\sigma}_s$ in the sample (Figure 8), which are also shown in Table 2.

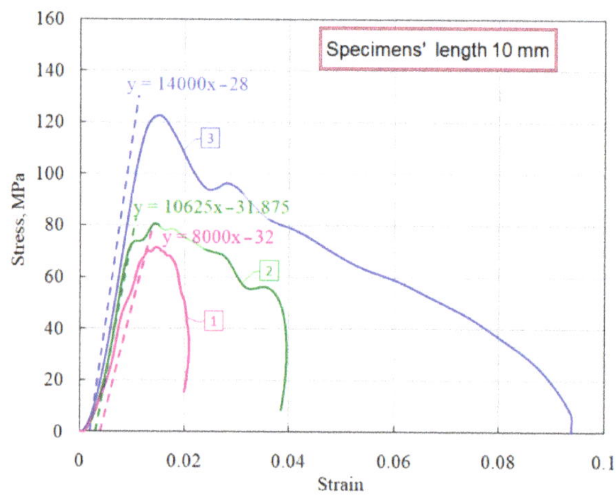

Figure 7. Definition of modules of load branches.

Table 2. Results of dynamic tests of ceramics.

Curve Number in the Diagram	The Module of the Load Branch, MPa	Average Strain Rate, 1/s	Stress Growth Rate, MPa/μs	Destruction Start Point			
				Strength, MPa	Strain, %	Time, μs	Energy Capacity, MJ/m^3
1	8000	220	1.9	71	1.5	82	0.93
2	10,625	400	6.5	80	1.4	40	2.42
3	14,000	1040	10.0	113	1.4	27	6.18

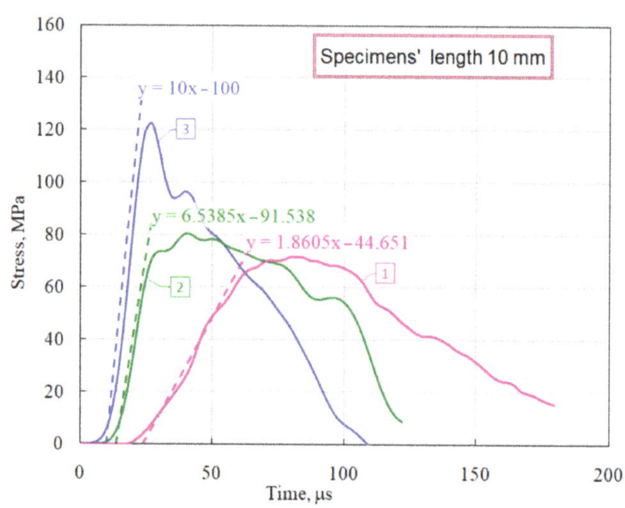

Figure 8. Definition the stress growth rate.

An important characteristic of ceramics working as a protective (energy-absorbing) material is their energy capacity, calculated as the area under the curve σ~ε (Figure 9). This characteristic is also given in Table 2.

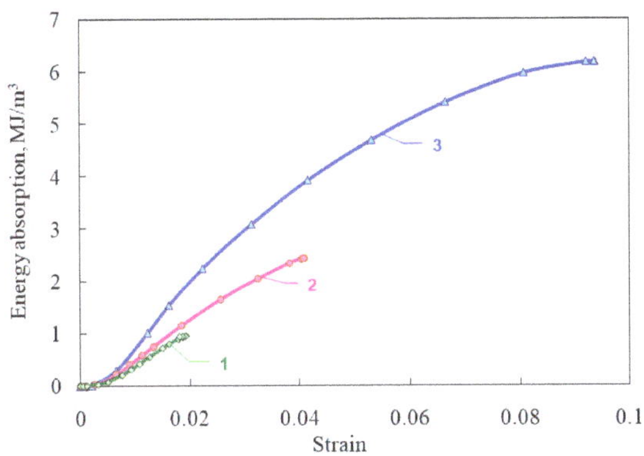

Figure 9. Energy absorption of ceramics.

As noted earlier [6,7], the thickness of the specimen has a significant effect on the course of the obtained dynamic stress–strain curves. In this regard, a comparative study of the influence of various geometries of ceramic specimens on mechanical characteristics was carried out. Figure 10 shows the average diagrams during compression of specimens with a diameter of 20 mm and a thickness of 10 mm, 5 mm, and 2 mm.

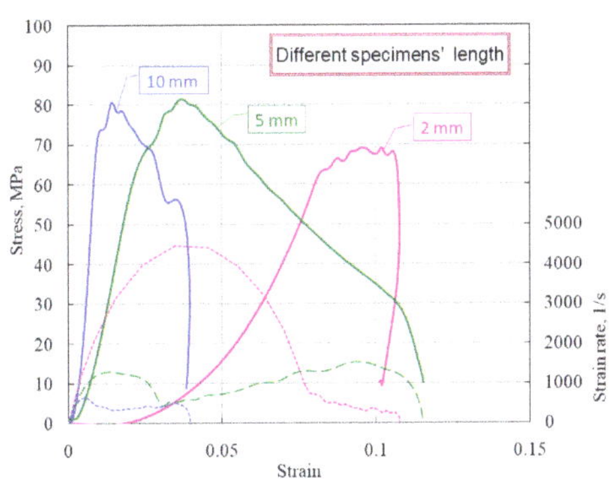

Figure 10. Effect of specimens' length on ceramic properties.

It is clearly seen that short specimens have a much more sloping area of active loading. Using the obtained stress–strain diagrams, the average values of the deformation module $d\sigma/d\varepsilon$ (steepness of the active loading section) were measured (Figure 11). The dashed lines show the linear approximations of the sections of active loading with the corresponding equations.

A similar trend was noted earlier in the study of ceramics [6] and fine-grained concrete [7]. The reason for the greater deformability of specimens of brittle media of shorter length can be as follows. Specimens whose length is 5 times different have approximately the same porosity and grain size. In addition, they have similar roughness characteristics of the end surfaces and possible deviations in their parallelism. When the incident strain (stress) pulse reaches the end of the specimen, it is divided into reflected and transmitted. These two pulses are the responses of the material to the applied load. In accordance with the formulas of the Kolsky method (1)–(3), the strain rate is directly proportional to the reflected pulse, and the specimen deformation is proportional to the integral of the reflected pulse. The voltage in the specimen is directly proportional to the transmitted pulse. The partition of the fractions of the reflected and transmitted pulses in the incident pulse is determined by a number of factors. The magnitude of the reflected wave is determined primarily by the ratio of impedances ρC of the incident bar and the specimen. In addition, in general, the ratio of reflected and transmitted waves is determined by the mechanical rigidity of the specimen material as well as its porosity. However, at the initial moment of exposure of the incident pulse to the specimen, this ratio is also determined by the roughness of the surface of the specimen, the non-flatness of its end and the possible non-parallelism of the end surface of the incident bar and the loading end surface of the specimen. Naturally, the shorter the length of the specimen, the greater the contribution of these defects to the reflected pulse, which determines the deformation of the specimen at the initial stage of loading, and, accordingly, the less the deformation modulus of the loading branch of the stress–strain diagram. Since, in accordance with Equation (2), the deformation of the specimen is inversely proportional to its length, then, for the same amplitude of the reflected pulse, the deformation of the 2-mm long specimen will be five times larger than for the 10-mm long specimen 10. This fact determines the corresponding difference obtained in the deformation modules of the sections of active loading of specimens of various lengths.

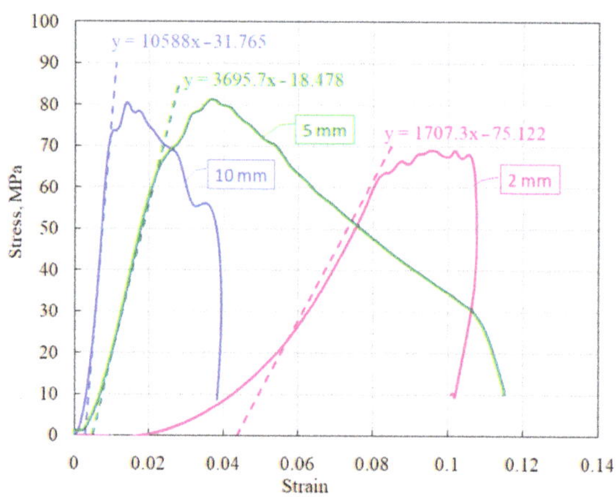

Figure 11. Definition of modules of load branches of ceramic diagrams for specimens of different lengths.

However, it should be noted that specimens of tested ceramics with a length of 2 mm cannot be considered representative, since, as follows from the granulometric composition of the ceramic (Table 1), fractions of 0.63–1 mm in size are present in it, in an amount of 40 wt.%. When choosing the geometry of specimens of brittle porous media, the structure of the material should be taken into account so that the size of the specimen (both length and diameter) exceed the size of the internal fractions of the material by at least five times.

Thus, in dynamic tests of brittle porous media, the optimal geometry of the test specimens, determined due to the Davies–Hunter recommendation [12], is preferable when the contributions of axial and radial components of inertia are mutually compensated, and the contribution of the effects of friction to the resulting diagram is minimal.

5. Conclusions

In dynamic tests of brittle porous media, special attention should be paid to quality control of the end surfaces of the specimen: to ensure minimal roughness, as well as parallelism of the ends and their flatness. The thickness of the specimen has a significant effect on the course of the obtained dynamic stress–strain diagrams: short specimens have a much more sloping area of active loading. The main contribution to the modulus of the load branch resulting from tests of brittle porous media is made by the geometry of the specimens and the porosity of the material. When choosing the length of specimens for dynamic tests, the optimal geometry of the tested specimens, determined due to the Davies–Hunter recommendation [12], is preferable when the contributions of axial and radial components of inertia are mutually compensated, and the contribution of the effects of friction in the resulting diagram is minimal.

Author Contributions: Conceptualization, A.M.B. and L.A.I.; methodology, A.K.L. and A.A.B.; formal analysis, A.M.B.; investigation, A.K.L.; data curation, A.M.B.; writing—original draft preparation, A.K.L.; writing—review and editing, L.A.I. and V.A.E.; visualization, A.A.B.; supervision, V.A.E.; project administration, A.M.B. All authors have read and agreed to the published version of the manuscript.

Funding: Production of samples and carrying out a set of dynamic tests were performed with the financial support of the Russian Science Foundation (task 21-19-00283). Analysis of the effect of the geometry of samples on the accuracy of the results obtained was carried out with financial support by the Scientific and Education Mathematical Center "Mathematics for Future Technologies" (Project No. 075-02-2021-1394).

Institutional Review Board Statement: Not applicable.

Informed Consent Statement: Not applicable.

Data Availability Statement: Not applicable.

Conflicts of Interest: The authors declare no conflict of interest. The funders had no role in the design of the study; in the collection, analyses, or interpretation of data; in the writing of the manuscript, or in the decision to publish the results.

References

1. Kolsky, H. An investigation of the mechanical properties of material at very high rates of loading. *Proc. Phys. Soc.* **1949**, *62*, 676–700. [CrossRef]
2. Frew, D.J.; Forrestal, M.J.; Chen, W. A split Hopkinson pressure bar technique to determine compressive stress–strain data for rock materials. *Exp. Mech.* **2001**, *41*, 40–46. [CrossRef]
3. Frew, D.J.; Forrestal, M.J.; Chen, W. Pulse shaping techniques for testing brittle materials with a split Hopkinson pressure bar. *Exp. Mech.* **2002**, *42*, 93–106. [CrossRef]
4. Frantz, C.E.; Follansbee, P.S.; Wright, W.J. New experimental techniques with the split Hopkinson pressure bar. In Proceedings of the 8th International Conference on High Energy Rate Fabrication, San Antonio, TX, USA, 17–21 June 1984; pp. 17–21.
5. Bragov, A.M.; Lomunov, A.K. Methodological aspects of studying dynamic material properties using the Kolsky method. *Int. J. Imp. Eng.* **1995**, *16*, 321–335. [CrossRef]
6. Bragov, A.; Kruszka, L.; Lomunov, A.; Konstantinov, A.; Lamzin, D.; Filippov, A. High-speed deformation and fracture of the dioxide-zirconium ceramics and zirconium alumina concrete. *EPJ Web Conf.* **2012**, *26*, 01055. [CrossRef]
7. Bragov, A.; Karihaloo, B.; Konstantinov, A.; Kruszka, L.; Lamzin, D.; Lomunov, A.; Petrov, Y. High-speed deformation and destruction of concrete and brick. In Proceedings of the 7th International Conference "Progress in Mechanics and Materials in Design", Albufeira, Portugal, 11–15 June 2017; pp. 381–392.
8. Chen, W.W.; Song, B. *Split Hopkinson (Kolsky) Bar: Design, Testing and Applications*; Springer Science & Business Media: Boston, MA, USA, 2011.
9. Bragov, A.M.; Igumnov, L.A.; Konstantinov, A.Y.; Lomunov, A.K.; Rusin, E.E.; Eremeyev, V.A. Experimental analysis of wear resistance of compacts of fine-dispersed iron powder and tungsten monocarbide nanopowder produced by impulse pressing. *Wear* **2020**, *456*, 203358. [CrossRef]
10. Bragov, A.M.; Lomunov, A.K.; Sergeichev, I.V. Modification of the Kolsky method for studying properties of low-density materials under high-velocity cyclic strain. *J. Appl. Mech. Tech. Phys.* **2001**, *42*, 1090–1094. [CrossRef]
11. Bragov, A.M.; Konstantinov, A.Y.; Lamzin, D.A.; Lomunov, A.K.; Filippov, A.R. Dynamic deformation and fracture of heterogeneous brittle media. *Probl. Strength Plast.* **2012**, *74*, 59–67. (In Russian) [CrossRef]
12. Davies, E.D.H.; Hunter, S.C. Dynamic compression testing of solids by the method of the split Hopkinson pressure bar. *J. Mech. Phys. Solids* **1963**, *11*, 155–179. [CrossRef]
13. Zlatin, N.A.; Pugachev, G.S.; Stepanov, V.A. On Breaking Pressures under Short-Time Impact Loading. *J. Teh. Fiz.* **1979**, *8*, 1786–1788. (In Russian)
14. Bragov, A.; Lomunov, A.; Kruszka, L. Study of Dry and Wet Cement Mortar Dynamic Properties. *Strength Mater.* **2002**, *34*, 37–42. [CrossRef]

materials

Article

Autonomous Design of Photoferroic Ruddlesden-Popper Perovskites for Water Splitting Devices

Alexandra Craft Ludvigsen, Zhenyun Lan * and Ivano E. Castelli *

Department of Energy Conversion and Storage, Technical University of Denmark, DK-2800 Kgs. Lyngby, Denmark; s184365@student.dtu.dk
* Correspondence: zhenlan@dtu.dk (Z.L.); ivca@dtu.dk (I.E.C.)

Abstract: The use of ferroelectric materials for light-harvesting applications is a possible solution for increasing the efficiency of solar cells and photoelectrocatalytic devices. In this work, we establish a fully autonomous computational workflow to identify light-harvesting materials for water splitting devices based on properties such as stability, size of the band gap, position of the band edges, and ferroelectricity. We have applied this workflow to investigate the Ruddlesden-Popper perovskite class and have identified four new compositions, which show a theoretical efficiency above 5%.

Keywords: light-harvesting; water splitting; photoferroics; high-throughput screening; Ruddlesden-Popper perovskites

Citation: Ludvigsen, A.C.; Lan, Z.; Castelli, I.E. Autonomous Design of Photoferroic Ruddlesden-Popper Perovskites for Water Splitting Devices. *Materials* **2022**, *15*, 309. https://doi.org/10.3390/ma15010309

Academic Editors: Michele Bacciocchi and Abbas S. Milani

Received: 22 November 2021
Accepted: 28 December 2021
Published: 2 January 2022

1. Introduction

The development of novel energy devices is required to meet the challenges of increasing energy demand and dependence on fossil fuels. The conversion of solar energy into electricity, using a photovoltaic (PV) device, or fuels, e.g., hydrogen and oxygen from water [1], by means of a photoelectrochemical (PEC) cell, are among the most promising solutions to achieve a green future. Both of these technologies rely on materials that show high stability, optimal light-harvesting properties, and low electron-hole recombination rates. The maximum theoretical efficiency obtainable from a single photoactive material in a PV cell is ≈33% (Schockley–Queisser limit), which corresponds to a material with a band gap around 1.3 eV, under 1.5 G solar irradiation and including all possible losses [2]. The efficiency is much lower for PEC devices, where the minimum required band gap is above 2 eV to overcome the bare energy to split water (1.23 eV), the reaction overpotentials (≈0.1 and ≈0.4 eV for the hydrogen and oxygen evolution [3]), and the Quasi Fermi-level (≈0.25 eV per band edge) [4]. The maximum theoretical efficiency is thus not larger than 7% [5]. Different solutions have been suggested to increase the PV and PEC efficiencies [6], both at the device level, by using solar concentrators and multi-junctions, and at the material level [7], by discovering novel compounds with supreme properties. Two new classes of materials have shown great potential to improve the solar conversion efficiency. (1) Organometal halide perovskites, where organic molecules are embedded in an inorganic crystal, have superior light absorption properties, high electron-hole mobility, and long lifetime, i.e., low electron-hole recombination rate [8]. However, they also show low stability and contain Pb, which can cause health issues. (2) Ferroelectric semiconductors (photoferroics) have two properties that make them very interesting for a new generation of solar energy conversion materials [9]. On one side, they can generate photovoltages larger than the band gap, and from the other side, they show an intrinsic polarization, which spontaneously separates the electrons and holes without the need of a $p-n$ junction or co-catalysts, in PV and PEC devices, respectively [10,11]. By generating photovoltages larger than the band gap, photoferroic materials would be able to easily provide the driving force (reaction overpotentials) necessary to run the hydrogen and, especially, oxygen evolution reactions. This could allow us to use materials with band gaps smaller than the 2 eV, mentioned above; thus, drastically increasing the efficiency of PEC devices. Moreover, the spontaneous separation of the photogenerated charges could solve some of the issues related to low mobility and high recombination rates, which is often solved with the use of co-catalysts, making the device easier and cheaper to produce. Despite the high potential, this technology is still at its early stages. The literature reports only a few photoferroic perovskite materials useful for PV and PEC devices, such as the oxides $BiFeO_3$ and its derivates [12–14], $KBiFe_2O_5$ [15], $Ba_2Bi^{3+}Bi^{5+}O_6$ [16], and chalcogenides [17], but the optimal materials have not been discovered yet. Moreover, recently, we have shown the polarization of $InSnO_2N$ can be switched during the oxygen evolution reaction (OER) to reduce the overpotential and thus increase the sun-to-chemical conversion efficiency [18].

In this work, we establish an autonomous workflow in the framework of Density Functional Theory (DFT) calculations to discover new photoferroic materials for PEC devices. This workflow is based on the calculation of stability, electronic, and ferroelectric properties and then applied to the class of Ruddlesden-Popper oxide and chalcogenide perovskites. Four new photoferroic materials have been identified to absorb at least 5% of the incident photons and be promising for one-photon water splitting applications.

2. Autonomous Workflow and Computational Methods

Thanks to methodological improvements [19–21] and an increase in the computational power, computational methods have been successfully used to design novel materials with desired functionalities and improved performance. Among others [22], high-throughput approaches and autonomous workflows have been used, in combination with DFT calculations, to design better catalysts [23], batteries [24], novel 2D and 1D materials [25–27], and solar energy conversion devices [4,28–30].

Starting from the properties of the constituent elements, a good photoferroic material should be formed by abundant, cheap, and non-toxic chemicals. The material should then be stable, absorb a good fraction of the solar spectrum, show good intrinsic polarization, suitable electron-hole mobility, and have good photoelectrocatalytic properties. These properties are calculated thanks to descriptors, which are easy to calculate and, at the same time, provide a good estimation of the quantity under investigation. For example, the stability is calculated using a convex hull analysis. The convex hull is constructed considering all the possible competing phases (constituent atoms, binary, and ternary compounds), taken from the Materials Project database, in which the candidate material can be separated [31]. The heat of formation is calculated as the difference between the DFT total energy of the candidate material and the energy of the convex hull at that particular composition. To include metastability [32], we consider a material thermodynamically stable when its heat of formation is up to 0.1 eV/atom. Furthermore, calculations of the mechanical and dynamic stability could be useful to confirm whether the candidate material could be synthesized or not. The light-harvesting efficiency is often estimated by the size of the band gap or full absorption spectrum and the photocatalytic properties with the position of the band edges [4,10,28,33].

The workflow established here to identify photoferroic materials is shown in Figure 1. All calculations are performed using the GPAW code and the Atomistic Simulation Environment (ASE) [34–36]. The workflow is implemented in the framework of MyQueue [37]. After having selected an appropriate chemical space, we use a structure prototype approach, in which all possible combinations obtained by decorating the prototype with the different chemicals are calculated. We then reduce the possible pool of candidate materials by considering simple structural and chemical rules, such as the sum of the electrons should be even to ensure that no bands are crossing the Fermi level, the sum of the possible oxidation states should be equal to zero to ensure a charge balance in the unit cell, and the size of the A and B-cations [38]. This reduces the original search space to around 30% of it. For these possible combinations, we calculate the relaxed structures (until the forces are below 0.05 eV/Å) of the different prototypes, their energies and band gaps. These calculations are performed in the framework of the Generalized Gradient Approximation (GGA) using PBEsol as the exchange-correlation function [39]. The simulations are performed in the Plane Waves (PW) mode with an energy cutoff of 800 eV and a K-point density equal to 3 Å^{-1}. We then compare the energies of the different prototypes and if the most stable one is non-centrosymmetric, has a convex hull energy below 0.1 eV/atom, and shows a band gap, we calculate its electronic properties, such as band gap, band structure, and density of states, using more accurate methods, in this case using the GLLB-SC exchange-correlation function [40–42]. For a better description of the electron density, we use a k-point density of 5 Å^{-1} and include the spin-orbit coupling (SOC) correction. To be able to evolve oxygen and hydrogen, the band edges need to straddle the redox levels of water. This is estimated using the geometrical average of the Mulliken electronegativities of the constituent atoms [28,43,44]. For a general $A_aB_bX_x$ compound, the position of the valence and conduction band edges, $E_{VB,CB}$, are thus given by

$$E_{VB,CB} = E_0 + \sqrt[a+b+x]{\chi_A^a \chi_B^b \chi_X^x} \pm E_{gap}/2\,, \tag{1}$$

where E_0 is the difference between the normal hydrogen electrode (NHE) and the vacuum level ($E_0 = -4.5$ eV) and χ_I is the electronegativity of the neutral I atom in the Mulliken scale. If the band gap is in the visible range, i.e., with a gap between 1.5 and 3 eV and the band edges straddle the redox levels of water, we proceed to calculate the absorption spectrum using Time-Dependent DFT, which gives a more accurate estimation of the light-harvesting efficiency. The theoretical efficiency, η, is calculated as

$$\eta = \frac{1}{n_{tot}} \int_{gap_d}^{\infty} ph_{abs}(E) n_{ph}(E) dE\,, \tag{2}$$

where n_{tot} is the total number of photons emitted by the sun at AM1.5, $ph_{abs}(E)$ is the photon absorptivity of the material, and $n_{ph}(E)$ the number of sun photons at the energy, E, in eV. We assume that no absorption takes place below the direct band gap, gap_d. Here, since we do not consider phonons, which are required to change the momentum in indirect transitions, we assume that no photons are absorbed below the direct gap. This approach is explained in detail elsewhere in the literature and has been used to estimate the light-harvesting properties in perovskites [45]. Moreover, the ferroelectricity/spontaneous polarization using the Berry phase approximation is estimated [10]. We note here that the indirect band gap materials are relevant only if phonons are involved in the absorption process. If that is not the case, e.g., for thin-films, the relevant gap to consider is the direct value. A material is considered a candidate only if it shows a spontaneous polarization and an efficiency of at least 5%.

Although this workflow can be applied to any crystal structure, we use it here to investigate the perovskite family. Perovskite compounds have shown a manifold of properties from efficient light-harvesting and high stability, superconductivity, and photo and ferroelectricity [46].

Moreover, the perovskite structure is able to host almost all elements from the periodic table, which allows for a very wide range of combinations, optimal for a screening project. Conventional cubic perovskites are, however, centrosymmetric so they will not show any polarization. On the other hand, perovskites exist in many different symmetries, such as double and layered, which can show an intrinsic polarization [47].

In this work, we consider the Ruddlesden-Popper (RP) layered perovskite phase, with formula $A_3B_2X_7$, where A and B are cations (A = Ba, Ca, Mg, or Sr; B = Ge, Hf, Pb, Sn, Ti, or Zr; $r_A > r_B$ and $\dagger_A \leq \dagger_B$, where r is the radius and $\dagger$ is the oxidation number) and X is an anion (X = O, S, or Se). An RP is formed by alternately stacked rock-salt layers (AX) and two perovskite-like layers (with formula ABX_3) along the c-axis of the crystal. Therefore, the RP phase shows some similar properties to the cubic perovskites, and, simultaneously, because of the two different chemical environments (rock salt and perovskite-like layers), more unique properties, including a polar structure, which can cause the generation of intrinsic ferroelectric behavior, as it has been recently shown for selected chalcogenide RP [17,48]. We investigate here the five most common RP prototypes, two centrosymmetric (space groups *Ccca* and $I4/mmm$, which by definition cannot show a polarization) and three non-centrosymmetric ($Cmc2_1$, *Pbcn*, and $P4_2/mnm$), as shown in Figure 1.

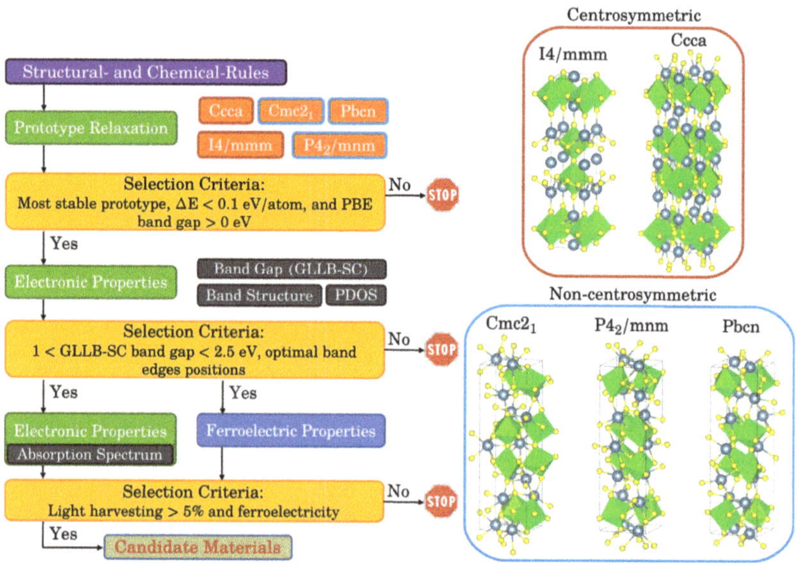

Figure 1. Workflow to autonomously identify photoferroic materials (**left**). The five most common Ruddlesden-Popper prototypes considered in this work (**right**). The A-cation is shown in blue, the B-cation in green, and the X-anion in yellow.

3. Results

Figure 2 shows the heat of formation and the band gap of all the calculated compositions. A material that shows good stability and band gap in the desired range is indicated in red. Overall, looking at the heat of formation, most of the investigated materials are stable, or at least metastable. We note that, despite having good stabilities, most oxides show very large band gaps (larger than 3 eV), which remove the majority of them from the pool of candidate, as we are considering only materials with a gap in the visible range. The wide band gaps of the oxides are a result of the large electronegativities difference between metals and oxygen [28,49]. Moreover, the very large electronegativity of oxygen has the effect of generating materials with rather deep bands at lower energies compared to the oxygen evolution potential. This, combined with a band gap in the visible range, makes the band edges not well-aligned with the redox levels of water, causing either large energy loss or making the material not suitable for evolving oxygen and hydrogen simultaneously. Sulfides behave differently from oxides. Firstly, almost all of them are more stable in the $Cmc2_1$ prototype, which is non-polar and thus allows for spontaneous polarization. Most of the materials have heat of formations below the metastability threshold, except for the compounds formed by Ge and Pb, which are very unstable. Secondly, the band gaps are smaller than the ones of the oxides and are within the visible light range due to the fact that sulfur is less electronegative than oxygen, which also impacts the position of the band edges with respect to the redox levels of water. Selenides seem even more promising than sulfides. The most stable prototypes are non-polar, the heats of formation are more negative than the ones of the sulfides, and the band gaps are smaller.

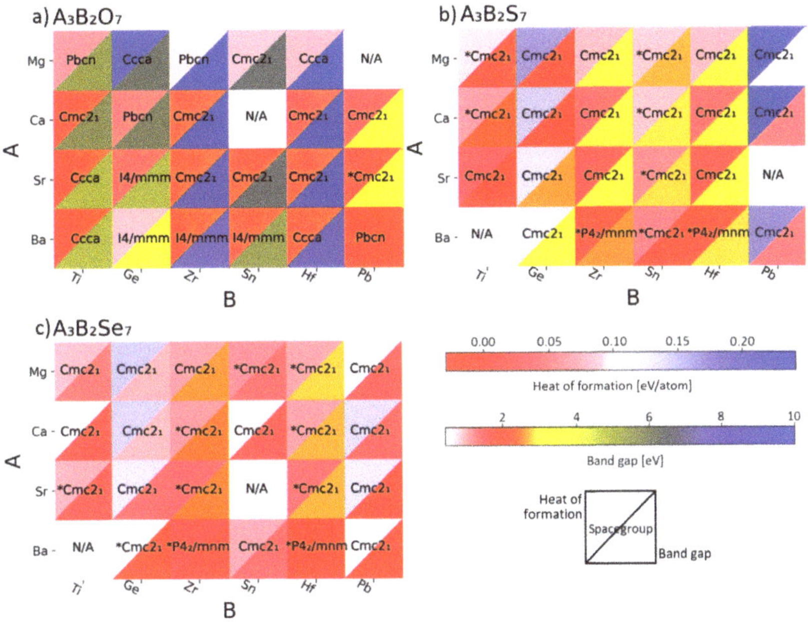

Figure 2. Heat maps showing the heat of formation (top **left** triangle) and band gap (bottom **right** triangle) for the oxide (**a**), sulfide (**b**), and selenide-based perovskites (**c**). A completely red square indicates a stable compound with good electronic properties, which is thereby considered a potential candidate. The space group of the most stable prototype is indicated in each square and stars (*) mark the materials that show an intrinsic polarization.

A total of 25 compositions survive the criteria based on stability and band gap, as well as having a non-centrosymmetric most stable prototype. Out of these, only 19 show a spontaneous polarization, as indicated in Table 1. The materials that show polarization in the Z-direction have the $Cmc2_1$ space group, while the ones where the polarization is in the X/Y-direction have the $P4_2/mnm$ space group.

The polarization direction becomes important to construct the water-splitting device. The absorption of light should happen in the thickest direction of the material to allow for an increased light-absorption ratio, while the splitting of the photogenerated charges should occur in the thinnest direction, to avoid their recombination. If the polarization is used to enhance the splitting of the charges, it should point along the thinnest direction and the light-absorption in the perpendicular direction. In practice, if the polarization points towards the Z-direction, then the absorption should happen in the XY-plane, and vice versa.

To be considered as candidate materials for water splitting, the band edges of a material should straddle the redox levels of water, which is a condition to allow the evolution of hydrogen and oxygen from water. Figure 3 shows the position of the band edges of these 19 candidate materials. Only 10 of them straddle both the hydrogen and oxygen evolution potentials, while 9 only straddle the hydrogen level. While the former can be used to run an overall (one-photon) water-splitting reaction, the latter can be used in a tandem device (two-photons) to evolve hydrogen [50,51].

Table 1. Calculated spontaneous polarization of the stable candidate materials with a band gap in the visible range. * indicates which materials also have well-positioned band edges, according to Figure 3.

Formula	Pol. ($\mu C/m^2$)	Direction
$Mg_3Hf_2Se_7$ *	23.97	Z
$Mg_3Sn_2Se_7$	46.56	Z
$Mg_3Sn_2S_7$ *	31.24	Z
$Mg_3Ti_2S_7$ *	51.12	Z
$Ca_3Zr_2Se_7$	20.36	Z
$Ca_3Hf_2Se_7$	3.72	Z
$Ca_3Sn_2S_7$ *	34.29	Z
$Sr_3Zr_2Se_7$	8.12	Z
$Ca_3Ti_2S_7$	15.72	Z
$Sr_3Hf_2Se_7$	24.19	Z
$Sr_3Sn_2S_7$ *	8.99	Z
$Sr_3Pb_2O_7$ *	31.74	Z
$Ba_3Sn_2S_7$ *	10.64	Z
$Ba_3Ge_2Se_7$ *	20.58	Z
$Sr_3Ti_2Se_7$	24.23	Z
$Ba_3Zr_2Se_7$	28.07; 1.12	X; Y
$Ba_3Hf_2Se_7$	19.89; 10.16	X; Y
$Ba_3Zr_2S_7$ *	24.84; 0.14	X; Y
$Ba_3Hf_2S_7$ *	17.51; 7.76	X; Y

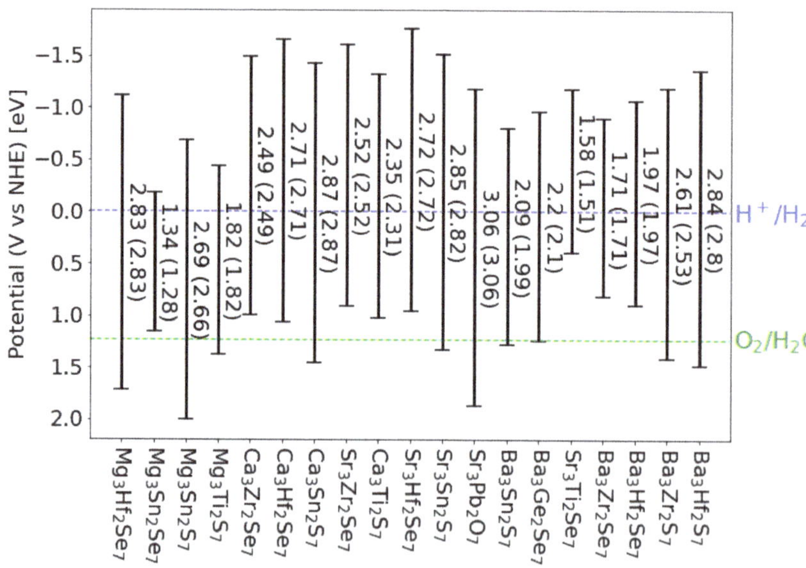

Figure 3. Position of the band edges, calculated for direct gaps, for all the materials that show stability and optimal size of the band gap. The values of the direct (indirect in parentheses) gap is indicated for each composition. The oxygen and hydrogen evolution potentials are also indicated.

The band gap is the most simple descriptor for the light-harvesting efficiency. This, however, does not take into account the kind of transition or its strength. For this reason, we calculate the absorption spectrum and calculate the number of absorbed photons. This procedure and its details have already been used to estimate the light-harvesting efficiency in cubic and layered perovskites [45]. The theoretical photon-absorption efficiencies are calculated in the direction perpendicular to the polarization as the ratio between the number of absorbed photons and total amount of photons from the Sun (AM1.5), and they are plotted as a function of the band gap in Figure 4. Four compositions ($Mg_3Ti_2S_7$, $Ba_3Sn_2S_7$, $Ba_3Ge_2Se_7$, and $Mg_3Sn_2S_7$) can be used for one-photon water splitting, with a theoretical capacity above 5%. To our knowledge, only $Ba_3Sn_2S_7$ has been previously synthesized, however, in a different space group [52]. In addition, we have identified a handful of materials, which can be used for the hydrogen evolution in a two-photon water splitting device, with an efficiency well-above 10%. We note that this estimation of the efficiency does not include recombination losses but is still an improvement of the efficiency obtained from the band gap values.

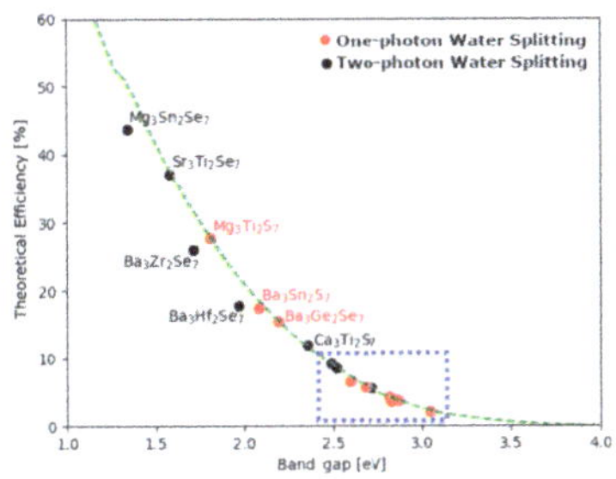

Formula	Theoretical efficiency (%)
$Ca_3Zr_2Se_7$	9.16%
$Sr_3Zr_2Se_7$	8.42%
$Ba_3Zr_2S_7$	6.49%
$Mg_3Sn_2S_7$	5.78%
$Ca_3Hf_2Se_7$	5.50%
$Sr_3Hf_2Se_7$	5.49%
$Mg_3Hf_2Se_7$	4.24%
$Sr_3Sn_2S_7$	3.96%
$Ca_3Sn_2S_7$	3.69%
$Ba_3Hf_2S_7$	3.68%
$Sr_3Pb_2O_7$	2.16%

Figure 4. Efficiencies of the 19 candidate materials. The materials indicated in red show potential for one-photon water splitting, while all the others could be used for a two-photon water splitting device. The green line represents the maximum theoretical efficiency. The figure reports materials with efficiency above 10%, while below 10% (enclosed in the dashed box) are summarized in the table.

4. Conclusions

In this work, we have described an autonomous workflow to identify new photoferroic materials for light-harvesting in a photoelectrochemical water splitting device. Our workflow has been applied to investigate oxide and chalcogenide Ruddlesden-Popper perovskites. Based on descriptors, such as stability, size of the band gap, position of the band edges, absorption spectrum, and ferroelectrocity, we have identified four new compounds that have a theoretical light-harvesting efficiency above 5%. Five other compositions could be used for a two-photon water splitting device, with an efficiency above 10%. Beyond perovskite structures, this workflow can now be used to investigate any crystal structure both using a similar structure prototype approach and investigating known materials, for example, from the Inorganic Crystal Structure Database (ICSD) [53].

Author Contributions: Conceptualization, I.E.C.; methodology, I.E.C.; formal analysis, A.C.L. and Z.L.; investigation, A.C.L. and Z.L.; data curation, A.C.L. and Z.L.; writing—original draft preparation, A.C.L., Z.L. and I.E.C.; writing—review and editing, Z.L. and I.E.C.; visualization, A.C.L. and I.E.C.; supervision, Z.L. and I.E.C.; project administration, I.E.C.; funding acquisition, Z.L. and I.E.C. All authors have read and agreed to the published version of the manuscript.

Funding: This research was funded by the China Scholarship Council (CSC) and the Department of Energy Conversion and Storage, Technical University of Denmark, through the Special Competence Initiative "Autonomous Materials Discovery (AiMade)" [54].

Data Availability Statement: The raw data generated for this work, the workflow for running the calculations, and the script produced to analyse the data are available upon reasonable request and can be obtained contacting I.E.C.

Acknowledgments: I.E.C. would like to thank A.T. Las for the useful discussions on this topic.

Conflicts of Interest: The authors declare no conflict of interest.

References

1. Sivula, K.; van de Krol, R. Semiconducting materials for photoelectrochemical energy conversion. *Nat. Rev. Mater.* **2016**, *1*, 1–16. [CrossRef]
2. Shockley, W.; Queisser, H.J. Detailed balance limit of efficiency of p-n junction solar cells. *J. Appl. Phys.* **1961**, *32*, 510–519. [CrossRef]
3. Trasatti, S. Surface chemistry of oxides and electrocatalysis. *Croat. Chem. Acta* **1990**, *63*, 313–329.
4. Castelli, I.E.; Landis, D.D.; Thygesen, K.S.; Dahl, S.; Chorkendorff, I.; Jaramillo, T.F.; Jacobsen, K.W. New cubic perovskites for one- and two-photon water splitting using the computational materials repository. *Energy Environ. Sci.* **2012**, *5*, 9034. [CrossRef]
5. Weber, M.; Dignam, M. Splitting water with semiconducting photoelectrodes—Efficiency considerations. *Int. J. Hydrog. Energy* **1986**, *11*, 225–232. [CrossRef]
6. Nayak, P.K.; Mahesh, S.; Snaith, H.J.; Cahen, D. Photovoltaic solar cell technologies: Analysing the state of the art. *Nat. Rev. Mater.* **2019**, *4*, 269–285. [CrossRef]
7. Wong, L.H.; Zakutayev, A.; Major, J.D.; Hao, X.; Walsh, A.; Todorov, T.K.; Saucedo, E. Emerging inorganic solar cell efficiency tables (Version 1). *J. Phys. Energy* **2019**, *1*, 032001. [CrossRef]
8. Chiarella, F.; Zappettini, A.; Licci, F.; Borriello, I.; Cantele, G.; Ninno, D.; Cassinese, A.; Vaglio, R. Combined experimental and theoretical investigation of optical, structural, and electronic properties of $C_3NH_3SnX_3$ thin films (X = Cl, Br). *Phys. Rev. B* **2008**, *77*, 045129. [CrossRef]
9. Fridkin, V.M. *Ferroelectric Semiconductors*; Springer: New York, NY, USA, 1980.
10. Castelli, I.E.; Olsen, T.; Chen, Y. Towards photoferroic materials by design: Recent progress and perspectives. *J. Phys. Energy* **2019**, *2*, 011001. [CrossRef]
11. Wallace, S.K.; Svane, K.L.; Huhn, W.P.; Zhu, T.; Mitzi, D.B.; Blum, V.; Walsh, A. Candidate photoferroic absorber materials for thin-film solar cells from naturally occurring minerals: Enargite, stephanite, and bournonite. *Sustain. Energy Fuels* **2017**, *1*, 1339–1350. [CrossRef]
12. Choi, T.; Lee, S.; Choi, Y.J.; Kiryukhin, V.; Cheong, S.W. Switchable ferroelectric diode and photovoltaic effect in $BiFeO_3$. *Science* **2009**, *324*, 63–66. [CrossRef]
13. Nie, R.; Yun, H.S.; Paik, M.J.; Mehta, A.; Park, B.W.; Choi, Y.C.; Seok, S.I. Efficient solar cells based on light-harvesting antimony sulfoiodide. *Adv. Energy Mater.* **2017**, *8*, 1701901. [CrossRef]
14. Nechache, R.; Harnagea, C.; Li, S.; Cardenas, L.; Huang, W.; Chakrabartty, J.; Rosei, F. Bandgap tuning of multiferroic oxide solar cells. *Nat. Photonics* **2014**, *9*, 61–67. [CrossRef]
15. Zhang, G.; Wu, H.; Li, G.; Huang, Q.; Yang, C.; Huang, F.; Liao, F.; Lin, J. New high Tc multiferroics $KBiFe_2O_5$ with narrow band gap and promising photovoltaic effect. *Sci. Rep.* **2013**, *3*, 1–9. [CrossRef]
16. Tang, J.; Zou, Z.; Ye, J. Efficient photocatalysis on $BaBiO_3$ driven by visible light. *J. Phys. Chem. C* **2007**, *111*, 12779–12785. [CrossRef]
17. Wang, H.; Gou, G.; Li, J. Ruddlesden–Popper perovskite sulfides $A_3B_2S_7$: A new family of ferroelectric photovoltaic materials for the visible spectrum. *Nano Energy* **2016**, *22*, 507–513. [CrossRef]
18. Lan, Z.; Småbråten, D.R.; Xiao, C.; Vegge, T.; Aschauer, U.; Castelli, I.E. Enhancing oxygen evolution reaction activity by using switchable polarization in ferroelectric $InSnO_2N$. *ACS Catal.* **2021**, *11*, 12692–12700. [CrossRef]
19. Marzari, N. Materials modelling: The frontiers and the challenges. *Nat. Mater.* **2016**, *15*, 381–382. [CrossRef]
20. Seminario, J. (Ed.) *Recent Developments and Applications of Modern Density Functional Theory (Theoretical and Computational Chemistry)*; Elsevier Science: New York, NY, USA, 1996.
21. Lejaeghere, K.; Bihlmayer, G.; Björkman, T.; Blaha, P.; Blügel, S.; Blum, V.; Caliste, D.; Castelli, I.E.; Clark, S.J.; Dal Corso, A.; et al. Reproducibility in density functional theory calculations of solids. *Science* **2016**, *351*, 1415. [CrossRef]
22. Curtarolo, S.; Hart, G.L.W.; Nardelli, M.B.; Mingo, N.; Sanvito, S.; Levy, O. The high-throughput highway to computational materials design. *Nat. Mater.* **2013**, *12*, 191–201. [CrossRef]
23. Back, S.; Tran, K.; Ulissi, Z.W. Toward a design of active oxygen evolution catalysts: Insights from automated density functional theory calculations and machine learning. *ACS Catal.* **2019**, *9*, 7651–7659. [CrossRef]
24. Bölle, F.T.; Mathiesen, N.R.; Nielsen, A.J.; Vegge, T.; Garcia-Lastra, J.M.; Castelli, I.E. Autonomous discovery of materials for intercalation electrodes. *Batte. Supercaps* **2020**, *3*, 488–498. [CrossRef]
25. Haastrup, S.; Strange, M.; Pandey, M.; Deilmann, T.; Schmidt, P.S.; Hinsche, N.F.; Gjerding, M.N.; Torelli, D.; Larsen, P.M.; Riis-Jensen, A.C.; et al. The Computational 2D Materials Database: High-throughput modeling and discovery of atomically thin crystals. *2D Mater.* **2018**, *5*, 042002. [CrossRef]

26. Mounet, N.; Gibertini, M.; Schwaller, P.; Campi, D.; Merkys, A.; Marrazzo, A.; Sohier, T.; Castelli, I.E.; Cepellotti, A.; Pizzi, G.; et al. Two-dimensional materials from high-throughput computational exfoliation of experimentally known compounds. *Nat. Nanotechnol.* **2018**, *13*, 246–252. [CrossRef]
27. Bölle, F.T.; Mikkelsen, A.E.G.; Thygesen, K.S.; Vegge, T.; Castelli, I.E. Structural and chemical mechanisms governing stability of inorganic Janus nanotubes. *NPJ Comput. Mater.* **2021**, *7*, 1–8. [CrossRef]
28. Castelli, I.E.; Olsen, T.; Datta, S.; Landis, D.D.; Dahl, S.; Thygesen, K.S.; Jacobsen, K.W. Computational screening of perovskite metal oxides for optimal solar light capture. *Energy Environ. Sci.* **2012**, *5*, 5814–5819. [CrossRef]
29. Wu, Y.; Lazic, P.; Hautier, G.; Persson, K.; Ceder, G. First principles high throughput screening of oxynitrides for water-splitting photocatalysts. *Energy Environ. Sci.* **2013**, *6*, 157–168. [CrossRef]
30. Kuhar, K.; Crovetto, A.; Pandey, M.; Thygesen, K.S.; Seger, B.; Vesborg, P.C.K.; Hansen, O.; Chorkendorff, I.; Jacobsen, K.W. Sulfide perovskites for solar energy conversion applications: Computational screening and synthesis of the selected compound LaYS3. *Energy Environ. Sci.* **2017**, *10*, 2579–2593. [CrossRef]
31. Materials Project—A Materials Genome Approach. Available online: https://materialsproject.org (accessed on 30 December 2021).
32. Esposito, V.; Castelli, I.E. Metastability at defective Metal oxide interfaces and nanoconfined structures. *Adv. Mater. Interfaces* **2020**, *7*, 1902090. [CrossRef]
33. Castelli, I.E.; Kuhar, K.; Pandey, M.; Jacobsen, K.W. Chapter 3. Computational screening of light-absorbing materials for photoelectrochemical water splitting. In *Advances in Photoelectrochemical Water Splitting*; Royal Society of Chemistry: London, UK, 2018; pp. 62–99. [CrossRef]
34. Mortensen, J.J.; Hansen, L.B.; Jacobsen, K.W. Real-space grid implementation of the projector augmented wave method. *Phys. Rev. B* **2005**, *71*, 035109. [CrossRef]
35. Enkovaara, J.; Rostgaard, C.; Mortensen, J.J.; Chen, J.; Dułak, M.; Ferrighi, L.; Gavnholt, J.; Glinsvad, C.; Haikola, V.; Hansen, H.A.; et al. Electronic structure calculations with GPAW: A real-space implementation of the projector augmented-wave method. *J. Phys. Condens. Matter* **2010**, *22*, 253202. [CrossRef] [PubMed]
36. Larsen, A.H.; Mortensen, J.J.; Blomqvist, J.; Castelli, I.E.; Christensen, R.; Dułak, M.; Friis, J.; Groves, M.N.; Hammer, B.; Hargus, C.; et al. The atomic simulation environment—A Python library for working with atoms. *J. Phys. Condens. Matter* **2017**, *29*, 273002. [CrossRef]
37. Mortensen, J.; Gjerding, M.; Thygesen, K. MyQueue: Task and workflow scheduling system. *J. Open Source Softw.* **2020**, *5*, 1844. [CrossRef]
38. Castelli, I.E.; Jacobsen, K.W. Designing rules and probabilistic weighting for fast materials discovery in the Perovskite structure. *Model. Simul. Mater. Sci. Eng.* **2014**, *22*, 055007. [CrossRef]
39. Perdew, J.P.; Ruzsinszky, A.; Csonka, G.I.; Vydrov, O.A.; Scuseria, G.E.; Constantin, L.A.; Zhou, X.; Burke, K. Restoring the Density-Gradient Expansion for Exchange in Solids and Surfaces. *Phys. Rev. Lett.* **2008**, *100*, 136406. [CrossRef]
40. Gritsenko, O.; van Leeuwen, R.; van Lenthe, E.; Baerends, E.J. Self-consistent approximation to the Kohn-Sham exchange potential. *Phys. Rev. A* **1995**, *51*, 1944–1954. [CrossRef]
41. Kuisma, M.; Ojanen, J.; Enkovaara, J.; Rantala, T.T. Kohn-Sham potential with discontinuity for band gap materials. *Phys. Rev. B* **2010**, *82*, 115106. [CrossRef]
42. Castelli, I.E.; Hüser, F.; Pandey, M.; Li, H.; Thygesen, K.S.; Seger, B.; Jain, A.; Persson, K.A.; Ceder, G.; Jacobsen, K.W. New light-harvesting materials using accurate and efficient bandgap calculations. *Adv. Energy Mater.* **2014**, *5*, 1400915. [CrossRef]
43. Butler, M.A.; Ginley, D.S. Prediction of flatband potentials at semiconductor-electrolyte interfaces from atomic electronegativities. *J. Electrochem. Soc.* **1978**, *125*, 228–232. [CrossRef]
44. Xu, Y.; Schoonen, M.A. The absolute energy positions of conduction and valence bands of selected semiconducting minerals. *Am. Mineral.* **2000**, *85*, 543–556. [CrossRef]
45. Castelli, I.E.; Thygesen, K.S.; Jacobsen, K.W. Calculated optical absorption of different perovskite phases. *J. Mater. Chem. A* **2015**, *3*, 12343–12349. [CrossRef]
46. Ishihara, T. *Perovskite Oxide for Solid Oxide Fuel Cells*; Springer: Boston, MA, USA, 2009.
47. Benedek, N.A.; Rondinelli, J.M.; Djani, H.; Ghosez, P.; Lightfoot, P. Understanding ferroelectricity in layered perovskites: New ideas and insights from theory and experiments. *Dalton Trans.* **2015**, *44*, 10543–10558. [CrossRef]
48. Zhang, Y.; Shimada, T.; Kitamura, T.; Wang, J. Ferroelectricity in Ruddlesden–Popper chalcogenide perovskites for photovoltaic application: The role of tolerance factor. *J. Phys. Chem. Lett.* **2017**, *8*, 5834–5839. [CrossRef]
49. Aguiar, R.; Logvinovich, D.; Weidenkaff, A.; Rachel, A.; Reller, A.; Ebbinghaus, S.G. The vast colour spectrum of ternary metal oxynitride pigments. *Dye. Pigment.* **2008**, *76*, 70–75. [CrossRef]
50. Grätzel, M. Photoelectrochemical cells. *Nature* **2001**, *414*, 338–344. [CrossRef]
51. Seger, B.; Castelli, I.E.; Vesborg, P.C.K.; Jacobsen, K.W.; Hansen, O.; Chorkendorff, I. 2-Photon tandem device for water splitting: Comparing photocathode first versus photoanode first designs. *Energy Environ. Sci.* **2014**, *7*, 2397–2413. [CrossRef]
52. Jumas, J.; Olivier Fourcade, J.; Vermot-Gaud-Daniel, F.; Ribes, M.; Philippot, E.; Maurin, M. Etude structurale de thiocomposes a groupements anioniques de type -pyro-, $Na_6X_2S_7$ (X = Ge, Sn) et $Ba_3Sn_2S_7$. *Rev. Chim. Miner.* **1974**, *11*, 13–26.
53. ICSDWeb. Available online: http://www.fiz-karlsruhe.de/icsd_web.html (accessed on 30 December 2021).
54. Autonomous Materials Discovery (AiMade). Available online: http://www.aimade.org (accessed on 30 December 2021).

materials MDPI

Article

The Use of a High-Pressure Water-Ice Jet for Removing Worn Paint Coating in Renovation Process

Grzegorz Chomka [1], Jarosław Chodór [1,*], Leon Kukiełka [2] and Maciej Kasperowicz [2]

1 Faculty of Mechanical Engineering, Branch of the Koszalin University of Technology in Szczecinek, Koszalin University of Technology, Waryńskiego 1 Street, 78-400 Szczecinek, Poland; grzegorz.chomka@tu.koszalin.pl
2 Faculty of Mechanical Engineering, Koszalin University of Technology, Racławicka 15-17 Street, 75-620 Koszalin, Poland; leon.kukielka@tu.koszalin.pl (L.K.); maciej.kasperowicz@tu.koszalin.pl (M.K.)
* Correspondence: jaroslaw.chodor@tu.koszalin.pl

Abstract: The paper presents the results of investigations into the possibility of using ahigh-pressure water-ice jet as a new method for removing a worn-out paint coating from the surface of metal parts (including those found in means of transportation) and for preparing the base surface for the application of renovation paint coating. Experimental investigations were carried out in four stages, on flat specimens, sized S × H = 75 × 115 mm, cut from sheet metal made of various materials such as steel X5CrNi18-10, PA2 aluminium alloy and PMMA polymethyl methacrylate (plastic). In the first stage, the surfaces of the samples were subjected to observation of surface morphology under a scanning electron microscope, and surface topography (ST) measurements were made on a profilographometer. Two ST parameters were analysed in detail: the maximum height of surface roughness Sz and the arithmetic mean surface roughness Sa. Next, paint coatings were applied to the specimens as a base. In the third stage, the paint coating applied was removed by means of a high-pressure water-ice jet (HPWIJ) by changing the values of the technological parameters, i.e., water jet pressure p_W, dry ice mass flow rate $\dot{m}_L$, distance between the sprinkler head outlet and the surface being treated (the so-called working jet length) l_2 and spray angle κ for the following constants: the number of TS = 4 holes, water hole diameter φ = 1.2 mm and sprinkler head length L_k = 200 mm. Afterwards, the surface morphology was observed again and the surface topography of the specimen was investigated by measuring selected 3D parameters of the ST structure, Sz and Sa. The results of investigations into the influence of selected HPWIJ treatment parameters on the surface Q_F removal efficiency obtained are also presented. Univariate regression functions were developed for the mean stripping efficiency based on the following: dry ice mass flow rate $\dot{m}_L$, working jet length l_2 and spray angle κ. Based on these functions, the values of optimal parameters were determined that allow the maximum efficiency of the process to be obtained. A 95% confidence region for the regression function was also developed. The results demonstrated that HPWIJ treatment does not interfere with the geometric structure of the base material, and they confirmed the possibility of using this treatment as an efficient method of removing a worn paint layer from bases made of various metal and plastic materials, and preparing it for applying a new layer during renovation.

Keywords: high-pressure water jet; high-pressure water-ice jet; paint coating; geometric structure of the surface; base; surface efficiency of the process; renovation

Citation: Chomka, G.; Chodór, J.; Kukiełka, L.; Kasperowicz, M. The Use of a High-Pressure Water-Ice Jet for Removing Worn Paint Coating in Renovation Process. *Materials* **2022**, *15*, 1168. https://doi.org/10.3390/ma15031168

Academic Editors: Carmine Maletta and Michele Bacciocchi

Received: 1 December 2021
Accepted: 29 January 2022
Published: 3 February 2022

1. Introduction

The continuous dynamic development of the means of transport requires the use of various materials in their manufacture. One of the main elements of vehicles of various types is their body, which is usually protected against corrosion with a paint layer. However, in order for the paint coating to fulfil its function, it is necessary, among other things, to properly prepare the surface of the base for the coatings to be applied. This refers both to factory and refinish coatings. During vehicle transport operations, factory paint coating may be damaged for various reasons; therefore, the preparation of the base for the application of the refinish paint coating becomes of particular importance.

Most frequently, damage to the factory paintwork in the vehicle body occurs as a result of natural wear and tear, such as erosion or ultraviolet radiation [1]. Other factors that contribute to damage of the factory paint coating include aggressive media (e.g., aqueous solutions of salts, acids, bases; tree sap; bird droppings; acid rain), which have a negative impact on the paint coating throughout the entire vehicle service life. Thermal factors causing the so-called temperature shock, and, in particular, extremely high or low temperatures, constitute yet another group [2]. However, it is a synergy of the factors mentioned above that is one of the most common causes of damage to the paint coating. Damage may also occur as a result of road traffic incidents. Damaged factory paintwork is most often replaced with refinish coatings. High demands are placed on these coatings as they must fulfil their protective and decorative functions as close as possible to those of factory coatings. Many technical and technological factors must be met for this to be possible. One of the most important requirements is the correct preparation of the base material [3,4]. The correct temperature as well as the quantity and cleanliness of the air flowing through the spray booth are also important [2].

Prior to regeneration of paintwork, a thorough assessment of the condition of the factory paintwork is carried out. The size and amount of visible damage as well as the type and origin of discoloration occurring on the car body are assessed. It is only then that a decision is made as to the method of paint stripping and preparation of the base for renovation. Old paintwork is usually removed using mechanical methods. Their advantage is high efficiency and the possibility to remove traces of corrosion occurring under the paint coating. The disadvantage of these methods is interference in the geometric structure of the base material prepared by the manufacturer. Therefore, when there is no need to interfere with the topography of the base material, chemical methods (alkaline, acid or solvent-based preparations) or thermal methods (pyrolysis furnaces) are used for paint removal. However, it is not always possible to use one of these methods, for example due to the necessity to disassemble the car body or the proximity of components that are sensitive to temperature or chemical agents. In such cases, sandblasting in hermetic chambers can be used. However, in this case, the topography of the base material is damaged after sandblasting [5–7]. In addition, abrasive particles often remain in the base material, and they form the so-called surface reinforcement following treatment, which leads to the formation of corrosion centres and defects in the refinish coating being applied.

In the search for surface treatment methods that do not damage the base, more and more attention is being paid to air-ice [8,9] or water-ice [10] jets. Most often, the carrier jet is air to which particles of crushed carbon dioxide ice are added. Such a tool for removing paint from aircraft is described in [11]. The authors of the study [12] performed the optimization of the cleaning process for circular and flat nozzles. They also presented experimental results related to the removal of paint from sheet metal and the removal of silicon seals from aluminum-magnesium alloys. Paper [13] describes the removal of particulate contaminants adhering to a surface, a process investigated using a dry ice blasting system. The experimental results showed that, for surface cleaning, dry ice blasting performs well, which is attributed to the collision of the dry ice particles with the contaminants. For submicron-sized contaminants, a lower temperature jet was required in order to produce a larger number of dry ice particles to enhance the removal efficiency.

Very few publications have investigated the use of a water-ice jet where carbon dioxide ice is used as an abrasive. The effectiveness of the surface treatment in removing paint layers using a high-pressure water-ice jet, using different nozzle variants, is presented in [14]. Frozen gas processing methods are also gaining importance, e.g., for machining of nuclear fuel pins [15]. Paper [16] describes the results of particle removal mechanisms in cryogenic surface cleaning. The agglomeration process of dry ice particles produced by expansion of liquid carbon dioxide was presented by the authors [17]. In the experiments, the temperatures of the jet flow and the tube wall were measured by thermocouples, and dry ice particles in the jet flow were observed by a high speed camera with a zoom lens. It was found that two stages of temperature reduction occurred in the jet flow, corresponding to the agglomeration process. It was also found that the particle size of the agglomerates increased and the particle velocity decreased with increasing tube diameter.

Processing can also involve a cryogenic jet of liquid nitrogen [18]. The possibility of using a jet of liquid nitrogen in combination with an abrasive (garnet) to cut metals and brittle materials was described in [19]. The results of subsequent studies [20] allowed the identification of areas of perspective application of cryogenic jets (surface cleaning, disinfection, cutting materials of different strength, cutting explosives, medicine).

The vast majority of tests are conducted on prototype test benches. The difference between the high-pressure water-ice jet (HPWIJ) presented in this paper and the ice jet technology, known as ice abrasive water jet (IAWJ), is the pressure and the abrasive used. The IAWJ method uses liquid nitrogen at the appropriate pressure and the abrasive is water ice. In the presented paper, the carrier jet is water and crushed carbon dioxide ice is used as the abrasive. A significant advantage of treatment with a high-pressure water-ice jet is the removal of surface layers without introducing surface tension [21]. In the literature, the most commonly reported results are the use of high-pressure waterjet for cutting [22,23] rather than cleaning, where the abrasive is usually garnet, corundum, or olivine. There are also published papers addressing the disintegration intensity of abrasives in the high-speed abrasive water jet (AWJ) cutting process [24–28]. New research has also been devoted to optimizing the cutting parameters of SiC-reinforced aluminum composite [29], and titanium alloy [30].

The possibility of using a high-pressure water jet without any additives seems to be safe for metal substrate materials and plastic parts. High-pressure water jet has previously been used to process such delicate materials as fish muscle and skin [31,32]. The necessary jet pressure to completely cut through the muscle and skin of rainbow trout was determined. The possibility also needs to be mentioned to utilise modern computer methods in further research in the aspect of modelling and simulating the aforementioned material processing methods. Experimental studies and numerous simulations of various other types of processing [33–37] support the validity of using such techniques in modelling the high-pressure water-ice jet.

High-pressure water jet treatment possesses an unquestionable advantage: after treatment, water is filtered and reused, thus protecting natural resources [38]. In order to increase the efficiency of the removal of different types of coatings, the water jet is mixed with CO_2 dry ice particles. After treatment, the ice particles sublimate into the atmosphere and filtered water is reused. The hardness of dry ice particles is estimated at 2 on the Mohs scale, making them comparable to rock salt or calcite. These properties of dry ice increase the erosivity of the water jet, but there is no negative impact on the topography of the base material.

An experimental study was conducted to see if the application of high-pressure water-ice jet (HPWIJ) causes changes in the substrate material after paint removal. The influence of selected processing parameters (water jet pressure p_w, dry ice mass flow rate $\dot{m}_L$, working jet length l_2i and spray angle) on the surface removal efficiency Q_F was also investigated. Regression functions were developed for the average paint removal efficiency as a function of: dry ice mass flow rate $\dot{m}_L$, working jet length l_2 and spray angle κ. Based on these functions, the values of the optimal parameters to obtain the maximum process efficiency were determined.

2. Materials and Methods

The experimental tests were carried out on a test bench constructed for this purpose. The prototype test stand (Figure 1) includes major components in the form of a stationary hydromonitor 3 and a sprinkler 7. Water is the medium that creates the so-called "jet stream" for the dry ice particles. Its source is the city water system. Water through valve 1 flows into cooler 2 where it is pre-cooled. The reduced temperature water is then compressed to the required pressure by the high-pressure pump 3. The maximum pressure of the water jet reached p_W = 57 MPa and the flow rate $\dot{V} = 80$ dm^3·min^{-1}. The water jet pressure was stabilized by the control system 4. After compression, the temperature of the water jet rises and it needs to be cooled down again in the radiator 5. After the temperature has been reduced, the water jet enters the high-pressure gun 6 and the sprinkler 7. A high-pressure jet of water flowing through the sprinkler 7 creates a vacuum in the conduit 9, drawing ice particles from the reservoir 10 located in the temperature-controlled room. The ice particles sucked from the reservoir 10 into the sprinkler nozzle 7 are accelerated by the high-pressure water jet and shaped into a final water-ice jet sprayed onto the treated surface 8.

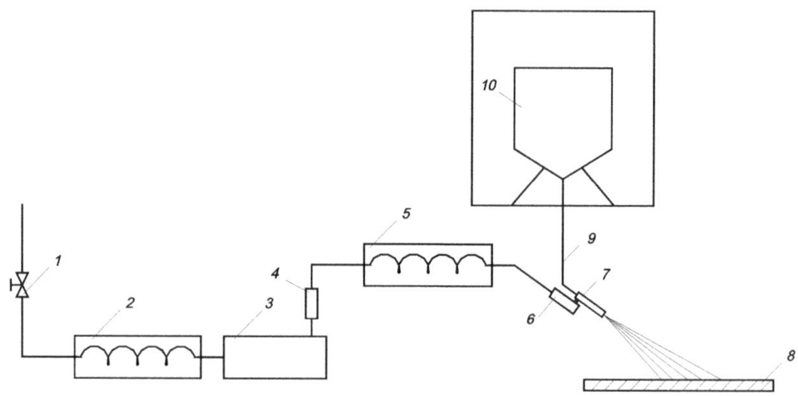

Figure 1. Scheme of the experimental system: 1—valve, 2—radiator, 3—water pump, 4—control system, 5—radiator, 6—high-pressure gun, 7—sprinkler, 8—work surface, 9—suction hose, 10—CO_2 dry ice particle tank.

The test specimens were prepared in an identical manner. Once rectangular specimens sized S × H = 75 × 115 mm from a sheet had been cut out, they were first degreased and dried. Next, they were covered with two layers of phthalic primer with the trade name of Nobikor manufactured by Nobiles Włocławek (one layer along the specimens and the second across the specimens). Following the guidelines from the Nobikor manufacturer, the surfaces painted were seasoned for 24 h. Once the primer had dried, the specimens were additionally coated with a non-metallic base paint: phthalic carbamide enamel with the trade name of Autorenolak F manufactured by Polifarb Cieszyn (two coats at an interval of 24 h). The coat obtained was dried at ambient temperature (ca. 20 °C) for 10 days. Then, one layer of Spectral colourless paint was applied and seasoned for 7 days.

Experimental investigations were carried out to determine the impact of high-pressure water-ice jet treatment on the surface morphology and the surface topography structure (ST). It was also important to obtain reliable information on the selection of appropriate hydraulic and technological parameters of the cleaning process to ensure maximum surface paint removal efficiency while not interfering with the substrate material. Several substrate materials were selected for testing and paint coating were applied to them. The main consideration was their application in various means of transportation vehicle (cars, aeroplanes, light vehicle bodies, etc.). The mechanical properties of the materials used were also taken into account, as they determine the degree of damage to their geometric surface structure in the process of removing paint coatings with a water-ice jet (Table 1). The first material used is chromium-nickel stainless steel of the X5CrNi18-10 grade, which is hard yet flexible at the same time. The second base material tested is an aluminium alloy with magnesium and manganese (PA2 grade) admixtures. It is a soft material with a high susceptibility to surface damage. The third base material used for the paint coating was polymethyl methacrylate (PMMA), i.e., a macromolecular plastic characterised by good mechanical properties as well as high brittleness and susceptibility to scratching.

Table 1. Mechanical and physical properties of materials used in the research.

Material Used in the Research	Material According to the Norm (EN)	Density ρ [kg × m$^{-\rho}$]	Tensile Strength Rm [MPa]	Young's Modulus E [GPa]	Yield Point Re [MPa]
X5CrNi18-10	1.4301	7900	540	200	230
PA2	AW-5251	2700	130	70	60
PMMA	-	1180	75	3	0.06

Paint stripping was carried out using a four-hole helical nozzle with a water hole diameter of φ = 1.2 mm (TS = 4 × 1.2 mm). The appropriate shaping of the water-ice jet was carried out in the sprinkler head with an experimentally pre-determined length L_k = 200 mm. Water jet pressure p_w = 20 MPa and p_w = 35 MPa was used during the tests. The dry ice flow rate was altered in the range $\dot{m}_L = 52 \div 260$ kg·h^{-1} with a step of 26 kg·h^{-1}. The distance between the sprinkler head outlet and the surface treated (the so-called working jet length) was $l_2 = 150 \div 400$ mm and it was altered every 50 mm. The spray angle was changed in the range $\kappa = 60 \div 90°$ with a step of 15°. Each test was repeated three times. Variants of the tests performed for X5CrNi18-10 steel substrate specimens are shown in Tables 2 and 3.

Table 2. Sample test variants for p_w = 20 MPa, κ = 90°.

p_w [MPa]							20							
l_2 [mm]				150							200			
$\dot{m}_L$ [kg·h^{-1}]	52	78	104	130	156	182	208	52	78	104	130	156	182	208
l_2 [mm]				250							300			
$\dot{m}_L$ [kg·h^{-1}]	52	78	104	130	156	182	208	52	78	104	130	156	182	208
l_2 [mm]				350							400			
$\dot{m}_L$ [kg·h^{-1}]	52	78	104	130	156	182	208	52	78	104	130	156	182	208

Table 3. Sample test variants for p_w = 35 MPa, κ = 90°.

p_w [MPa]										35								
l_2 [mm]					150										200			
$\dot{m}_L$ [kg·h^{-1}]	52	78	104	130	156	182	208	234	260	52	78	104	130	156	182	208	234	260
l_2 [mm]					250										300			
$\dot{m}_L$ [kg·h^{-1}]	52	78	104	130	156	182	208	234	260	52	78	104	130	156	182	208	234	260
l_2 [mm]					350										400			
$\dot{m}_L$ [kg·h^{-1}]	52	78	104	130	156	182	208	234	260	52	78	104	130	156	182	208	234	260

An analogous study was conducted for samples with PA2 aluminum alloy substrates and polymethyl methacrylate (PMMA). After determining the central values (allowing to obtain the highest processing efficiency) for the dry ice output and the working length of the stream and the angle of spray, tests were carried out for the stream with pressure p_w = 25 MPa and p_w = 30 MPa.

Experimental investigations were carried out to determine the impact of the water-ice jet on the base material. For this reason, the macro- and microstructure of the surfaces of both untreated samples and samples after removal of the varnish coating with a high-pressure water-ice jet were evaluated. A JOEL JSM-5500LV scanning electronmicroscope was used to evaluate the surface morphology of the specimens examined. It was equipped with a computer system for recording and measurements that allows archiving the results obtained. The surface topography (ST) of the specimens was assessed on the basis of measurements made by means of a Talysurf CLI 2000 spatial profilometer manufactured by Taylor-Hobson. The measurement results recorded were processed and analysed using TalyMap software. The samples for measurement were 3 × 3 mm in size. Two hundred and one profiles were recorded during the measurement. The distance between the profiles was 15 µm. On one profile 1501 points were registered. The distance between the points of the profile was 2 µm. Each measurement was performed in single-pass mode. Measurements were made at the same sample location, i.e., the center of the sample. The maximum height of surface roughness (Sz) and the arithmetic mean surface roughness (Sa) were recorded during testing. The definitions and notation of spatial parameters are well known [39,40]. For each group of three samples, the average $\bar{z}$ mean value of Sa and Sz, as well as the standard deviation s(z) and the spread R(z) were calculated. The assessment of the quality of removing a worn-out paint coating from the surface of metal parts was also additionally verified with the use of the Kestler optical microscope—Vision Engineering Dynascope Ltd., Emmering, Deutschland.

A summary of the measurements of the arithmetic mean surface roughness Sa and the maximum surface roughness height Sz of the X5CrNi18-10 steel samples before paint coating is presented in Table 4. Table 5 presents the results of the same samples after removing the paint layer with a water-ice jet with pressure p_W = 20 MPa. Table 6 presents the results after the treatment with water jet at pressure p_W = 35.

Table 4. Results of Sa and Sz measurements of X5CrNi18-10 steel samples before paint coating.

Sample No.	Sa [µm]	$\bar{z}$ [µm]	s(z) [µm]	R(z) [µm]	Sz [µm]	$\bar{z}$ [µm]	s(z) [µm]	R(z) [µm]
1 2 3	0.77 0.75 0.76	0.76	0.01	0.02	7.3 7.4 7.2	7.30	0.10	0.20
4 5 6	0.78 0.76 0.78	0.77	0.01	0.02	7.4 7.3 7.5	7.40	0.10	0.20
7 8 9	0.77 0.78 0.78	0.78	0.01	0.01	7.4 7.5 7.6	7.50	0.10	0.20
10 11 12	0.77 0.76 0.77	0.77	0.01	0.01	7.2 7.3 7.4	7.30	0.10	0.20
13 14 15	0.78 0.77 0.77	0.77	0.01	0.01	7.4 7.5 7.3	7.40	0.10	0.20
16 17 18	0.74 0.76 0.77	0.76	0.02	0.03	7.6 7.5 7.6	7.57	0.06	0.10
19 20 21	0.78 0.75 0.77	0.77	0.02	0.03	7.2 7.3 7.6	7.37	0.21	0.40
22 23 24	0.78 0.74 0.76	0.76	0.02	0.04	7.5 7.1 7.4	7.33	0.21	0.40
25 26 27	0.78 0.76 0.75	0.76	0.02	0.03	7.3 7.2 7.6	7.37	0.21	0.40

Table 4. *Cont.*

Sample No.	Sa [µm]	$\bar{z}$ [µm]	s(z) [µm]	R(z) [µm]	Sz [µm]	$\bar{z}$ [µm]	s(z) [µm]	R(z) [µm]
28	0.78				7.4			
29	0.74	0.76	0.02	0.04	7.5	7.47	0.06	0.10
30	0.76				7.5			
31	0.76				7.3			
32	0.75	0.76	0.02	0.03	7.3	7.40	0.17	0.30
33	0.78				7.6			
34	0.77				7.4			
35	0.74	0.76	0.02	0.03	7.2	7.37	0.15	0.30
36	0.77				7.5			
37	0.77				7.3			
38	0.75	0.77	0.02	0.03	7.6	7.33	0.25	0.50
39	0.78				7.1			
40	0.77				7.4			
41	0.78	0.77	0.02	0.03	7.5	7.43	0.06	0.10
42	0.75				7.4			
43	0.77				7.6			
44	0.77	0.76	0.01	0.02	7.4	7.37	0.25	0.50
45	0.75				7.1			
46	0.76				7.2			
47	0.75	0.76	0.01	0.02	7.3	7.30	0.10	0.20
48	0.77				7.4			

Table 5. Results of Sa and Sz measurements of X5CrNi18-10 steel specimens after removal of paint coating (p_w = 20 MPa, l_2 = 250 mm, κ = 90°).

Sample No.	$\dot{m}_L$ [kg·h^{-1}]	Sa [µm]	$\bar{z}$ [µm]	s(z) [µm]	R(z) [µm]	Sz [µm]	$\bar{z}$ [µm]	s(z) [µm]	R(z) [µm]
1		0.77				7.4			
2	52	0.76	0.76	0.01	0.02	7.4	7.37	0.06	0.10
3		0.75				7.3			
4		0.77				7.3			
5	78	0.77	0.77	0.01	0.01	7.2	7.37	0.21	0.40
6		0.76				7.6			
7		0.77				7.3			
8	104	0.79	0.77	0.02	0.03	7.6	7.50	0.17	0.30
9		0.76				7.6			
10		0.77				7.3			
11	130	0.74	0.76	0.02	0.03	7.2	7.33	0.15	0.30
12		0.76				7.5			
13		0.77				7.4			
14	156	0.77	0.77	0.01	0.01	7.4	7.43	0.06	0.10
15		0.76				7.5			
16		0.75				7.5			
17	182	0.75	0.76	0.02	0.03	7.4	7.50	0.10	0.20
18		0.78				7.6			
19		0.75				7.3			
20	208	0.75	0.75	0.01	0.01	7.4	7.40	0.10	0.20
21		0.76				7.5			

Table 6. Results of Sa and Sz measurements of X5CrNi18-10 steel specimens after removal of paint coating (p_W = 35 MPa, l_2 = 250 mm, κ = 90°).

Sample No.	$\dot{m}_L$[kg·h^{-1}]	Sa [µm]	$\bar{z}$ [µm]	s(z) [µm]	R(z) [µm]	Sz [µm]	$\bar{z}$ [µm]	s(z) [µm]	R(z) [µm]
22 23 24	52	0.77 0.76 0.79	0.77	0.02	0.03	7.5 7.2 7.4	7.37	0.15	0.30
25 26 27	78	0.79 0.78 0.76	0.78	0.02	0.03	7.4 7.1 7.4	7.30	0.17	0.30
28 29 30	104	0.8 0.76 0.78	0.78	0.02	0.04	7.3 7.4 7.4	7.37	0.06	0.10
31 32 33	130	0.79 0.76 0.77	0.77	0.02	0.03	7.3 7.2 7.5	7.33	0.15	0.30
34 35 36	156	0.78 0.76 0.79	0.78	0.02	0.03	7.6 7.1 7.4	7.37	0.25	0.50
37 38 39	182	0.79 0.77 0.8	0.79	0.02	0.03	7.2 7.5 7.3	7.33	0.15	0.30
40 41 42	208	0.78 0.79 0.77	0.78	0.01	0.02	7.3 7.3 7.4	7.33	0.06	0.10
43 44 45	234	0.79 0.78 0.76	0.78	0.02	0.03	7.5 7.2 7.3	7.33	0.15	0.30
46 47 48	260	0.77 0.76 0.78	0.77	0.01	0.02	7.3 7.5 7.3	7.37	0.12	0.20

A summary of the measurements of the arithmetic mean surface roughness Sa and the maximum surface roughness height Sz of the PA2 aluminium alloysamples before paint coating is presented in Table 7. Table 8 presents the results of the same samples after removing the varnish coating with the water-ice jet with pressure p_W = 20 MPa. Table 9 presents the results after the treatment with water jet at pressure p_W = 35 MPa.

A summary of the measurements of the arithmetic mean surface roughness Sa and the maximum surface roughness height Sz of the PMMA polymethyl methacrylate samples before paint coating is presented in Table 10. Table 11 presents the results of the same samples after removing the paint coating with a water-ice jet at pressure p_W = 20 MPa. Table 12 shows the results after the treatment with water jet at pressure p_W = 35 MPa.

Table 7. Results of Sa and Sz measurements of PA2 aluminum alloy samples before paint coating.

Sample No.	Sa [μm]	$\bar{z}$ [μm]	s(z) [μm]	R(z) [μm]	Sz [μm]	$\bar{z}$ [μm]	s(z) [μm]	R(z) [μm]
1	1.04				10.6			
2	1.05	1.04	0.01	0.01	10.5	10.67	0.21	0.40
3	1.04				10.9			
4	1.07				10.7			
5	1.05	1.06	0.01	0.02	10.8	10.80	0.10	0.20
6	1.06				10.9			
7	1.05				10.6			
8	1.03	1.05	0.02	0.03	10.5	10.60	0.10	0.20
9	1.06				10.7			
10	1.06				10.5			
11	1.04	1.04	0.02	0.03	10.7	10.53	0.15	0.30
12	1.03				10.4			
13	1.05				10.8			
14	1.04	1.04	0.01	0.02	10.5	10.60	0.17	0.30
15	1.03				10.5			
16	1.02				10.6			
17	1.05	1.03	0.02	0.03	10.8	10.60	0.20	0.40
18	1.03				10.4			
19	1.04				10.3			
20	1.06	1.04	0.02	0.03	10.8	10.53	0.25	0.50
21	1.03				10.5			
22	1.02				10.5			
23	1.05	1.04	0.02	0.03	10.4	10.60	0.26	0.50
24	1.04				10.9			
25	1.03				10.8			
26	1.05	1.04	0.01	0.02	10.4	10.63	0.21	0.40
27	1.05				10.7			
28	1.06				10.4			
29	1.03	1.05	0.02	0.03	10.7	10.47	0.21	0.40
30	1.05				10.3			
31	1.07				10.7			
32	1.02	1.04	0.03	0.05	10.4	10.50	0.17	0.30
33	1.04				10.4			
34	1.04				10.6			
35	1.05	1.04	0.01	0.02	10.4	10.53	0.12	0.20
36	1.03				10.6			
37	1.04				10.5			
38	1.04	1.05	0.01	0.02	10.7	10.67	0.15	0.30
39	1.06				10.8			
40	1.04				10.6			
41	1.03	1.04	0.02	0.03	10.4	10.57	0.15	0.30
42	1.06				10.7			
43	1.05				10.2			
44	1.04	1.05	0.01	0.02	10.6	10.33	0.23	0.40
45	1.06				10.2			
46	1.05				10.2			
47	1.04	1.04	0.01	0.01	10.5	10.43	0.21	0.40
48	1.04				10.6			

Table 8. Results of Sa and Sz measurements of PA2 aluminum alloy samples before paint coating (p_w = 20 MPa, l_2 = 250 mm, κ = 90°).

Sample No.	$\dot{m}_L$ [kg·h^{-1}]	Sa [µm]	$\bar{z}$ [µm]	s(z) [µm]	R(z) [µm]	Sz [µm]	$\bar{z}$ [µm]	s(z) [µm]	R(z) [µm]
1		1.05				10.5			
2	52	1.05	1.05	0.01	0.01	10.2	10.43	0.21	0.40
3		1.06				10.6			
4		1.08				10.6			
5	78	1.04	1.06	0.02	0.04	10.9	10.77	0.15	0.30
6		1.06				10.8			
7		1.07				10.6			
8	104	1.05	1.07	0.02	0.03	10.4	10.53	0.12	0.20
9		1.08				10.6			
10		1.08				10.7			
11	130	1.05	1.06	0.02	0.04	10.6	10.60	0.10	0.20
12		1.04				10.5			
13		1.04				10.7			
14	156	1.07	1.05	0.02	0.03	10.5	10.53	0.15	0.30
15		1.05				10.4			
16		1.04				10.5			
17	182	1.07	1.05	0.02	0.03	10.7	10.50	0.20	0.40
18		1.04				10.3			
19		1.03				10.5			
20	208	1.08	1.05	0.03	0.05	10.5	10.50	0.00	0.00
21		1.04				10.5			

Table 9. Results of Sa and Sz measurements of PA2 aluminum alloy specimens after removal of paint coating (p_w = 35 MPa, l_2 = 250 mm, κ = 90°).

Sample No.	$\dot{m}_L$ [kg·h^{-1}]	Sa [µm]	$\bar{z}$ [µm]	s(z) [µm]	R(z) [µm]	Sz [µm]	$\bar{z}$ [µm]	s(z) [µm]	R(z) [µm]
22		1.02				10.4			
23	52	1.06	1.05	0.02	0.04	10.2	10.43	0.25	0.50
24		1.06				10.7			
25		1.05				10.7			
26	78	1.08	1.06	0.02	0.03	10.5	10.63	0.12	0.20
27		1.06				10.7			
28		1.06				10.4			
29	104	1.04	1.06	0.02	0.03	10.7	10.60	0.17	0.30
30		1.07				10.7			
31		1.06				10.5			
32	130	1.04	1.05	0.01	0.02	10.6	10.43	0.21	0.40
33		1.05				10.2			
34		1.06				10.6			
35	156	1.08	1.06	0.02	0.04	10.2	10.53	0.31	0.60
36		1.04				10.8			
37		1.05				10.8			
38	182	1.04	1.05	0.01	0.01	10.5	10.73	0.21	0.40
39		1.05				10.9			
40		1.06				10.6			
41	208	1.04	1.06	0.02	0.04	10.5	10.63	0.15	0.30
42		1.08				10.8			
43		1.06				10.3			
44	234	1.03	1.05	0.02	0.04	10.9	10.53	0.32	0.60
45		1.07				10.4			
46		1.07				10.5			
47	260	1.05	1.06	0.01	0.02	10.6	10.67	0.21	0.40
48		1.06				10.9			

Table 10. Sa and Sz measurement results of polymethyl methacrylate PMMA samples before paint coating.

Sample No.	Sa [μm]	$\bar{z}$ [μm]	s(z) [μm]	R(z) [μm]	Sz [μm]	$\bar{z}$ [μm]	s(z) [μm]	R(z) [μm]
1	0.48				5.48			
2	0.47	0.48	0.01	0.02	5.42	5.48	0.07	0.13
3	0.49				5.55			
4	0.5				5.49			
5	0.48	0.48	0.02	0.03	5.42	5.49	0.08	0.15
6	0.47				5.57			
7	0.48				5.43			
8	0.5	0.49	0.01	0.02	5.56	5.49	0.07	0.13
9	0.48				5.49			
10	0.5				5.48			
11	0.47	0.49	0.02	0.03	5.41	5.47	0.06	0.12
12	0.49				5.53			
13	0.49				5.47			
14	0.5	0.50	0.01	0.02	5.5	5.48	0.02	0.04
15	0.51				5.46			
16	0.49				5.49			
17	0.5	0.49	0.01	0.02	5.43	5.46	0.03	0.06
18	0.48				5.46			
19	0.49				5.47			
20	0.47	0.48	0.01	0.02	5.43	5.47	0.04	0.07
21	0.49				5.5			
22	0.51				5.49			
23	0.5	0.50	0.02	0.03	5.47	5.46	0.03	0.06
24	0.48				5.43			
25	0.49				5.46			
26	0.47	0.48	0.01	0.02	5.44	5.46	0.03	0.05
27	0.48				5.49			
28	0.5				5.51			
29	0.49	0.49	0.02	0.03	5.48	5.48	0.03	0.05
30	0.47				5.46			
31	0.49				5.57			
32	0.48	0.49	0.01	0.02	5.43	5.49	0.07	0.14
33	0.5				5.46			
34	0.49				5.5			
35	0.5	0.49	0.02	0.03	5.47	5.47	0.03	0.06
36	0.47				5.44			
37	0.47				5.47			
38	0.5	0.49	0.02	0.03	5.46	5.48	0.03	0.05
39	0.5				5.51			
40	0.49				5.47			
41	0.5	0.49	0.01	0.01	5.5	5.47	0.04	0.07
42	0.49				5.43			
43	0.49				5.44			
44	0.47	0.49	0.02	0.04	5.51	5.47	0.04	0.07
45	0.51				5.45			
46	0.49				5.41			
47	0.51	0.50	0.01	0.02	5.43	5.43	0.02	0.04
48	0.5				5.45			

Table 11. Sa and Sz measurement results of polymethyl methacrylate PMMA samples after paint removal (p_W = 20 MPa, l_2 = 250 mm, κ = 90°).

Sample No.	$\dot{m}_L$ [kg·h^{-1}]	Sa [μm]	$\bar{z}$ [μm]	s(z) [μm]	R(z) [μm]	Sz [μm]	$\bar{z}$ [μm]	s(z) [μm]	R(z) [μm]
1		0.5				5.5			
2	52	0.48	0.50	0.02	0.03	5.44	5.49	0.05	0.09
3		0.51				5.53			
4		0.51				5.54			
5	78	0.5	0.49	0.02	0.04	5.43	5.51	0.07	0.13
6		0.47				5.56			
7		0.49				5.46			
8	104	0.51	0.49	0.02	0.04	5.54	5.50	0.04	0.08
9		0.47				5.49			
10		0.5				5.46			
11	130	0.49	0.50	0.01	0.01	5.43	5.47	0.04	0.08
12		0.5				5.51			
13		0.5				5.49			
14	156	0.52	0.50	0.02	0.03	5.52	5.48	0.05	0.10
15		0.49				5.42			
16		0.52				5.49			
17	182	0.49	0.50	0.02	0.03	5.51	5.48	0.04	0.07
18		0.49				5.44			
19		0.5				5.42			
20	208	0.49	0.50	0.01	0.02	5.49	5.48	0.05	0.10
21		0.51				5.52			

Table 12. Sa and Sz measurement results of polymethyl methacrylate PMMA samples after paint removal (p_W = 35 MPa, l_2 = 250 mm, κ = 90°).

Sample No.	$\dot{m}_L$ [kg·h^{-1}]	Sa [μm]	$\bar{z}$ [μm]	s(z) [μm]	R(z) [μm]	Sz [μm]	$\bar{z}$ [μm]	s(z) [μm]	R(z) [μm]
22		1.45				38.7			
23	52	1.42	1.44	0.02	0.04	38.2	38.40	0.26	0.50
24		1.46				38.3			
25		1.48				38.3			
26	78	1.47	1.45	0.04	0.07	38.8	38.43	0.32	0.60
27		1.41				38.2			
28		1.5				38.8			
29	104	1.48	1.47	0.04	0.07	38.2	38.43	0.32	0.60
30		1.43				38.3			
31		1.51				38.6			
32	130	1.42	1.47	0.05	0.09	38.3	38.37	0.21	0.40
33		1.49				38.2			
34		1.49				38.1			
35	156	1.5	1.49	0.02	0.03	38.6	38.40	0.26	0.50
36		1.47				38.5			
37		1.49				38.6			
38	182	1.51	1.50	0.01	0.02	38.7	38.50	0.26	0.50
39		1.5				38.2			
40		1.48				38.4			
41	208	1.44	1.48	0.04	0.07	38.6	38.40	0.20	0.40
42		1.51				38.2			
43		1.5				38.2			
44	234	1.44	1.48	0.03	0.06	38.8	38.43	0.32	0.60
45		1.49				38.3			
46		1.52				38.3			
47	260	1.5	1.50	0.02	0.03	38.7	38.40	0.26	0.50
48		1.49				38.2			

Experimental tests were carried out in accordance with the static factor-selection programme shown in Figure 2. Three-fold repeatability of the tests was used. The results of the experiments are recorded in Tables 13–18. The surface treatment capacity for each sample $Q_F m^2 \cdot h^{-1}$ was determined during testing. The mean value of the surface treatment efficiency $\overline{Q}_F$ $m^2 \cdot h^{-1}$ and the standard deviation and spread were calculated for three samples treated with identical parameters.

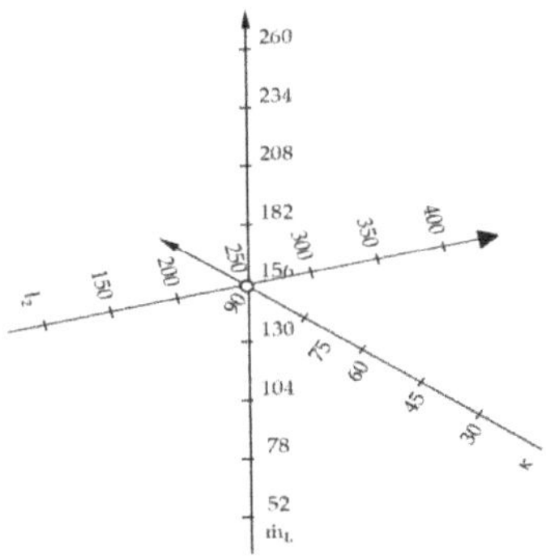

Figure 2. Graphical illustration of the test programme.

Table 13. Measuring results of average surface paint coat stripping efficiency $\overline{Q}_F$ [$m^2 \cdot h^{-1}$] from the surface of X5CrNi18-10 steel specimen as a function of the dry ice mass flow rate $\dot{m}_L$ [$kg \cdot h^{-1}$].

Dry ice Mass Flow Rate $\dot{m}_L$[$kg \cdot h-^k$]	Water Pressure p_W = 20 MPa p_W = 35 MPa Average Surface Efficiency $\overline{Q}_F$ [$m^2 \cdot h^{-h}$]	
52	0.014	0.025
78	0.025	0.040
104	0.050	0.077
130	0.086	0.110
156	0.129	0.157
182	0.125	0.200
208	0.116	0.228
234	-	0.210
260	-	0.176

Table 14. Measuring results of average surface paint coat stripping efficiency $\overline{Q}_F$ [$m^2 \cdot h^{-1}$] from the surface of PA2.

Dry ice Mass Flow Rate $\dot{m}_L$[$kg \cdot h-^k$]	Water Pressure p_W = 20 MPa p_W = 35 MPa Average Surface Efficiency $\overline{Q}_F$ [$m^2 \cdot h^{-h}$]	
52	0.018	0.033
78	0.040	0.060
104	0.070	0.110
130	0.120	0.153
156	0.175	0.22
182	0.171	0.262
208	0.160	0.307
234	-	0.296
260	-	0.273

Table 15. Measuring results of average surface paint coat stripping efficiency $\overline{Q}_F$ [$m^2 \cdot h^{-1}$] from the surface of specimen made of X5CrNi18-10 steel depending on jet working length l_2 [mm].

Working Jet Lengthl_2[mm]	Water Pressure p_w = 20 MPa	p_w = 35 MPa Average Surface Efficiency $\overline{Q}_F$ [$m^2 \cdot h^{-h}$]
150	0.103	0.126
200	0.122	0.149
250	0.129	0.157
300	0.113	0.141
350	0.090	0.126
400	0.060	0.102

Table 16. Measuring results of average surface paint coat stripping efficiency $\overline{Q}_F$ [$m^2 \cdot h^{-1}$] of the PA2 aluminium alloy specimen depending on the working length of the spray l_2 [mm].

Working Jet Lengthl_2[mm]	Water Pressure p_w = 20 MPa	p_w = 35 MPa Average Surface Efficiency $\overline{Q}_F$ [$m^2 \cdot h^{-h}$]
150	0.140	0.176
200	0.166	0.209
250	0.175	0.220
300	0.153	0.198
350	0.119	0.176
400	0.082	0.143

Table 17. Measuring results of average surface paint coat stripping efficiency $\overline{Q}_F$ [$m^2 \cdot h^{-1}$] of PA2 aluminium alloy specimen depending from spray angle κ [°].

Spray Angle κ [°]	Water Pressure p_w = 20 MPa	p_w = 35 MPa Average Surface Efficiency $\overline{Q}_F$ [$m^2 \cdot h^{-h}$]
30	0.016	0.021
45	0.075	0.096
60	0.160	0.200
75	0.172	0.218
90	0.175	0.220

Table 18. Measuring results of average surface paint coat stripping efficiency $\overline{Q}_F$ [$m^2 \cdot h^{-1}$] for central values $\dot{m}_L$ = 156 [$kg \cdot h^{-1}$], l_2 = 250 [mm], κ = 90 [°].

Water Pressure p_w [MPa]	PMMA	PA2 Average Surface Efficiency $\overline{Q}_F$[$m^2 \cdot h^{-1}$]	X5CrNi18-10
20	0.210	0.175	0.129
25	0.228	0.189	0.138
30	0.250	0.201	0.150
35	0.264	0.220	0.157

The mean values of the object's outputs $\overline{Q}_F$ were approximated by a third degree polynomial obtaining a regression equation as one-parameter functions:

$$\hat{\overline{Q}}_F = b_o + b_1\overline{x} + b_2\overline{x}^2 + b_3\overline{x}^3, \ \left[m^2 \cdot h^{-1}\right] \quad (1)$$

where:

b_o, b_1, b_2 and b_3—unknown coefficients of the regression equation,

$\overline{x}$—input variables: $\overline{x} = \dot{m}_L$ [kg·h^{-1}] or $\overline{x} = l_2$ [mm] or $\overline{x} = \kappa$ [°] or $\overline{x} = p_w$ [MPa].

Using matrix calculus, the column vector **{b}** of the unknown coefficients in Equation (5) was calculated from the matrix formula:

$$\{b\} = \left([X]^T[X]\right)^{-1}[X]^T\{Y\}, \quad (2)$$

where:

$[X]$–input variable matrix of dimension N × L. For data N = 5, 6, 7 and 9 and L = 4, the following matrix forms $[X]$ were developed that are presented in Table 19:

Table 19. Matrix forms $[\mathbf{X}]$ for N = 5 and L = 4; N = 6 and L = 4; N = 7 and L = 4 as well as N = 9 and L = 4.

	Matrix Forms [X]			
N = 5 and L = 4	1	30	900	27,000
	1	45	2025	91,125
	1	60	3600	216,000
	1	75	5625	421,875
	1	90	8100	729,000
N = 6 and L = 4	1	150	22,500	3,375,000
	1	200	40,000	8,000,000
	1	250	62,500	1,562,500
	1	300	90,000	27,000,000
	1	350	122,500	4,2875,000
	1	400	160,000	64,000,000
N = 7 and L = 4	1	52	2704	140,608
	1	78	6084	474,552
	1	104	10,816	1,124,864
	1	130	16,900	2,197,000
	1	156	24,336	3,796,416
	1	182	33,124	602,8568
	1	208	43,264	8,998,912
N = 9 and L = 4	1	52	2704	140,608
	1	78	6084	47,4552
	1	104	10,816	1,124,864
	1	130	16,900	2,197,000
	1	156	24,336	3,796,416
	1	182	33,124	6,028,568
	1	208	43,264	8,998,912
	1	234	54,756	12,812,904
	1	260	67,600	17,576,000

$[X]^T$—transposed matrix $[X]$,

$\left([X]^T[X]\right)^{-1}$—covariance matrix,

$\{Y\}$—column vector of the average values of the experimental results (Tables 13–18).

The boundaries of the confidence region for Regression Function (1) were determined from the following formula:

$$\hat{\overline{Q}}_F \pm t_{kr(\alpha;f=N-L)} \cdot \frac{S_R}{\sqrt{N-L-1}} \cdot \sqrt{\{\overline{x}\}^T\left([X]^T[X]\right)^{1}\{\overline{x}\}},\ \left[m^2 \cdot h^{-1}\right] \quad (3)$$

where:

$\hat{\overline{Q}}_F$—regression equation according to Formula (1),

$t_{kr(\alpha;f=N-L)}$—critical value of Student t test for significance level $\alpha = 0.05$ and the number of the degrees of freedom f = N-L (here, f = 1, 2, 3 or 5),

N—number of measurement points in the experimental design N = 5, 6, 7 or 9,

L—number of unknown coefficients in Regression Equation (1); here, L = 4,

$\{\overline{x}\}$ and $\{\overline{x}\}^T$—column vector of the functions of input variables (test factors in real form) and its transposition: $\{\overline{x}\}^T = \left[1\ \overline{x}\overline{x}^2\overline{x}^3\right]$,

$S_R = \sum_{i=1}^{i=N} \left(\hat{\overline{y}}_i - \overline{y}_i\right)^2$—residual variance,

$\hat{\overline{y}}_i$—average values of model outputs for plan points calculated from Equation (1) $(\overline{y}_i = \hat{\overline{Q}}_{Fi})$,

$\overline{y}_i$— average values of experimental results (Tables 13–18).

The test results after statistical processing according to the algorithm presented in the study [41] were used to develop regression equations. The results of single-factor tests are successive functions in the form of a third degree polynomial that defines the influence of the factor covered by the tests: the mass flow rate of dry ice, the working jet length and the spray angle on the surface average paint stripping rate as well as the 95% confidence region for the regression function (Figure 3).

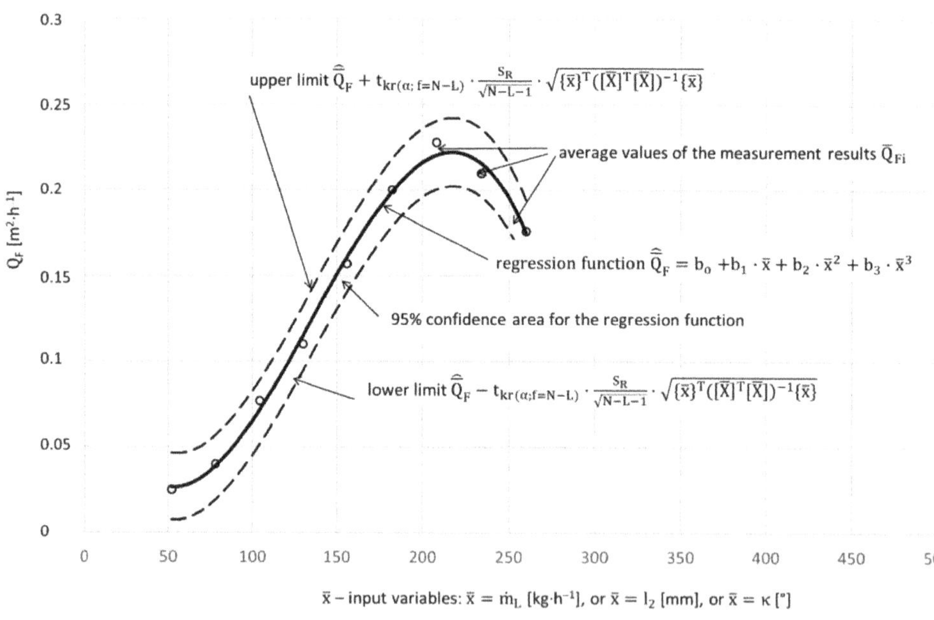

Figure 3. Illustration of the influence of factor x on the average paint removal efficiency, lower and upper limits, and the 95% confidence region for the regression function.

The regression equations developed take the following form.

Dependence of the average surface efficiency $\hat{Q}_F$ [$m^2 \cdot h^{-1}$] of the removal of the paint coating from the surface of the X5CrNi18-10 steel specimen from the dry ice mass flow rate $\dot{m}_L$ [$kg \cdot h^{-1}$]:

$$\hat{Q}_F = -1214 \cdot 10^{-10} \cdot (\dot{m}_L)^3 + 4327 \cdot 10^{-8} \cdot (\dot{m}_L)^2 - 372 \cdot 10^{-5} \cdot \dot{m}_L + 0.1077,\ R = 0.9944,\ p_w = 20\ [MPa] \tag{4}$$

$$\hat{Q}_F = -884 \cdot 10^{-10} \cdot (\dot{m}_L)^3 + 358 \cdot 10^{-7} \cdot (\dot{m}_L)^2 - 305 \cdot 10^{-5} \cdot \dot{m}_L + 0.1004,\ R = 0.9982, p_w = 35\ [MPa] \tag{5}$$

Dependence of the average surface efficiency $\hat{Q}_F$ [$m^2 \cdot h^{-1}$], of the removal of paint from the surface of the PA2aluminiumalloy specimen from the dry ice mass flow rate $\dot{m}_L$ [$kg \cdot h^{-1}$]:

$$\hat{Q}_F = -1478 \cdot 10^{-10} \cdot (\dot{m}_L)^3 + 519 \cdot 10^{-7} \cdot (\dot{m}_L)^2 - 4217 \cdot 10^{-6} \cdot \dot{m}_L + 0.1187,\ R = 0.9942,\ p_w = 20\ [MPa] \tag{6}$$

$$\hat{Q}_F = -878 \cdot 10^{-10} \cdot (\dot{m}_L)^3 + 35 \cdot 10^{-6} \cdot (\dot{m}_L)^2 - 242 \cdot 10^{-5} \cdot \dot{m}_L + 0.0661,\ R = 0.9999,\ p_w = 35\ [MPa] \tag{7}$$

Dependence of the average surface efficiency $\hat{Q}_F$ [$m^2 \cdot h^{-1}$], of the removal of the paint coating from the surface of the X5CrNi18-10 steel specimen from the working jet length l_2 [mm]:

$$\hat{Q}_F = 54 \cdot 10^{-10} \cdot (l_2)^3 - 7078 \cdot 10^{-9} \cdot (l_2)^2 + 24 \cdot 10^{-4} \cdot l_2 - 0.1272,\ R = 0.9982,\ p_w = 20\ [MPa] \tag{8}$$

$$\hat{Q}_F = 93 \cdot 10^{-10} \cdot (l_2)^3 - 10^{-5} \cdot (l_2)^2 + 315 \cdot 10^{-5} \cdot l_2 - 0.1602,\ R = 0.9999,\ p_w = 35\ [MPa] \tag{9}$$

Dependence of the average surface efficiency $\hat{Q}_F$ [$m^2 \cdot h^{-1}$], of the removal of the paint coating from the surface of the PA2 aluminium alloy specimen from the working jet length l_2 [mm]:

$$\hat{Q}_F = 94 \cdot 10^{-10} \cdot (l_2)^3 - 112 \cdot 10^{-7} \cdot (l_2)^2 + 367 \cdot 10^{-5} \cdot l_2 - 0.1903,\ R = 1,\ p_w = 20\ [MPa] \tag{10}$$

$$\hat{Q}_F = 114 \cdot 10^{-10} \cdot (l_2)^3 - 127 \cdot 10^{-7} \cdot (l_2)^2 + 4096 \cdot 10^{-6} \cdot l_2 - 0.1912,\ R = 0.9999,\ p_w = 35\ [MPa] \tag{11}$$

Dependence of the average surface efficiency $\hat{Q}_F$ [$m^2 \cdot h^{-1}$], of the removal of the paint coating from the surface of the PA2 aluminium alloy specimen from the spray angle κ [°]:

$$\hat{Q}_F = -864 \cdot 10^{-9} \cdot (\kappa)^3 + 968 \cdot 10^{-7} \cdot (\kappa)^2 + 114 \cdot 10^{-5} \cdot \kappa - 0.0844,\ R = 1,\ p_w = 20\ [MPa] \tag{12}$$

$$\hat{Q}_F = -106 \cdot 10^{-8} \cdot (\kappa)^3 + 12 \cdot 10^{-5} \cdot (\kappa)^2 + 1097 \cdot 10^{-6} \cdot \kappa - 0.095,\ R = 1,\ p_w = 35\ [MPa] \tag{13}$$

Dependence of the average surface efficiency $\hat{Q}_F$ [$m^2 \cdot h^{-1}$] of the removal of the paint coating from the surface specimen from the water-ice jet pressure p_w [MPa]:

- X5CrNi18-10 steel

$$\hat{Q}_F = 0.001932 \cdot p_w + 0.09032,\ R = 0.995 \tag{14}$$

- PA2 aluminium alloy

$$\hat{Q}_F = 0.00294 \cdot p_w + 0.1154,\ R = 0.995 \tag{15}$$

- polymethyl methacrylate PMMA

$$\hat{Q}_F = 0.00368 \cdot p_w + 0.1368,\ R = 0.995 \tag{16}$$

3. Results and Discussion

3.1. The Surface Quality of the Base

The surface quality of the base after treatment plays an important role in the paint stripping process. In order to verify whether the High-Pressure Water Ice Jet (HPWIJ) treatment does not damage the base, the results of 3D surface topography (ST) and morphology of the surface of the specimens before and after paint stripping were compared.

An example of a scanning electron microscope SEM image of the surface morphology of an X5CrNi18-10 steel specimen before paint coating is presented in Figure 4a. It shows a characteristic fine-grained structure with clearly visible dark traces of grain boundaries. The isometric appearance (that is a topographic image) of the surface of the same steel is shown in Figure 4b. In this case, irregularly distributed protrusions and indentations were observed. As a result of the measurements carried out, the maximum height of the surface roughness was found to be Sz = 7.4 μm, while the arithmetic mean surface roughness was found to be Sa = 0.77μm.

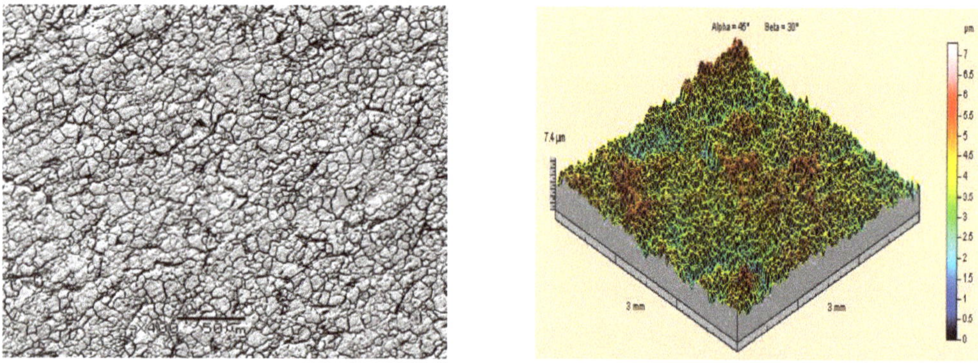

(**a**) (**b**)

Figure 4. Surface topography of X5CrNi18-10 steel specimens prior to the application of paint coating and prior to ice-water jet treatment: (**a**) Grey scale map under scanning electron microscope (400× magnification); (**b**) Isometric image.

After removal of the paint coating using a high-pressure water-ice jet, carried out with different treatment parameters (p_W = 20 MPa and p_W = 35 MPa), scanning electron microscope images of the surface were observed, an example of which is shown in Figure 5a. You can see the very fine grain structure with its characteristic dark borders. To a small extent, you can see the individual primer paint particles remaining at the grain boundaries. They are only discernible at 400× magnification. Comparing the crystalline structure of the surface of the specimen before (Figure 4a) and after treatment (Figure 5a), it was found that they are similar, no significant differences were found.

Similar conclusions were drawn with regard to the measurement of ST parameters. Here, too, almost identical results to those recorded for unpainted specimens were found. The surface topography, consisting of hills and depressions, is also almost identical. Their occurrence is rather irregular and similar to the results of specimens that are not subjected to the water-ice jet. The maximum height of the surface roughness of the specimens treated with the high-pressure water-ice jet was Sz = 7.3 μm, while the arithmetic mean surface roughness was Sa = 0.78 μm. The results obtained do not differ significantly at the adopted level of significance α = 0.05 from the results obtained for those specimens that were not painted.

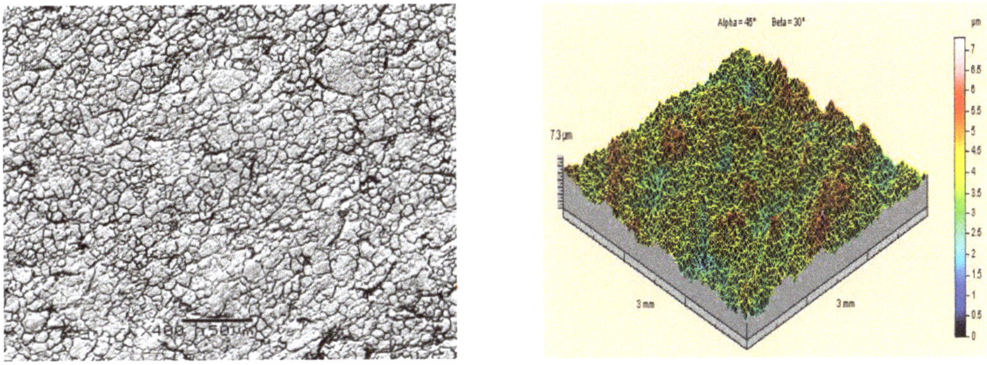

(**a**) (**b**)

Figure 5. Surface topography of X5CrNi18-10 steel specimens after water-ice jet treatment (p_W = 35 MPa, $\dot{m}_L = 208$ kg·h^{-1}, l_2 = 250, κ = 90°): (**a**) Grey scale map under a scanning electron microscope (400× magnification); (**b**) Isometric image.

Similar tests as for steel base specimens were carried out for PA2 aluminium alloy specimens. An example of a scanning image of the surface morphology and an isometric image of the surface of PA2 aluminium alloy specimens prior to paint coating application is presented in Figure 6. It was established that the surface of the PA2 aluminium alloy specimen is covered with parallel traces of unidirectional periodic structure formed during the rolling process. On the surface of the material, small scratches and spot indentations can also be seen as residues from previous machining processes. Based on the measurements of the SGS parameters, the height of the irregularities was found to be Sz = 10.6 µm. The second important ST parameter, i.e., the arithmetic mean surface roughness, for the PA2 aluminium alloy base is Sa = 1.04 µm.

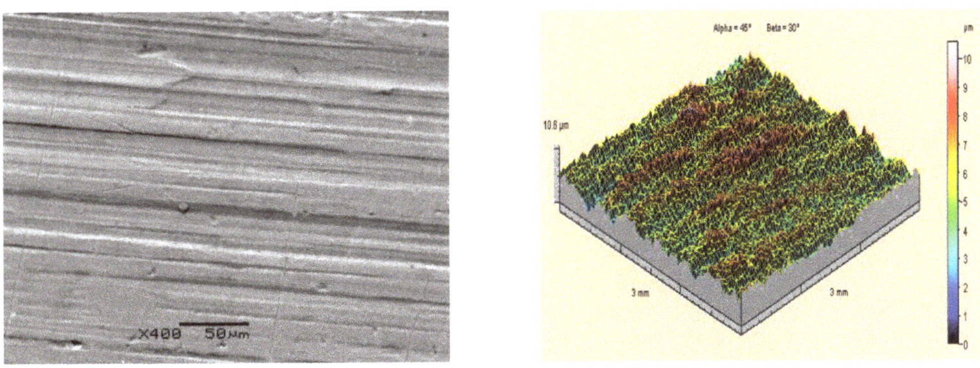

(**a**) (**b**)

Figure 6. Surface topography of a PA2 aluminium alloy specimen prior to the application of paint coating and before high-pressure water-ice jet treatment: (**a**) Scanning electron microscope image (400× magnification); (**b**) Isometric image.

Figure 7 presents the results of surface topography observations under a scanning electron microscope and of the measurements of ST parameters of PA2 aluminium alloy base material, from which the paint coating was removed by a high-pressure water-ice jet with pressure p_W = 35 MPa and dry ice mass flow rate $\dot{m}_L = 208$ kg·h^{-1} using the distance between the sprinkler nozzle outlet, the surface treated l_2 =250 mm and the jet spray angle κ = 90°. In the microscopic image, alternating protrusions and indentations are observed, ones which indicate the application of the base rolling operation. Fine scratches and indentations can also be seen. Analogous traces were observed for all those specimens from which paint coatings were removed when the maximum pressure of the water-ice jet p_W = 35 MPa was applied and with other process parameters being variable. The measurements of the ST parameters (Figure 7b) demonstrated that the height of the roughness, which is identical to previous studies for those specimens that were not painted, is Sz = 10.6 μm. Similar conclusions were drawn based on the measurements of the arithmetic mean deviation of the surface roughness. In this case, it was also observed that the results obtained with the value Sa = 1.06 μm are almost identical to those obtained for the specimens that were not painted. By analysing all the results obtained, it was found that the use of a high-pressure water-ice jet to remove paint coatings from PA2 aluminium alloy does not cause any changes in the ST parameters. There were also no significant differences in the appearance of the base morphology on the scanning images. The assessment of the quality of removing a worn-out paint coating from the surface of metal parts was also additionally verified with the use of the Kestler optical microscope—Vision Engineering Dynascope Ltd., Emmering, Deutschland.

(a)

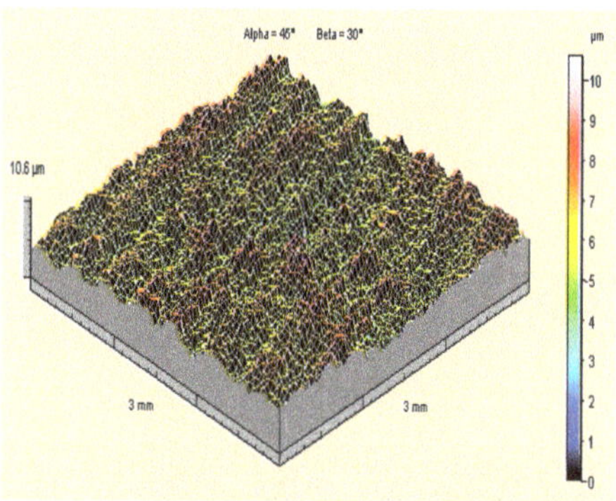

(b)

Figure 7. Surface topography of PA2 aluminium alloy specimens after ice-water jet treatment (p_W = 35 MPa, $\dot{m}_L = 208$ kg·h^{-1}, l_2 = 250 mm, κ = 90°): (**a**) Scanning electron microscope image (400× magnification); (**b**) Isometric image.

A microscopic photograph of the surface of a specimen made of PMMA polymethyl methacrylate, at 50 times magnification, is shown in Figure 8a. This is a uniform, smooth and even surface. For specimens made of PMMA, the geometric structure of their surfaces before the application of paint coatings is shown in the example of Figure 8b.

Based on the image provided and the measurement results obtained, it was found that the surface of the specimens made of PMMA polymethyl methacrylate possesses the lowest roughness in comparison with PA2 aluminium alloy and X5CrNi18-10 steel. The arithmetic mean surface roughness of the PMMA prior to treatment was Sa = 0.49 μm, while the height of the roughness was Sz = 5.47 μm.

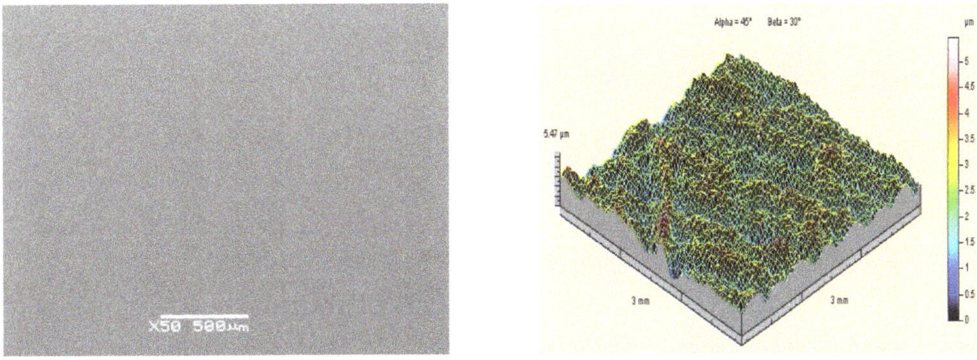

(**a**) (**b**)

Figure 8. View of the surface of a polymethyl methacrylate PMMA specimen prior to the application of a paint coating and prior to high-pressure water-ice jet treatment: (**a**) Scanning electron microscope image (50× magnification); (**b**) Isometric image.

A scanning electron microscope image of a PMMA polymethyl methacrylate specimen, from which the paint coating was removed by means of a high-pressure water-ice jet is presented in Figure 9a. It shows clear signs of surface chipping. Similar defects are observed during ST measurements. The surface morphology contains numerous "craters", which are the effect of the water-ice jet impact. As a result of the formation of micro-chips in the PMMA surface, the arithmetic mean deviation of its surface roughness increases to Sa = 1.48 µm. Damage to the base material as a result of the application of the water-ice jet is also evidenced by a significant increase in the height of the irregularities amounting to a maximum of as much as Sz = 38.4 µm.

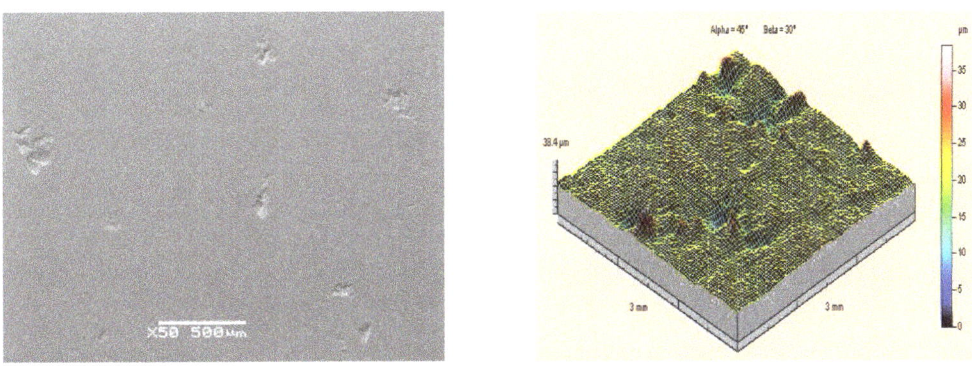

(**a**) (**b**)

Figure 9. View of the surface of polymethyl methacrylate PMMA specimen after high-pressure water-ice jet treatment (p_W = 35 MPa, $\dot{m}_L = 208$ kg·h^{-1}, l_2 = 250 mm, κ = 90°): (**a**) Scanning electron microscope image (50× magnification); (**b**) Isometric image.

Reducing the working pressure of the jet to 20 MPa and the dry ice output to a maximum of $\dot{m}_L = 156$ kg·h^{-1} enabled an effective paint stripping tool to be obtained. At the same time, the applied treatment did not cause damage to the substrate surface. A scanning electron microscope image of the surface of the PMMA polymethyl methacrylate sample from which the paint layer was removed is shown in Figure 10a, while the surface topography of the substrate is shown in Figure 10b.

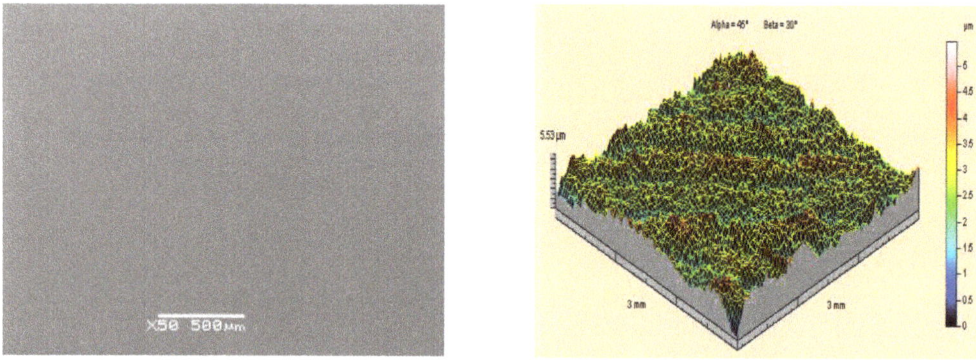

(**a**) (**b**)

Figure 10. View of the surface of PMMA polymethyl methacrylate specimens after high-pressure water-ice jet treatment (p_w = 20 MPa, $\dot{m}_L$ = 156 $kg \cdot h^{-1}$, l_2 = 250 mm, κ = 90°): (**a**) Scanning electron microscope image (50× magnification); (**b**) Isometric image.

Based on the results obtained from the conducted research, it was concluded that the removal of paint layers with a high-pressure water-ice jet is possible without disturbing the geometric structure of the substrate material. This includes substrates made from X5CrNi18-10 steel, PA2 aluminum alloy, and polymethyl methacrylate (PMMA).

3.2. Influence of Selected Treatment Parameters on the Surface Performance of the HPWIJ Process

On the basis of the results obtained from the research conducted, it was established that the removal of paint coatings from various base materials is possible without disturbing the structure of the base material. However, in order to accomplish this, it is necessary to know the values of the technological parameters which enable the cleaning process to be carried out correctly. From the perspective of practical applications, for operators that employ a high-pressure water-ice jet for removing paint coatings, it is not only the effect of such technology on the base material that is of great importance but, above all, this is the impact on the surface treatment efficiency achieved during work. First, it should be noted that paint stripping with a high-pressure water jet cannot be compared to conventional abrasive blasting methods [42–47]. This is not a competitive method considering the hardness and density of the abrasive that is admixed. However, the most important advantage of machining using an ice-water jet is the absence of the traces of impact on the base material. Increasingly, manufacturing companies are looking for technologies of this type, e.g., for cleaning injection moulds and ship hulls. It is therefore important to consciously use new technologies in industry. For this purpose, it is necessary to determine the values of technological and hydraulic parameters in order to achieve the maximum treatment efficiency while ensuring that there is no destructive effect on the base.

Figure 11 presents the effect of CO_2 dry ice particle output on the surface paint coat stripping efficiency of X5CrNi18-10 steel for jet pressures p_w = 20 MPa and p_w = 35 MPa. As the pressure of the water jet increases, its velocity increases, and so does the velocity of CO_2 dry ice particles that are accelerated by it. The higher kinetic energy of dry ice particles results in a stronger mechanical (impact) effect on the surface being treated. The sublimation of CO_2 particles in the treatment region is also more intense. In addition, a higher-pressure water jet exerts a greater force on the paint coating to be removed. This is why the microcracks in the top layers of the paint film are separated and torn apart more quickly. Consequently, a higher pressure water-ice jet treatment leads to an increase in the surface paint coat stripping efficiency. At the same time, it should be noted that a higher pressure water jet, with its higher kinetic energy, is able to accelerate a larger mass of dry ice to the maximum speed than a lower pressure jet.

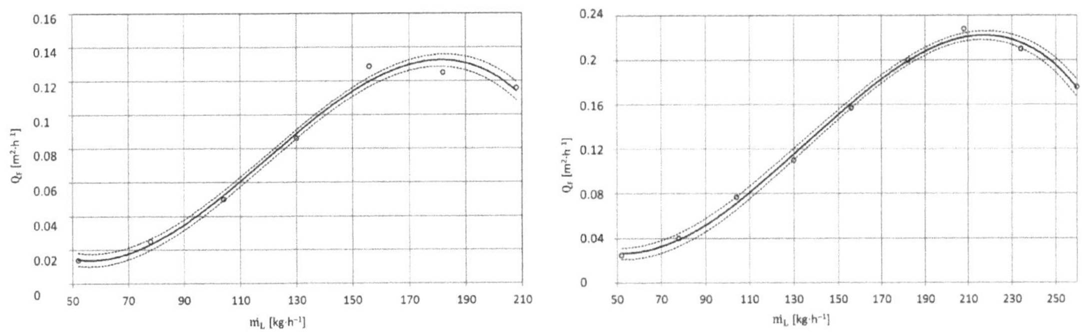

(**a**) (**b**)

Figure 11. Influence of CO_2 $\dot{m}_L$ [kg·h^{-1}] dry ice output on the maximum surface efficiency Q_F [m^2·h^{-1}] of paint stripping by high-pressure water-ice jet from X5CrNi18-10 steel surfaces (TS = 4 × 1.2 mm, L_k = 200 mm, l_2 = 250 mm, κ = 90°): (**a**) jet pressure p_w = 20 MPa; (**b**) p_w = 35 MPa.

In the case of X5CrNi18-10 steel specimens, a minimum surface paint coat stripping efficiency of $\overline{Q}_F = 0.014$ m^2·h^{-1} is obtained with dry ice mass flow rate $\dot{m}_L = 52$ kg·h^{-1} and water jet pressure p_w = 20 MPa. An increase in dry ice flow rate results in a higher number of CO_2 particles hitting the surface being cleaned. Weakened by the impact of the dry ice particles and of the water jet, the outer layer of the paint coatis removed more quickly. Doubling the CO_2 dry ice output ($\dot{m}_L = 104$ kg·h^{-1}) (Figure 11a) leads to a more than threefold increase in the surface paint coat stripping efficiency ($\dot{Q}_F$ = 0.05 m^2·h^{-1}). The optimum dry ice particle output to be used for stripping with water-ice jet pressure p_w = 20 MPa is $\dot{m}_L = 182$ kg·h^{-1}. This being the case, surface paint coat stripping efficiency reaches the value ($\dot{Q}_F$ = 0.125 m^2·h^{-1}). With an increasing CO_2 output, the decreasing number of dry ice particles to be used to remove the paint coating on the "unit surface" indicates that the "supercooled" water jet fully accelerating dry ice particles added to it is a tool with maximum erosivity. A flow rate of CO_2 particles that is too high causes the nozzle pass to be "choked" with excess ice. The water jet is then unable to accelerate dry ice particles so that they could achieve maximum velocity. Therefore, with a flow rate being too high, dry ice particles have less kinetic energy compared to particles that are fully accelerated. The end result is a decrease in surface paint coat removal efficiency. In addition, using CO_2 mass flow rate that is too high results in increased treatment costs, which is undesirable from a practical point of view. For example, with dry ice mass flow rate of $\dot{m}_L = 208$ kg·h^{-1}, surface paint coat stripping efficiency decreased to Q_F = 0.116 m^2·h^{-1}. Under these conditions, more than 75 CO_2 dry ice particles have to be used to obtain 1 mm^2 of cleaned surface.

The use of a high-pressure water-ice jet with water jet pressure p_w = 35 MPa (Figure 11b) ensures maximum surface paint coat stripping efficiency $\dot{Q}_F$= 0.228 m^2·h^{-1} with dry ice flow rate $\dot{m}_L = 216$ kg·h^{-1}. This is an increase in the maximum surface paint coat stripping efficiency $\Delta\dot{Q}_F$= 0.112 m^2·h^{-1} compared to that obtained for a jet with pressure p_w = 20 MPa, while maintaining the remaining processing parameters at the same level.

The effect of dry ice output on the surface stripping efficiency of paint coatings from PA2 aluminium alloy specimens for water jet pressure p_w = 20 MPa and p_w = 35 MPa is presented in Figure 12a,b. Similarly as for the specimens with X5CrNi18-10 steel base, also in this case, an increase of water jet pressure is accompanied by an increase of surface paint coat stripping efficiency. It needs to be noted that for the removal of the paint coating from PA2 aluminium alloy, it is most beneficial to use a water-ice jet with pressure p_W = 35 MPa and CO_2 output $\dot{m}_L = 225$ kg·h^{-1} because it is then that the highest surface treatment efficiency is obtained.

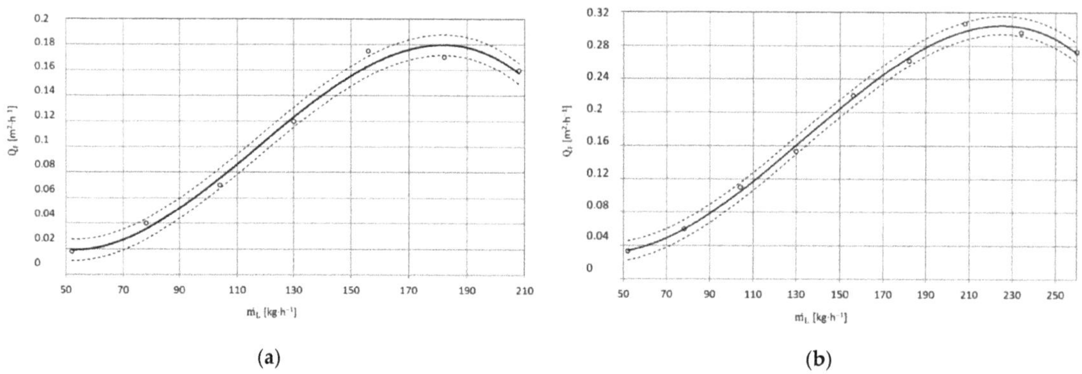

Figure 12. Influence of dry ice CO_2 output $\dot{m}_L$ [kg·h^{-1}] on maximum surface efficiency $\overline{Q}_F$ [m^2·h^{-1}] of paint coat stripping by high-pressure water-ice jet from PA2 aluminium alloy surface (TS = 4 × 1.2 mm, L_k = 200 mm, l_2 = 250 mm, κ = 90°): (**a**) Jet pressure p_W = 20 MPa; (**b**) p_W = 35 MPa.

When using a jet with pressure p_W = 20 MPa, the best treatment effects, based on experimental research results, are obtained using dry ice flow ranging from $\dot{m}_L = 156$ kg·h^{-1} to $\dot{m}_L = 182$ kg·h^{-1} (Table 3). With these treatment parameters, as a result of a cumulative interaction of dry ice particles (i.e., mechanical interaction and interaction from their sublimation) and the water jet, a water-ice jet of the highest erosivity is obtained. When dry ice output $\dot{m}_L = 234$ kg·h^{-1}(for p_W = 35 MPa) (Figure 12b) and $\dot{m}_L = 182$ kg·h^{-1} (for p_W = 20 MPa) (Figure 12a) is exceeded, the water jet is unable to accelerate the increased CO_2 mass to its maximum velocity, the result being decreased treatment efficiency.

Figures 13 and 14 show the effect of the distance between the sprinkler nozzle head outlet l_2 (also known as the working jet length) and the surface being treated on the maximum surface efficiency Q_F of paint coat stripping. It should be noted that when using a working jet length that is too small, high erosivity is achieved but not the highest treatment efficiency is obtained. In the case of the base made of X5CrNi18-10 steel, an increase in the working length of the water-ice jet from l_2 = 150 mm to l_2 = 200 mm is accompanied by an increase in the surface paint coat stripping efficiency by 18 % on average when using a water-ice jet with pressure p_W = 20 MPa (Figure 13a). The maximum increase in stripping efficiency of just under 26% is achieved by changing the working jet length from l_2 = 150 mm to l_2 = 250 mm. Any further increase in the working jet length from l_2 = 150 mm to l_2 = 300 mm only results in a 9% increase in stripping efficiency. The use of length l_2 = 350 mm leads to a 13% reduction (compared to that obtained with l_2 = 150 mm) in the surface paint coat stripping efficiency. For the greatest spray length used of l_2 = 400 mm, the reduction in treatment efficiency is just over 41%. The results from the use of a water-ice jet with pressure p_W = 35 MPa (Figure 13b) are similar to those obtained with lower pressure (p_W = 20 MPa).

The surface paint coat stripping efficiency as a function of the working jet length for PA2 aluminium alloy specimens using a water-ice jet created in a sprinkler head with length L_k = 200 mm is presented in Figure 14a,b. It was found that an increase in the working jet length from l_2 = 150 mm to l_2 = 200 mm is accompanied by an increase in the surface paint coat stripping efficiency caused by an increase in the impact area of the water-ice jet.

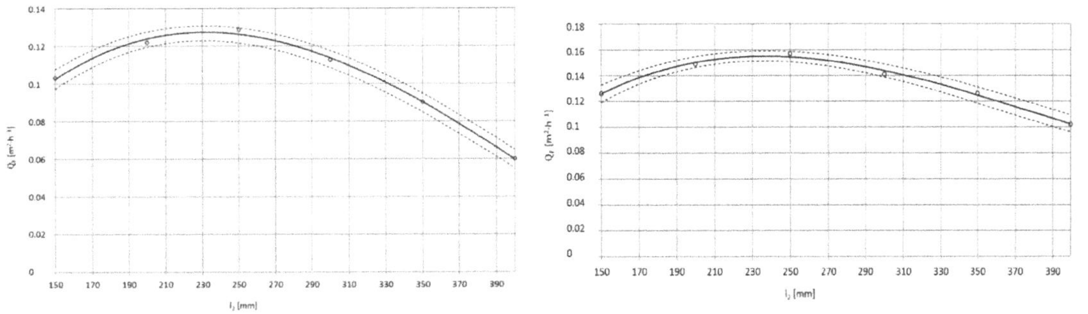

(a) (b)

Figure 13. Influence of jet working length l_2 [mm] on maximum surface efficiency $\overline{Q}_F$ [$m^2 \cdot h^{-1}$] of coat paint stripping with high-pressure water-ice jet from X5CrNi18-10 steel surface (TS = 4 × 1.2 mm, L_k = 200 mm, κ = 90°, $\dot{m}_L = 156\ kg \cdot h^{-1}$): (**a**) Jet pressure p_w = 20 MPa; (**b**) p_w = 35 MPa.

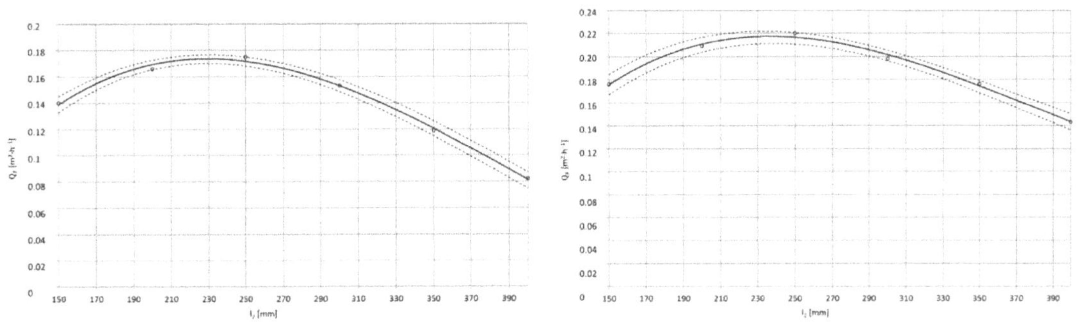

(a) (b)

Figure 14. Influence of jet working length l_2 [mm] on maximum surface efficiency $\overline{Q}_F$ [$m^2 \cdot h^{-1}$] of high-pressure water-ice jet paint coat stripping from PA2 aluminium alloy surface (TS = 4 × 1.2 mm, L_k = 200 mm, κ = 90°, $\dot{m}_L = 156\ kg \cdot h^{-1}$): (**a**) Jet pressure p_w = 20 MPa; (**b**) p_w = 35 MPa.

For a water-ice jet with pressure p_w = 20 MPa (Figure 14a), this increase is 9% on average. The maximum increase in treatment efficiency of less than 26% is obtained when the working jet length is changed from l_2 = 150 mm to l_2 = 250 mm. Further increase of the working length from l_2 = 150 mm to l_2 = 300 mm only results in a 9% increase in surface paint coat stripping efficiency. The use of l_2 = 350 mm leads to a 15% reduction (compared to that obtained with l_2 = 150 mm) in surface paint coat stripping efficiency. For the largest working jet length used l_2 = 400 mm, the decrease in treatment efficiency is just over 41% (Figure 14a). The results from the use of a water-ice jet with pressure p_w = 35 MPa (Figure 14b) are similar to those obtained with lower pressure (p_w = 20 MPa).

In the paint stripping process, apart from dry ice output and the distance between the sprinkler head outlet, the spray angle is also of great importance (Figure 15a,b). This is connected both with the so-called jet reflection and the preservation of the mechanical properties of the coating during the treatment process. In natural conditions, the paint coating is characterised by elasticity, but during an interaction of a large number of CO_2 dry ice particles with the temperature of 194.6 K, a local spot depression of its temperature occurs, which leads to the formation of thermal stress that causes microcracks, which initiate the erosion process [10]. The effect of the jet spray angle on the surface paint coat stripping efficiency from PA2 aluminium alloy when using a water-ice jet with pressure p_w = 20 MPa and p_w = 35 MPa is presented in Figure 15a,b.

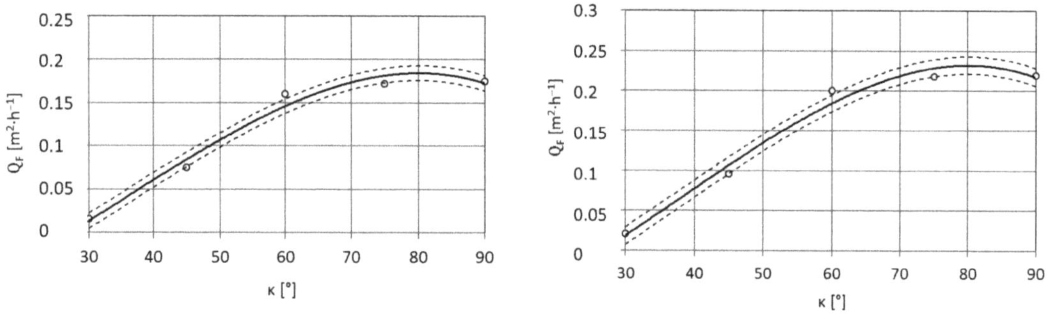

(**a**) (**b**)

Figure 15. Impact of spray angle κ [°] on maximum surface efficiency $\overline{Q}_F$ [$m^2 \cdot h^{-1}$] of high-pressure water-ice jet paint coat stripping from PA2 aluminium alloy surface (TS = 4 × 1.2 mm, L_k = 200 mm, l_2 = 250 mm, $\dot{m}_L = 156$ kg·h^{-1}): (**a**) Jet pressure p_w = 20 MPa; (**b**) p_w = 35 MPa.

With an increase of the jet spray angle κ = 30 ÷ 60°, there is an almost linear increase in surface paint coat stripping efficiency. Further increase of the spray angle up to κ = 90° causes an increase of the treatment efficiency yet that is not as sharp as observed with smaller spray angles. It can therefore be accepted that the best machining results are obtained with a spray angle of κ = 75 ÷ 90°.

The effect of water jet pressure on paint removal efficiency is shown in Figure 16. As the pressure increases, the processing efficiency increases. For example, increasing the pressure from p_w = 20 MPa to p_w = 25 MPa increased the treatment efficiency to $\Delta\dot{Q}_F$ = 0.014 $m^2 \cdot h^{-1}$ for the PA2 aluminum alloy. Further increase of the jet pressure to p_w = 30 MPa, with unchanged other processing parameters, resulted in paint removal efficiency of $\dot{Q}_F$ = 0.201 $m^2 \cdot h^{-1}$. This is an outgrowth of the treatment efficiency by $\Delta\dot{Q}_F$ = 0.012 $m^2 \cdot h^{-1}$ compared to that obtained for a jet with pressure p_w = 25 MPa.

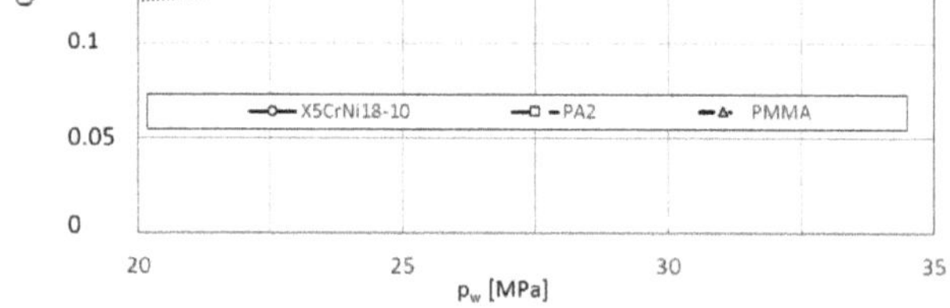

Figure 16. Influence of water-ice jet pressure p_w [MPa] on the maximum surface stripping efficiency Q_F [$m^2 \cdot h^{-1}$] (TS = 4 × 1.2 mm, L_k = 200 mm, l_2 = 250 mm, $\dot{m}_L = 156$ kg·h^{-1}, κ = 90°).

4. Conclusions

The optical microscope by Kestler—Vision Engineering Dynascope Ltd. with the ND 1300 Quadra-Chek measuring system and the sensitivity of ±0.001 was used to measure the surface quality of the processed materials. Based on the research and analysis of the results, it was found that:

1. The use of a high-pressure water-ice jet with maximum pressure p_W = 35 MPa does not cause any changes in the geometric structure of the surface of X5CrNi18-10 steel and PA2 aluminium alloy.
2. In the case of brittle materials, such as polymethyl methacrylate PMMA, which form bases for lacquer coatings, it is necessary to use a water-ice jet with the maximum working pressure of p_W = 20 MPa and dry ice flow rate of $\dot{m}_L = 156$ kg·h^{-1}. The use of a water jet with pressure p_W = 35 MPa and dry ice flow rate $\dot{m}_L = 208$ kg·h^{-1} causes traces of jet impact in the form of small chipping of the base material (PMMA polymethyl methacrylate). The highest surface treatment efficiency is obtained with a higher water jet pressure and an appropriately increased dry ice particle flow rate. For water jet pressure p_W = 20 MPa, the best results are obtained when dry ice flow rate is $\dot{m}_L = 156 \div 182$ kg·h^{-1}, while for water jet pressure p_W = 35 MPa, this output should not exceed $\dot{m}_L = 234$ kg·h^{-1}.
3. In order to obtain the maximum paint coat stripping efficiency irrespective of the water jet pressure and the dry ice particle flow rate, the working length of the jet needs to be l_2 = 250 mm and the spray angle needs to be $\kappa = 75 \div 90°$.
4. Increasing the pressure of the water jet results in higher efficiency of paint removal. The increase of processing efficiency is proportional in the tested range of jet pressure $p_W = 20 \div 35$ MPa.
5. The above research proves that one of the effective methods of removing paint coatings is the use of a high-pressure stream of water with ice. This method of processing allows for effective preparation of the substrate for the application of a renovation paint coating, while simultaneously maintaining appropriate processing parameters in that it does not change the geometric structure of the substrate.

Author Contributions: The individual contributions of the Authors: conceptualization, G.C. and J.C.; methodology, G.C., J.C and L.K.; investigation, G.C., J.C. and M.K.; software, G.C., J.C., L.K. and M.K.; data curation, G.C.; validation, J.C. and L.K.; writing—original draft preparation, G.C., J.C., L.K. and M.K.; writing—review and editing, G.C., J.C and L.K.; visualization, G.C., J.C. and M.K.; supervision, L.K.; All authors have read and agreed to the published version of the manuscript.

Funding: This research received no external funding.

Institutional Review Board Statement: Not applicable.

Informed Consent Statement: Not applicable.

Data Availability Statement: Data sharing is not applicable to this article.

Conflicts of Interest: The authors declare no conflict of interest.

References

1. Kotnarowska, D. Destruction kinds of car body coating. *J. Logistyka* **2010**, *6*.
2. Kotnarowska, D.; Kotnarowski, A. Influence of ageing on kinetics of epoxy coatings erosive wear. *Int. J. Appl. Mech. D Eng.* **2004**, *9*, 53–58.
3. Kittel, J.; Celati, N.; Keddam, M.; Takenouti, H. Influence of the coating–substrate interactions on the corrosion protection: Characterisation by impedance spectroscopy of the inner and outer parts of a coating. *Prog. Org. Coat.* **2003**, *46*, 135–147. [CrossRef]
4. Vesga, L.F.; Vera, E.; Panqueva, J.H. Use of the electrochemical impedance spectroscopy to evaluate the performance of a primer applied under different surface preparation methods. *Prog. Org. Coat.* **2000**, *39*, 61–65. [CrossRef]
5. Santágata, D.M.; Seré, P.R.; Elsner, C.I.; Di Sarli, A.R. Evaluation of the surface treatment effect on the corrosion performance of paint coated carbon steel. *Prog. Org. Coat.* **1998**, *33*, 44–54. [CrossRef]

6. Geng, S.; Sun, J.; Guo, L. Effect of sandblasting and subsequent acid pickling and passivation on the microstructure and corrosion behavior of 316L stainless steel. *Mater. Des.* **2015**, *88*, 1–7. [CrossRef]
7. Barranco, V.; Onofre, E.; Escudero, M.L.; García-Alonso, M.C. Characterization of roughness and pitting corrosion of surfaces modified by blasting and thermal oxidation. *Surf. Coat. Technol.* **2010**, *204*, 3783–3793. [CrossRef]
8. Geskin, E.S.; Shishkin, D.; Babets, K. Application of ice particles for precision cleaning of sensitive surfaces. In Proceedings of the 10th American Waterjet Conference, Houston, TX, USA, 14–17 August 1999; Volume 1.
9. Dong, S.J.; Song, B.; Hansz, B.; Liao, H.L.; Coddet, C. Modelling of dry ice blasting and its application in thermal spray. *Mater. Res. Innov.* **2012**, *16*, 61–66. [CrossRef]
10. Borkowski, P.J. *Theoretical and Experimental Basis of Hydrostreaming Surface Treatment*; University of Koszalin University of Technology: Koszalin, Poland, 2004.
11. Ivey, R.B. Carbon dioxide pellets blasting paint removal for potential application on Warner Robins Managed Air Force Aircraft. In Proceedings of the 1st Annual International Workshop on Solvent Substitution, Phoenix, AZ, USA, 4–7 December 1990.
12. Spur, G.; Uhlmann, E.; Elbing, F. Dry-ice blasting for cleaning: Process, optimization and application. *Wear* **1999**, *233–235*, 402–411. [CrossRef]
13. Liu, Y.H.; Maruyama, H.; Matsusaka, S. Effect of particle impact on surface cleaning using dry ice jet. *Aerosol Sci. Technol.* **2011**, *45*, 1519–1527. [CrossRef]
14. Chomka, G.; Chudy, J. Analysis and interpretation of measurements of surface machining effectiveness in the process of varnish removal by a water-ice jet. *Teh.-Tech. Gazette* **2013**, *20*, 847–852.
15. Dunsky, C.M.; Hashish, M. Feasibility study of the use of ultrahigh-pressure liquefied gas jets for machining of nuclear fuel pins. In Proceedings of the 8th American Water Jet Conference, Houston, TX, USA, 26–29 August 1995.
16. Toscano, C.; Ahmadi, G. Particle removal mechanisms in cryogenic surface cleaning. *J. Adhes.* **2003**, *79*, 175–201. [CrossRef]
17. Liu, Y.H.; Maruyama, H.; Matsusaka, S. Agglomeration process of dry ice particles produced by expanding liquid carbon dioxide. *Adv. Powder Technol.* **2010**, *21*, 652–657. [CrossRef]
18. Liu, H.T.; Fang, S.; Hibbard, C.; Maloney, J. Enhancement of ultrahigh-pressure technology with LN2 cryogenic jets. In Proceedings of the 10th American Waterjet Conference, Houston, TX, USA, 14–17 August 1999.
19. Dunsky, C.M.; Hashish, M. Observations on cutting with abrasive-cryogenic jets. In Proceedings of the 13th International Conference on Jetting Technology—Applications and Opportunities, Sardinia, Italy, 29–31 October 1996.
20. Hashish, M.; Dunsky, C.M. The formation of cryogenic and abrasive-cryogenic jets. In Proceedings of the 14th International Conference on Jetting Technology, Brugge, Belgium, 21–23 September 1998.
21. Myslinski, P.; Precht, W.; Kukiełka, L.; Kamasa, P.; Pietruszka, K. A possibility of application of MTDIL to the residual stresses analysis: The hard coating-substrate system. *J. Therm. Anal. Calorim.* **2004**, *77*, 253–258. [CrossRef]
22. Hlavacova, I.M.; Geryk, V. Abrasives for water-jet cutting of high-strength and thick hard materials. *Int. J. Adv. Manuf. Technol.* **2017**, *90*, 1217–1224. [CrossRef]
23. Krajcarz, D. Comparison metal water jet cutting with laser and plasma cutting. *Procedia Eng.* **2014**, *69*, 838–843. [CrossRef]
24. Perec, A. Research into the disintegration of abrasive materials in the abrasive water jet machining process. *Materials* **2021**, *14*, 3940. [CrossRef]
25. Perec, A. Disintegration and recycling possibility of selected abrasives for water jet cutting. *DYNA* **2017**, *84*, 249–256. [CrossRef]
26. Perec, A. Investigation of limestone cutting efficiency by the Abrasive Water Suspension Jet. In *Advances in Manufacturing Engineering and Materials*; Lecture Notes in Mechanical Engineering; Springer Nature: Cham, Switzerland, 2019; pp. 124–134. [CrossRef]
27. Perec, A.; Radomska-Zalas, A. Modeling of abrasive water suspension jet cutting process using response surface method. *AIP Conf. Proc.* **2019**, *2078*, 0200511–0200518. [CrossRef]
28. Perec, A. Environmental aspects of abrasive water jet cutting. *Ann. Set Environ. Prot.–Rocz.* **2018**, *20*, 258–274.
29. Thangaraj, M.; Ahmadein, M.; Alsaleh, N.A.; Elsheikh, A.H. Optimization of Abrasive Water Jet Machining of SiC Reinforced Aluminum Alloy Based Metal Matrix Composites Using Taguchi–DEAR Technique. *Materials* **2021**, *14*, 6250. [CrossRef] [PubMed]
30. Muthuramalingam, T.; Vasanth, S.; Vinothkumar, P.; Geethapriyan, T.; Rabik, M.M. Multi Criteria Decision Making of Abrasive Flow Oriented Process Parameters in Abrasive Water Jet Machining Using Taguchi–DEAR Methodology. *Silicon* **2018**, *10*, 2015–2021. [CrossRef]
31. Kasperowicz, M.; Chomka, G.; Chudy, J. Determining the supply pressure depending on the feed speed and the diameter of the nozzle. *Carpathian J. Food Sci. Technol.* **2018**, *10*, 17–23.
32. Kasperowicz, M.; Chomka, G.; Bil, T. Determination of supply pressure during cutting fish using high-pressure water stream taking into account the place and diameter of the water nozzle. *DEGRUYTER Cutting. Int. J. Food Eng.* **2019**, *16*, 1–9. [CrossRef]
33. Bohdal, L. Application of a SPH Coupled FEM Method for Simulation of Trimming of Aluminum Autobody Sheet. *Acta Mech. Et Autom.* **2016**, *10*, 56–67. [CrossRef]
34. Bohdal, L.; Kukielka, L.; Radchenko, A.M.; Patyk, R.; Kułakowski, M.; Chodór, J. Modelling of guillotining process of grain oriented silicon steel using FEM. *Comput. Technol. Eng.* **2019**, *2078*, 020080.
35. Chodor, J.; Zurawski, L. Investigations on the chip shape and its upsetting and coverage ratios in partial symmetric face milling process of aluminium alloy AW-7075 and the simulation of the process with the use of FEM. In *Advances in Mechanics: Theoretical, Computational and Interdisciplinary Issues*; Taylor & Francis Ltd.: Gdansk, Poland, 2016; pp. 121–124.

36. Florianowicz, M.; Bohdal, L. Modeling of the Refrigerants Condensation in the Superheated Vapor Area. *Annu. Set Environ. Prot.* **2012**, *14*, 393–406.
37. Patyk, R.; Kukielka, L.; Kaldunski, P.; Bohdal, L.; Chodor, J.; Kulakowska, A.; Kukielka, K.; Nagnajewicz, S. Experimental and numerical researches of duplex burnishing process in aspect of achieved productive quality of the product. *ESAFORM* **2018**, *1960*, 070021.
38. McGeough, J.A. Cutting of Food Products by Ice-particles in a Water-jet. *Procedia CIRP* **2016**, *42*, 863–865. [CrossRef]
39. Mathia, T.G.; Pawlus, P.; Wieczorowski, M. Recent trends in surface metrology. *Wear* **2011**, *3–4*, 494–508. [CrossRef]
40. Wieczorowski, M. Use of Topographic Analysis in Surface Roughness Measurements. Habilitation Thesis, Poznan University of Technology, Poznań, Poland, 2009; 429p.
41. Kukiełka, L. *Fundamentals of Engineering Studies*; National Scientific Publisher: Warsaw, Poland, 2002; 273p. (In Polish)
42. Rosenberg, B.; Yuan, L.; Fulmer, S. Ergonomics of abrasive blasting: A comparison of high pressure water and steel shot. *Appl. Ergon.* **2006**, *37*, 659–667. [CrossRef]
43. Schnakovszky, C.; Herghelegiu, E.; Radu, C.; Cristea, I. *The influence of the feed rate on the quality of surfaces processed by AWJ at high pressures. InAdvanced Materials Research*; Trans Tech Publications: Bäch, Switzerland, 2014.
44. Xiong, S.; Jia, X.; Wu, S.; Li, F.; Ma, M.; Wang, X. Parameter optimization and effect analysis of low-pressure abrasive water jet (LPAWJ) for paint removal of remanufacturing cleaning. *Sustainability* **2021**, *13*, 2900. [CrossRef]
45. Yang, M.; Choi, J.; Lee, J.; Hur, N.; Kim, D. Wet blasting as a deburring process for aluminum. *J. Mater. Process. Technol.* **2014**, *214*, 524–530. [CrossRef]
46. Yu, Y.; Sun, T.; Yuan, Y.; Gao, H.; Wang, X. Experimental investigation into the effect of abrasive process parameters on the cutting performance for abrasive waterjet technology: A case study. *Int. J. Adv. Manuf. Technol.* **2020**, *107*, 2757–2765. [CrossRef]
47. Zhong, Z.W.; Han, Z.Z. Performance comparison of four waterjet nozzles. *Mater. Manuf. Process.* **2003**, *18*, 965–978. [CrossRef]

materials MDPI

Article

Can a Black-Box AI Replace Costly DMA Testing?—A Case Study on Prediction and Optimization of Dynamic Mechanical Properties of 3D Printed Acrylonitrile Butadiene Styrene

Ronak Vahed, Hamid R. Zareie Rajani and Abbas S. Milani *

School of Engineering, University of British Columbia, Kelowna, BC V1V 1V7, Canada; ronak.vahed@gmail.com (R.V.); hamid.r.zareie@gmail.com (H.R.Z.R.)
* Correspondence: abbas.milani@ubc.ca

Abstract: The complex and non-linear nature of material properties evolution during 3D printing continues to make experimental optimization of Fused Deposition Modeling (FDM) costly, thus entailing the development of mathematical predictive models. This paper proposes a two-stage methodology based on coupling *limited* data experiments with black-box AI modeling and then performing heuristic optimization, to enhance the viscoelastic properties of FDM processed acrylonitrile butadiene styrene (ABS). The effect of selected process parameters (including nozzle temperature, layer height, raster orientation and deposition speed) as well as their combinative effects are also studied. Specifically, in the first step, a Taguchi orthogonal array was employed to design the Dynamic Mechanical Analysis (DMA) experiments with a minimal number of runs, while considering different working conditions (temperatures) of the final prints. The significance of process parameters was measured using Lenth's statistical method. Combinative effects of FDM parameters were noted to be highly nonlinear and complex. Next, artificial neural networks were trained to predict both the storage and loss moduli of the 3D printed samples, and consequently, the process parameters were optimized via Particle Swarm Optimization (PSO). The optimized process of the prints showed overall a closer behavior to that of the parent (unprocessed) ABS, when compared to the unoptimized set-up.

Keywords: acrylonitrile butadiene styrene; additive manufacturing; artificial neural networks; dynamic mechanical properties; Particle Swarm Optimization

Citation: Vahed, R.; Zareie Rajani, H.R.; Milani, A.S. Can a Black-Box AI Replace Costly DMA Testing?—A Case Study on Prediction and Optimization of Dynamic Mechanical Properties of 3D Printed Acrylonitrile Butadiene Styrene. *Materials* **2022**, *15*, 2855. https://doi.org/10.3390/ma15082855

Academic Editor: Gabriele Milani

Received: 5 February 2022
Accepted: 3 April 2022
Published: 13 April 2022

1. Introduction

Fused Deposition Modeling (FDM) is increasingly being used as a reliable rapid prototyping tool in industries. However, an outstanding challenge in the field of additive manufacturing is yet to employ FDM to build high-quality end-use parts with minimal waste, while maintaining a high rate of production [1]. Exploiting the full potential of FDM for manufacturing requires the proper selection of process control factors through a good understanding of their nature [2]. Over years of evolving additive manufacturing techniques, several investigations have been performed to study the effect of FDM process parameters, e.g., [3–5]. In these studies, the road width (w), which is the width of the road deposited through the nozzle, layer thickness (t), which is the thickness of each 2D layer, feeding rate (v), which is the rate at which the thermoplastic filament is fed into the nozzle, nozzle temperature (NT), raster orientation (θ), which is the orientation of roads in each 2D layer, overlap (b), which is the amount of overlap between two adjacent roads, and nozzle diameter (d) have been introduced as the main effective process parameters [3–5]. Figure 1 schematically depicts the above-mentioned process parameters.

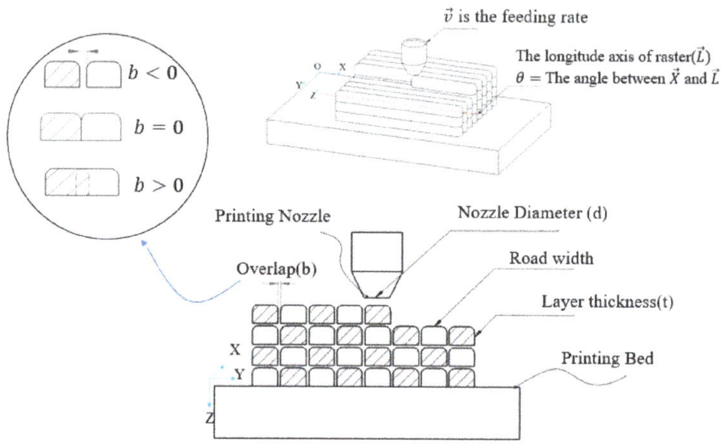

Figure 1. FDM main process parameters.

Assessment of FDM fabricated parts could be performed using several quality measures such as durability and static mechanical properties, dynamic mechanical properties, and manufacturing accuracy [6–17]. Despite the fact that thermoplastics are viscoelastic materials, due to the expensive and time-consuming nature of the dynamic mechanical analysis (DMA) tests, the time- and temperature-dependent mechanical properties of FDM processed parts have not been characterized as extensively as their static mechanical properties [3].

One of the earliest studies on dynamic mechanical properties of FDM processed parts goes back to the work of Chin Ang et al. [18], with respect to a few select process parameters including air gap, raster width, raster orientation, deposition profile, and layer height. In their study, the air gap and raster width were identified as the most effective process parameters to control the porosity and strength of processed parts. Furthermore, they claimed that there is a logarithmic relationship between mechanical properties and porosity, meaning that 3D printed scaffold parts with a lower porosity would show a higher strength. Later, Arivazhagan et al. [5] used the dynamic mechanical analysis (DMA) to examine the effects of road width, raster orientation, and nozzle temperature on viscosity and dynamic moduli of FDM processed Acrylonitrile Butadiene Styrene (ABS) samples. They showed that a raster orientation of 30°/60° and a road width of 0.454 mm improves the dynamic moduli of 3D-printed ABS. Mohamed et al. [19] considered layer thickness, overlap, raster angle, raster orientation, and road width as control factors to investigate the dynamic mechanical properties of FDM processed ABS. The results of their study indicated that the overlap and layer thickness are the most effective process parameters. Specifically, it was shown that a layer thickness of 0.3302 mm, a road width of 0.4572 mm, and an overlap of 0 mm with a raster angle of 0° can increase the dynamic moduli of ABS.

However, none of the above past studies on dynamic mechanical properties of FDM processed thermoplastics accounted for other important parameters including nozzle temperature and feeding rate. In addition, a lack of systematic experimental design (e.g., using the DOE methods) did not allow to fully account for the combined effects of various process parameters (most studies used a one-factor-at-a-time sensitivity analysis). This could lead to errors in analyzing the role of FDM parameters in the dynamic mechanical characteristics of 3D printing materials.

This paper presents a two-stage methodology to study the concurrent effects of multiple FDM process parameters including nozzle temperature, raster orientation, layer thickness, and feeding rate on dynamic mechanical properties of FDM processed ABS. Specifically, the variations of the dynamic moduli and glass transition temperature of ABS as a function of changes in four key FDM parameters (nozzle temperature, layer height, raster orientation and deposition speed) are analyzed using a Taguchi design of experiment (DOE). The latter not only minimizes the number of costly and time-consuming DMA tests, but also accounts for the combined effects of the FDM parameters. Once the required (limited) dataset was gathered through performing experiments, a series of artificial neural networks were developed and employed to predict the properties of the 3D prints, and consequently, the process parameters were optimized via a particle swarm optimization (PSO). Finally, the contribution and ranking of process parameters were identified.

2. Methods

2.1. Material and Fabrication

Owing to its relatively low melting temperature and high-quality surface finish, ABS is one of the most commonly used materials in 3D printing industries [20]. The initial ABS material in the current study was in the form of filaments, which were then fed into a 3D printer (MakerGear M2, Beachwood, OH, USA) to manufacture rectangular samples of 57 mm × 14 mm × 1.25 mm (length × width × thickness) for subsequent DMA characterization tests. The as-received filaments were extruded out of ABS POLYLAC® by CHIMEI Corp. (Tainan City, Taiwan), with specifications shown in Table 1.

Table 1. Properties of the ABS filaments.

Commercial code	CHIMEI PA-747S
Nominal diameter (mm)	1.75
Purity	>98%
Nominal Young's modulus (GPa)	2
Relative density—H_2O $\left(\frac{g}{cm^3}\right)$	1.03–1.10
Decomposition temperature (°C)	>310

2.2. Dynamic Mechanical Analysis

Dynamic mechanical analysis (DMA) [21] characterizes the viscoelastic properties of materials, described by the storage modulus (E'), loss modulus (E''), complex modulus (E^*), and $\tan\delta$ [22]; $\tan\delta$ is a dimensionless property, defined as the ratio of the loss modulus to the storage modulus. DMA also quantifies the glass transition temperature (T_g), which denotes the transition point between glassy and rubbery states [23]. In this study, the dynamic mechanical analyses of ABS 3D-printed samples were carried out using DMA Q800 (TA Instruments, New Castle, DE, USA). DMA testing was also conducted on original ABS filaments (length of 57 mm) prior to printing; this will be referred to as unprocessed material hereafter. The test specimens were prepared according to the DMA 800 manufacturer instructions.

2.3. Artificial Neural Network (ANN)

The Artificial Neural Network (ANN) modeling technique, inspired by the neurologic system of the brain, has received notable attention in recent years in the AI field. The ability to approximate complex non-linear relationships between input and output parameters in complex systems is the main advantage that has made the technique a useful predictive modeling tool in a wide range of applications [24]. In addition, ANNs are known to provide predictions with minimal prior assumptions, hence making them a useful black-box modeling means for unknown/complex systems, as opposed to more explicit, e.g., regression techniques that require a "pre-defined" form of input-output relationship. As depicted in Figure 2, an artificial neural network model is a collection of processing units called nodes (also called neurons). In a neural network, neurons are connected to each other by numerically assigned connections, known as weights (W_i) and are fed by a signal from each input (x_i). The neuron's scalar output (a) is the summation of the weighted inputs and the bias term (b) modified by a transfer function (f). Summation of weighted inputs is denoted by s and is presented as:

$$s = w_1x_1 + w_2x_2 + \ldots + w_nx_n + b = \sum_{i=1}^{n} w_ix_i + b \tag{1}$$

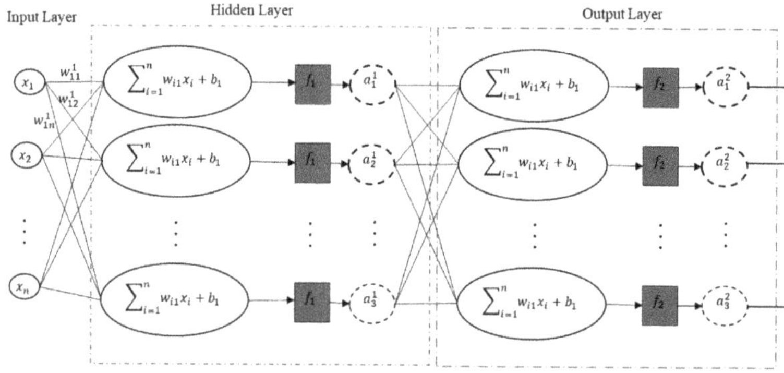

Figure 2. Architecture of artificial neural network with one hidden layer. The superscripts represent the layer number. The first and second subscripts, respectively, represent the input and neuron number that are associated together in each layer.

The bias term acts similar to an input with the value of 1 and its existence in a network is not mandatory. However, it often improves the performance of the model [25,26]. It must be noted that weights and biases are adjustable terms that are updated through the learning algorithm. The transfer function is fixed throughout the whole process.

An ANN architecture consists of three layers: input, hidden, and output layers. The input layer, which is statistically related to independent variables, contains no neurons. On the other hand, all the neurons in the hidden (which is responsible for the major mathematical process) and the output layers (which delivers the dependent variables) take the outputs of their preceding layer as their inputs. Figure 2 shows an example of an ANN architecture with one hidden layer.

For a designed network architecture, defining the initial weights and updating them would be the next step. This step, which is also known as the learning or training algorithm, is in essence the process of minimizing the network error. The training procedure starts by calculating the error with the initial weights and continues with adjusting the interconnecting weights until a maximum iteration level or an acceptable error level is achieved. The network performance can be evaluated, e.g., by the Mean Squared Error (MSE) between the desired and the predicted values of the output:

$$E = \frac{1}{2N}\sum_{i=1}^{N}[t_i - a_i]^2 \tag{2}$$

where N stands for the number of training sample points, t is the desired value and a is the predicted value for the output of the i-th sample point. As the network error is calculated, the weights and biases are updated through back propagation to reduce the error value. This process is repeated until the error becomes minimized. In each iteration, an adjusted weight is calculated based on:

$$W_{ji}^{(n)}(k) = W_{ji}^{(n)}(k-1) - \alpha\left(\frac{\partial E}{\partial W_{ji}^{(n)}}\right) \tag{3}$$

where $W_{ji}^{(n)}(k)$ is an updated weight of the n-th layer in the k-th iteration (also known as epoch). $\frac{\partial E}{\partial W_{ji}^{(n)}}$ is the partial derivative of the error. In this equation, α is the learning rate, which is less than 1. Calculation of the derivative part of the equation is normally achieved by the chain rule [27]. In order to develop a robust neural network, datasets are divided into three subsets—training, validation and testing—and the performance of the model is assessed through each of these subsets. The training dataset is employed to update the weights and bias terms, and in order to prevent overfitting of the network and test set, the validation set is employed [28].

2.4. Particle Swarm Optimization (PSO)

Particle Swarm Optimization is known as a powerful numerical algorithm to optimize complex functions by finding the best solution in a space of feasible solutions. This technique, which has been inspired by the social behavior of animals, was first introduced by Eberhart and Kennedy [29]. A simple interpretation of PSO is the behavior of a group of birds who are seeking food. None of the birds know where the single piece of food is located. However, they know their distance from the other birds. Therefore, the simplest and fastest way to achieve the food is to follow the closest bird to that food. In essence, it combines local search methods (through self-experience) with global search methods (through neighboring experience). Here, PSO was chosen merely as an example of global search methods (next to other heuristics methods such as genetics algorithm), owing to its simple and intuitive mathematical rules in finding the optimum solution in high-dimensional spaces.

In PSO, each candidate solution is called a "particle", which is part of a community known as a "swarm". PSO solves the optimization problem by moving the particles in a space of all feasible solutions, also known as the search space. Each particle has a memory to keep its best experience, and the cooperation of particles helps them find the best global solution in the search space.

As shown in Figure 3, the position of particle i at time step t, denoted by $\overrightarrow{X_i(t)}$, and its velocity, denoted by $\overrightarrow{V_i(t)}$, are the key properties to define a particle. The previous experience of each particle, $\overrightarrow{X_i(t)}$, its previous movement $\overrightarrow{V_i(t)}$, and the best experience of the whole swarm, $g(t)$, guide each particle moving towards its next position by $\overrightarrow{X_i(t+1)}$, which is probably a better experience. This process continues until the swarm meets its best experience (denoted by $P_i(t)$).

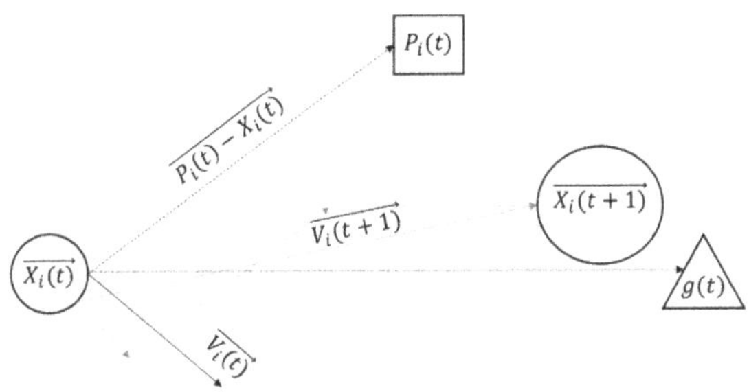

Figure 3. Movement of each particle toward its best position in particle swarm optimization.

In PSO, each particle obeys two rules to update its position and velocity vectors:

$$\overrightarrow{X_i(t+1)} = \overrightarrow{X_i(t)} + \overrightarrow{V_i(t+1)} \tag{4}$$

$$\overrightarrow{V_i(t+1)} = wV_i(t) + C_1(P_i(t) - X_i(t)) + C_2(g(t) - X_i(t)) \tag{5}$$

where w is the inertia coefficient and C_1 and C_2 are acceleration coefficients.

The j-th component of new position and speed vectors can be calculated as follows:

$$X_{ij}(t+1) = X_{ij}(t) + V_{ij}(t+1) \tag{6}$$

$$V_{ij}(t+1) = wV_{ij}(t) + r_1C_1\left(P_{ij}(t) - X_{ij}(t)\right) + r_2C_2\left(g_j(t) - X_{ij}(t)\right) \tag{7}$$

where $V_{ij}(t+1)$ is the j-th component of velocity of particle i at time step $(t+1)$. In Equation (7), the first component ($wV_{ij}(t)$) is known as inertia term. The second component is called the cognitive component, and the third term is the social component; r_1 and r_2 are the uniformly distributed numbers in the range of 0 and 1. $P_{ij}(t)$ is the j-th component of position that gives the best value ever experienced by particle i and $g_j(t)$ is that experienced by all particles in the swarm.

3. Results and Discussion

Although there are several process parameters controlling the properties of FDM fabricated parts, this study is focused on four selected factors including the nozzle temperature, the layer height, the raster orientation and the deposition speed. Based on the initial trial and error experiments to ensure that the printed parts have minimum acceptable quality with no visible defects, four levels were assigned to each process parameter. The selected factors and their corresponding levels are presented in Table 2. For this case, the total (full-factorial) number of experiments would be 4^4 or 256. However, assuming that 'the factor interactions are negligible', the Taguchi approach [30,31] was implemented to reduce the number of costly DMA experimental runs by means of using orthogonal arrays. Taguchi orthogonal arrays are often shown by "$L_n(x^y)$", where n stands for the total number of the experiments, x represents the levels, and y is the number of the controlling factors. Here, an $L_{16}(4^4)$ design was used (Table 3, where the levels 1–4 are representing the physical values in Table 2). Variations of the dynamic moduli and tan δ over working temperatures were measured for 16 designed samples of Table 3, and the results are presented in Figures 4–6.

Table 2. Control factors used in the experimental procedures with their assigned levels.

Control Factors	Level 1	Level 2	Level 3	Level 4
Raster orientation	0° 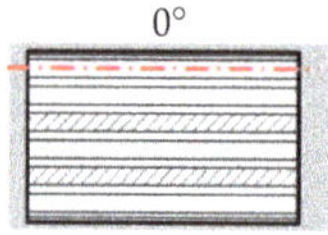	90° 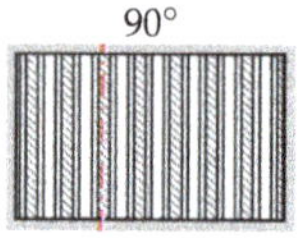	45° 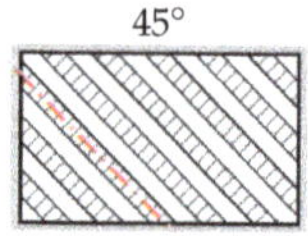	±45° 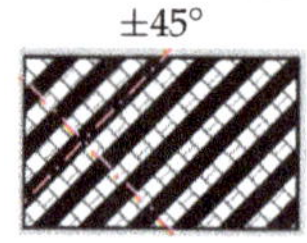
Layer height (µm)	50	130	210	300
Temperature (°C)	250	270	290	310
Feeding rate (mm/min)	1000	2000	3000	4000

According to the observed trends, when comparing the processed (3D printed) and unprocessed samples (as-received filaments), it can be concluded that all the processed samples generally follow a similar trend. However, the comparison between the group of the processed samples and unprocessed samples clearly shows that the FDM has reduced the magnitude of both storage and loss moduli regardless of the combination of process parameters used. The extent of the decrease, however, is highly correlated to the temperature at which the property has been measured in DMA (equivalent to the working temperature of the print in use). As an instance, according to Figure 4, an average reduction of 40% is observed on the storage modulus when pooling all processed samples. This reduction ranges from 15% to 62%, corresponding to samples 2 and 9, respectively. At a specific working temperature, e.g., 100 °C, there is an average reduction of 25% in the storage modulus due to the FDM process. Similarly, the same trends can be seen while measuring the loss modulus of 3D printed ABS samples. As shown in Figure 5, regardless of assigned values to process parameters, the FDM process decreases the loss modulus of fabricated parts compared to the unprocessed sample. For instance, at 40 °C the loss modulus has decreased 11–56% as a result of different FDM process conditions. The reduction at 40 °C (as a nominal working temperature example) has an average value of 33.5%. Nevertheless, the reduction increases drastically and reaches the average value of 60.7% at a working temperature of 100 °C. Regardless of the selected process parameters, on average it is confirmatory to notice that FDM fabricated samples heated up to, e.g., 100 °C show less viscous behavior (represented by the loss modulus) compared to samples tested at 40 °C. Although the FDM process seems to unavoidably decrease the storage and loss moduli of the printed parts, by selecting a suitable (optimized) set of process parameters this reduction may be minimized.

Table 3. Orthogonal array used to design the experimental layout with respect to configuration levels given in Table 2.

Sample #	Raster Orientation	Layer Height (μm)	Nozzle Temperature (°C)	Deposition Speed (mm/min)
1	1	1	1	1
2	1	2	2	2
3	1	3	3	3
4	1	4	4	4
5	2	1	2	3
6	2	2	1	4
7	2	3	4	1
8	2	4	3	2
9	3	1	3	4
10	3	2	4	3
11	3	3	1	2
12	3	4	2	1
13	4	1	4	2
14	4	2	3	1
15	4	3	2	4
16	4	4	1	3

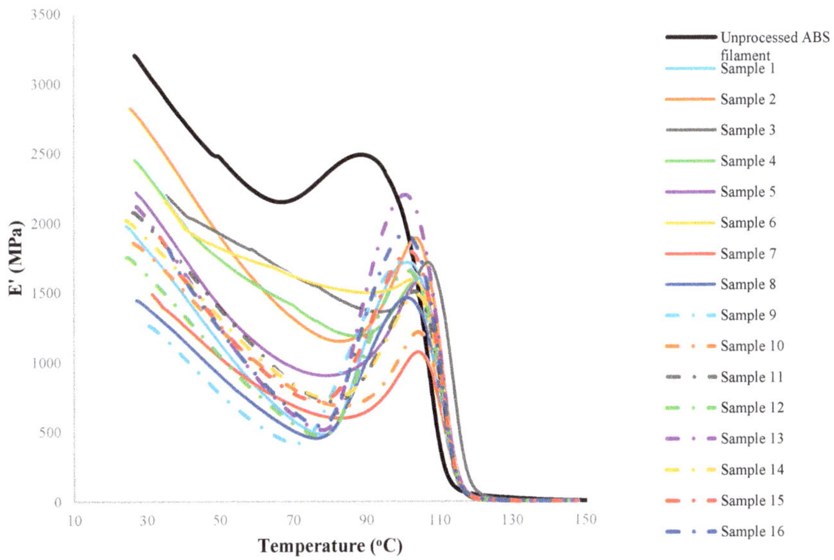

Figure 4. The variation of the storage modulus versus temperature for the test specimens. The overshoot close to glass transition temperature often occurs as the stresses trapped in the part during processing are relieved with an increase in temperature and rearrangement of polymer chains. Note that the sample numbers in the legend correspond to the setups in Table 3, and do not refer to repeats.

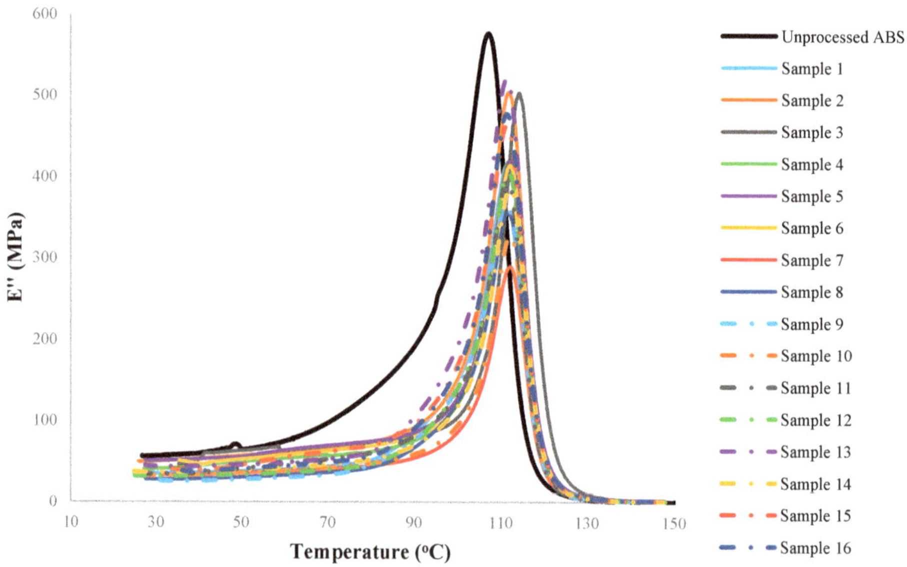

Figure 5. The variation of the loss modulus versus temperature for the test specimens.

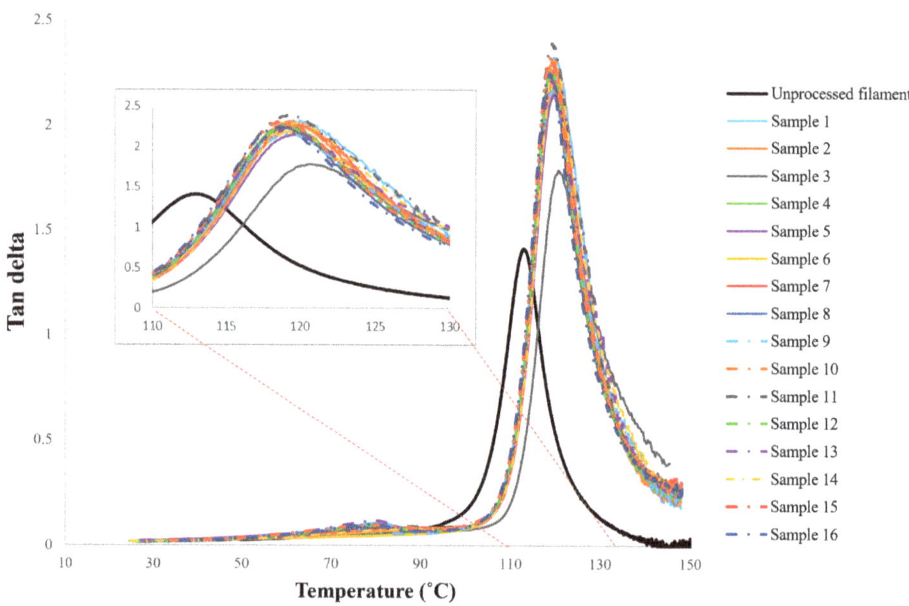

Figure 6. The variation of tan delta versus temperature for the test specimens, with the more detailed view of the variation in the range of 110 °C to 130 °C.

Generally, it is common to identify the glass transition temperature (T_g) of thermoplastics by measuring the peak of temperature-tan δ curve. Similarly, here Figure 6 was used along with its zoomed view, to find this transitional point of mechanical properties from glassy to rubbery behavior. The results are summarized in Table 4.

Table 4. Values of glass transition temperature measured by DMA.

Sample	T_g (°C)	Sample	T_g (°C)
1	119.466	9	119.268
2	119.363	10	119.682
3	120.578	11	119.131
4	119.156	12	118.986
5	119.608	13	118.918
6	119.178	14	119.771
7	119.711	15	117.984
8	118.667	16	118.514
		Unprocessed ABS filament	112.854

From the results in Table 4, it can be concluded that the FDM-processed parts have a higher glass transition temperature in comparison to unprocessed ABS; albeit the difference in the values among processed samples themselves is small. In other words, regardless of the combination of the assigned process parameters, FDM processed parts stay longer in the glassy region in comparison to unprocessed material. For instance, according to Table 3, the glass transition temperature has raised from 112.8 °C for unprocessed ABS filament to 120.5 °C under sample 3 (i.e., a 6% increase). Figure 7 shows the effect of each process parameter on the glass transition temperature of the FDM printed parts. In order to statistically rank the effect of process parameters on T_g, Lenth's method [32,33] was employed. This method is known as a powerful statistical tool to analyze costly experiments with single replicate factorial designs.

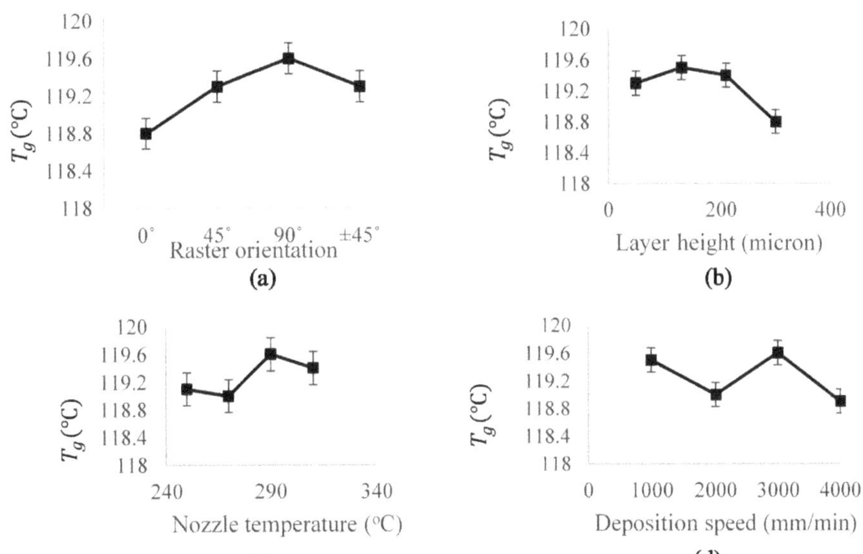

Figure 7. The variation of glass transition temperature of FDM processed ABS as a function of: (**a**) Raster orientation; (**b**) Layer height; (**c**) Nozzle temperature; (**d**) Deposition speed. Note that the presence of air gap (or inversely an overlap) between the printed roads would make a major difference in ensuing macro-level properties and non-linearities observed in the response above, and other mechanical properties as also reported by [19]. The curved shown are average values from all the performed 16 tests.

Assuming a design with m effects, considering both factors and interactions, denoted by $c_1, c_2, \ldots, c_m$, Lenth's method performs the effect analysis using a numerical value called the Pseudo Standard Error (PSE). For a 2^k design, m is equal to $2^k - 1$.

$$PSE = 1.5 * median(|c_j| : |c_j| < 2.5S_0 \quad (8)$$

where:

$$S_0 = 1.5median(|c_j|) \quad (9)$$

According to the method, when there is no sufficient information on repeats of a test, PSE is a reasonable measure to estimate the Standard Error [33]. The margin of Error (ME) is the final factor used to compare factor effects:

$$ME = t_{\frac{\alpha}{2},d}PSE \quad (10)$$

where $t_{\frac{\alpha}{2},d}$ is the t-distribution with the significance level of α and the degree of freedom of $d = \frac{m}{3}$. Given a specific factor, if the absolute value of the effect is greater than ME , that factor is considered effective (statistically significant). Table 5 represents the mean value of T_g under each studied factor, based on Figure 7. The Delta parameter is the difference between the maximum and minimum value of each data column and was considered as the factor effect. Lenth's parameter (ME threshold) has been calculated via Equations (8)–(10). Comparing the Delta values with each corresponding threshold, it can be concluded that the raster orientation is ranked above all other factors, followed by the feeding rate and layer height, and finally the nozzle temperature, to control glass transition temperature in FDM of ABS parts.

Table 5. Lenth's method of factor analysis for glass transition temperature; the values shown correspond to the average of response under each corresponding level of the process factors. The physical values of the levels were given in Table 2.

Level	Raster Orientation	Layer Height	Temperature	Deposition Speed
1	118.8	119.3	119.1	119.5
2	119.6	119.5	119	119
3	119.3	119.4	119.6	119.6
4	119.3	118.8	119.4	118.9
Delta	0.8	0.7	0.6	0.7
ME threshold	0.515	0.515	0.515	0.515

Figures 8 and 9 exemplify the relationship between process parameters and the reductions in dynamic mechanical moduli at two specific working temperatures, including 40 °C and 100 °C. As seen in these figures, the relationship between response modulus and process parameters is sizable, nonlinear and highly dependent on the working temperature.

Owing to the complex nature of the FDM process, the presence of combinative effects of the parameters necessitates the use of advanced modeling techniques to predict the material behavior. As addressed in Section 2.3, the Artificial Neural Network (ANN) is one of the known examples of powerful black-box modeling techniques to approximate such complex non-linear relationships between input and output parameters. In order to collect an adequate size of data to train the ANN model, the moduli values were taken at the working temperature steps of 5 °C from Figures 4 and 5, under each process condition. For each storage and loss modulus response, a separate multi-layer perceptron neural network architecture, including input layer, hidden layer and output layer was designed. After performing a series of training algorithms and testing various ANN architectures, the optimum model was selected for each modulus. Namely, to approximate the storage modulus, a 5-9-1 architecture was employed (five input neurons, nine hidden neurons and one output neuron), while a 5-7-1 architecture was used to simulate the loss modulus.

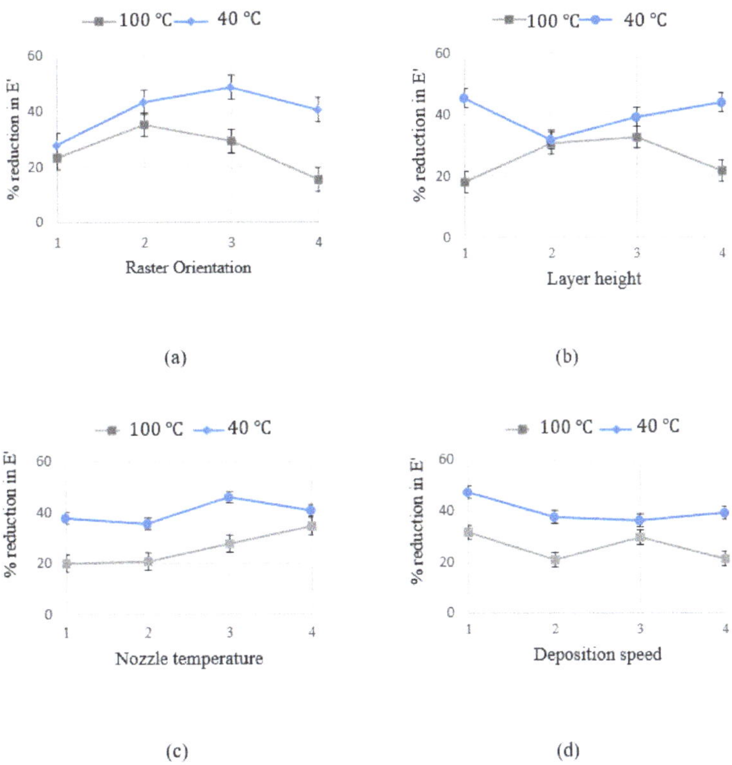

Figure 8. Percentage of reduction in storage modulus as a function of FDM process parameters: (**a**) Raster orientation; (**b**) Layer height; (**c**) Nozzle temperature; (**d**) Deposition speed. The blue and the grey lines represent the average reduction in the storage modulus at 40 °C and 100 °C respectively.

Both designed networks were trained using 60% of randomly selected data points via Levenberg- Marquardt algorithm. Then, the testing validation for each network was performed using a 20–20% portion of the remaining data. However, it must be mentioned that all data points, except one experimental run (number 9), were used to build the storage modulus and loss modulus networks. The latter experimental run was selected randomly not to be used in training, testing and validation steps. Instead, the data points corresponding to the response curves of test 9 were used to evaluate the robustness of the final developed model for each modulus.

The storage modulus ANN showed an acceptable performance represented by means of the coefficient of correlation (R). The R-values corresponding to training, validation, and testing, respectively, equaled 0.99317, 0-99256, and 0.99503. Figure 10 illustrates the network performance in detail. This designed network was then employed to predict the storage modulus under the fully unseen run #9. Figure 11 depicts the actual values versus the simulated values under this setup.

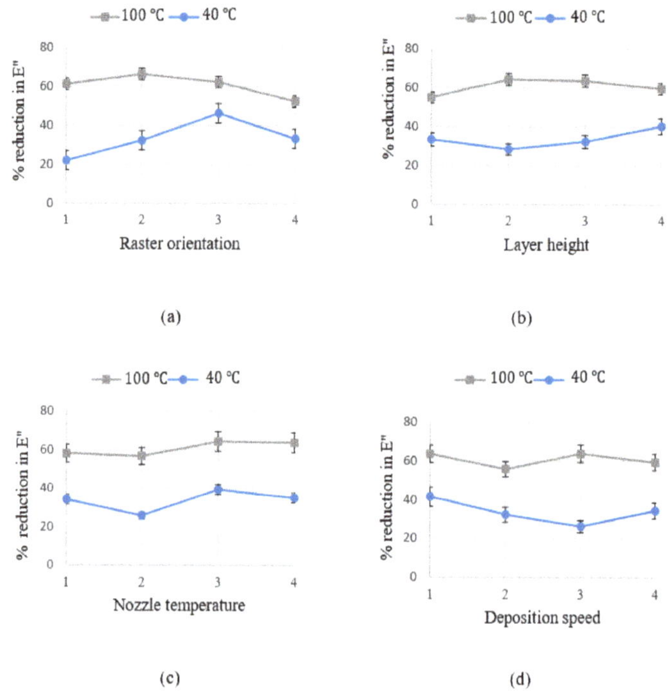

Figure 9. Percentage of reduction in loss modulus as a function of FDM process parameters: (**a**) Raster orientation; (**b**) Layer height; (**c**) Nozzle temperature; (**d**) Deposition speed. The blue and the grey lines represent the average reduction in the loss modulus at 40 °C and 100 °C respectively.

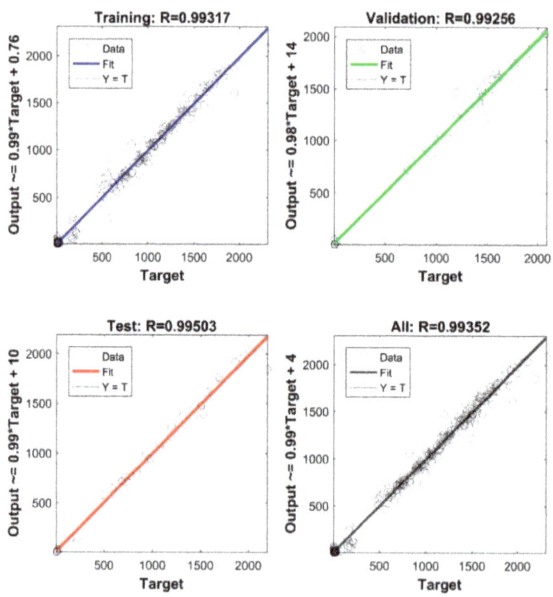

Figure 10. The performance of the developed 5-9-1 neural network to approximate the storage modulus of the 3D prints.

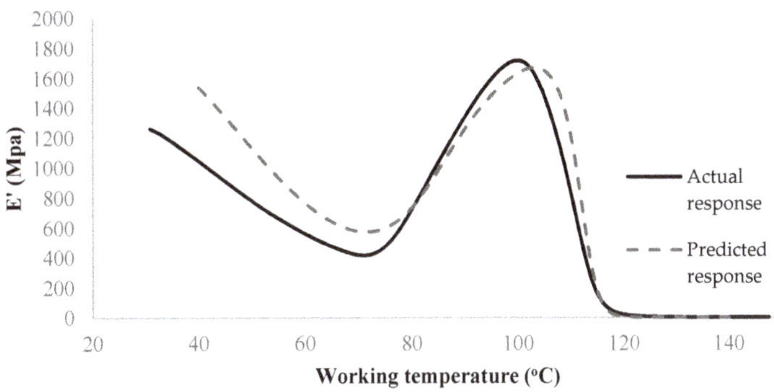

Figure 11. Comparison of the actual and predicted (virtual DMA test of) storage modulus, E', under run #9.

As shown in Figure 11, the developed network is usable to avoid actual experimentation and predict the storage modulus of the untested sample. The deviation between the experiment and model in the early stage of the DMA test can be linked to the large differences seen in the original data (Figure 4) in the same range when comparing different process conditions. Accordingly, as more DMA tests become available for training (here only 16 samples were used), the performance of ANN would have also been improved. Despite limited data, the overall validation R-score of the present model (i.e., considering all samples and all regions of response) is still >99% (Figure 10). Similarly, a 5-7-1 network architecture was selected to predict the loss modulus of FDM fabricated ABS samples. The network performance is illustrated in Figure 12. The final verification of the network performance was completed by predicting set-up condition #9 (Figure 13).

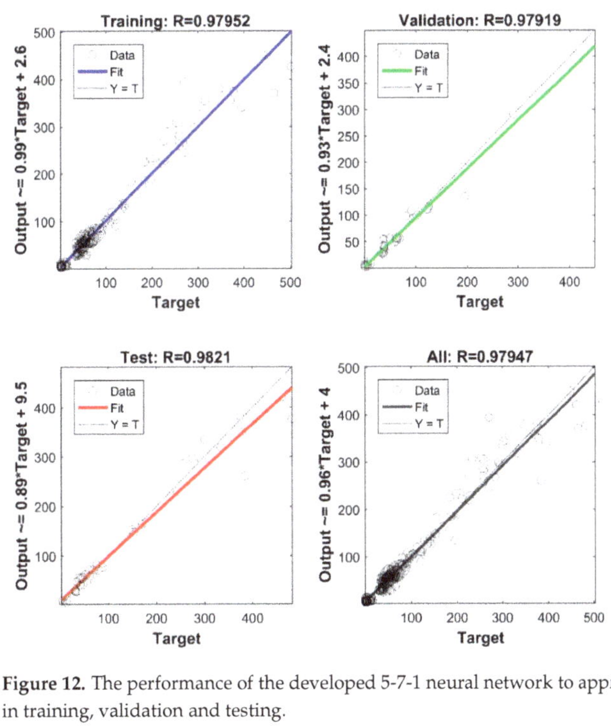

Figure 12. The performance of the developed 5-7-1 neural network to approximate the loss modulus in training, validation and testing.

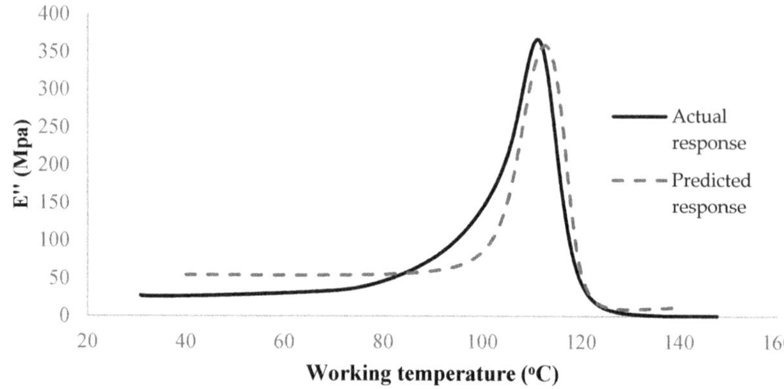

Figure 13. Comparison of the actual and predicted (virtual DMA test of) loss modulus, E'', under run #9.

Figure 14 represents an example of a simulated modulus as a function of process parameters, using the above AI model, at a given working temperature of 40 °C. According to Figure 14a, during printing of samples with 0° raster orientation, increasing the nozzle temperature decreases the general trend of storage modulus response. However, as shown in Figure 14b, increasing the nozzle temperature, first decreases and then increases the storage modulus. Moreover, both mentioned figures show that at higher values of the layer height, the storage modulus is higher. This trend is also visible for prints at 45° raster orientation, as depicted in Figure 14c. The effect of nozzle temperature on the general trend of storage modulus based on the layer height and deposition speed at raster orientation of 45° is highly nonlinear where the contours are crossing each other (indicating a high level of interaction of process parameters). Finally, as illustrated in Figure 14d, at the raster orientation of ±45°, regardless of the value assigned to the nozzle temperature, the higher the layer height, the higher the storage modulus. Factor interaction effects can also be clearly observed in the ±45° raster orientation case.

Noticing the above complex relationship between the process parameters and their interactions on the ensuing dynamic moduli of the prints, which in turn also vary at each given target working temperature, performing a nonlinear optimization of the process is deemed vital for FDM designers to ensure maximized performance of printed parts during use. Accordingly, the optimum set of process parameters would be the one at which the fabricated part shows the highest value of storage or loss moduli, i.e., as close as possible to the parent (unprocessed) material. Here, using the developed ANNs, such an optimum set of process parameters at each working temperature condition was obtained using the Particle Swarm Optimization (PSO) technique. As per Table 2, the layer height was allowed to change from 50 μm to 300 μm, the nozzle temperature from 250 °C to 310 °C and the deposition speed from 1000 mm/min to 4000 mm/min, at each level of raster orientation. The working temperature was varied between 40 °C and 140 °C. The PSO was performed in MATLAB (R2016b, MathWorks, Natick, MA, USA) using 100 particles in each swarm. In order to end the iterating process, the maximum number of iterations was set at 1000 and the function tolerance was defined to be 10^{-25}. The minimum and maximum inertia weights were chosen to be 0.1 and 1.1, respectively. It should be noted that prior to the optimization process, the data points were normalized to be between 0 and 1.

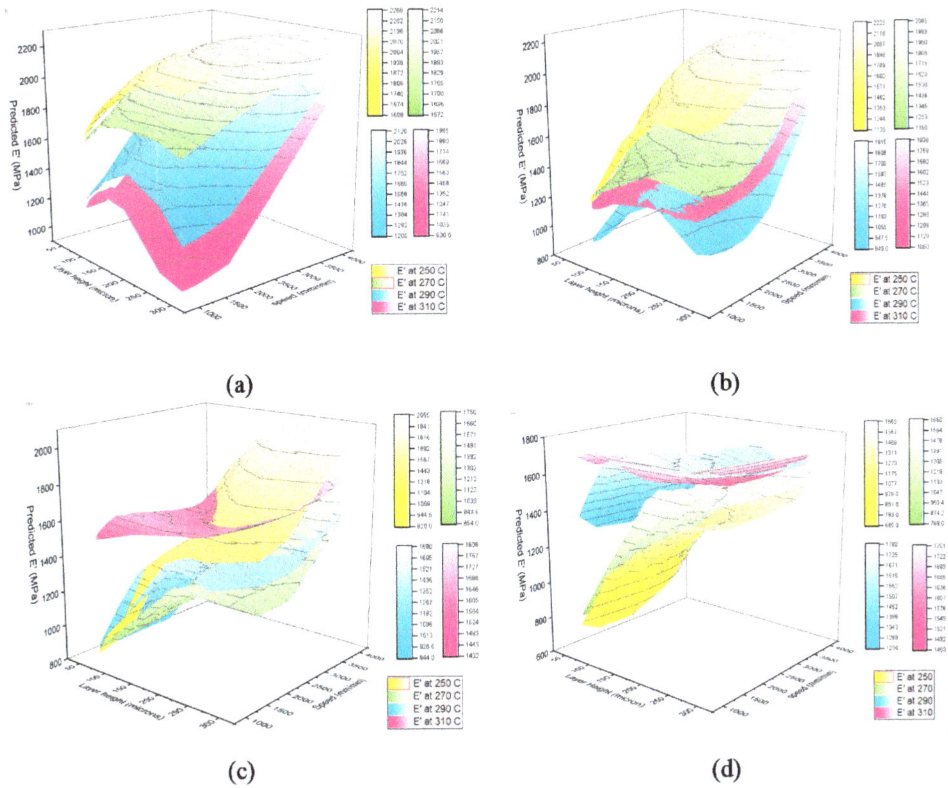

Figure 14. Simulated response of the storage modulus numeral network, at a working temperature of 40 °C and fixed nozzle temperatures (shown by different colors); (**a**) at 0° raster orientation, (**b**) at 90° raster orientation, (**c**) at 45° raster orientation, (**d**) at ±45° raster orientation.

The obtained optimum values of the process parameters along with the maximum achievable storage modulus are presented in Table A1. Table A2 similarly illustrates the optimization results for the loss modulus. The latter optimum values were reported for each working temperature. Results clearly show that there is no single optimum process condition that can optimize the printed part performance at all working temperature conditions. In practice, such look-up tables may be used by a designer, e.g., to choose the best FDM condition given the target operating temperature (application) of the 3D printed part. To better visualize the efficiency of the optimization, Figure 15 shows the optimized moduli of prints compared to the unprocessed material. When comparing Figure 15a with Figure 4, it can be noticed that the optimized prints exhibit closer behavior to that of the unprocessed material, albeit the storage modulus of the unprocessed material prior to T_g is still relatively higher than the prints. A similar effect of optimization, though to a lesser extent, can be noticed in regard to the loss moduli of the prints (compare Figures 5 and 15b). From Figure 15, it can be induced that the optimum moduli values for raster orientations 0°, 90°, and 45° are comparable, but at the orientation ±45° a distinct behavior is evidenced, namely providing a much lower storage modulus (pre-glass transition) but a higher storage modulus upon process optimization.

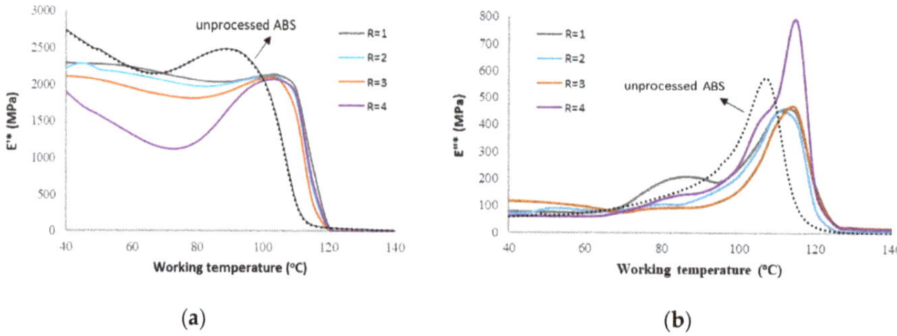

Figure 15. Comparison of the optimized (**a**) storage and (**b**) loss moduli of the prints, as compared to the parent (unprocessed) material, at different working temperatures and raster orientations (R). 4 levels provided for R are corresponding to the levels presented in Table 2.

4. Conclusions

In this paper, first, the variation of the viscoelastic response of 3D printed ABS samples in a wide range of working temperatures (the room condition to 140 °C) was studied as a function of four FDM process parameters (raster orientation, layer height, nozzle temperature, and deposition speed). The experimental layout was designed via a Taguchi orthogonal array in order to minimize the size of data required for subsequent ANN training. Dependency of the viscoelastic properties of the printed samples on the process control factors was shown via Lenth's method. Consequently, the optimum parameter values corresponding to each working condition were obtained using the ANN models (for storage and loss moduli), integrated with the PSO algorithm. The optimum values were reported for various working temperatures for both moduli. Results clearly showed that there is no single optimum process condition that can optimize the printed part performance at all working temperature conditions. However, specific conclusions could be drawn from the observations, as follows.

- The FDM process condition could directly affect the maximum allowable working temperature (represented by glass transition temperature) for the 3D printed thermoplastic.
- Based on Lenth's statistical analysis, among the process parameters, raster orientation was the most effective factor to increase the glass transition temperature of the 3D printed parts. Subsequently, the deposition speed and the layer height were ranked second, followed by the nozzle temperature.
- Distinct trends between viscoelastic responses of unprocessed and processed ABS filaments under various process conditions pointed to the fact that all FDM process conditions significantly (on average 40%) lowered the magnitude of viscoelastic moduli regardless of a specific combination of process parameters, which is also in agreement with the earlier study [5]. This effect is deemed critical for designers to consider for the reliable application of 3D printed parts, especially at high temperatures.
- Although it was shown that there are distinct trends between the behavior of processed and unprocessed ABS samples, the exact change in the moduli was highly dependent on the working temperature, at which the part viscoelastic properties were measured. For instance, at a working temperature of 100 °C, there was an average reduction of 25% in storage modulus when compared to the unprocessed sample. On the other hand, this reduction at a 40 °C working temperature was about 33.5%. The reduction increased drastically and reached as high as 60.7% at high working temperatures >100 °C.
- It was validated that the developed neural network architectures are capable of predicting the entire DMA curve of 3D printed parts, including for samples that were fully unseen to the original model. Using such networks, optimum values of process parameters can be obtained via global search methods such as particle swarm

optimization for each given target working temperature (Tables A1 and A2). The optimized prints indicated a closer behavior to that of the parent material.

- The optimized prints with orientation ±45° showed clearly a distinct behavior compared to the 0°, 90°, and 45° orientations.

Further study may be worthwhile to test the presented AI modeling approach for other thermoplastics, and possibly for improving the permeance of the predictions by employing and comparing other high-fidelity machine learning methods. Furthermore, the observed variation of dynamic mechanical properties can be further supported with, e.g., a molecular weight analysis of samples.

Author Contributions: R.V.: Conceptualization, Methodology, Software, Formal analysis, Writing—original draft. H.R.Z.R. and A.S.M.: Investigation, Resources. A.S.M.: Supervision, Writing—review & editing. All authors have read and agreed to the published version of the manuscript.

Funding: This research received no external funding.

Institutional Review Board Statement: Not applicable.

Informed Consent Statement: Not applicable.

Data Availability Statement: Data is contained within the article.

Acknowledgments: The authors gratefully acknowledge the contributions of K. Yesilcimen, B. Crawford and M. Takaffoli for their assistance in different stages of this work.

Conflicts of Interest: The authors declare no conflict of interest.

Appendix A. Optimization Results Raw Data

Table A1. The optimum values (look-up table) of the process parameters using PSO on E′ at each raster orientation level and under each working temperature.

R	LH* (μm)	NT* (°C)	DS* ($\frac{mm}{min}$)	WT (°C)	E′* (MPa)	R	LH* (μm)	NT* (°C)	DS* ($\frac{mm}{min}$)	WT (°C)	E′* (MPa)
1	300	310	1275	40	2292	2	270	250	3374	40	2227
1	300	310	1281	45	2282	2	279	250	3357	45	2285
1	300	310	1295	50	2277	2	286	250	3338	50	2199
1	50	310	1464	55	2263	2	288	250	3320	55	2172
1	50	310	1527	60	2234	2	292	250	3314	60	2136
1	50	310	1567	65	2193	2	299	250	3318	65	2093
1	50	310	1575	70	2141	2	300	250	3332	70	2046
1	50	310	1548	75	2099	2	300	250	3344	75	2003
1	50	310	1484	80	2058	2	300	250	3341	80	1972
1	50	310	1378	85	2033	2	300	250	3314	85	1973
1	76	310	1211	90	2039	2	300	250	3268	90	2011
1	87	310	1000	95	2080	2	300	250	3229	95	2066
1	94	310	1000	100	2125	2	300	252	3263	100	2105
1	102	310	1000	105	2128	2	300	253	3081	105	2088
1	118	310	1000	110	1945	2	300	253	2841	110	1817
1	50	268	3956	115	922	2	300	251	2564	115	607
1	50	269	3941	120	71	2	300	252	2601	120	37
1	175	310	1237	125	6	2	300	250	2894	125	4
1	300	297	1000	130	2	2	300	250	3215	130	2
1	300	296	1000	135	2	2	300	250	3452	135	2
1	300	298	1048	140	2	2	300	250	3538	140	2
3	300	310	4000	40	2111	4	300	310	2023	40	1887
3	235	250	4000	45	2092	4	177	296	1000	45	1692
3	242	250	4000	50	2056	4	194	297	1000	50	1573
3	251	250	4000	55	2004	4	207	297	1000	55	1443
3	265	250	4000	60	1944	4	220	298	1000	60	1312
3	287	250	4000	65	1886	4	239	297	1000	65	1196
3	300	250	4000	70	1848	4	300	292	1000	70	1131
3	300	250	4000	75	1817	4	300	292	1000	75	1130
3	300	250	4000	80	1810	4	300	293	1000	80	1222

Table A1. *Cont.*

R	LH* (μm)	NT* (°C)	DS* $(\frac{mm}{min})$	WT (°C)	E′* (MPa)	R	LH* (μm)	NT* (°C)	DS* $(\frac{mm}{min})$	WT (°C)	E′* (MPa)
3	300	250	4000	85	1842	4	300	294	1000	85	1423
3	300	250	4000	90	1910	4	167	310	1000	90	1696
3	300	250	3774	95	2008	4	167	310	1000	95	1931
3	300	250	3639	100	2086	4	173	310	1000	100	2051
3	300	250	3577	105	2056	4	190	310	1000	105	2060
3	300	250	3548	110	1620	4	229	310	1000	110	1873
3	300	310	1000	115	414	4	223	310	1000	115	704
3	300	310	1000	120	22	4	300	310	1000	120	46
3	300	250	3861	125	3	4	300	310	1000	125	4
3	300	250	4000	130	2	4	300	250	4000	130	2
3	300	250	4000	135	2	4	300	250	4000	135	2
3	300	250	4000	140	2	4	162	250	4000	140	2

* LH*, NT*, DS* stand for the optimum Layer height, Nozzle temperature, Deposition speed and given Raster orientation R and Working temperature WT. E′* is the corresponding optimum storage modulus.

Table A2. The optimum values (look-up table) of the process parameters using PSO on E″ at each raster orientation level and under each working temperature.

R	LH* (μm)	NT* (°C)	DS* $(\frac{mm}{min})$	WT (°C)	E″* (MPa)	R	LH* (μm)	NT* (°C)	DS* $(\frac{mm}{min})$	WT (°C)	E″* (MPa)
1	50	310	4000	40	79.8	2	54	277	4000	40	74.1
1	50	310	4000	45	78.3	2	50	276	4000	45	73.9
1	50	310	4000	50	76.7	2	300	250	1000	50	90.5
1	50	310	4000	55	75.3	2	300	250	1000	55	91.2
1	50	310	4000	60	73.4	2	278	250	1000	60	84.4
1	50	273	1000	65	80	2	300	310	4000	65	80.2
1	300	250	1000	70	99.3	2	300	310	4000	70	85.6
1	300	250	1000	75	146.7	2	50	250	4000	75	98.2
1	300	250	1000	80	188.8	2	50	250	4000	80	107.6
1	75	250	1000	85	208.7	2	50	250	4000	85	107.4
1	300	250	1000	90	204.5	2	300	296	4000	90	129.6
1	300	250	1000	95	189.6	2	300	297	4000	95	163.1
1	50	250	3548	100	239.6	2	300	298	4000	100	213.8
1	98	250	3967	105	334	2	300	298	4000	105	313.5
1	110	250	4000	110	448.5	2	50	279	4000	110	448.9
1	115	254	4000	115	442.2	2	50	282	4000	115	407.6
1	284	278	4000	120	176.4	2	300	310	4000	120	89.4
1	300	283	4000	125	34	2	300	310	3860	125	13.6
1	300	285	4000	130	17.2	2	300	271	1000	130	7.7
1	186	280	4000	135	14.7	2	50	310	1586	135	8.2
1	148	281	4000	140	14.4	2	50	310	1895	140	9.9
3	50	250	4000	40	117.7	4	300	310	1000	40	63.7
3	50	250	4000	45	114.9	4	300	310	1000	45	62.6
3	50	250	4000	50	111	4	133	250	1000	50	61.9
3	50	250	4000	55	105.2	4	148	250	1000	55	61.6
3	50	250	4000	60	97	4	170	250	1000	60	60.8
3	50	250	1000	65	86.2	4	170	310	4000	65	64.7
3	50	250	1000	70	76	4	150	310	4000	70	79.6
3	50	250	1000	75	86.1	4	155	250	4000	75	99.3
3	50	250	1000	80	92.4	4	140	250	4000	80	125
3	50	250	1000	85	93.9	4	134	250	4000	85	141.8
3	300	250	3364	90	96.9	4	131	250	4000	90	149.9
3	258	310	1000	95	116.7	4	50	310	1000	95	181.2
3	255	310	1000	100	158.2	4	50	310	1000	100	252.5
3	251	310	1000	105	251.9	4	50	250	1000	105	408.6
3	102	250	4000	110	412.5	4	50	250	1000	110	508.5
3	67	250	4000	115	458.6	4	300	250	2278.8	115	783.6
3	50	250	4000	120	164.1	4	50	250	1000	120	155.1
3	50	250	4000	125	37.3	4	50	250	1000	125	34.2
3	50	250	4000	130	21.6	4	50	250	1000	130	19.5
3	50	250	4000	135	18.2	4	50	250	1000	135	15.8
3	50	250	3911	140	16.7	4	50	250	1000	140	14.2

* LH*, NT*, DS* stand for the optimum Layer height, Nozzle temperature, Deposition speed and given Raster orientation R and Working temperature WT. E″* is the corresponding optimum loss modulus.

References

1. Huang, Y.; Leu, M.C.; Mazumder, J.; Donmez, A. Additive manufacturing: Current state, future potential, gaps and needs, and recommendations. *J. Manuf. Sci. Eng.* **2015**, *137*, 014001. [CrossRef]
2. Bagsik, A.; Schöppner, V.; Klemp, E. FDM part quality manufactured with ultem* 9085. In Proceedings of the 14th International Scientific Conference on Polymeric Materials, Halle (Saale), Germany, 15–17 September 2010.
3. Doubrovski, Z.; Verlinden, J.C.; Geraedts, J.M. Optimal design for additive manufacturing: Opportunities and challenges. In Proceedings of the ASME 2011 International Design Engineering Technical Conferences and Computers and Information in Engineering Conference, Washington, DC, USA, 28–31 August 2011; pp. 635–646.
4. Gibson, I.; Rosen, D.W.; Stucker, B. *Additive Manufacturing Technologies*; Springer: Berlin/Heidelberg, Germany, 2010; p. 238.
5. Arivazhagan, A.; Masood, S. Dynamic mechanical properties of ABS material processed by fused deposition modelling. *Int. J. Eng. Res. Appl.* **2012**, *2*, 2009–2014.
6. Weeren, R.V.; Agarwala, M.; Jamalabad, V.R.; Bandyopadhyay, A.; Vaidyanathan, R.; Langrana, N.; Safari, A.; Whalen, P.; Danforth, S.C.; Ballard, C. Quality of parts processed by fused deposition. In Proceedings of the Solid Freeform Fabrication Symposium, Austin, TX, USA, 7–9 August 1995; pp. 314–321.
7. Bossett, E.; Rivera, L.; Qiu, D.; McCuiston, R.; Langrana, N.; Rangarajan, S.; Venkataraman, N.; Danforth, S.; Safari, A. Real time video microscopy for the fused deposition method. In Proceedings of the Solid Freeform Fabrication Symposium, Austin, TX, USA, 11–13 August 1998; pp. 113–120.
8. Gray, R.W.; Baird, D.G.; Bøhn, J.H. Effects of processing conditions on short TLCP fiber reinforced FDM parts. *Rapid Prototyp. J.* **1998**, *4*, 14–25. [CrossRef]
9. Rodríguez, J.F.; Thomas, J.P.; Renaud, J.E. Mechanical behavior of acrylonitrile butadiene styrene (ABS) fused deposition materials. Experimental investigation. *Rapid Prototyp. J.* **2001**, *7*, 148–158. [CrossRef]
10. Es-Said, O.S.; Foyos, J.; Noorani, R.; Mendelson, M.; Marloth, R.; Pregger, B.A. Effect of layer orientation on mechanical properties of rapid prototyped samples. *Mater. Manuf. Process.* **2000**, *15*, 107–122. [CrossRef]
11. Ahn, S.-H.; Montero, M.; Odell, D.; Roundy, S.; Wright, P.K. Anisotropic material properties of fused deposition modeling ABS. *Rapid Prototyp. J.* **2002**, *8*, 248–257. [CrossRef]
12. Lee, B.; Abdullah, J.; Khan, Z. Optimization of rapid prototyping parameters for production of flexible ABS object. *J. Mater. Process. Technol.* **2005**, *169*, 54–61. [CrossRef]
13. Sood, A.K.; Ohdar, R.K.; Mahapatra, S.S. Parametric appraisal of fused deposition modelling process using the grey Taguchi method. *Proc. Inst. Mech. Eng. Part B J. Eng. Manuf.* **2010**, *224*, 135–145. [CrossRef]
14. Onwubolu, G.C.; Rayegani, F. Characterization and Optimization of Mechanical Properties of ABS Parts Manufactured by the Fused Deposition Modelling Process. *Int. J. Manuf. Eng.* **2014**, *2014*, 598531. [CrossRef]
15. Liu, X.; Zhang, M.; Li, S.; Si, L.; Peng, J.; Hu, Y. Mechanical property parametric appraisal of fused deposition modeling parts based on the gray Taguchi method. *Int. J. Adv. Manuf. Technol.* **2017**, *89*, 2387–2397. [CrossRef]
16. Mohamed, O.A.; Masood, S.; Bhowmik, J.L. Optimization of fused deposition modeling process parameters: A review of current research and future prospects. *Adv. Manuf.* **2015**, *3*, 42–53. [CrossRef]
17. Domingo-Espin, M.; Borros, S.; Agullo, N.; Garcia-Granada, A.A.; Reyes, G. Influence of Building Parameters on the Dynamic Mechanical Properties of Polycarbonate Fused Deposition Modeling Parts. *3D Print. Addit. Manuf.* **2014**, *1*, 70–77. [CrossRef]
18. Ang, K.C.; Leong, K.F.; Chua, C.K.; Chandrasekaran, M. Investigation of the mechanical properties and porosity relationships in fused deposition modelling-fabricated porous structures. *Rapid Prototyp. J.* **2006**, *12*, 100–105.
19. Mohamed, O.A.; Masood, S.H.; Bhowmik, J.L.; Nikzad, M.; Azadmanjiri, J. Effect of Process Parameters on Dynamic Mechanical Performance of FDM PC/ABS Printed Parts Through Design of Experiment. *J. Mater. Eng. Perform.* **2016**, *25*, 2922–2935. [CrossRef]
20. Kulich, D.M.; Gaggar, S.K.; Lowry, V.; Stepien, R. Acrylonitrile–butadiene–styrene (ABS) polymers. In *Kirk-Othmer Encyclopedia of Chemical Technology*; Wiley: NewYork, NY, USA, 2004.
21. Menard, K.P. *Dynamic Mechanical Analysis: A Practical Introduction*; CRC Press: Boca Raton, FL, USA, 2008.
22. Ferry, J.D. *Viscoelastic Properties of Polymers*; John Wiley & Sons: Hoboken, NJ, USA, 1980.
23. Murayama, T. *Dynamic Mechanical Analysis of Polymeric Material*; Elsevier Scientific Pub. Co.: New York, NY, USA, 1978.
24. Lek, S.; Delacoste, M.; Baran, P.; Dimopoulos, I.; Lauga, J.; Aulagnier, S. Application of neural networks to modelling nonlinear relationships in ecology. *Ecol. Model.* **1996**, *90*, 39–52. [CrossRef]
25. Rafiq, M.; Bugmann, G.; Easterbrook, D. Neural network design for engineering applications. *Comput. Struct.* **2001**, *79*, 1541–1552. [CrossRef]
26. Kartalopoulos, S.V.; Kartakapoulos, S.V. *Understanding Neural Networks and Fuzzy Logic: Basic Concepts and Applications*; Wiley-IEEE Press: New York, NY, USA, 1997.
27. Hagan, M.; Demuth, H.; Beale, M. Neural Network Design. *Boston Mass. PWS* **1996**, *2*, 734.
28. Qi, M.; Zhang, G.P. An investigation of model selection criteria for neural network time series forecasting. *Eur. J. Oper. Res.* **2001**, *132*, 666–680. [CrossRef]
29. Eberhart, R.; Kennedy, J. A new optimizer using particle swarm theory. In *Micro Machine and Human Science, 1995. MHS'95, Proceedings of the Sixth International Symposium on, Nagoya, Japan, 4–6 October 1995*; IEEE: Piscataway, NJ, USA, 1995; pp. 39–43.
30. Menard, K.P.; Menard, N.R. Dynamic mechanical analysis in the analysis of polymers and rubbers. *Encycl. Polym. Sci. Technol.* **2002**, 1–33. [CrossRef]

31. Roy, R.K. *Design of Experiments Using the Taguchi Approach: 16 Steps to Product and Process Improvement*; John Wiley & Sons: Hoboken, NJ, USA, 2001.
32. Lenth, R.V. Lenth's method for the analysis of unreplicated experiments. *Encycl. Stat. Qual. Reliab.* **2008**. [CrossRef]
33. Komeili, M.; Milani, A. The effect of meso-level uncertainties on the mechanical response of woven fabric composites under axial loading. *Comput. Struct.* **2012**, *90*, 163–171. [CrossRef]

materials

MDPI

Article

On the Numerical Modeling of Flax/PLA Bumper Beams

Liu Jiao-Wang, José A. Loya and Carlos Santiuste *

Department of Continuum Mechanics and Structural Analysis, Universidad Carlos III de Madrid, Avenida de la Universidad 30, 28911 Leganés, Spain
* Correspondence: csantius@ing.uc3m.es; Tel.: +34-91-624-99-20

Abstract: Significant progress has been made in green composites developing fully biodegradable composites made of microbially degradable polymers reinforced with natural fibers. However, an improvement in the development of numerical models to predict the damage of green composites is necessary to extend their use in industrial applications of structural responsibility. This paper is focused on developing a numerical model that can predict the failure modes of four types of bumper beams made of flax/PLA green composites with different cross sections. The predictions regarding energy absorption, contact force history, and extension of delamination were compared with experimental results to validate the FEM model, and both results revealed a good agreement. Finally, the FEM model was used to analyze the failure modes of the bumper beams as a function of the impact energy and cross-section roundness. The impact energy threshold defined as the maximum absorbed-energy capability of the beam match with the impact energy that produces delaminations extended through all the cross sections. Experimental and numerical results revealed that the threshold energy, where the maximum energy-absorption capability is reached, for Type A is over 60 J; for Type B and C is around 60 J; and for Type D is at 50 J. Since delamination is concentrated at the cross-section corners, the threshold energy decreases with the cross-section roundness because the higher the roundness ratio, the wider the delamination extension.

Keywords: green composite; bumper beam; numerical modeling; energy absorption; natural fibers

Citation: Jiao-Wang, L.; Loya, J.A.; Santiuste, C. On the Numerical Modeling of Flax/PLA Bumper Beams. *Materials* **2022**, *15*, 5480. https://doi.org/10.3390/ma15165480

Academic Editor: Gaetano Giunta

Received: 14 July 2022
Accepted: 3 August 2022
Published: 9 August 2022

1. Introduction

Plastics and composites are ideal materials in the automotive industry for interior, exterior, and structural applications [1]. Composites present advantages in terms of specific mechanical properties and they are competitive in manufacturing cost; however, composites are difficult to recycle. The creation of the fully biodegradable composite material, also called "green composites", can counter this disadvantage of composites with respect to metals. Green composites are essentially composites made of natural fibers and microbially degradable polymers. Their mechanical properties are quite different from those of carbon- or glass-fiber-reinforced polymer composite materials. Elastic nonlinearity and strain-rate dependency features make this green material more difficult to analyze [2]. Therefore, more comprehensive research on automotive components made of fully biodegradable composite becomes essential.

Among the vehicle protective components, bumper beams remarkably affect the structural energy-absorption capacity to protect the occupant and pedestrian under low-speed impacts [3]. Bumper beams can be one of the most suitable car components to be manufactured using natural fiber-reinforced polymers (NFRP). There are many types of green composites and NFRPs that can be applied to the automotive industry. Some of the most popular green composites studied in scientific literatures are reinforced with, e.g., flax, hemp, jute, and sisal fiber, and sustainable polymers such as polylactic acid (PLA), polyhydroxybutyrate (PHB), and poly(butylene succinate) (PBS) are typically used as the matrix [4]. Some examples of natural fiber-reinforced composites are PP reinforced with hemp, flax, and kenaf fibers [5], Supersap CLR matrix reinforced with sisal fibers [6], PLA/flax composites [7], PHB/piassava composites [8], and PLA/kenaf composites [9].

Experimental methods can be used to study new materials in the automotive industry. However, simulation tools based on the finite element method (FEM) as Abaqus, Ansys, and LS-DYNA have emerged as computer-aided engineering tools due to the elevated cost and time consumption of the experiments.

The usefulness of the FEM models to analyze the mechanical behavior of components made from biocomposites is unquestionable. In the past few years, the mechanical engineering field has been inseparable from such effective and powerful software. The theoretical principles behind it allow to for the development of new theories or criteria and ensure the accuracy of the simulation results, as Jalón et al. [10] did in their study about the impact behavior of flax fiber-reinforced polylactic acid (PLA) composite. They created the FEM model, simulated the impact scenario, and validated the constitutive model developed for natural fiber-based composites.

Moreover, FEM has also been further applied by many researchers in the new composite materials and automobiles field, in order to predict the failure mechanism, deformation behavior, and mechanical characteristics of automobile bumper beams made of natural fiber composites more effectively under various impact experiments. Hassan et al. [11,12] conducted their work on the low-speed impact study of an oil palm empty fruit bunch (OPEFB) fiber-reinforced composite bumper beam compared with a conventional aluminum one. Their numerical simulations (LS-DYNA) showed that the deformation was more severe in the natural fiber composite bumper than in the aluminum one, and 56% lighter. Therefore, they concluded that this biocomposite has the potential to replace aluminum in bumper beam applications. Arbintarson et al. [13] used an FEM model based on Solidworks software to study the behavior of a vehicle front bumper made of pineapple leaf fiber-reinforced composite in the front collision to a wall. They varied the impact velocity and concluded that the composite bumper could bear collision speeds up to 70 km/h. Ramasubbu and Madasamy [14] designed a sisal/kenaf fiber-reinforced-composite car bumper with a FEM model developed in Solidworks to simulate the drop-weight test. Their analysis of the NFRP bumper beam mechanical behavior was similar to a polycarbonate bumper. Rubio-López et al. [15] created a numerical FEM model in Abaqus/Explicit to predict the crashworthiness of all-cellulose composite (ACC) plates under low-velocity impact. After the simulation, the damage modes of the plate were given, and the energy threshold that led to its complete failure was obtained.

There are other studies in which researchers analyze the use of natural fiber-reinforced composites to enhance the chemical, functional, and mechanical performance of traditional composites such as those reinforced with glass fibers. Ghani et al. [16] studied the combination of glass/jute fiber-reinforced PBS composite material and the influence of the stacking sequence on the mechanical, thermal, and water-resistance properties of this composite. They found that the hybrid composite showed better mechanical performance than the pure composite, and they mentioned that the glass fiber layer on the surface could improve the impact performance of the combination. Murugu et al. [17] studied the quasistatic behavior of hybridized glass-/hemp-fiber-reinforced composite bumper using an FEM model developed in the Ansys. They concluded that this hybrid composite showed prominent properties to replace the conventional glass-fiber-reinforced composites.

Fiber-reinforced composites present distinct failure mechanisms to traditional metallic materials due to their anisotropy and heterogeneity. The different failure criteria in the literature consider three main failure modes to predict the failure of a fiber-reinforced composite: matrix cracking, fiber breakage, and delamination. However, many works have demonstrated that flax-fiber-reinforced composites are failed mainly by fiber breakage. Qian et al. [18] claimed that when the processing temperature exceeds a critical point, flax-fiber-reinforced composite fails during impacts, and the primary failure mode appears to be the fiber breakage of the material. Flax-fiber-reinforced composite thermally treated below 180 °C presents better fatigue and impact resistance. However, when the temperature exceeds this value, the mechanical properties of the composite are degraded, and the failure morphology will be changed. The work carried out by Díaz-Álvarez et al. [19] also reported that the fiber breakage occurs in the flax-fiber composite when impact energies are high, and the after-impact behavior of the structure presents promising results due to the absence of the delamination phenomenon. In addition, they also demonstrated that for low impact energies, the material fails due to matrix cracking.

Numerous studies show that the bumper beam geometry remarkably influences the structure impact behavior. Significantly, it depends on the structure curvature, cross-section geometry, and fiber orientation. For instance, in a previous work [20], four types of bumper beams that differed in the roundness of the cross section were studied under low-velocity impact test and bending after impact test. The results showed that the cross-section selection depends on whether the purpose is to achieve a better impact or postimpact performance. Additionally, the biocomposite-made bumper beams all show low peak force, which means that using green material can effectively reduce the acceleration during impact.

The main novelty of the present study is the development of an effective FEM model for predicting the damage behavior of bumper beams made entirely with fully-biodegradable flax-fiber-reinforced PLA composite. Four cross-sections were modeled to evaluate the prediction capacity of the model. First, tensile and peeling tests were carried out to obtain the composite mechanical properties and fracture toughness. Then, the properties were introduced in a numerical model developed to reproduce the low-velocity impact tests. Subsequently, the model was validated by comparing numerical predictions and experimental tests published in the previous work [20] in terms of absorbed energy and contact force history. Moreover, some specimens were inspected using X-ray scan tomography to validate the accuracy of the delamination extension predicted by the FEM model. Finally, the model results are used to analyze the different failure modes and energy absorption mechanisms as a function of the impact energy and cross-section roundness.

2. Methodology

The finite element method was used to create numerical models in Abaqus/Explicit for the mechanical response prediction of flax/PLA bumper beams under low-velocity impact tests. The experimental results needed for the numerical model calibration were published in previous work [20]. Four models were created to reproduce the geometry of the corresponding different bumper beams (named Type A–D). The methodology was divided into four steps:

1. Definition of the numerical model. It includes an intralaminar failure model, which considers nonlinear viscoplastic behavior and the influence of strain rate, and an interlaminar failure model based on cohesive interaction.
2. Most of the material parameters were obtained through experimental characterization tests. However, some parameters included in the cohesive interaction are difficult to obtain experimentally. Thus, they were fitted using the bumper beam Type A results.
3. Once the cohesive parameters were fitted using only the experimental results of the bumper beam Type A, the experimental results of the other bumper beam types (B, C, and D) were used to validate the prediction accuracy of the numerical model.
4. Finally, the validated numerical models are used to analyze the failure modes and the influence of the cross section on the absorbed energy.

2.1. Numerical Model

Before the impact test, every specimen's weight and dimension was measured for the modeling in Abaqus/Explicit. The density of the flax/PLA composite was equal to 1.14 ton/m^3. According to the previous studies of flax/PLA, this biocomposite presents nonlinear elastic viscoplastic behavior. Therefore, the traditional mechanical behavior model for composites considering linear elastic behavior up to failure cannot be used to reproduce the behavior of green composites. In the present model, two regions were considered in the stress–strain curve: elastic and viscoplastic. As shown in Figure 1, the first elastic stage is assimilated to linear elastic with mechanical properties of E = 5 GPa and Poisson ratio = 0.3; the second region is viscoplastic and was defined using experimental strain-rate-dependent data. Tensile tests were carried out in the laboratory at different velocities to characterize the influence of strain rate. A constitutive viscoplastic model was defined in a previous work [21], which was implemented in an FEM model to reproduce the mechanical behavior of flax/PLA plates either under low-velocity impacts [10] or machining operations scenarios [22,23].

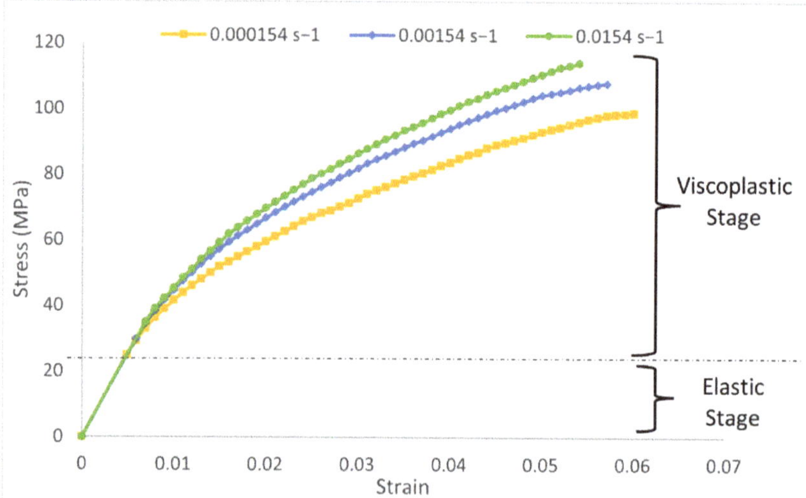

Figure 1. Stress-strain curves of flax/PLA biocomposite obtained in tensile tests at different strain rates.

Finally, a VUSDFLD user subroutine was used to define an ultimate strain criterion to remove elements that reach the ultimate strain. Figure 2 shows the function of the ultimate strain as a function of the strain rate implemented in this subroutine. For strain rates greater than 0.0154 s^{-1}, the ultimate strain was fixed as 0.056; otherwise, the ultimate strain obeys a linear function. These values are based on the results obtained in the experimental tensile tests; more detail can be found in [21]. Therefore, the subroutine calculates the ultimate strain as a function of the strain rate and deletes elements that exceed the ultimate strain.

The model developed in Abaqus/Explicit is composed of three solids to mimic the actual impact scenario: striker, supports, and bumper beam, as shown in Figure 3, to reproduce the low-velocity impact tests performed [20]. The experimental tests were performed in three-point bending configuration set in a drop-weight tower to analyze the impact behavior of four geometrically different flax/PLA bumper beams. An INSTRON-CEAST Fractovist 6785 drop-weight tower (High Wycombe, Buckinghamshire, UK) was used; the instrumented striker that moved along the vertical guide rail had customized features: Charpy nose, the weight of 5.93 kg and diameter of 20 mm.

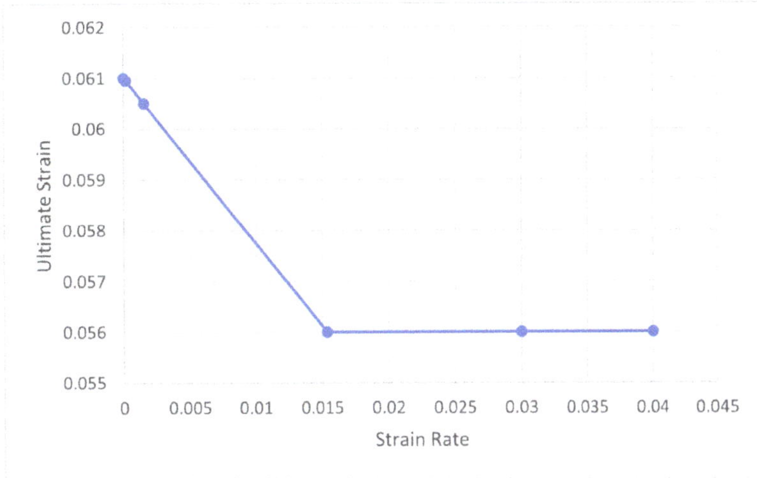

Figure 2. The ultimate strain as a function of the strain rate. The element-deletion criterion is implemented in the VUSDFLD subroutine.

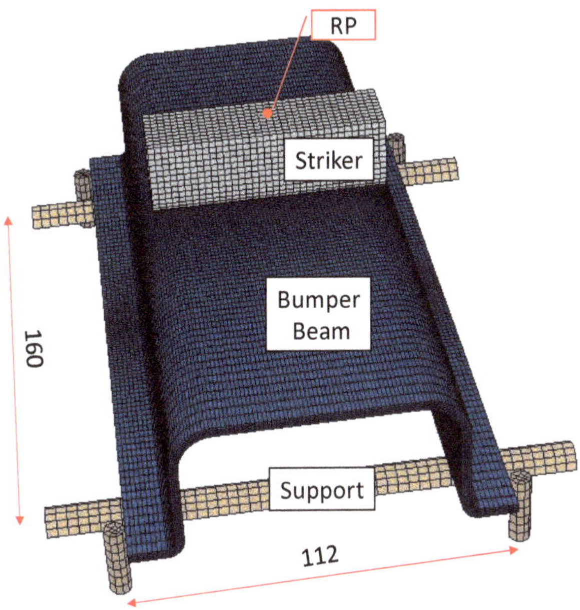

Figure 3. Numerical model of the bumper beam under low-velocity impact test. The overall setting and dimension.

The striker was modeled as a 3D discrete rigid shell, element type R3D4, U-form, with overall length of 70 mm, overall height 25 mm, and radius of the bottom surface of 10 mm. It has a point mass of 5.93 Kg applied in the up-center point of the structure, called the reference point (RP), see Figure 3), where a multipoint constraint (MPC) was created to tie all the nodes of the striker to the RP, constraining the motion of the slave nodes to the motion of a single point. This simplification was assumed to reduce the computational cost because the stiffness of the steel striker was much higher than the bumper beam. An additional advantage of this simplification is that the striker results (velocity, acceleration, reaction forces ...) can be registered in a single node. All the degrees of freedom of the striker nodes were restricted except the vertical displacement. Moreover, a predefined velocity field was defined with values between 1.006 m/s to 4.859 m/s (equivalent to 3–70 J of kinetic energy).

The supports are composed of two cylinders that hold the bumper beam and four pins that limit its lateral movement. Since no permanent deformations after impact were found, the supportive structures were modeled as homogeneous solids with linear elastic properties of ordinary steels (E = 210 GPa and Poisson Ratio = 0.3), and the element type is C3D8R. In addition, their boundary conditions were *Encastre*, see Figure 3.

The actual beam is modeled as a four-layer 3D deformable body, including cohesive contact properties between layers to reproduce delamination failure. All four bumper beam types were modeled having the same average thickness (t = 2.5 mm), width (w = 106 mm), and length (L = 210 mm). However, they differ in the cross-section geometry, as depicted in Figure 4. Their fillet radius increases from Type A to Type D according to the bumper beam geometries used in the experimental tests of a previous study [20]. The maximum radius corresponds to a semicircular geometry and the minimum radius is determined by the compression-molding manufacturing method. The characteristics of the four cross sections are summarized in Table 1.

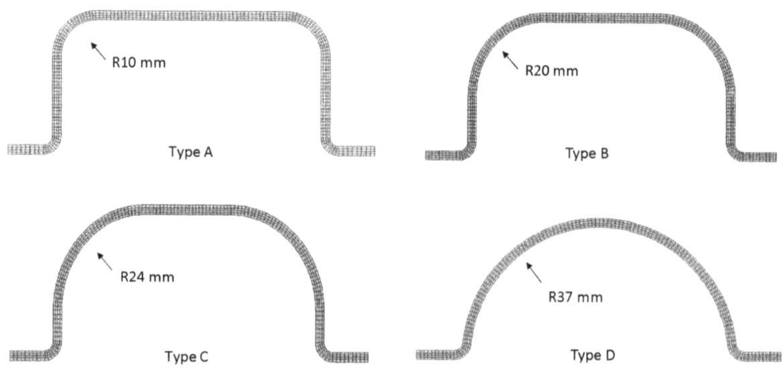

Figure 4. The cross-section geometry of the bumper beams.

Table 1. Main geometric variables of the four bumper beam cross sections.

Bumper Beam	Radius (mm)	Height (mm)	Weight (g)
Type A	10	40	97.4
Type B	20	40	96.8
Type C	24	40	95.7
Type D	37	40	84.6

The FEM model was verified using the experimental results of low-velocity impacts. The model mesh should be sufficiently refined to ensure that the results are reliable and accurate, considering that the computational cost rises according to the model refinement level. Thus, it was essential to find the optimal mesh to acquire a balance between the accuracy of the results and the computational cost.

Two variables: the absorbed energy (Eabs) for the whole model and the peak force (Fmax), were compared with the experimental values to verify the model prediction capacity. Three meshes (coarse, fine, and very fine mesh) were chosen to analyze the convergence of results for an impact energy of 60 J. The three meshes included a nonuniform distribution of elements along the beam length to reduce the processing time without compromising simulation accuracy. The double-bias seeding changed the mesh density progressively from the coarsest end to the finest center area of the bumper beam, emphasizing the reliability of the impact zone; see Figure 3.

For a very fine mesh (101,000 elements), the model lasted 18 h to complete the simulation and the predicted results were 57.66 J of absorbed energy and 2466.6 kN of peak force. On the other hand, for the fine mesh (65,000 elements), the simulation lasted 11 h and the results were similar: 58.84 J of absorbed energy and 2490.1 kN of peak force. Therefore, the fine mesh of approximately 65,000 elements was selected with sizes from 4 mm in beam ends to 1 mm in the contact area.

2.2. Fitting

The fitting process was conducted on the type A bumper beam, while other bumper beam types were used in the validation process. As mentioned in the previous section, cohesive interaction was used to simulate the interlaminar damage onset and propagation. Polylactic acid (PLA) thermoplastic was used to bind the different layers of the bumper beam, working as the cohesive element for the entire system; thus, its mechanical properties were taken as references for the cohesive properties in Abaqus/Explicit (see Table 2).

Table 2. Mechanical properties of PLA, data from [24].

Mechanical Properties	PLA
Tensile strength (MPa)	54.27
Young modulus (MPa)	3180

As the interface thickness of the bonded layers was negligible, a cohesive constitutive response based on traction–separation law was used for the numerical model. Abaqus offers the traction–separation model that assumes an initial linear elastic behavior followed by two stages, the initiation of damage (the onset of degradation) and the evolution of damage (the propagation of the damage up to failure). The elastic behavior is defined as an elastic matrix that relates the nominal and shear stress to the nominal and shear separation [25]. Therefore, the stiffness coefficients ($K_{nn} = K_{ss} = K_{tt}$) were specified as the Young modulus of PLA for uncoupled traction–separation behavior (Table 2). These values were not modified during the fitting process.

To define the damage initiation stage, the maximum nominal stress criterion was considered, i.e., when the maximum contact stress ratio reaches one, as shown in Equation (1), the damage of the instances will initiate.

$$max\left\{\frac{t_n}{t_n^o}, \frac{t_s}{t_s^o}, \frac{t_t}{t_t^o}\right\} = 1 \tag{1}$$

where t_n, t_s, and t_t are the normal and the two shear tractions from the nominal traction stress vector, respectively.

The peak values of the contact stress (t_n^o, t_s^o, t_t^o) were established during the fitting process. The initial value for these three parameters was equal to the PLA tensile strength, 54.27 MPa (Table 2). However, low absorbed energy and poor delamination response were generally detected in the simulations. The comparison with experimental results showed that these parameters required modification to increase the model accuracy. By the trial-and-error method, lower values seemed to give more reasonable results of energy and force, so the final attempt led to final stress of $t_n^o = t_s^o = t_t^o = 30$ MPa.

For the damage evolution specification, the approach used in this study was based on the dissipated energy due to failure (fracture energy), which is equal to the area under the traction–separation curve, and a linear softening behavior was considered to simplify the scenario. Once that, Benzeggagh–Kenane (BK) fracture criterion was chosen to compute the damage evolution of the cohesive interface, and the formulation is given by

$$G_{IC} + (G_{IIC} - G_{IC})\left\{\frac{G_S}{G_T}\right\}^{\eta} = G_C \tag{2}$$

where $G_{IIC} = G_{IIIC}$, $G_S = G_{IIC} + G_{IIIC}$, $G_T = G_{IC} + G_S$, and η is an empiric value called cohesive property parameter. According to Riccio et al. [26], the cohesive property parameter comes from experimental tests, ranging between 1 and 1.6. In order to simplify the study, η was equal to 1 for the numerical model.

Since green composites are still in their early stages of development, these materials are not commonly tested through the double cantilever beam test method (DCB) to determine the interlaminar fracture toughness energy (G_{Ic}), and few experimental tests can be found in the literature. To obtain the interlaminar fracture toughness energy in Mode I (G_{Ic}), double cantilever beam (DCB) tests according to the ASTM D5528-13 standard [27] were carried out, Figure 5a). It should be noticed that this standard was developed for carbon-fiber-reinforced plastics. However, it was used in this study because no specific standard for natural fiber-reinforced composites was found. According to the ASTM D 5528 standard, G_{IC} can be found using the area of the force–displacement curve, Figure 5b), using Equation (3):

$$G_{Ic} = \frac{A}{a \cdot w} \tag{3}$$

where A represents the area below the load–displacement diagram, a indicates the propagated crack length, and w is the width of the specimen, Figure 5c).

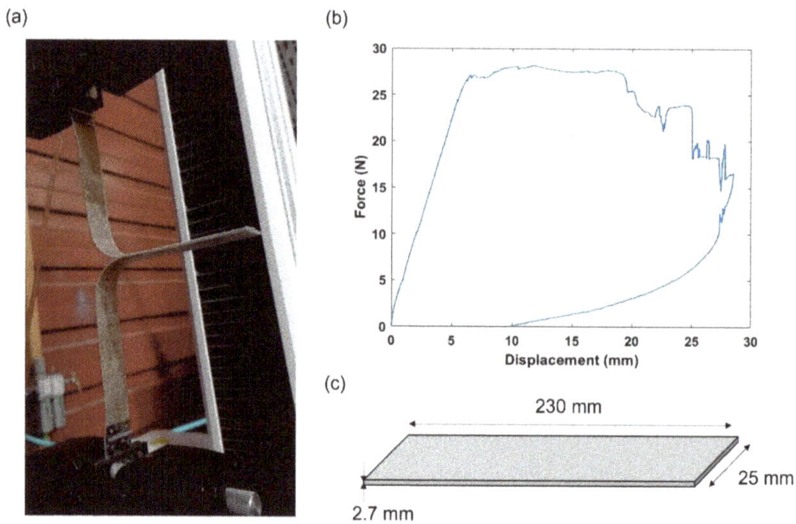

Figure 5. Experimental tests conducted to determine the interlaminar fracture toughness energy in Mode I. (**a**) Experimental setup; (**b**) load–displacement curve; (**c**) geometry of the specimens. Dimension not to scale.

The G_{Ic} value obtained in the experimental tests conducted on flax/PLA specimens was 2 mJ/mm^2. However, no experimental tests were conducted on the flax/PLA specimens to determine the fracture toughness energy mode II (G_{IIc}) and mode III (G_{IIIc}). The reason is the lack of specific standards and the difficulty of applying standards developed for CFRPs because the stiffness of natural fibers is much lower than that of carbon fibers. Therefore, to fit G_{IIc} and G_{IIIc}, values between 2 and 3 mJ/mm^2 were checked. This range was selected because the fracture toughness energy modes II and III are equal to or higher than mode I in composites. After the fitting process, values of fracture toughness energies were fixed at $G_{Ic} = G_{IIc} = G_{IIIc} = 2$ mJ/mm^2.

As a result of the fitting process, the model reasonably predicted the contact force history, the evolution of absorbed energy, and the delamination extension for different impact energies on bumper beam Type A.

2.2.1. Contact Force History

Figure 6 compares the experimental results and the numerical predictions regarding the contact-force history for impact energies equal to 30 J, 50 J, and 70 J. These impact energies were selected because they were representative of the three typical impacts: damage located in contact area at 30 J, complete failure of the bumper beam at 70 J, and intermediate case at 50 J. It can be seen that the model can predict not only the peak forces but also the different trends of contact force history and the duration of the impact event. The FEM model can predict the force history recorded in experimental results for 30 J and 70 J with a great accuracy; however, the accuracy in the 50 J impact test is lower. There are two possible reasons for this lack of precision: first, there is an intrinsic scattering in experimental results for composites reinforced with natural fibers; and second, 70 J is a case of high impact energy that produces the total failure of the beam, and 30 J is a case of low impact energy with localized damage, while 50 J is an intermediate impact energy, and the transition energies are more difficult to reproduce with numerical of theoretical models.

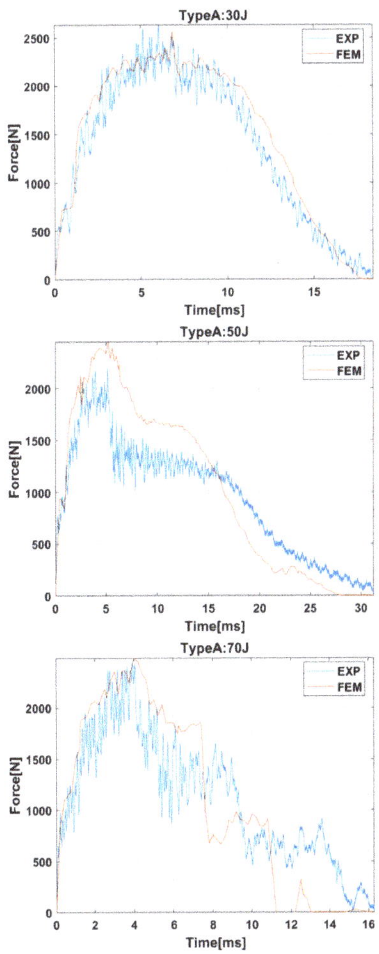

Figure 6. Force history of bumper beam type A under impact energies of 30 J, 50 J, and 70 J.

2.2.2. Absorbed Energy

The evolution of the absorbed energy is shown in Figure 7. For impact energies of 30 J or 50 J, the curves can be divided into two stages. First, absorbed energy increases until it reaches the maximum value, the impact energy. Then, the absorbed energy decreases until the final value. The decrease in the second stage is produced because part of the absorbed energy was elastic energy that the striker recovered in the form of kinetic energy.

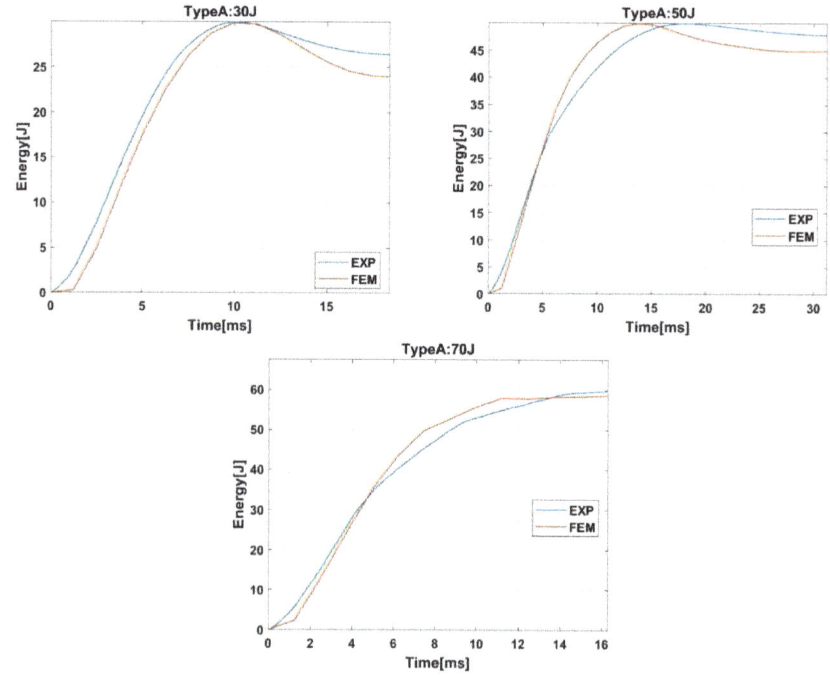

Figure 7. Energy history of bumper beam type A under impact energies of 30 J, 50 J, and 70 J.

On the other hand, for impact energy of 70 J, the absorbed energy increases until a maximum value of around 60 J because that is the maximum capacity of energy absorption of the bumper beam. When that energy is achieved, a complete failure is produced in the bumper beam, and the contact forces drop to zero, see Figure 6. The model can reproduce the absorbed energy evolution and predict the bumper beam total absorbed energy.

2.2.3. Delamination Damage Extension

Finally, three Type A specimens were subjected to X-ray scan tomography to evaluate the damage after impact. The equipment used—a Phoenix v/tomex of GE Sensing and Inspection Technologies X Ray company—has an x-ray tube with a 140 kV nanofocus. Figures 8–10 show a comparison between the tomography results and the numerical model predictions. At impact energy of 30 J, Figure 8, the predicted extension of the delamination in the top corner of the middle section (section B) is quite similar to the experimental scenario, which indicates a good accuracy of this FEM model. The numerical model also predicted slight delamination at the bottom corners of the sections near the supports (sections A and C). However, these delaminations were not detected in the tomography inspections; this discrepancy can be explained by the boundary conditions that cannot reproduce the complex nature of experimental impact test. The delaminations seem to occur only at the impact zone of the bumper beam.

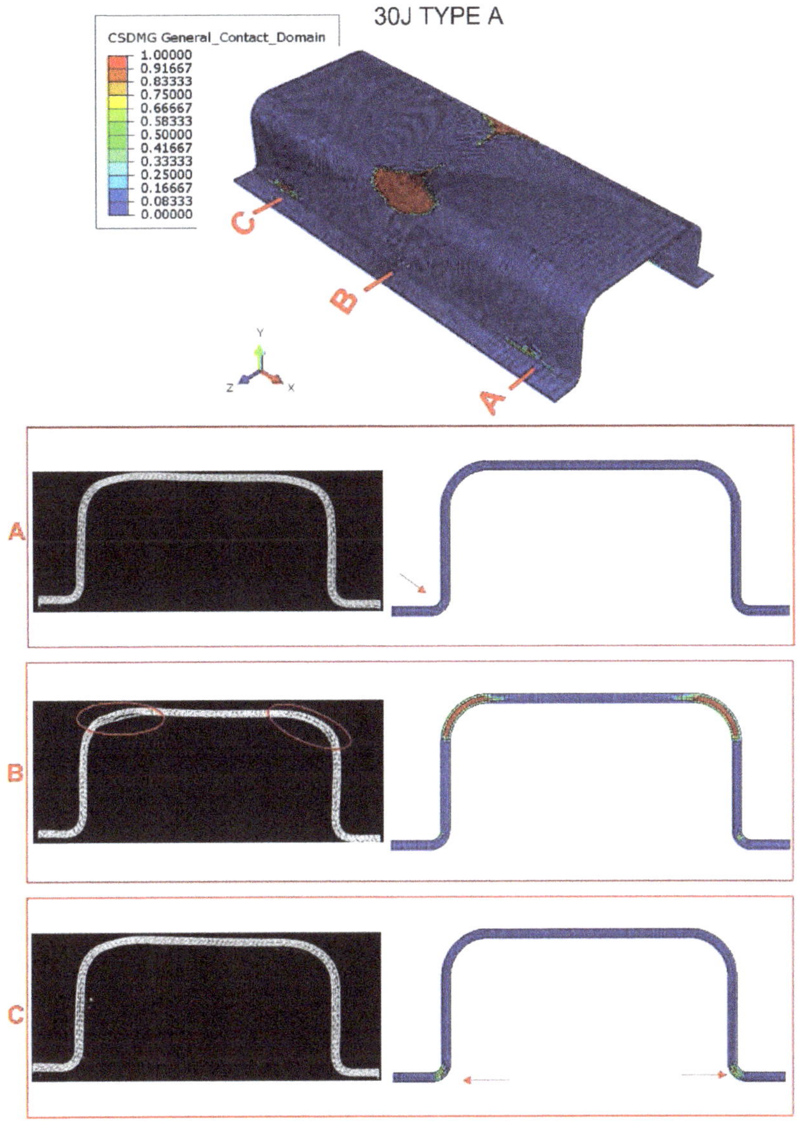

Figure 8. Tomography of bumper beam type A, impact energy of 30 J; and comparison with numerical results. Observation of the interlaminar failure in the ends of the beam (sections A and C) and in the impacted section (section B). Red arrows point the appearance of failure in sections A and C in the numerical screenshot, red ovals point to failure at fillet radius in section B.

When impact energy increases to 50 J (Figure 9), the damage in the central section also increases considerably. Delaminations can be found at the top corners and on the web. Moreover, a permanent deflection can be appreciated in the top flange. The model can reproduce this increase in delamination and permanent deformations due to plastic strains. For cross sections near the supports, both experimental and numerical results agree on the presence of small delaminations in the web and the bottom corners. The rest of the bumper beam sections do not present delaminations.

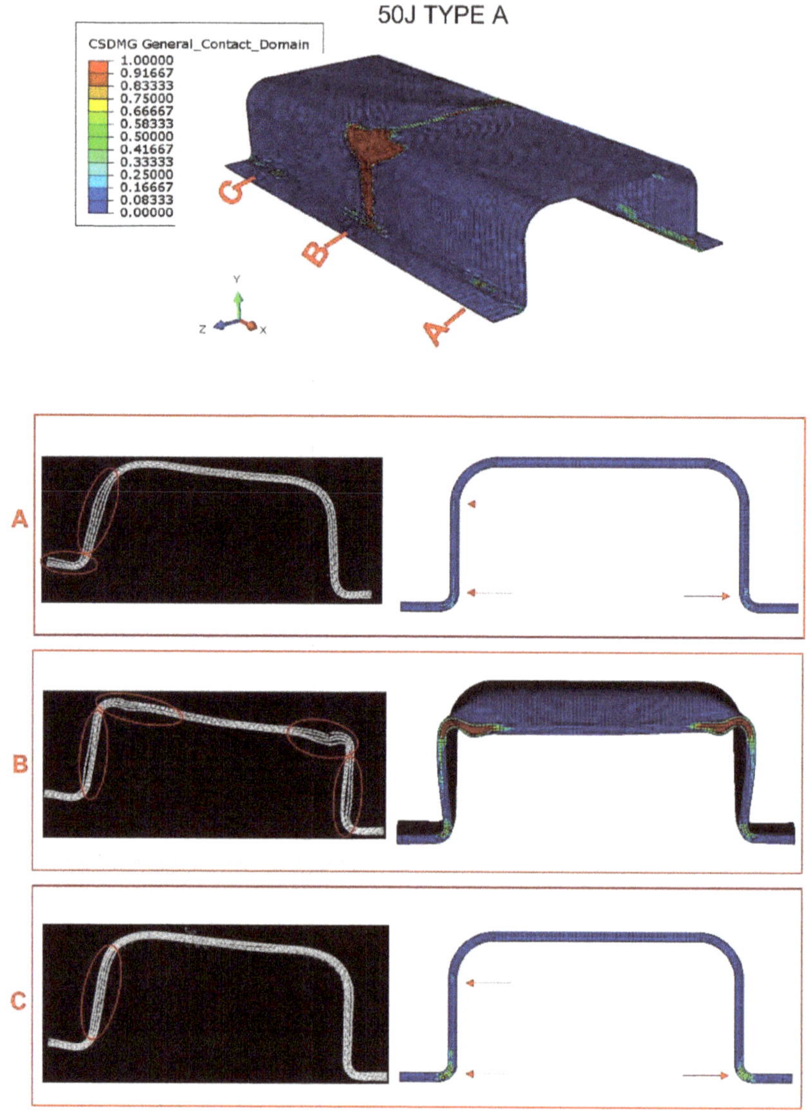

Figure 9. Tomography of bumper beam type A, impact energy of 50 J; comparison with numerical results. Observation of the interlaminar failure in the ends of the beam (sections A and C) and the center of impact (section B). The red arrows in the numerical screenshot point the appearance of failure in sections A and C.

If impact energy increases to 60 J (see Figure 10), the numerical model again accurately reproduces the damage in the impact section of the structure. Moreover, smaller delaminations appear in the top flange and the web of the cross sections near the support. Finally, considering the numerical model prediction capability in terms of contact force, absorbed energy, and damage extension, the fitted parameters were fixed, and the results in bumper beam types B, C, and D were used to validate the numerical model.

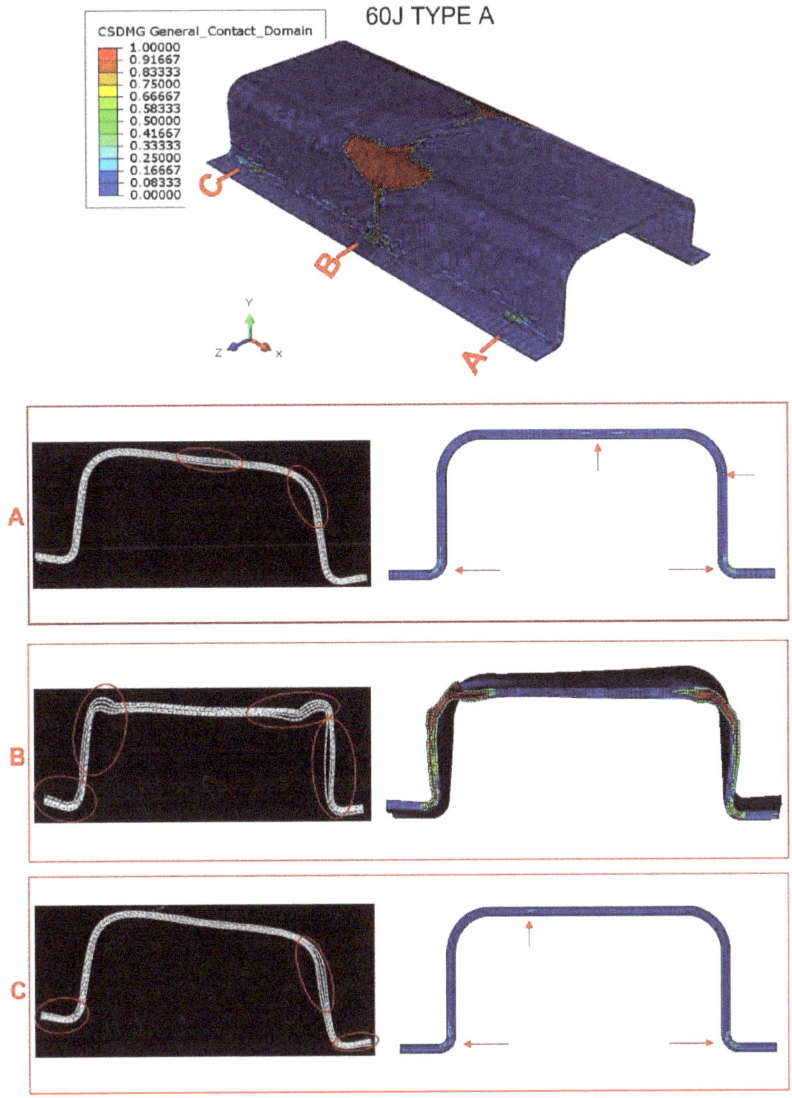

Figure 10. Tomography of bumper beam type A, impact energy of 60 J; comparison with numerical results. Observation of the interlaminar failure in the ends of the beam (sections A and C) and the center of impact (section B). The red arrows in the numerical screenshot point the appearance of failure in sections A and C.

3. Model Validation: Comparison with Experimental Data

After the fitting process, the same loading and boundary conditions were used for the remaining types of bumper beams by only changing the cross-section geometry. Validation of the models is studied in this section by comparing numerical and experimental results.

Figure 11 shows the absorbed energy versus the impact energy curves (Eabs-Eimp) of the four bumper beams in which the numerical prediction is compared with the experimental data. Generally, the numerical results show excellent agreement with the experimental data, except for 30–50 J. These disagreement produces different slopes in numerical and experimental curves. The main reason for this underestimation of absorbed energy is that the threshold energy of the bumper beams occurs within 30–50 J. For impact energies near the threshold energy, the behavior is intermediate between local damage in the contact area and complete failure of the bumper beam. This intermediate behavior is the most difficult to reproduce with a numerical model. Moreover, as massive damage (matrix cracking, fiber breaking, and delamination) initiates at this range, more uncertainty makes the scenario harder to predict and explains the difference between the experimental and numerical values. Nevertheless, the numerical model can reproduce the tendency and the values of absorbed energy with reasonable accuracy for the four bumper beam types.

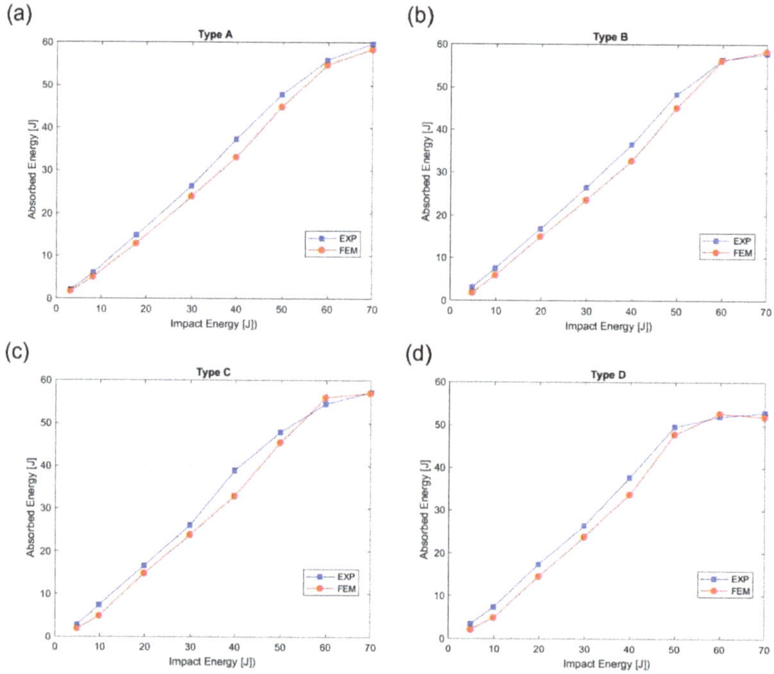

Figure 11. Absorbed energy as a function of impact energy. Experimental and numerical results comparison for all the bumper beam types. (**a**) E_{abs}-E_{imp} for Type A; (**b**) E_{abs}-E_{imp} for Type B; (**c**) E_{abs}-E_{imp} for Type C; and (**d**) E_{abs}-E_{imp} for Type D.

Figure 12 shows the contact force versus displacement curves for all the bumper beam types. Three impact energies were selected because they are representative of the three different impact behaviors. The lower impact energy, 30 J, represents a typical low-velocity impact with localized damage in the contact area. The force increases with displacement until the peak force is reached, then both force and displacement decrease until force drops to zero, but there is a permanent displacement around 10–15 mm. The impact energy of 70 J is the highest impact energy analyzed in this work and represents the complete failure of the bumper beam. The force increases with displacement until the peak force is reached and then drops to zero, but the displacements always increase, indicating that the striker does not rebound. The impact energy of 50 J represents an intermediate behavior with significant damage in the bumper beam but without a complete failure. The force–displacement curve at this energy is also a transitional curve between low and high impact energy. The force increases with displacement until peak force is reached, then the force decreases, but the displacement keeps increasing until maximum displacement is reached; finally, both force and displacement decrease. It can be observed that the numerical model can predict the experimental trends for the three impact energies, even the unique case of Type C at 50 J

(Figure 12c), where there is a second peak force. This unusual appearance might point out that the cracks start to appear in the beam structure, and this tendency grows more remarkable from Type A (Figure 12a) to Type D (Figure 12d). Therefore, the numerical model showed an excellent prediction capability in terms of contact force, displacement, and energy absorption.

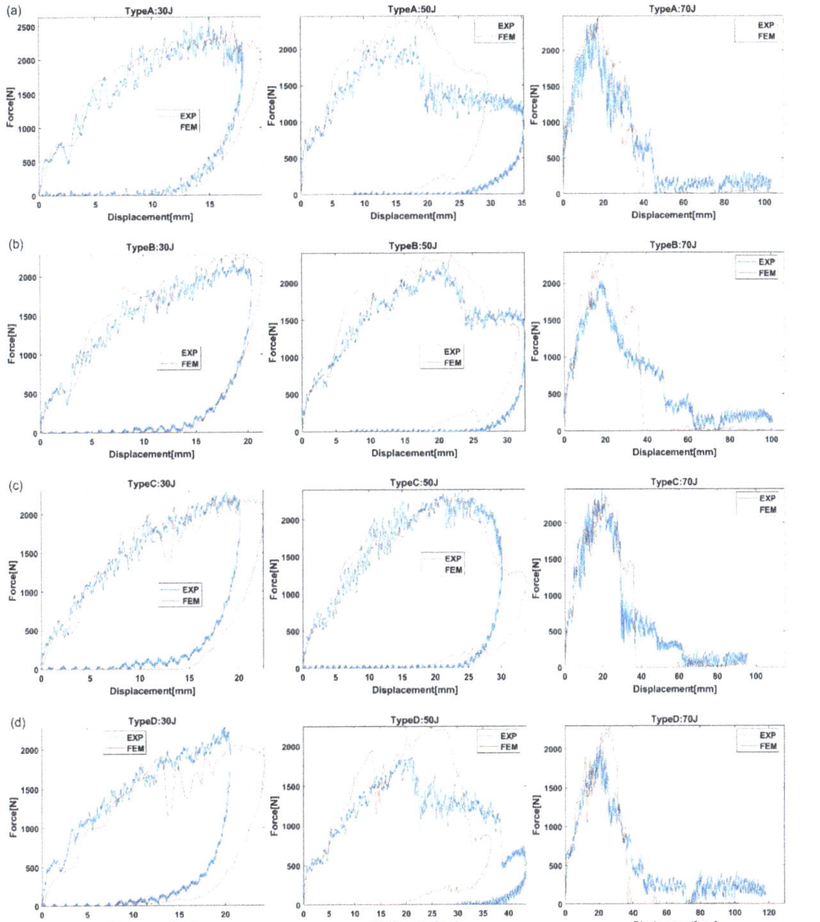

Figure 12. Force versus displacement curves of all the bumper beam types. Comparison between the experimental and the numerical results. (**a**) Curves for Type A; (**b**) Curves for Type B; (**c**) Curves for Type C; and (**d**) Curves for Type D.

4. Damage Analysis

The validated numerical model was used to analyze the damage evolution and better understand the energy-absorption mechanisms of the biocomposite bumper beams. Figures 13–16 show the delamination pattern in the middle cross section for all the bumper beam types at different impact energies. The goal of the analysis of damage in the middle cross section is the explanation of the results shown in Figure 11. The energy-absorption capability of the bumper beam Type A is not reached because the absorbed energy at 70 J is clearly higher than that at 60 J. In bumper beam types B and C, the absorbed energy at 60 J is almost equal to at 70 J, indicating that 60 J is the threshold energy where the maximum energy absorption capability is reached. This maximum energy-absorption capability was reached at 50 J for bumper beam type D because the absorbed energy does not increase more at 60 J or 70 J.

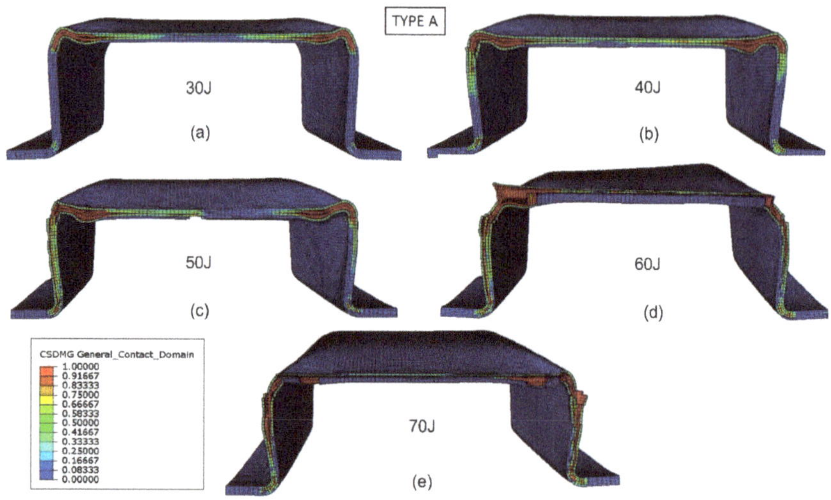

Figure 13. Delamination initiation criterion at the middle cross section of bumper beam Type A. Impact energy: 30 J (**a**), 40 J (**b**), 50 J (**c**), 60 J (**d**), and 70 J (**e**).

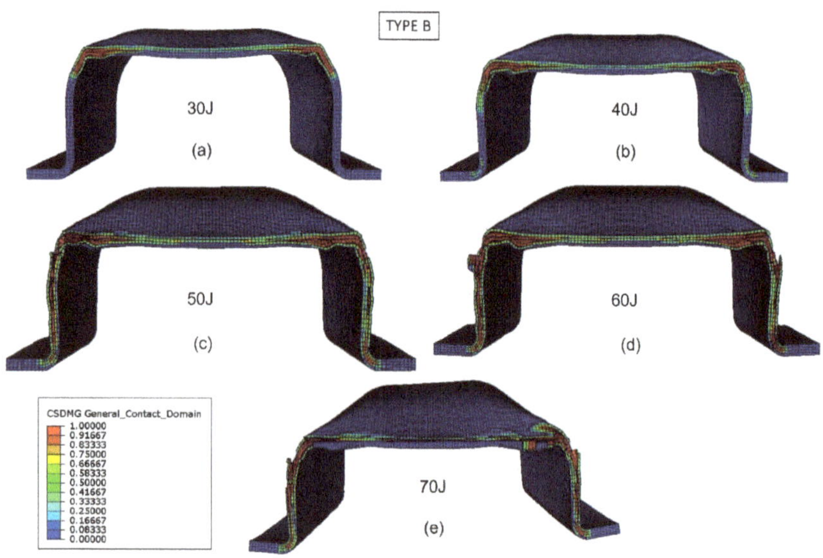

Figure 14. Delamination initiation criterion at the middle cross section of bumper beam Type B. Impact energy: 30 J (**a**), 40 J (**b**), 50 J (**c**), 60 J (**d**), and 70 J (**e**).

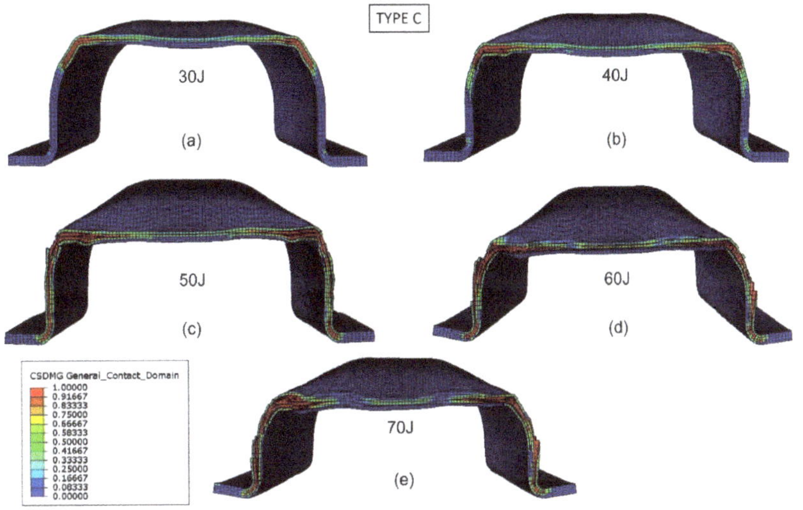

Figure 15. Delamination initiation criterion at the middle cross section of bumper beam Type C. Impact energy: 30 J (**a**), 40 J (**b**), 50 J (**c**), 60 J (**d**), and 70 J (**e**).

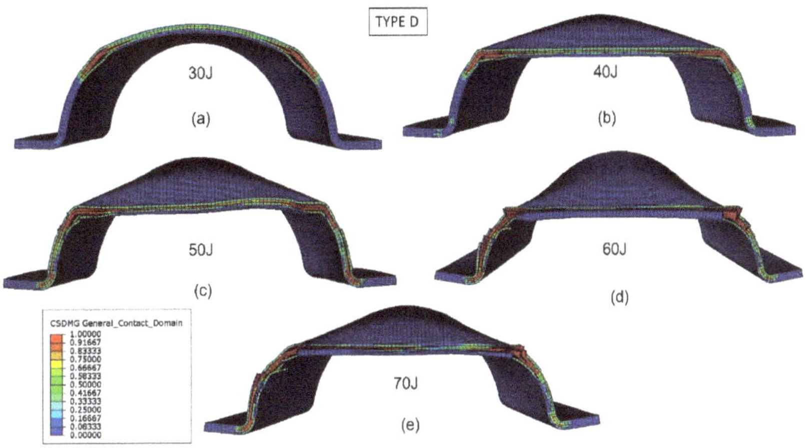

Figure 16. Delamination initiation criterion at the middle cross section of bumper beam Type D. Impact energy: 30 J (**a**), 40 J (**b**), 50 J (**c**), 60 J (**d**), and 70 J (**e**).

The evolution of delamination with impact energy in bumper beam Type A can be observed in Figure 13. For the impact energy of 30 J (Figure 13a), delamination is localized at the top corners, but the center of the top flange and the webs are free of delaminations. When impact energy increases to 40 J and 50 J (Figure 13b,c), delaminations propagate along the top flange and the webs but do not entirely fail. It can also be seen that the cracking of the section begins from the corner at 40 J and 50 J, which corresponds exactly to the anomaly that appears in the force-displacement at 50 J of Figure 12. When impact energy is equal to 60 J (Figure 13d), delamination is propagated through the top flange, and still, part of the webs is free of delamination. Only at the impact energy of 70 J, (Figure 13e), all the webs and top flange are delaminated; thus, the stiffness of the cross section drops to zero, and the absorbed energy capability of the bumper beam Type A is reached. This tendency agrees with the results shown in Figure 11, indicating that absorbed energy increases with impact energy in the range of impact energy studied in this work.

Figure 14 shows the delamination of bumper beam Type B for different impact energies at the middle cross section. For impact energy of 30 J (Figure 14a), delaminations are located at the top corners. When impact energy increases to 40 J (Figure 14b), there is a slight propagation of delaminations, but they are located around the top corners. If impact energy increases to 50 J or 60 J (Figure 14c,d), delaminations propagate in the top flange and the webs, but some parts of the webs and top flange are still free of delamination. For impact energy of 70 J (Figure 14e), delaminations are extended through the whole top flange and webs; thus, the stiffness of the cross section is lost, and the energy-absorption capability of the bumper beam Type B is reached. Figure 11 shows that the absorbed energy of bumper beam Type B increases with impact energies up to 70 J. However, the increment from 60–70 J is almost negligible, indicating that the threshold energy is probably between 60 J and 70 J.

The predictions of delaminations in the bumper beam Type C under different impact energies at the middle cross-section are shown in Figure 15. Delaminations are located at the top corner for the impact energy equal to 30 J (Figure 15a). For impact energies of 40 J and 50 J (Figure 15b,c), delaminations propagate through the top flange and webs, but with some parts of them free of delaminations. Furthermore, for impact energies of 60 J and 70 J (Figure 15d,e), delaminations are entirely extended in the top flange and webs, indicating that the cross-section stiffness drops to zero and the maximum energy absorption capability of the bumper Type C is reached at 60 J. These results agree with those shown in Figure 11 because the absorbed energy of bumper beam Type C increases with impact energy until 60 J, and the absorbed energy at 70 J is almost equal to that of 60 J.

Figure 16 shows the bumper beam Type D delamination at the middle cross section for different impact energies. For impact energy of 30 J (Figure 16a), delaminations are concentrated at the $\pm 45^{\circ}$ angles of the arc. If the impact energy increases to 40 J (Figure 16b), delaminations propagate toward the arc center, but the arc bottom part is free of interlaminar damage. For impact energies equal to or higher than 50 J (Figure 16c), the whole arc of the cross section is delaminated; thus, the cross section loses its stiffness, and the energy-absorption capability of the bumper beam Type D is reached. The results shown in Figure 11 confirm that 50 J is the impact energy threshold because the absorbed energy increases with impact energy until 50 J, and the absorbed energy at 60 J and 70 J is almost equal to that of 50 J.

The comparison of the different bumper beam types indicates that the lower the roundness of the cross section, the higher the energy-absorption capability. The squarest cross section, Type A, is the bumper beam with the highest energy absorption capability. On the other hand, the hemi-circumference cross-section-shape bumper beam, Type D, with the highest roundness, presents the lowest energy-absorption capability. Finally, the energy absorbed by types B and C is intermediate because their roundness is also transitional between types A and D. The main reason for this phenomenon is that the moment of inertia of the cross section decreases with the roundness. Therefore, contact forces also decrease with the cross-section roundness, as shown in Figure 12. The absorbed energy increases with the contact force; thus, bumper beam Type A shows the highest energy-absorption capability, and Type D the lowest.

Another reason behind this phenomenon is that the threshold energy is associated with the damage extended to the whole cross section. The delamination onset is located at the cross-section corners. If the corner radius is small, as in Type A, delaminations are more concentrated around the corners. In contrast, a high radius implies that delaminations are more prone to be distributed through the entire cross section. In other words, delamination can propagate prematurely through the cross section at lower impact energies.

These results are in clear contradiction with previous works on impact on flat plates of NFRPs [18,19], where the main failure mode observed is fiber breakage, and delaminations are almost negligible. This work, and a previous study on bumper beams manufactured with green composites [20], demonstrate that the impact behavior of curved specimens as bumper beams are strongly dominated by delaminations, being the delamination effect negligible only on flat plates.

To obtain a better understanding of the differences between the impact behavior of the four cross sections, a comparison of the failure modes at 60 J was performed using a high-speed camera to record the impact tests. Figure 17 compares the bumper beam deformed shape predicted by the numerical model with the experimental results at the impact energy of 60 J. It can be seen in Figure 17a) that types A and B do not fracture at 60 J. The bumper beam absorbed most of the kinetic energy, and the rest was returned to the striker with the elastic recovery of the bumper. Therefore, as the impact performance of bumper beam types A and B has not reached the limit, they can still have residual properties in postimpact testing. However, bumper beams C and D have already experienced a complete failure at this impact energy.

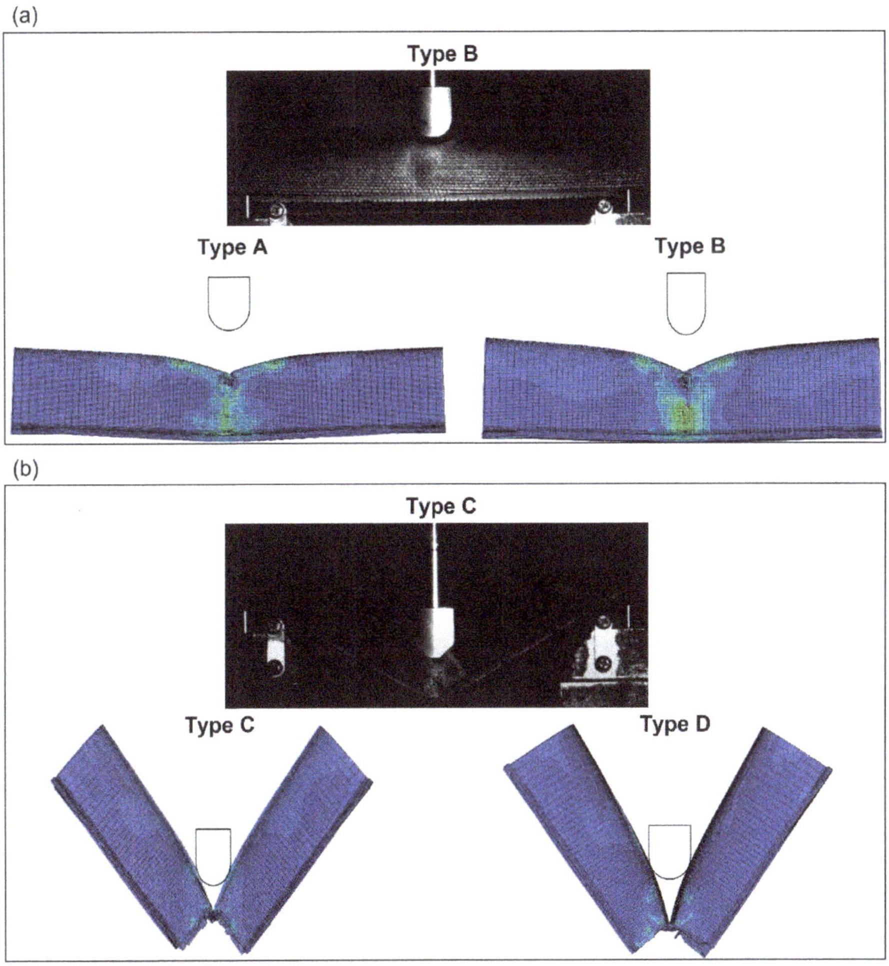

Figure 17. Side view of the bumper beams impacted at 60 J. (**a**) Photograph of Type B and numerical screenshots of Type A and B; (**b**) photograph of Type C and numerical screenshots of Type C and D.

5. Conclusions

The numerical model developed in Abaqus/Explicit shows accurate predictions on the damage behavior of the four types of bumper beams made of flax/PLA green composites. Its validation was carried out by comparing with the experimental low-velocity impact results: the energy absorption, contact-force history, and extension of delamination. Moreover, the model was also used to analyze the damage evolution and the energy-absorption mechanisms of the biocomposite bumper beams as a function of the impact energy and cross-section roundness. The reliability of the numerical model in predicting the growth of the delamination phenomenon was verified by comparing the damage with the X-ray scan tomography of Type A.

From the study, several conclusions are drawn:

- Regarding the prediction of the absorbed energy of the bumper beam, the numerical results show excellent agreement with the experimental data. The model is able to predict the different behavior as a function of impact energy, from localized damage to complete failure.
- Analysis of the force–displacement curves at impact energies of 30 J, 50 J, and 70 J shows that the permanent deformations of the bumper beam initiate at 30 J; the damage on the structure becomes significant and shows different severity for the four types at 50 J; and the complete failure of the beams is produced at 70 J.
- Experimental and numerical results revealed that the threshold energy, where the maximum energy absorption capability is reached, for Type A is over 60 J; for Type B and C is around 60 J; and for Type D is at 50 J.
- The damage evolution showed that the delamination manifests initially in the section corners and then spreads further. It implies that delaminations are more prone to propagate through the entire cross section and more prematurely for rounder types.
- Adding the fact that the rounder section presents smaller peak force and threshold energy. Therefore, the squarest cross section, Type A, is the bumper beam with the highest energy-absorption capability. Conversely, the roundest one, Type D, presents the lowest energy-absorption capability. The lower the roundness of the cross section, the higher the energy-absorption capability.
- Lastly, the numerical model predicted the same deformed shape of the four bumper beams under 60 J as the experimental scenario.

The development of a numerical model that can accurately predict the impact behavior of structures manufactured with green composites opens new lines of research in the use of green composites in structures designed to absorb energy from impacts. However, to facilitate the use of green composites in industry, further research must be focused on the impact behavior of green composites after ageing tests.

Author Contributions: Conceptualization, J.A.L. and C.S.; Funding acquisition, J.A.L. and C.S.; Investigation, L.J.-W.; Software, L.J.-W.; Supervision, C.S.; Validation, L.J.-W.; Writing—original draft, L.J.-W.; Writing—review & editing, J.A.L. and C.S. All authors have read and agreed to the published version of the manuscript.

Funding: This research was funded by the Ministry of Economy and Competitiveness of Spain grant number RTC-2015-3887-8.

Data Availability Statement: The data presented in this study are available on request from the corresponding author. The data are not publicly available due to privacy.

Acknowledgments: The authors acknowledge the Ministry of Economy and Competitiveness of Spain under the Project RTC-2015-3887-8 for the partial financial support of the work. In addition, authors are indebted to the CENIEH for carrying out the CT scans.

Conflicts of Interest: The authors declare no conflict of interest.

References

1. Greene, J.P. Introduction. In *Automotive Plastics and Composites*; Elsevier: Amsterdam, The Netherlands, 2021; pp. 1–15.
2. Poilâne, C.; Cherif, Z.E.; Richard, F.; Vivet, A.; ben Doudou, B.; Chen, J. Polymer Reinforced by Flax Fibres as a Viscoelastoplastic Material. *Compos. Struct.* **2014**, *112*, 100–112. [CrossRef]
3. López-Alba, E.; Schmeer, S.; Díaz, F. Energy Absorption Capacity in Natural Fiber Reinforcement Composites Structures. *Materials* **2018**, *11*, 418. [CrossRef] [PubMed]

4. Loureiro, N.C.; Esteves, J.L. Green Composites in Automotive Interior Parts. In *Green Composites for Automotive Applications*; Koronis, G., Silva, A., Eds.; Elsevier: Amsterdam, The Netherlands, 2019; pp. 81–97.
5. Hoffmann, K.G.; Haag, K.; Müssig, J. Biomimetic Approaches towards Lightweight Composite Structures for Car Interior Parts. *Mater. Des.* **2021**, *212*, 110281. [CrossRef]
6. Militello, C.; Bongiorno, F.; Epasto, G.; Zuccarello, B. Low-Velocity Impact Behaviour of Green Epoxy Biocomposite Laminates Reinforced by Sisal Fibers. *Compos. Struct.* **2020**, *253*. [CrossRef]
7. Nickels, L. Car with a Biodegradable Core. *Reinf. Plast.* **2017**, *61*, 332–334. [CrossRef]
8. Santos, E.B.C.; Barros, J.J.P.; de Moura, D.A.; Moreno, C.G.; Fim, F.; da Silva, L.B. Rheological and thermal behavior of PHB/piassava fiber residue-based green composites modified with warm water. *J. Mater. Res. Technol.* **2019**, *8*, 531–540. [CrossRef]
9. Azlin, M.N.M.; Sapuan, S.M.; Zuhri, M.Y.M.; Zainudin, E.S. Effect of Stacking Sequence and Fiber Content on Mechanical and Morphological Properties of Woven Kenaf/Polyester Fiber Reinforced Polylactic Acid (PLA) Hybrid Laminated Composites. *J. Mater. Res. Technol.* **2022**, *16*, 1190–1201. [CrossRef]
10. Jalón, E.; Hoang, T.; Rubio-López, A.; Santiuste, C. Analysis of Low-Velocity Impact on Flax/PLA Composites Using a Strain Rate Sensitive Model. *Compos. Struct.* **2018**, *202*, 511–517. [CrossRef]
11. Hassan, C.S.; Pei, Q.; Sapuan, S.M.; Abdul Aziz, N.; Mohamed Yusoff, M.Z. Crash Performance of Oil Palm Empty Fruit Bunch (OPEFB) Fibre Reinforced Epoxy Composite Bumper Beam Using Finite Element Analysis. *Int. J. Automot. Mech. Eng.* **2018**, *15*, 5826–5836. [CrossRef]
12. Hassan, C.S.; Qiang, P.; Sapuan, S.M.; Nuraini, A.A.; Zuhri, M.Y.M.; Ilyas, R.A. Unidirectional Oil Palm Empty Fruit Bunch (OPEFB) Fiber Reinforced Epoxy Composite Car Bumper Beam—Effects of Different Fiber Orientations on Its Crash Performance. In *Biocomposite and Synthetic Composites for Automotive Applications*; Sapuan, S.M., Ilyas, R.A., Eds.; Elsevier: Amsterdam, The Netherlands, 2021; pp. 233–253.
13. Arbintarso, E.S.; Muslim, M.; Rusianto, T. Simulation and Failure Analysis of Car Bumper Made of Pineapple Leaf Fiber Reinforced Composite. In *IOP Conference Series: Materials Science and Engineering*; IOP Publishing: Bristol, UK, 2018; Volume 306, p. 012038. [CrossRef]
14. Ramasubbu, R.; Madasamy, S. Fabrication of Automobile Component Using Hybrid Natural Fiber Reinforced Polymer Composite. *J. Nat. Fibers* **2022**, *19*, 736–746. [CrossRef]
15. Rubio-López, A.; Olmedo, A.; Santiuste, C. Modelling Impact Behaviour of All-Cellulose Composite Plates. *Compos. Struct.* **2015**, *122*, 139–143. [CrossRef]
16. Ghani, M.U.; Siddique, A.; Abraha, K.G.; Yao, L.; Li, W.; Khan, M.Q.; Kim, I.-S. Performance Evaluation of Jute/Glass-Fiber-Reinforced Polybutylene Succinate (PBS) Hybrid Composites with Different Layering Configurations. *Materials* **2022**, *15*, 1055. [CrossRef] [PubMed]
17. Murugu Nachippan, N.; Alphonse, M.; Bupesh Raja, V.K.; Palanikumar, K.; Sai Uday Kiran, R.; Gopala Krishna, V. Numerical Analysis of Natural Fiber Reinforced Composite Bumper. *Mater. Today Proc.* **2021**, *46*, 3817–3823. [CrossRef]
18. Li, Q.; Li, Y.; Ma, H.; Cai, S.; Huang, X. Effect of Processing Temperature on the Static and Dynamic Mechanical Properties and Failure Mechanisms of Flax Fiber Reinforced Composites. *Compos. Commun.* **2020**, *20*, 100343. [CrossRef]
19. Díaz-Álvarez, A.; Jiao-Wang, L.; Feng, C.; Santiuste, C. Energy Absorption and Residual Bending Behavior of Biocomposites Bumper Beams. *Compos. Struct.* **2020**, *245*, 112343. [CrossRef]
20. Jiao-Wang, L.; Loya, J.A.; Santiuste, C. Influence of Cross-Section on the Impact and Post-Impact Behavior of Biocomposites Bumper Beams. *Mech. Adv. Mater. Struct.* **2022**, 1–12. [CrossRef]
21. Rubio-López, A.; Hoang, T.; Santiuste, C. Constitutive Model to Predict the Viscoplastic Behaviour of Natural Fibres Based Composites. *Compos. Struct.* **2016**, *155*, 8–18. [CrossRef]
22. Díaz-Álvarez, A.; Díaz-Álvarez, J.; Cantero, J.L.; Santiuste, C. Analysis of Orthogonal Cutting of Biocomposites. *Compos. Struct.* **2020**, *234*, 111734. [CrossRef]
23. Díaz-Álvarez, A.; Díaz-Álvarez, J.; Feito, N.; Santiuste, C. Drilling of Biocomposite Materials: Modelling and Experimental Validation. *Simul. Model. Pr. Theory* **2021**, *106*, 102203. [CrossRef]
24. Porras, A.; Maranon, A. Development and Characterization of a Laminate Composite Material from Polylactic Acid (PLA) and Woven Bamboo Fabric. *Compos. Part B Eng.* **2012**, *43*, 2782–2788. [CrossRef]
25. © Dassault Systèmes Abaqus Analysis User's Guide: Surface-Based Cohesive Behavior 2014. Available online: http://130.149.89.49:2080/v6.14/books/usb/default.htm (accessed on 23 June 2022).
26. Riccio, A.; Saputo, S.; Raimondo, A.; Sellitto, A. *Modeling Low Velocity Impact Phenomena on Composite Structures*; Elsevier: Amsterdam, The Netherlands, 2017.
27. *ASTM D5528-13*; Standard Test Method for Mode I Interlaminar Fracture Toughness of Unidirectional Fiber-Reinforced Polymer Matrix Composites. ASTEM International: West Conshohocken, PA, USA, 2022.

materials

Article

Nonlocal Free Vibration of Embedded Short-Fiber-Reinforced Nano-/Micro-Rods with Deformable Boundary Conditions

Ömer Civalek [1,2], Büşra Uzun [3,*] and Mustafa Özgür Yaylı [3]

[1] Civil Engineering Department, Akdeniz University, 07058 Antalya, Turkey
[2] Research Center for Interneural Computing, China Medical University, Taichung 40447, Taiwan
[3] Division of Mechanics, Civil Engineering Department, Bursa Uludag University, 16059 Bursa, Turkey
* Correspondence: buzun@uludag.edu.tr; Tel.: +90-224-294-26-36

Abstract: An efficient eigenvalue algorithm is developed for the axial vibration analysis of embedded short-fiber-reinforced micro-/nano-composite rods under arbitrary boundary conditions. In the formulation, nonlocal elasticity theory is used to capture the size effect, and the deformable boundary conditions at the ends are simulated using two elastic springs in the axial direction. In addition, to determine the reinforcing effect of restrained nano-/micro-rods, a new system of linear equations with the concept of the infinite power series is presented. After performing the mathematical processes known as Fourier sine series, Stokes' transformation and successive integration, we finally obtain a coefficient matrix in terms of infinite series for various rigid or deformable boundary conditions. Some accurate eigenvalue solutions of the free axial vibration frequencies of the short-fiber-reinforced micro-/nano-composite rods with and without being restrained by the means of elastic springs are given to show the performance of the present method. The presence of the elastic spring boundary conditions changes the axial vibration frequencies and corresponding mode shapes.

Keywords: axial vibration; short-fiber-reinforced; Fourier series; nonlocal elasticity

Citation: Civalek, Ö.; Uzun, B.; Yaylı, M.Ö. Nonlocal Free Vibration of Embedded Short-Fiber-Reinforced Nano-/Micro-Rods with Deformable Boundary Conditions. *Materials* **2022**, *15*, 6803. https://doi.org/10.3390/ma15196803

Academic Editors: Michele Bacciocchi, Abbas S. Milani and Panagiotis G. Asteris

Received: 3 September 2022
Accepted: 26 September 2022
Published: 30 September 2022

Publisher's Note: MDPI stays neutral with regard to jurisdictional claims in published maps and institutional affiliations.

1. Introduction

Due to their unique qualities, composite materials have garnered increasing attention over the past few decades. Composites are materials formed from at least two different components with different chemical and physical properties. These special materials have a more advanced structure than their constituent components thanks to the combination of different properties.

There are various types of composite materials used in different applications in the field of engineering. They can be constituted in different forms in accordance with the intended use. One of these is short-fiber-reinforced composite materials. These materials are formed by placing fibers of short length into a matrix in various arrangements.

Recently, studies involving the analysis of reinforced structures have gained momentum. Fiber-reinforced [1,2], carbon nanotube-reinforced [3–10] and graphene platelet-reinforced [11–15] structures can be found in the recent literature. However, other composite structures like functionally graded [16–20], sandwich [21,22] and porous functionally graded structures [23–28] have also attracted considerable interest. As can be seen, composite structures have attracted attention not only at a macro scale but also at nano and micro scales. While analyzing structures at nano and micro scales, we see that many of them are presented with theories based on the size effect. When we look at the governing equations represented by these theories, it is understood that they contain one or more parameters in addition to the classical constants. These parameters, generally called small-scale parameters, make it possible to investigate the size effect. Researchers working on the analysis of various very small-scale nano- and micro-structures have used these size effect theories, such as doublet mechanics theory [29–33], modified couple stress theory [34–39], nonlocal elasticity theory [1,40–45], nonlocal strain gradient theory [46–49] and strain gradient theory [50–53]. Researchers use a variety of techniques to analyze an element or structure, such as a rod, beam, plate, frame, etc., whether at the macro, nano, or micro level. Some of these techniques are: the finite element method [1,33,41,42,54,55], artificial neural networks [56], the Laplace transform [57], Stokes' transformation [18,28,43,45], the perturbation technique [5] and the Chebyshev–Ritz method [19]. This is the first work to investigate the longitudinal vibration behavior of short-fiber-reinforced micro-/nano-rods embedded in an elastic medium via Fourier sine series with Stokes' transformation.

This paper presents the free axial vibrational response of a restrained and size-dependent micro- and nano-scale rod embedded in an elastic medium based on Eringen's nonlocal elasticity theory [58]. The size dependency of the material characteristics are modeled according to the short-fiber-reinforced micro- and nano-composite rods. In addition, displacements at the ends are defined based on classical rod theory. This paper presents for the first time a solution based on the Stokes' transformation and Fourier sine series for the axial vibration of short-fiber-reinforced nano-/micro-rods with arbitrary boundary conditions in the presence of an elastic medium. The contribution of this work is that it provides an approach to study the effect of both an elastic medium and arbitrary boundary conditions on the axial vibration of short-fiber-reinforced micro-/nano-rods. Fourier sine series are also utilized to define the axial deflection function. Nonlocal force boundary conditions are utilized to derive the systems of linear equations for specifying the elastic foundation, nonlocal and short fiber parameters. The linear system of equations obtained is discretized with the help of Stokes' transformation. A coefficient matrix and the corresponding eigenvalue problem is constructed for longitudinal dynamic analysis of the short-fiber-reinforced micro- and nano-composite rods under rigid or restrained boundary conditions. This coefficient matrix includes nonlocal parameter boundary conditions, a short fiber constant and an elastic foundation coefficient.

2. Nonlocal Elasticity

Nonlocal elasticity theory is the most preferred continuum mechanics approach for nano-sized structures. This continuum mechanics theory was introduced by Eringen [58]. In accordance with nonlocal elasticity theory, the stresses and strains of one location inside a structure are related to the stresses and strains of other locations that are close to the reference point.

For the axial vibration behavior of a composite micro-/nano-sized rod, the constitutive equation based on the nonlocal elasticity theory is written as follows [1]:

$$\left[1-\mu\frac{\partial^2}{\partial x^2}\right]\tau_{xx}=E_c\varepsilon \tag{1}$$

where $\mu=e_0a^2$, τ_{xx} denotes the nonlocal stress, E_c specifies the Young's modulus of the composite, and ε represents the axial strain. In addition, e_0a is called the nonlocal parameter. This nonlocal parameter allows us to investigate the size effect on the composite micro-/nano-rod. Furthermore, a defines the internal characteristic length and e_0 specifies a material constant. The equation of motion for the longitudinal vibration of the composite micro- and nano-rods can be described by [44]:

$$\frac{\partial N^l}{\partial x}+f=\rho_cA\frac{\partial^2u(x,t)}{\partial t^2} \tag{2}$$

in which N^l denotes the axial force of local (classical) elasticity, f specifies the distributed axial force acting on the composite rod, ρ_c represents the density of the composite, A is the cross-sectional area, $u(x,t)$ is the axial displacement and t is the time. In addition, the axial force of local elasticity N^l can be defined as:

$$N^l = \int_A \sigma_{xx} dA \tag{3}$$

In Equation (3), σ_{xx} is the component of local stress. By integrating Equation (1) with respect to the cross-sectional area of the composite, the following relation is obtained:

$$N - \mu \frac{\partial^2 N}{\partial t^2} = N^l \tag{4}$$

Here, N represents the axial force of nonlocal elasticity theory and is defined as:

$$N = \int_A \tau_{xx} dA \tag{5}$$

One can derive the equation of motion for longitudinal vibration of the composite micro- and nano-rods via Equations (2)–(5):

$$E_c A \frac{\partial^2 u(x,t)}{\partial x^2} + f - \mu \frac{\partial^2 f}{\partial x^2} = \rho_c A \frac{\partial^2 u(x,t)}{\partial t^2} - \mu \rho_c A \frac{\partial^4 u(x,t)}{\partial x^2 \partial t^2} \tag{6}$$

In this paper, the influence of an elastic medium on the longitudinal vibration frequencies of the short-fiber-reinforced composite micro- and nano-rods is investigated for the first time. For this purpose, the force based on the elastic medium is considered in the following form [44]:

$$f = -ku(x,t) \tag{7}$$

In this study, the composite micro-/nano-rod is considered to be surrounded by an elastic medium. As expected, this elastic medium has a stiffness. In Equation (7), k denotes the stiffness of the elastic medium. One can derive the equation of motion for an embedded composite micro-/nano-rod by inserting Equation (7) into Equation (6) as follows:

$$E_c A \frac{\partial^2 u(x,t)}{\partial x^2} - ku(x,t) + \mu k \frac{\partial^2 u(x,t)}{\partial x^2} - \rho_c A \frac{\partial^2 u(x,t)}{\partial t^2} + \mu \rho_c A \frac{\partial^4 u(x,t)}{\partial x^2 \partial t^2} = 0 \tag{8}$$

It should be highlighted here that if the parameter μ is set to zero, the equation is simplified to the equation of the classical embedded composite rod as follows:

$$E_c A \frac{\partial^2 u(x,t)}{\partial x^2} - ku(x,t) - \rho_c A \frac{\partial^2 u(x,t)}{\partial t^2} = 0 \tag{9}$$

The other point that should be highlighted here is that if the elastic medium stiffness k is set to zero, the equation is reduced to the equation of the un-embedded composite micro-/nano-rod as follows [1]:

$$E_c A \frac{\partial^2 u(x,t)}{\partial x^2} - \rho_c A \frac{\partial^2 u(x,t)}{\partial t^2} + \mu \rho_c A \frac{\partial^4 u(x,t)}{\partial x^2 \partial t^2} = 0 \tag{10}$$

3. Material Properties of Short-Fiber-Reinforced Composite

In this study, the free longitudinal vibration behavior of the embedded short-fiber-reinforced composite micro-/nano-rod with deformable boundary conditions is investigated via Eringen's nonlocal elasticity theory [58]. Furthermore, an aligned composite micro-/nano-rod and a randomly oriented composite micro-/nano-rod with elastic springs are shown in Figures 1 and 2, respectively. In addition, an illustration of a randomly oriented composite material may be seen in Figure 3. As can be seen in the governing equation of embedded nano-rods given in Equation (10), Young's modulus (E_c) and density (ρ_c) are properties of short-fiber-reinforced micro-/nano-rods that should be defined. In this section, these properties are described. This study adopts the Halpin–Tsai equations [59]. These simple and easy-to-use equations are quite reasonable as they give accurate predictions, as long as the fiber volume fraction does not approach one [60]. Via Halpin–Tsai equations, the longitudinal and transverse Young's moduli of aligned short-fiber-reinforced composite materials are written as follows [60]:

$$E_L = E_m\left(\frac{1+\left(\frac{2l}{d}\right)\eta_L V_f}{1-\eta_L V_f}\right) \tag{11}$$

$$E_T = E_m\left(\frac{1+2\eta_T V_f}{1-\eta_T V_f}\right) \tag{12}$$

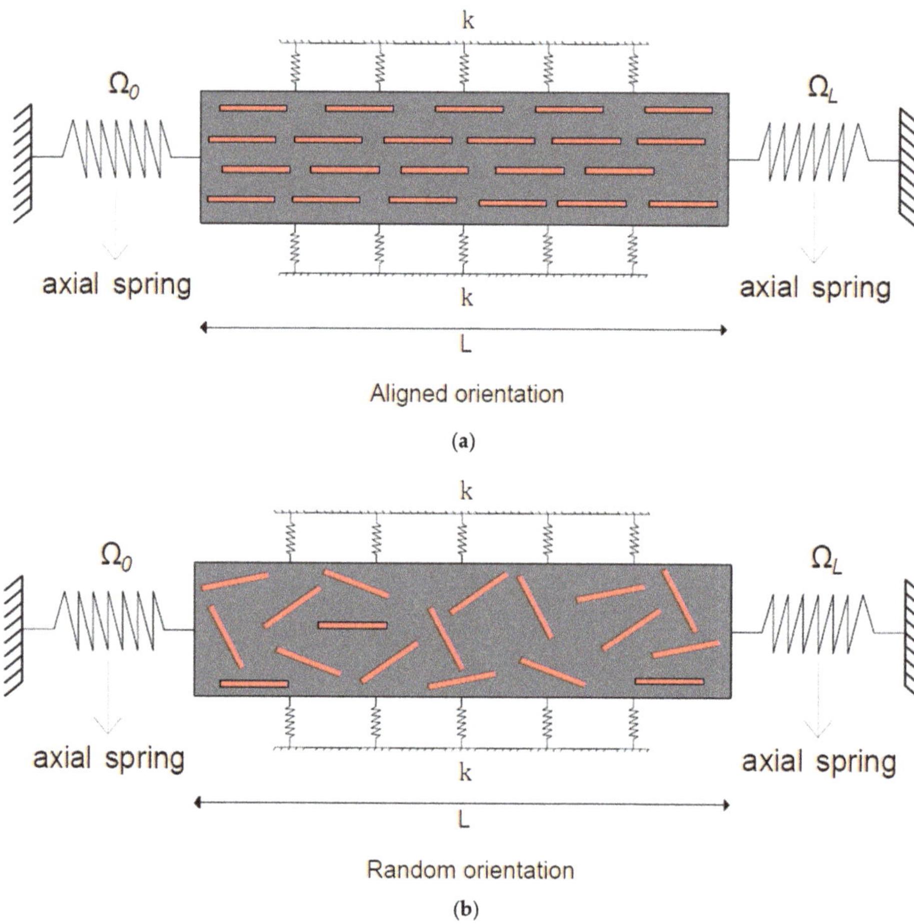

Figure 1. Illustrations of composite micro-/nano-rods with elastic springs: (**a**) aligned composite micro-/nano-rod; (**b**) randomly oriented composite micro-/nano-rod.

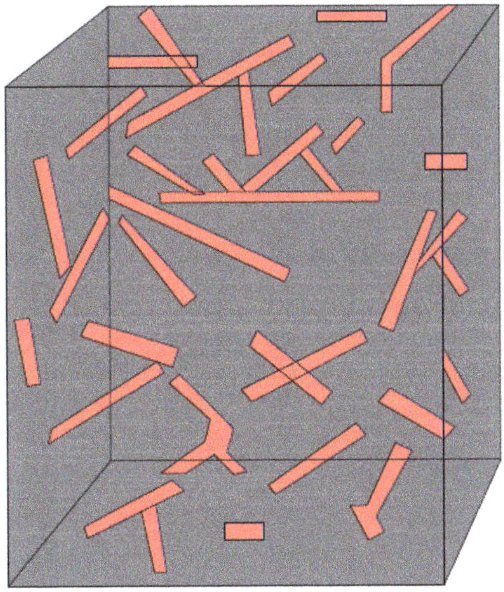

Figure 2. Illustration of a randomly oriented composite material.

By arranging short fibers in a matrix in different ways, short-fiber-reinforced composite materials are formed. In the above equations, E_m represents the Young's modulus of the matrix, d is the diameter of the fiber, l denotes the length of the fiber and V_f is the volume fraction of the fiber. In addition, η_L and η_T seen in Equations (11) and (12) are described as:

$$\eta_L = \left(\frac{\frac{E_f}{E_m} - 1}{\frac{E_f}{E_m} + 2\left(\frac{l}{d}\right)} \right) \tag{13}$$

$$\eta_T = \left(\frac{\frac{E_f}{E_m} - 1}{\frac{E_f}{E_m} + 2} \right) \tag{14}$$

in which E_f specifies the Young's modulus of the fiber of the composite. Moreover, the Young's modulus of a randomly oriented short-fiber-reinforced composite is described as follows [60]:

$$E_c = E_{random} = \frac{3}{8}E_L + \frac{5}{8}E_T \tag{15}$$

Lastly, the density of the short-fiber-reinforced composite should be defined. The density of the short-fiber-reinforced composite is given as follows [1]:

$$\rho_c = \rho_m\left(1 - V_f\right) + \rho_f V_f \tag{16}$$

In the above equation, ρ_c, ρ_m and ρ_f are the density of the short-fiber-reinforced composite, matrix and fiber, respectively. Short-fiber-reinforced composites include two different components, the matrix and the fibers, with different material properties. Depending on these components' properties, the composite's material properties are calculated with the equations given above.

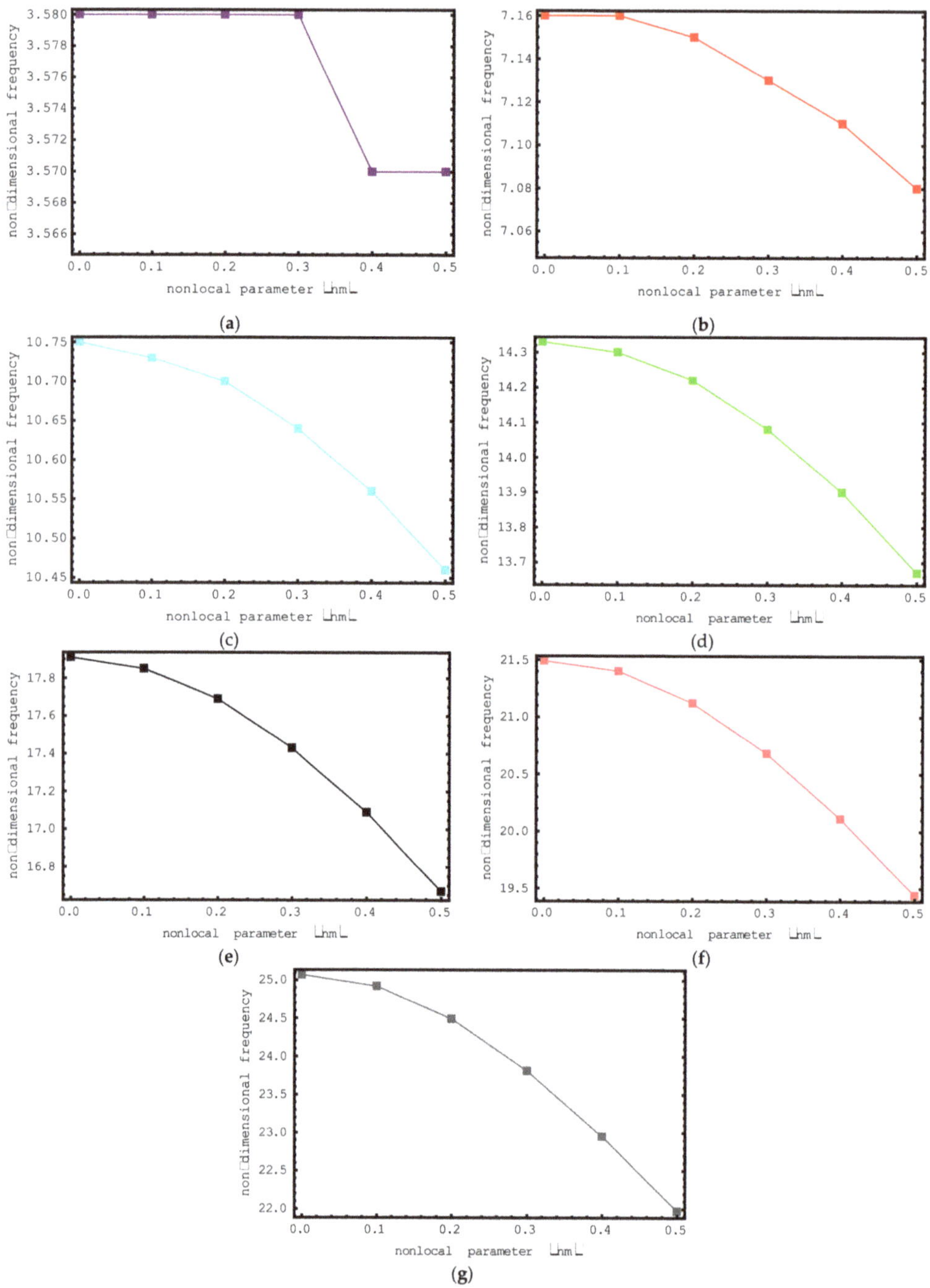

Figure 3. The variations of non-dimensional frequencies of short-fiber-reinforced composite nanorods versus nonlocal parameter: (**a**) 1st mode (**b**) 2nd mode (**c**) 3rd mode (**d**) 4th mode (**e**) 5th mode (**f**) 6th mode (**g**) 7th mode.

4. Fourier Infinite Series with Stokes' Transformation

In this section of the study, the adopted solution procedure is applied to the longitudinal vibration of embedded short-fiber-reinforced composite micro-/nano-rods restrained with axial elastic springs at both ends. Assuming harmonic vibrations, $u(x, t)$ may be represented by:

$$u(x,t) = \Psi(x)\cos(\omega t) \tag{17}$$

One can get the following expression by substituting Equation (17) into Equation (8):

$$-E_c A\frac{d^2\Psi(x)}{dx^2} - k\Psi(x) - \mu k\frac{d^2\Psi(x)}{dx^2} - \rho_c A\omega^2\Psi(x) + \mu\rho_c A\omega^2\frac{d^2\Psi(x)}{dx^2} = 0 \tag{18}$$

in which ω specifies the natural frequency of the composite micro-/nano-rod in terms of rad/s and $\Psi(x)$ defines the modal displacement function, and $\Psi(x)$ can be written in three separate regions as below [43,45]:

$$\Psi(x) = \Psi_0 \; x = 0, \tag{19}$$

$$\Psi(x) = \Psi_L \; x = L, \tag{20}$$

$$\Psi(x) = \sum_{j=1}^{\infty} H_j \sin\left(\frac{j\pi x}{L}\right) 0 < x < L \tag{21}$$

H_j in Equation (21) is defined as:

$$H_j = \frac{2}{L}\int_0^L \Psi(x)\sin\left(\frac{j\pi x}{L}\right)dx \tag{22}$$

The first derivative of Equation (21) leads to:

$$\Psi'(x) = \sum_{j=1}^{\infty} \frac{j\pi}{L} H_j \cos\left(\frac{j\pi x}{L}\right) \tag{23}$$

Furthermore, we can write Equation (23) as a Fourier cosine infinite series as follows:

$$\Psi'(x) = \frac{h_0}{L} + \sum_{j=1}^{\infty} h_j \cos\left(\frac{j\pi x}{L}\right) \tag{24}$$

The coefficients h_0 and h_j can be expressed as:

$$h_0 = \frac{2}{L}\int_0^L \Psi'(x)dx = \frac{2}{L}[\Psi(L) - \Psi(0)] \tag{25}$$

$$h_j = \frac{2}{L}\int_0^L \Psi'(x)\cos\left(\frac{j\pi x}{L}\right)dx \; j = 1,\, 2,\, \ldots \tag{26}$$

If we apply integration by parts, we obtain the following expressions:

$$h_j = \frac{2}{L}\left[\Psi(x)\cos\left(\frac{j\pi x}{L}\right)\right]_0^L + \frac{2}{L}\left[\frac{j\pi}{L}\int_0^L \Psi(x)\sin\left(\frac{j\pi x}{L}\right)dx\right] \tag{27}$$

$$h_j = \frac{2}{L}\left[(-1)^j\Psi(L) - \Psi(0)\right] + \frac{j\pi}{L}H_j \tag{28}$$

To continue the mathematical steps, we should find the first two derivatives of the displacement function $\Psi(x)$. The first two derivatives of $\Psi(x)$ are calculated as [43,45]:

$$\frac{d\Psi(x)}{dx} = \frac{\Psi_L - \Psi_o}{L} \sum_{j=1}^{\infty} \cos(\xi_j x)\left(\frac{2\left((-1)^j\Psi_L - \Psi_0\right)}{L} + \xi_j H_j\right) \tag{29}$$

$$\frac{d^2\Psi(x)}{dx^2} = \sum_{j=1}^{\infty} \xi_j \sin(\xi_j x)\left(\frac{2\left((-1)^j\Psi_L - \Psi_0\right)}{L} + \xi_j H_j\right) \tag{30}$$

in which ξ_j is defined as:

$$\xi_j = \frac{j\pi}{L} \tag{31}$$

In this step of the solution, we should find the Fourier coefficient H_j. To find H_j, we substitute Equations (21), (30) and (31) into Equation (18):

$$H_j = \frac{2j\pi(-\gamma^2\lambda^2 + \gamma^2 K + 1)\left(\Omega_0 - (-1)^j\Omega_L\right)}{-\lambda^2 + K^2 + \pi^2 j^2\gamma^2 K - \pi^2 j^2\gamma^2\lambda^2 + \pi^2 j^2} \tag{32}$$

Here,

$$\gamma^2 = \frac{\mu}{L^2} \tag{33}$$

$$K = \frac{kL^2}{E_m A} \tag{34}$$

$$\lambda^2 = \frac{\rho_c A\omega^2 L^2}{E_c A} \tag{35}$$

The axial displacement for the axial vibration of a composite micro-/nano-sized rod embedded in an elastic medium yields:

$$u(x,t) = \sum_{j=1}^{\infty} \frac{2j\pi(-\gamma^2\lambda^2 + \gamma^2 K + 1)\left(\Omega_0 - (-1)^j\Omega_L\right)}{-\lambda^2 + K^2 + \pi^2 j^2\gamma^2 K - \pi^2 j^2\gamma^2\lambda^2 + \pi^2 j^2} \sin\left(\frac{j\pi x}{L}\right)\cos(\omega t) \tag{36}$$

The above equation is the more general axial displacement equation, consisting of the elastic medium effect and small size effect for a composite micro-/nano-rod.

5. Frequency Determinants for the Short-Fiber-Reinforced Micro-/Nano-Rods

In this section of the study, a number of eigenvalue problems for the various degenerated cases of short-fiber-reinforced micro-/nano-rods based on nonlocal elasticity are set up. Via these eigenvalue problems, the axial vibration frequencies of the short-fiber-reinforced micro-/nano-rods are found.

5.1. General Case

To obtain the free axial frequencies of embedded short-fiber-reinforced micro-/nano-rods, size-dependent boundary conditions should be written in terms of elastic axial springs at both ends.

$$E_c A\frac{\partial u}{\partial x} + \mu\rho_c A\frac{\partial^3 u}{\partial x\partial t^2} = \Omega_0\Psi_0, \qquad x = 0 \tag{37}$$

$$E_c A\frac{\partial u}{\partial x} + \mu\rho_c A\frac{\partial^3 u}{\partial x\partial t^2} = \Omega_L\Psi_L, \qquad x = L \tag{38}$$

In the above equations, Ω_0 and Ω_L define the axial spring stiffnesses of the short-fiber-reinforced nano-rod. By inserting Equations (29) and (36) into Equations (37) and (38), the following two homogeneous equations are found:

$$\begin{gathered}\left(\gamma^2\lambda^2 - \overline{\Omega_0} - 1 + \sum_{j=1}^{\infty}\frac{2(-\gamma^2\lambda^4+\lambda^2+\gamma^2\lambda^2K-K)}{-\lambda^2+\pi^2j^2\gamma^2K+K-\pi^2j^2\gamma^2\lambda^2+\pi^2j^2}\right)\Psi_0 + \\ \left(-\gamma^2\lambda^2 + 1 - \sum_{j=1}^{\infty}\frac{2(-1)^j(-\gamma^2\lambda^4+\lambda^2+\gamma^2\lambda^2K-K)}{-\lambda^2+\pi^2j^2\gamma^2K+K-\pi^2j^2\gamma^2\lambda^2+\pi^2j^2}\right)\Psi_L = 0\end{gathered} \tag{39}$$

$$\left(-\gamma^2\lambda^2+1-\sum_{j=1}^{\infty}\frac{2(-1)^j\left(-\gamma^2\lambda^4+\lambda^2+\gamma^2\lambda^2K-K\right)}{-\lambda^2+\pi^2j^2\gamma^2K+K-\pi^2j^2\gamma^2\lambda^2+\pi^2j^2}\right)\Psi_0+ \left(\gamma^2\lambda^2-\overline{\Omega_L}-1+\sum_{j=1}^{\infty}\frac{2\left(-\gamma^2\lambda^4+\lambda^2+\gamma^2\lambda^2K-K\right)}{-\lambda^2+\pi^2j^2\gamma^2K+K-\pi^2j^2\gamma^2\lambda^2+\pi^2j^2}\right)\Psi_L=0 \tag{40}$$

In the above equations, $\overline{\Omega_0}$ and $\overline{\Omega_L}$ are the non-dimensional forms of the stiffnesses of axial springs and they are defined by:

$$\overline{\Omega_0}=\frac{\Omega_0 L}{E_m A} \tag{41}$$

$$\overline{\Omega_L}=\frac{\Omega_L L}{E_m A} \tag{42}$$

Via Equations (39) and (40), the following eigenvalue problem is constructed to be resolved for the constants Ψ_0 and Ψ_L:

$$\begin{bmatrix}\Gamma_{11} & \Gamma_{12}\\ \Gamma_{21} & \Gamma_{22}\end{bmatrix}\begin{bmatrix}\Psi_0\\ \Psi_L\end{bmatrix}=0 \tag{43}$$

The elements of the coefficient matrix are given as:

$$\Gamma_{11}=\gamma^2\lambda^2-\overline{\Omega_0}-1+\sum_{j=1}^{\infty}\frac{2\left(-\gamma^2\lambda^4+\lambda^2+\gamma^2\lambda^2K-K\right)}{-\lambda^2+\pi^2j^2\gamma^2K+K-\pi^2j^2\gamma^2\lambda^2+\pi^2j^2} \tag{44}$$

$$\Gamma_{12}=-\gamma^2\lambda^2+1-\sum_{j=1}^{\infty}\frac{2(-1)^j\left(-\gamma^2\lambda^4+\lambda^2+\gamma^2\lambda^2K-K\right)}{-\lambda^2+\pi^2j^2\gamma^2K+K-\pi^2j^2\gamma^2\lambda^2+\pi^2j^2} \tag{45}$$

$$\Gamma_{21}=-\gamma^2\lambda^2+1-\sum_{j=1}^{\infty}\frac{2(-1)^j\left(-\gamma^2\lambda^4+\lambda^2+\gamma^2\lambda^2K-K\right)}{-\lambda^2+\pi^2j^2\gamma^2K+K-\pi^2j^2\gamma^2\lambda^2+\pi^2j^2} \tag{46}$$

$$\Gamma_{22}=\gamma^2\lambda^2-\overline{\Omega_L}-1+\sum_{j=1}^{\infty}\frac{2\left(-\gamma^2\lambda^4+\lambda^2+\gamma^2\lambda^2K-K\right)}{-\lambda^2+\pi^2j^2\gamma^2K+K-\pi^2j^2\gamma^2\lambda^2+\pi^2j^2} \tag{47}$$

Free vibration frequencies of embedded short-fiber reinforced nano-rods are found by the eigenvalues of the coefficient matrix in Equation (43).

$$\left|\Gamma_{\varphi\tau}\right|=0\ (\varphi,\tau=1,2) \tag{48}$$

The above solution covers the impacts of the elastic medium, the nonlocal parameter and axial spring parameters. Nano-/micro-rods are one of the most important elements in various engineering applications. When these elements are used as a component, they need to be fixed to a place or another element. In theoretical studies of nano-/micro-rods, the fixation patterns are often investigated with the same idealized combinations. The analysis considers these combinations as clamped–clamped or clamped–free for rod elements. In these combinations, the boundary conditions are considered to be and investigated as fully rigid. On the other hand, during the realization of the engineering applications mentioned, it may not be possible to give full rigidity to the boundaries of the rods. This leads to a situation where the boundary conditions allow deformation contrary to what is assumed. This paper presents an approach to investigate the deformation-permitting boundary conditions of short-fiber-reinforced micro-/nano-rods. In applications where short-fiber-reinforced micro-/nano-rods are used or likely to be used and subjected to vibration, the dynamic behavior of these elements is important. With the solution approach presented in this study, inferences regarding the vibration behavior of short-fiber-reinforced micro-/nano-rods can be found for any desired boundary condition. For this purpose, it is sufficient to input the desired stiffness values of the axial springs attached to the ends of the micro-/nano-rod.

5.2. Without Elastic Medium Effect

To compute the free axial frequencies of short-fiber-reinforced micro-/nano-rods without the elastic medium effect, the non-dimensional elastic medium parameter in Equations (39) and (40) is set to zero. If we adjust the non-dimensional elastic medium parameter K in Equations (39) and (40) to zero, we obtain the following equations:

$$\left(\gamma^2\lambda^2-\overline{\Omega_0}-1+\sum_{j=1}^{\infty}\frac{2(-\gamma^2\lambda^4+\lambda^2)}{-\lambda^2-\pi^2j^2\gamma^2\lambda^2+\pi^2j^2}\right)\Psi_0+\left(-\gamma^2\lambda^2+1-\sum_{j=1}^{\infty}\frac{2(-1)^j(-\gamma^2\lambda^4+\lambda^2)}{-\lambda^2-\pi^2j^2\gamma^2\lambda^2+\pi^2j^2}\right)\Psi_L=0 \quad (49)$$

$$\left(-\gamma^2\lambda^2+1-\sum_{j=1}^{\infty}\frac{2(-1)^j(-\gamma^2\lambda^4+\lambda^2)}{-\lambda^2-\pi^2j^2\gamma^2\lambda^2+\pi^2j^2}\right)\Psi_0+\left(\gamma^2\lambda^2-\overline{\Omega_L}-1+\sum_{j=1}^{\infty}\frac{2(-\gamma^2\lambda^4+\lambda^2)}{-\lambda^2-\pi^2j^2\gamma^2\lambda^2+\pi^2j^2}\right)\Psi_L=0 \quad (50)$$

Thus, the eigenvalue problem to be obtained from the above two equations will be as follows:

$$\begin{bmatrix}\chi_{11} & \chi_{12}\\ \chi_{21} & \chi_{22}\end{bmatrix}\begin{bmatrix}\Psi_0\\ \Psi_L\end{bmatrix}=0 \quad (51)$$

The elements of the coefficient matrix given above are given as:

$$\chi_{11}=\gamma^2\lambda^2-\overline{\Omega_0}-1+\sum\nolimits_{j=1}^{\infty}\frac{2(-\gamma^2\lambda^4+\lambda^2)}{-\lambda^2-\pi^2j^2\gamma^2\lambda^2+\pi^2j^2} \quad (52)$$

$$\chi_{12}=-\gamma^2\lambda^2+1-\sum\nolimits_{j=1}^{\infty}\frac{2(-1)^j(-\gamma^2\lambda^4+\lambda^2)}{-\lambda^2-\pi^2j^2\gamma^2\lambda^2+\pi^2j^2} \quad (53)$$

$$\chi_{21}=-\gamma^2\lambda^2+1-\sum\nolimits_{j=1}^{\infty}\frac{2(-1)^j(-\gamma^2\lambda^4+\lambda^2)}{-\lambda^2-\pi^2j^2\gamma^2\lambda^2+\pi^2j^2} \quad (54)$$

$$\chi_{22}=\gamma^2\lambda^2-\overline{\Omega_L}-1+\sum\nolimits_{j=1}^{\infty}\frac{2(-\gamma^2\lambda^4+\lambda^2)}{-\lambda^2-\pi^2j^2\gamma^2\lambda^2+\pi^2j^2} \quad (55)$$

The free vibration frequencies of embedded short-fiber-reinforced nano-rods are found by the eigenvalues of the coefficient matrix in Equation (51).

$$|\chi_{\varphi\tau}|=0\ (\varphi,\tau=1,2) \quad (56)$$

The above solution covers the influence of the nonlocal parameter and axial spring parameters.

5.3. Without Nonlocal Effect

To compute the free axial frequencies of embedded short-fiber-reinforced micro-/nano-rods without a size effect, the non-dimensional nonlocal parameter in Equations (39) and (40) is set to zero. If we adjust the non-dimensional nonlocal parameter β in Equations (39) and (40) to zero, we obtain the following equations:

$$\left(-\overline{\Omega_0}-1+\sum_{j=1}^{\infty}\frac{2(\lambda^2-K)}{-\lambda^2+K+\pi^2j^2}\right)\Psi_0+\left(1-\sum_{j=1}^{\infty}\frac{2(-1)^j(+\lambda^2-K)}{-\lambda^2+K+\pi^2j^2}\right)\Psi_L=0 \quad (57)$$

$$\left(1-\sum_{j=1}^{\infty}\frac{2(-1)^j(\lambda^2-K)}{-\lambda^2+K+\pi^2j^2}\right)\Psi_0+\left(-\overline{\Omega_L}-1+\sum_{j=1}^{\infty}\frac{2(\lambda^2-K)}{-\lambda^2+K+\pi^2j^2}\right)\Psi_L=0 \quad (58)$$

Thus, the eigenvalue problem to be obtained from Equations (57) and (58) will be as follows:

$$\begin{bmatrix} \phi_{11} & \phi_{12} \\ \phi_{21} & \phi_{22} \end{bmatrix} \begin{bmatrix} \Psi_0 \\ \Psi_L \end{bmatrix} = 0 \tag{59}$$

The elements of the coefficient matrix given in Equation (59) are defined as:

$$\phi_{11} = -\overline{\Omega_0} - 1 + \sum_{j=1}^{\infty} \frac{2(\lambda^2 - K)}{-\lambda^2 + K + \pi^2 j^2} \tag{60}$$

$$\phi_{12} = 1 - \sum_{j=1}^{\infty} \frac{2(-1)^j(+\lambda^2 - K)}{-\lambda^2 + K + \pi^2 j^2} \tag{61}$$

$$\phi_{21} = 1 - \sum_{j=1}^{\infty} \frac{2(-1)^j(+\lambda^2 - K)}{-\lambda^2 + K + \pi^2 j^2} \tag{62}$$

$$\phi_{22} = -\overline{\Omega_L} - 1 + \sum_{j=1}^{\infty} \frac{2(\lambda^2 - K)}{-\lambda^2 + K + \pi^2 j^2} \tag{63}$$

The free vibration frequencies of embedded short-fiber-reinforced nano-rods are found by the eigenvalues of the coefficient matrix in Equation (59).

$$\left|\phi_{\varphi\tau}\right| = 0 \ (\varphi, \tau = 1, 2) \tag{64}$$

The above solution includes the impacts of the elastic medium parameter and axial spring parameters.

5.4. Without Elastic Medium and Size-Effect

To obtain the free axial frequencies of short-fiber-reinforced micro-/nano-rods without a size effect, the non-dimensional nonlocal parameter and elastic foundation parameter in Equations (39) and (40) are adjusted to zero. If we set the non-dimensional nonlocal parameter β and elastic foundation parameter K in Equations (39) and (40) to zero, we derive the following expressions:

$$\left(-\overline{\Omega_0} - 1 + \sum_{j=1}^{\infty} \frac{2\lambda^2}{-\lambda^2 + \pi^2 j^2}\right)\Psi_0 + \left(1 - \sum_{j=1}^{\infty} \frac{2(-1)^j\lambda^2}{-\lambda^2 + \pi^2 j^2}\right)\Psi_L = 0 \tag{65}$$

$$\left(1 - \sum_{j=1}^{\infty} \frac{2(-1)^j\lambda^2}{-\lambda^2 + \pi^2 j^2}\right)\Psi_0 + \left(-\overline{\Omega_L} - 1 + \sum_{j=1}^{\infty} \frac{2\lambda^2}{-\lambda^2 + \pi^2 j^2}\right)\Psi_L = 0 \tag{66}$$

With the help of Equations (64) and (65), the following eigenvalue problem is derived:

$$\begin{bmatrix} \Lambda_{11} & \Lambda_{12} \\ \Lambda_{21} & \Lambda_{22} \end{bmatrix} \begin{bmatrix} \Psi_0 \\ \Psi_L \end{bmatrix} = 0 \tag{67}$$

The elements of the coefficient matrix given in Equation (67) are defined as:

$$\Lambda_{11} = -\overline{\Omega_0} - 1 + \sum_{j=1}^{\infty} \frac{2\lambda^2}{-\lambda^2 + \pi^2 j^2} \tag{68}$$

$$\Lambda_{12} = 1 - \sum_{j=1}^{\infty} \frac{2(-1)^j\lambda^2}{-\lambda^2 + \pi^2 j^2} \tag{69}$$

$$\Lambda_{21} = 1 - \sum_{j=1}^{\infty} \frac{2(-1)^j\lambda^2}{-\lambda^2 + \pi^2 j^2} \tag{70}$$

$$\Lambda_{22} = -\overline{\Omega_L} - 1 + \sum_{j=1}^{\infty} \frac{2\lambda^2}{-\lambda^2 + \pi^2 j^2} \tag{71}$$

The free vibration frequencies of short-fiber-reinforced classical rods are found by the eigenvalues of the coefficient matrix in Equation (67).

$$\left|\Lambda_{\varphi\tau}\right| = 0 \; (\varphi, \tau = 1, 2) \tag{72}$$

The above solution includes the effects of the axial spring parameters to examine the consequences of deformable boundary conditions on the axial vibration frequencies.

It should be noted here that there is wide-ranging, prominent knowledge on the effects of axial spring parameters. The main contribution of this study is that it presents the axial vibration behavior of embedded short-fiber-reinforced nano-rods with arbitrary boundary conditions. The boundary condition is one of the significant parameters affecting the vibration behavior of any element or structure. When looking at the boundary conditions studied in the literature, it is seen that most of them examine solutions performed under rigid boundaries (clamped at both ends or clamped–free for a nano-/micro-rod). In addition, the axial vibration behavior of short-fiber-reinforced nano-sized rods was examined by Gül and Aydoğdu [1] for the first time. Gül and Aydoğdu [1] considered clamped–clamped and clamped–free boundary conditions in their study. In the present paper, we examine the longitudinal vibration of short-fiber-reinforced nano-sized rods under arbitrary support conditions for the first time and include the impact of the elastic medium in the solutions.

6. Discussion

This section of the paper is dedicated to proving the correctness of the presented solution approach and presenting several numerical examples for the randomly oriented short-fiber-reinforced composite nano-rod. For this purpose, two comparison studies are first given for two different boundary conditions, with the results presented in the paper by Gül and Aydoğdu [1]. For these comparison studies, the material and geometrical properties of the composite nano-rod are considered as follows [1]: $\rho_f/\rho_m = 4$, $E_f/E_m = 10$, $l/d = 4$, $V_f = 0.5$ and $L = 20$ nm. Table 1 compares the non-dimensional axial frequencies in the first three modes of the randomly oriented composite nano-rod with two ends clamped, while Table 2 compares the dimensionless axial vibration frequencies of the randomly oriented composite nano-rod with one end clamped and the other end free. Gül and Aydoğdu have not examined the effect of elastic medium in their study. Therefore, the elastic medium parameter K is set equal to zero in the comparison studies.

In the following part of the study, various numerical studies are performed for randomly oriented short-fiber-reinforced composite nano-sized rods with axial springs of infinite stiffness at both ends ($\overline{\Omega_0} = \overline{\Omega_L} = \infty$). Also, all calculations are done by $j = 50$. These numerical studies are visualized with the help of a number of figures and the effects of various parameters are studied in detail. The frequencies examined in the study are in dimensionless form and the non-dimensional axial frequency of the composite nano-rod ($\overline{\lambda}$) is obtained in numerical studies as follows:

$$\overline{\lambda} = \omega L \sqrt{\frac{\rho_m A}{E_m A}} \tag{73}$$

First, the impacts of the nonlocal parameter e_0a on the dimensionless frequency values of the short-fiber-reinforced nano-rod are investigated. For this purpose, the dimensionless axial frequency values for nonlocal parameter values ranging from 0 nm to 0.5 nm are plotted for the first seven modes in Figure 3. The following properties are used for this figure: $\rho_f/\rho_m = 4$, $E_f/E_m = 10, l/d = 4$, $V_f = 0.5$ and $L = 20$ nm. Also, the elastic medium effect is omitted in this example. When we look at the changes in the dimensionless axial frequencies of the short-fiber-reinforced composite nano-rod with the help of the figure, we can say that a general decrease has occurred. In the first mode, when the nonlocal parameter is 0.0 nm, 0.1 nm, 0.2 nm and 0.3 nm, or when the nonlocal parameter is 0.0 nm and 0.1 nm in the second mode, there is no change in the dimensionless axial frequencies. This can be easily explained by the impact of the nonlocal parameter on the vibrational modes. It should be noted that the amount of the decrease in the non-dimensional frequencies increases with the increase in the vibration mode number. In the first and second modes the changes are especially negligible, while in the higher modes the differences become more pronounced. Thus, it can be concluded that the impact of the nonlocal parameter on the axial frequencies of the short-fiber-reinforced composite nano-rod in higher modes is more significant.

Table 1. Comparison of the first three non-dimensional axial frequencies of randomly oriented short-fiber-reinforced composite nano-rods for the clamped–clamped boundary condition.

Mode Number	Analytical Solution [1]	Present ($\overline{\Omega_0}$=$\overline{\Omega_L}$=∞)
e_0a = 0 nm		
1	3.5819	3.5819
2	7.1639	7.1639
3	10.7459	10.7459
e_0a = 0.5 nm		
1	3.5709	3.5709
2	7.0771	7.0771
3	10.4594	10.4594
e_0a = 1 nm		
1	3.5385	3.5385
2	6.8345	6.8345
3	9.7206	9.7206

Table 2. Comparison of the first three non-dimensional axial frequencies of randomly oriented short-fiber-reinforced composite nano-rods for the clamped–free boundary condition.

Mode Number	Analytical Solution [1]	Present ($\overline{\Omega_0}$=∞, $\overline{\Omega_L}$=0)
e_0a = 0 nm		
1	1.7909	1.7909
2	5.3729	5.3729
3	8.9549	8.9549
e_0a = 0.5 nm		
1	1.7896	1.7896
2	5.3360	5.3360
3	8.7871	8.7871
e_0a = 1 nm		
1	1.7854	1.7854
2	5.2297	5.2297
3	8.3352	8.3352

Secondly, the impacts of l/d ratios on the dimensionless frequency values of the short-fiber-reinforced nano-rod are examined. For this purpose, non-dimensional axial frequency values for l/d values ranging from one to seven are illustrated for the first seven modes via Figure 4. The following properties are utilized for this investigation: $\rho_f/\rho_m = 4$, $E_f/E_m = 10$, $e_0a = 0.2$ nm, $V_f = 0.5$, $K = 0$ and $L = 20$ nm. It can be clearly seen from Figure 4 that with increasing l/d values, the dimensionless frequency values of the short-fiber-reinforced nano-rod also increase. This increment is valid for all modes examined. It should be noted that at low l/d values, the change in the dimensionless frequencies of the short-fiber-reinforced composite nano-rod is more significant. As the l/d values increase, the change in the dimensionless frequencies decreases.

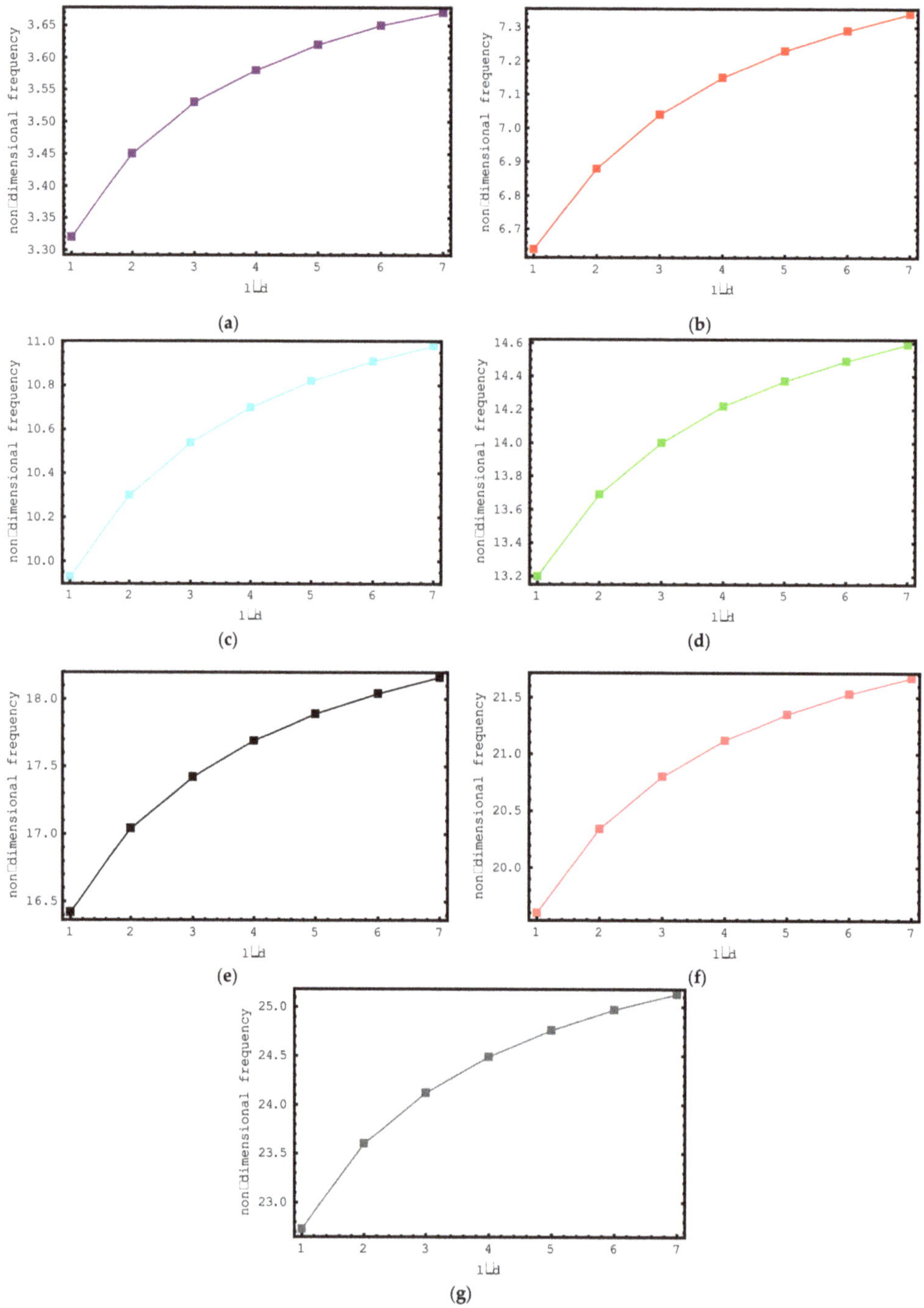

Figure 4. The variations of non-dimensional frequencies of short-fiber-reinforced composite nanorods versus l/d: (**a**) 1st mode (**b**) 2nd mode (**c**) 3rd mode (**d**) 4th mode (**e**) 5th mode (**f**) 6th mode (**g**) 7th mode.

Via Figure 5, the influence of composite nano-rod length on the dimensionless frequency values is discussed. In Figure 5, the variation in the non-dimensional axial frequency values of short-fiber-reinforced composite nano-rods versus length is plotted for the first seven modes. The length of the composite nano-rod ranges from 10 nm to 20 nm and the following properties are considered: $\rho_f/\rho_m = 2$, $E_f/E_m = 10$, $e_0a = 0.2$ nm, $V_f = 0.5$, $K = 0$ and $l/d = 2$. When we look at the changes in the dimensionless axial frequencies of the short-fiber-reinforced composite nano-rod via the figure, we can say that a general increase has occurred. In the first mode, when the length is 12 nm, 14 nm, 16 nm, 18 nm and 20 nm, there is no change in the dimensionless axial frequencies of the composite nano-rod. This can be explained by the impact of the length on the vibrational modes. It should be highlighted here that the amount of the increment in the non-dimensional frequencies increases with the increase in the vibration mode number. It may be said that the impact of nano-rod length on the non-dimensional frequencies in higher modes is more prominent.

In Figure 6, the influence of elastic foundation is investigated. For this aim, the dimensionless frequency values of the short-fiber-reinforced composite nano-rod are plotted against the dimensionless foundation parameter K for the first seven modes. The dimensionless foundation parameter impacting the composite nano-rod changes from zero to six and the following properties are considered in this investigation: $\rho_f/\rho_m = 2$, $E_f/E_m = 10$, $e_0a = 0.2$ nm, $V_f = 0.5$, $L = 20$ nm and $l/d = 2$. From this figure, the increase in the dimensionless frequencies of the short-fiber-reinforced nano-sized rod with increasing foundation parameter values can be clearly discerned. It should be noted here that if the dimensionless foundation parameter K is set to zero, the composite nano-sized rod becomes independent of the foundation effect. It should also be noted that the lowest frequency values are calculated at $K = 0$. From this it is clear that the presence of an elastic medium has a hardening effect on the short-fiber-reinforced composite nano-rod. Another important issue to be emphasized here is the influence of elastic foundation on the vibration modes. When Figure 6 is examined, it can be seen that the increases in the dimensionless frequency values are much higher in the first mode. If the modes are analyzed separately, it is clear that the lowest amount of change occurs in the seventh mode. Based on these results, it is possible to say that the elastic foundation effect is much more effective in lower modes.

Figure 7 aims to investigate the impacts of the foundation parameter and l/d ratio on the dimensionless frequencies of the short-fiber-reinforced composite nano-rod. For this purpose, Figure 7 demonstrates the alteration of the first mode dimensionless frequency values of the composite nano-rod versus l/d for various foundation parameters K. The dimensionless foundation parameter and l/d are changed from zero to six and from one to seven, respectively. In addition, the following properties are used in this investigation: $\rho_f/\rho_m = 4$, $E_f/E_m = 10$, $e_0a = 0.2$ nm, $V_f = 0.5$, $L = 20$ nm and $l/d = 2$. From here, the increase in frequencies caused by the foundation parameter and l/d can be observed again.

Figure 8 demonstrates the impacts of the foundation parameter and the E_f/E_m ratio on the first dimensionless frequencies of the embedded short-fiber-reinforced composite nano-rod. In this figure, the change in the first mode dimensionless frequencies of the composite nano-rod versus E_f/E_m for various foundation parameters K is plotted. The dimensionless foundation parameter and E_f/E_m are changed from zero to six and from 5 to 30, respectively. Furthermore, the following properties are assumed for this figure: $\rho_f/\rho_m = 4$, $l/d = 2$, $e_0a = 0.2$ nm, $V_f = 0.5$, $L = 20$ nm. It can be understood from this figure that an increment in the E_f/E_m value is accompanied by an increase in the first-mode axial frequencies.

The influence of E_f/E_m ratios on the dimensionless frequency values of the short-fiber-reinforced nano-rod are examined in Figure 9. For this purpose, the non-dimensional axial frequency values for E_f/E_m values ranging from 5 to 30 are demonstrated for the first seven modes. The following properties are considered for this study: $\rho_f/\rho_m = 4$, $e_0a = 0.2$ *nm*, $V_f = 0.5$, $K = 0$, $l/d = 2$, and $L = 20$ nm. It is clearly seen here that with increasing E_f/E_m values, the dimensionless frequencies of the composite nano-sized rod increase. It should be emphasized here that at low E_f/E_m values, the change in the dimensionless axial frequencies of the short-fiber-reinforced nano-rod is more prominent. As the E_f/E_m values increase, the change in the dimensionless axial frequencies of the composite nano-rod decreases.

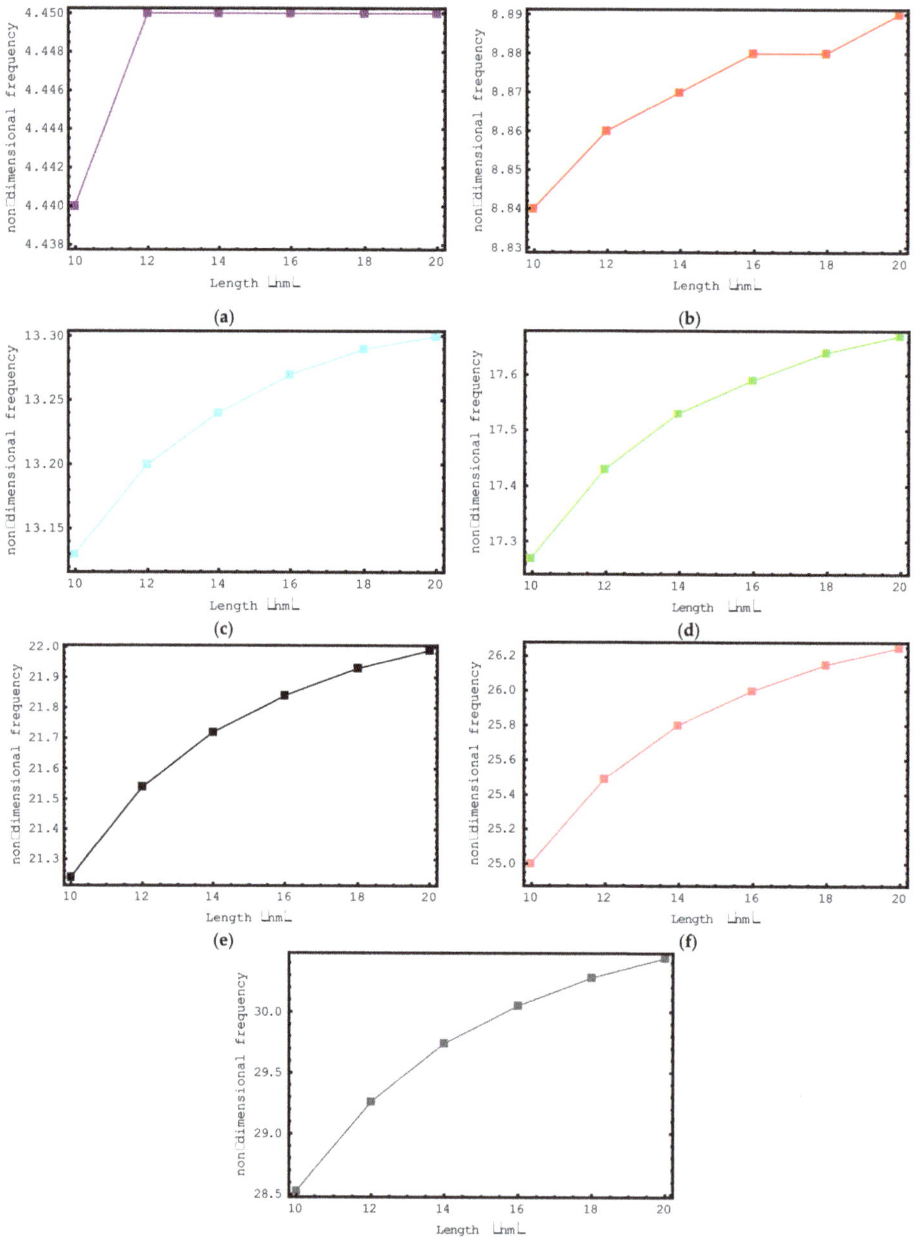

Figure 5. The variations of non-dimensional frequencies of short-fiber-reinforced composite nano-rods versus length: (**a**) 1st mode (**b**) 2nd mode (**c**) 3rd mode (**d**) 4th mode (**e**) 5th mode (**f**) 6th mode (**g**) 7th mode.

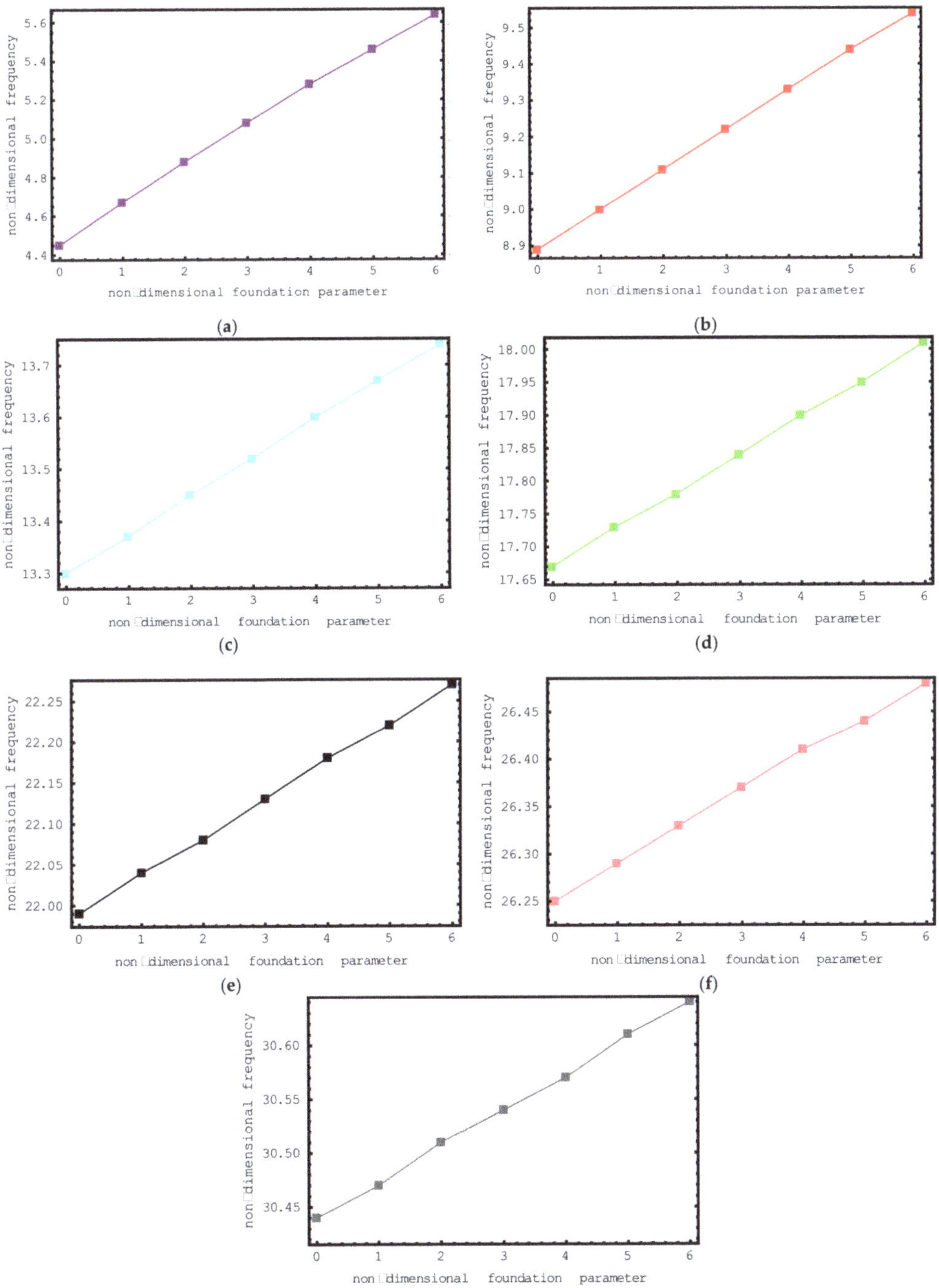

Figure 6. The variations of non-dimensional frequencies of short-fiber-reinforced composite nanorods versus the foundation parameter: (**a**) 1st mode (**b**) 2nd mode (**c**) 3rd mode (**d**) 4th mode (**e**) 5th mode (**f**) 6th mode (**g**) 7th mode.

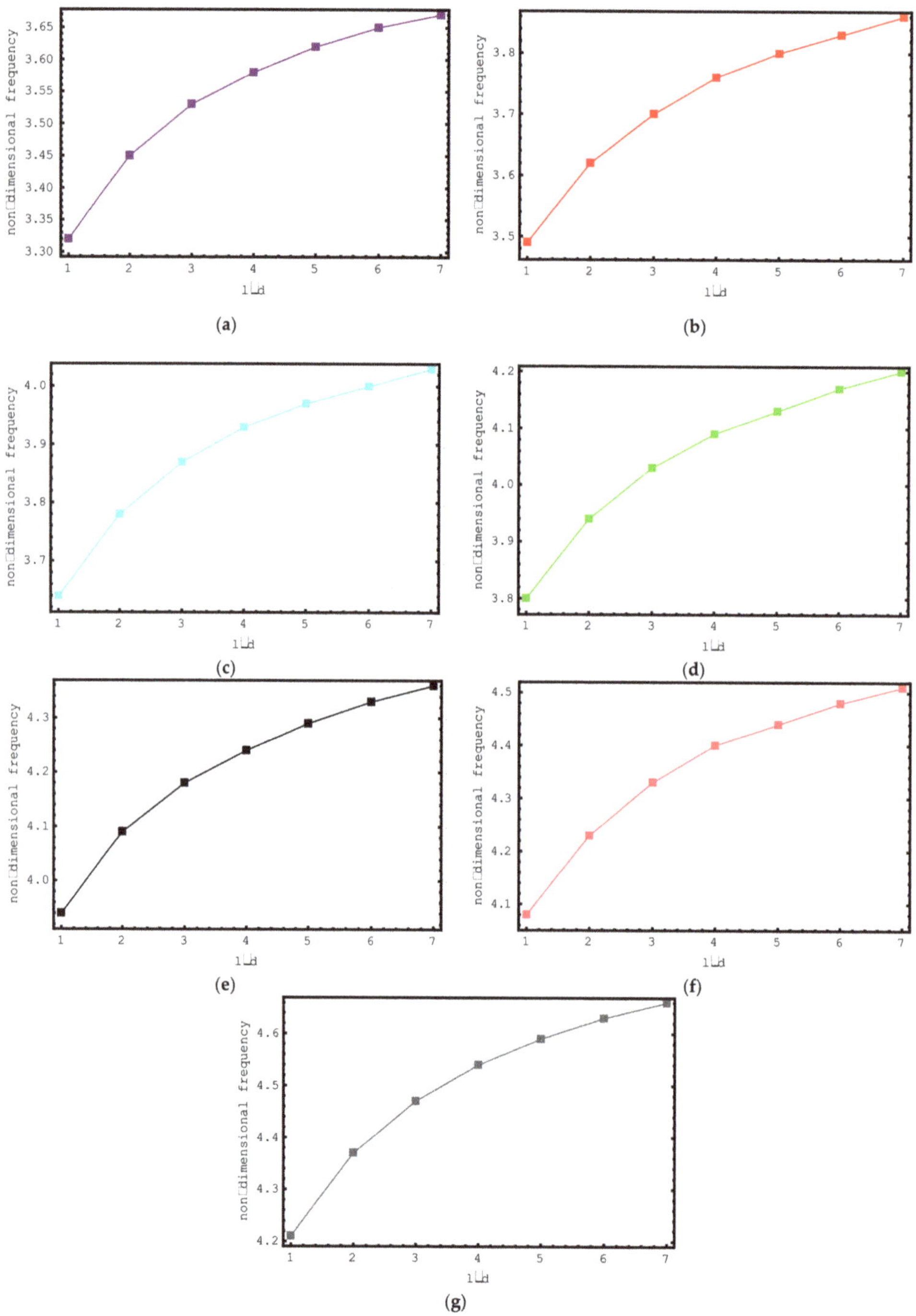

Figure 7. The variations of non-dimensional frequencies of short-fiber-reinforced composite nanorods versus l/d: (**a**) $K = 0$ (**b**) $K = 1$ (**c**) $K = 2$ (**d**) $K = 3$ (**e**) $K = 4$ (**f**) $K = 5$ (**g**) $K = 6$.

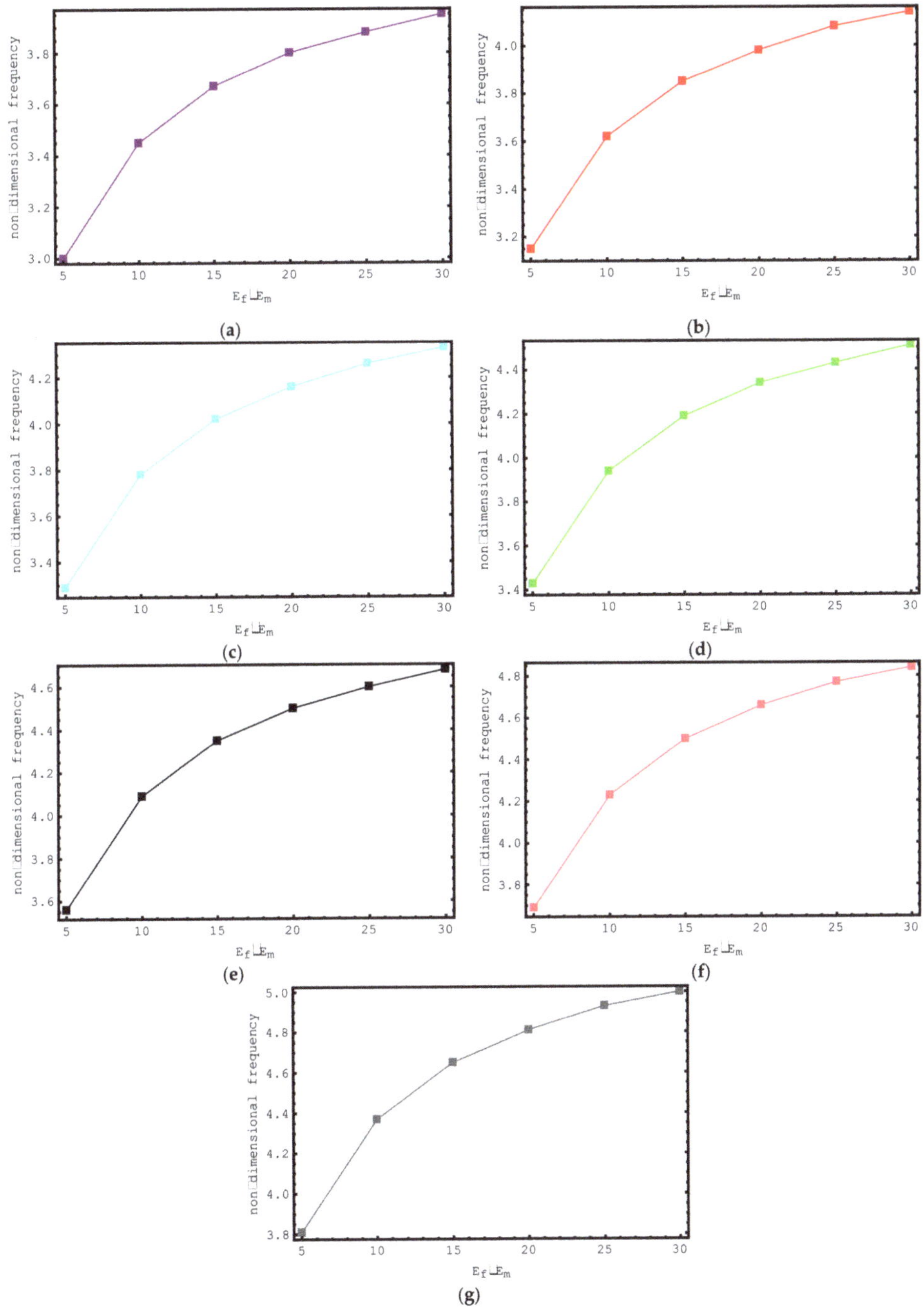

Figure 8. The variations of non-dimensional frequencies of short-fiber-reinforced composite nanorods versus E_f/E_m: (**a**) $K = 0$ (**b**) $K = 1$ (**c**) $K = 2$ (**d**) $K = 3$ (**e**) $K = 4$ (**f**) $K = 5$ (**g**) $K = 6$.

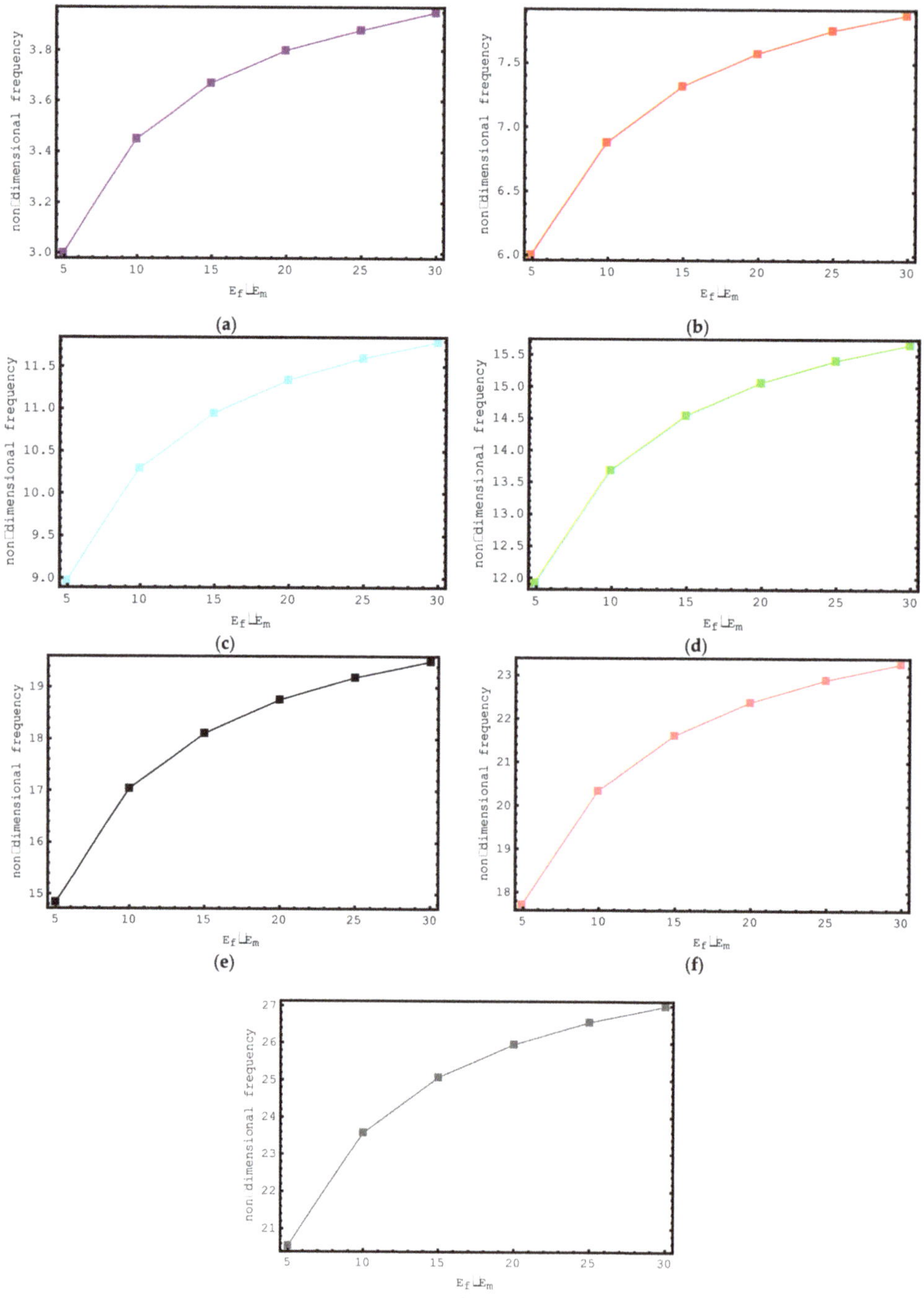

Figure 9. The variations of non-dimensional frequencies of short-fiber-reinforced composite nanorods versus E_f/E_m: (**a**) 1st mode (**b**) 2nd mode (**c**) 3rd mode (**d**) 4th mode (**e**) 5th mode (**f**) 6th mode (**g**) 7th mode.

Lastly, the impacts of ρ_f/ρ_m ratios on the dimensionless frequencies of the composite nano-sized rod are examined in Figure 10. For this purpose, non-dimensional axial frequency values for ρ_f/ρ_m values ranging from 2 to 12 are shown for the first seven modes. The following properties are considered for this study: $E_f/E_m = 10$, $e_0a = 0.2\ nm$, $V_f = 0.5$, $K = 0$, $l/d = 2$, and $L = 20$ nm. It is observed here that with increasing ρ_f/ρ_m values, the non-dimensional frequency values of the short-fiber-reinforced composite nano-rod decrease. This decrement in the frequencies is valid for all modes examined. It should be highlighted here that at low ρ_f/ρ_m values, the variation in the dimensionless axial frequencies of the short-fiber-reinforced nano-rod is more conspicuous.

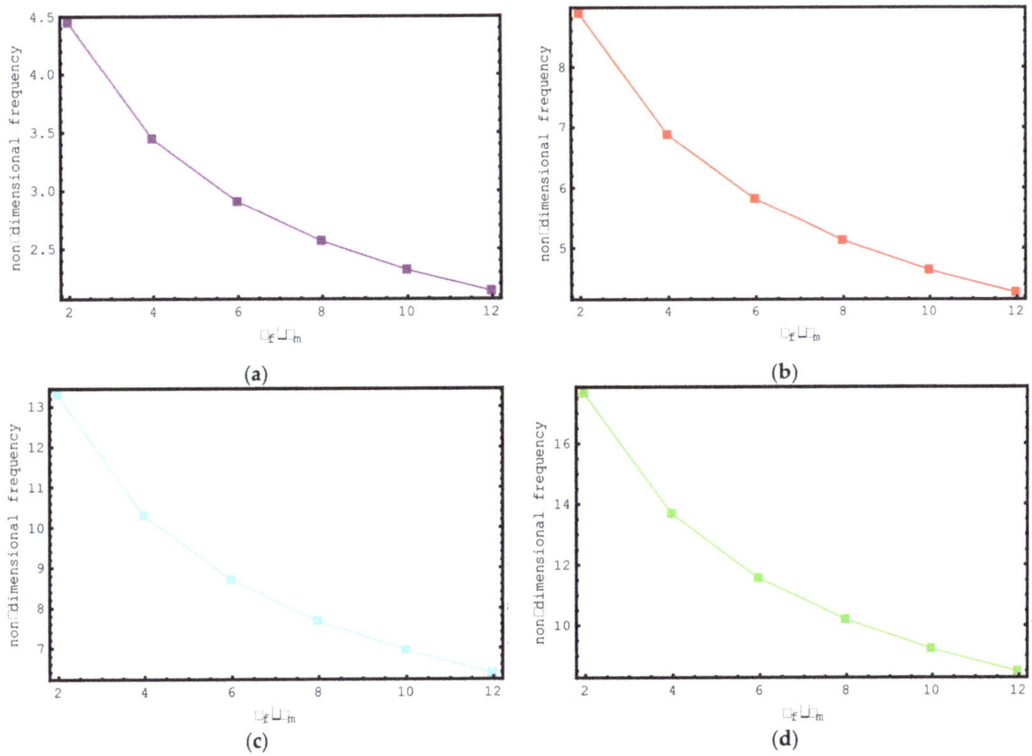

Figure 10. *Cont.*

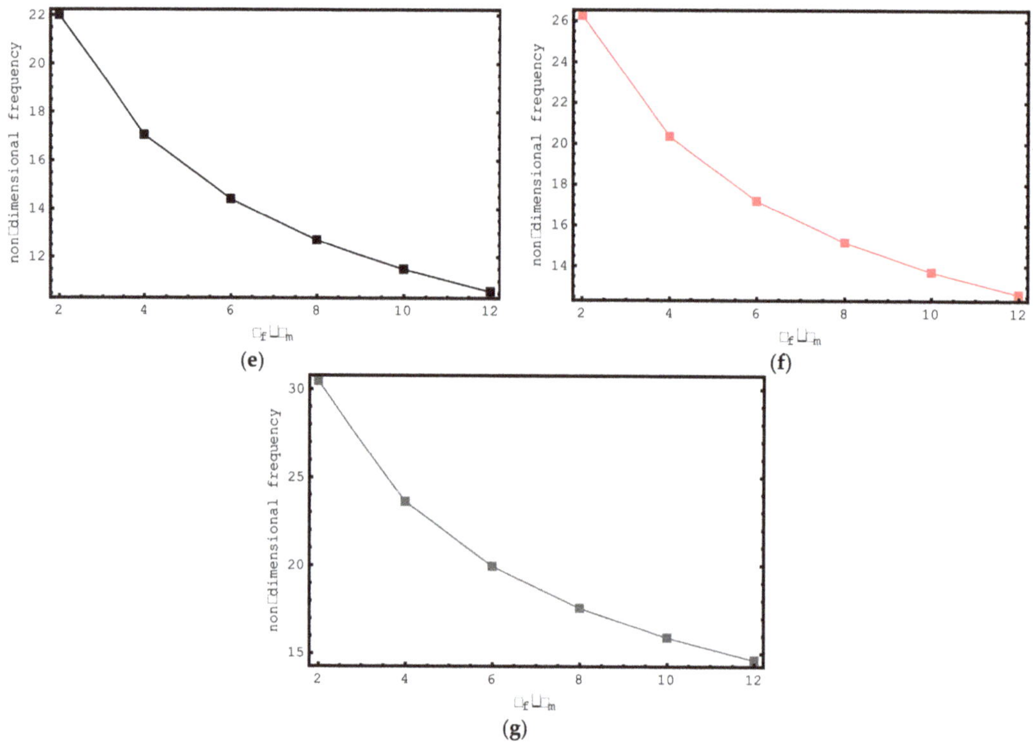

Figure 10. The variations of non-dimensional frequencies of short-fiber-reinforced composite nano-rods versus ρ_f/ρ_m: (**a**) 1st mode (**b**) 2nd mode (**c**) 3rd mode (**d**) 4th mode (**e**) 5th mode (**f**) 6th mode (**g**) 7th mode.

7. Conclusions

In this paper, the dynamics of embedded short-fiber-reinforced micro-/nano-rods have been investigated using Eringen's nonlocal elasticity theory. Based on this higher-order theory and the material properties of the nano-rods, a coefficient matrix including the elastic foundation, short fiber and the nonlocal parameter are obtained. The systems of linear equations including infinite power series are constructed by introducing the nonlocal force boundary conditions and with the help of the Stokes' transformation together with Fourier sine series. A precise and constant eigenvalue algorithm is applied to obtain the axial vibration frequencies of composite nano-rods with deformable and rigid boundary conditions. The present model is validated by comparing the analytical results with the results available in the scientific literature. Eringen's nonlocal small-scale parameter has a softening effect on the free axial vibration frequencies for all the boundary conditions (rigid or restrained), without observing the paradoxical response of nonlocal elasticity theory. The hardening effect of the short-fiber parameter is more pronounced for all the boundary conditions. Similar studies for other behaviors of short-fiber-reinforced nano-rods, such as buckling, bending, wave propagation and forced vibration propagation, can be conducted in future works.

Author Contributions: Conceptualization, Ö.C., B.U. and M.Ö.Y.; Investigation, Ö.C., B.U. and M.Ö.Y.; Methodology, Ö.C., B.U. and M.Ö.Y.; Writing—original draft, Ö.C., B.U. and M.Ö.Y.; Writing—review & editing, Ö.C., B.U. and M.Ö.Y. All authors have read and agreed to the published version of the manuscript.

Funding: This research received no external funding.

Conflicts of Interest: The authors declare no conflict of interest.

References

1. Gül, U.; Aydogdu, M. On the Axial Vibration of Viscously Damped Short-Fiber-Reinforced Nano/Micro-composite Rods. *J. Vib. Eng. Technol.* **2022**, 1–15. [CrossRef]
2. Akbaş, Ş.D. Forced vibration analysis of a fiber reinforced composite beam. *Adv. Mater. Res.* **2021**, *10*, 57–66.
3. Daikh, A.A.; Houari, M.S.A.; Belarbi, M.O.; Chakraverty, S.; Eltaher, M.A. Analysis of axially temperature-dependent functionally graded carbon nanotube reinforced composite plates. *Eng. Comput.* **2021**, *38*, 2533–2554. [CrossRef]
4. Esen, I.; Daikh, A.A.; Eltaher, M.A. Dynamic response of nonlocal strain gradient FG nanobeam reinforced by carbon nanotubes under moving point load. *Eur. Phys. J. Plus* **2021**, *136*, 458. [CrossRef]
5. Taati, E.; Borjalilou, V.; Fallah, F.; Ahmadian, M.T. On size-dependent nonlinear free vibration of carbon nanotube-reinforced beams based on the nonlocal elasticity theory: Perturbation technique. *Mech. Based Des. Struct. Mach.* **2022**, *50*, 2124–2146. [CrossRef]
6. Borjalilou, V.; Taati, E.; Ahmadian, M.T. Bending, buckling and free vibration of nonlocal FG-carbon nanotube-reinforced composite nanobeams: Exact solutions. *SN Appl. Sci.* **2019**, *1*, 1323. [CrossRef]
7. Lin, F.; Xiang, Y. Vibration of carbon nanotube reinforced composite beams based on the first and third order beam theories. *Appl. Math. Model.* **2014**, *38*, 3741–3754. [CrossRef]
8. Ke, L.L.; Yang, J.; Kitipornchai, S. Dynamic stability of functionally graded carbon nanotube-reinforced composite beams. *Mech. Adv. Mater. Struct.* **2013**, *20*, 28–37. [CrossRef]
9. Yas, M.H.; Samadi, N. Free vibrations and buckling analysis of carbon nanotube-reinforced composite Timoshenko beams on elastic foundation. *Int. J. Press. Vessel. Pip.* **2012**, *98*, 119–128. [CrossRef]
10. Madenci, E. Free vibration analysis of carbon nanotube RC nanobeams with variational approaches. *Adv. Nano Res.* **2021**, *11*, 157–171.
11. Chen, Q.; Zheng, S.; Li, Z.; Zeng, C. Size-dependent free vibration analysis of functionally graded porous piezoelectric sandwich nanobeam reinforced with graphene platelets with consideration of flexoelectric effect. *Smart Mater. Struct.* **2021**, *30*, 035008. [CrossRef]
12. Manickam, G.; Gupta, P.; De, S.; Rajamohan, V.; Polit, O. Nonlinear flexural free vibrations of size-dependent graphene platelets reinforced curved nano/micro beams by finite element approach coupled with trigonometric shear flexible theory. *Mech. Adv. Mater. Struct.* **2022**, *29*, 2489–2515. [CrossRef]
13. Sahmani, S.; Aghdam, M.M. Nonlocal strain gradient beam model for nonlinear vibration of prebuckled and postbuckled multilayer functionally graded GPLRC nanobeams. *Compos. Struct.* **2017**, *179*, 77–88. [CrossRef]
14. Yang, X.; Liu, H.; Ma, J. Thermo-mechanical vibration of FG curved nanobeam containing porosities and reinforced by graphene platelets. *Microsyst. Technol.* **2020**, *26*, 2535–2551. [CrossRef]
15. Zenkour, A.M.; Sobhy, M. Axial magnetic field effect on wave propagation in bi-layer FG graphene platelet-reinforced nanobeams. *Eng. Comput.* **2022**, *38*, 1313–1329. [CrossRef]
16. Shariati, A.; Jung, D.w.; Mohammad-Sedighi, H.; Żur, K.K.; Habibi, M.; Safa, M. On the vibrations and stability of moving viscoelastic axially functionally graded nanobeams. *Material* **2020**, *13*, 1707. [CrossRef] [PubMed]
17. Akgöz, B.; Civalek, O. Thermo-mechanical buckling behavior of functionally graded microbeams embedded in elastic medium. *Int. J. Eng. Sci.* **2014**, *85*, 90–104. [CrossRef]
18. Civalek, O.; Uzun, B.; Yaylı, M.Ö. An effective analytical method for buckling solutions of a restrained FGM nonlocal beam. *Comput. Appl. Math.* **2022**, *41*, 67. [CrossRef]
19. Ebrahimi, F.; Barati, M.R.; Civalek, O. Application of Chebyshev–Ritz method for static stability and vibration analysis of nonlocal microstructure-dependent nanostructures. *Eng. Comput.* **2020**, *36*, 953–964. [CrossRef]
20. Van Hieu, D.; Chan, D.Q.; Phi, B.G. Analysis of nonlinear vibration and instability of electrostatic functionally graded micro-actuator based on nonlocal strain gradient theory considering thickness effect. *Microsyst. Technol.* **2022**, *28*, 1845–1865. [CrossRef]
21. Eftekhari, S.A.; Toghraie, D. Vibration and dynamic analysis of a cantilever sandwich microbeam integrated with piezoelectric layers based on strain gradient theory and surface effects. *Appl. Math. Comput.* **2022**, *419*, 126867.
22. Fallah, Y.; Mohammadimehr, M. On the free vibration behavior of Timoshenko sandwich beam model with honeycomb core and nano-composite face sheet layers integrated by sensor and actuator layers. *Eur. Phys. J. Plus* **2022**, *137*, 741. [CrossRef]
23. Hong, J.; Wang, S.; Qiu, X.; Zhang, G. Bending and Wave Propagation Analysis of Magneto-Electro-Elastic Functionally Graded Porous Microbeams. *Crystals* **2022**, *12*, 732. [CrossRef]
24. Jalaei, M.H.; Thai, H.T.; Civalek, O. On viscoelastic transient response of magnetically imperfect functionally graded nanobeams. *Int. J. Eng. Sci.* **2022**, *172*, 103629. [CrossRef]
25. Kar, U.K.; Srinivas, J. Frequency analysis and shock response studies in bidirectional functionally graded microbeam with thermal effects. *J. Braz. Soc. Mech. Sci. Eng.* **2022**, *44*, 311. [CrossRef]
26. Faghidian, S.A.; Żur, K.K.; Reddy, J.N.; Ferreira, A.J.M. On the wave dispersion in functionally graded porous Timoshenko-Ehrenfest nanobeams based on the higher-order nonlocal gradient elasticity. *Compos. Struct.* **2022**, *279*, 114819. [CrossRef]

27. Tang, Y.; Qing, H. Size-dependent nonlinear post-buckling analysis of functionally graded porous Timoshenko microbeam with nonlocal integral models. *Commun. Nonlinear Sci. Numer. Simul.* **2022**, *116*, 106808. [CrossRef]
28. Uzun, B.; Yaylı, M.Ö. Porosity dependent torsional vibrations of restrained FG nanotubes using modified couple stress theory. *Mater. Today Commun.* **2022**, *32*, 103969. [CrossRef]
29. Eltaher, M.A.; Mohamed, N. Nonlinear stability and vibration of imperfect CNTs by doublet mechanics. *Appl. Math. Comput.* **2020**, *382*, 125311. [CrossRef]
30. Fatahi-Vajari, A.; Imam, A. Axial vibration of single-walled carbon nanotubes using doublet mechanics. *Indian J. Phys.* **2016**, *90*, 447–455. [CrossRef]
31. Gul, U.; Aydogdu, M. Noncoaxial vibration and buckling analysis of embedded double-walled carbon nanotubes by using doublet mechanics. *Compos. Part B Eng.* **2018**, *137*, 60–73. [CrossRef]
32. Gul, U.; Aydogdu, M.; Gaygusuzoglu, G. Vibration and buckling analysis of nanotubes (nanofibers) embedded in an elastic medium using Doublet Mechanics. *J. Eng. Math.* **2018**, *109*, 85–111. [CrossRef]
33. Karamanli, A. Structural behaviours of zigzag and armchair nanobeams using finite element doublet mechanics. *Eur. J. Mech.-A/Solids* **2021**, *89*, 104287. [CrossRef]
34. Alizadeh, A.; Shishehsaz, M.; Shahrooi, S.; Reza, A. Free vibration characteristics of viscoelastic nano-disks based on modified couple stress theory. *J. Strain Anal. Eng. Des.* **2022**, 03093247221116053. [CrossRef]
35. Babaei, A.; Arabghahestani, M. Free vibration analysis of rotating beams based on the modified couple stress theory and coupled displacement field. *Appl. Mech.* **2021**, *2*, 226–238. [CrossRef]
36. Hassannejad, R.; Hosseini, S.A.; Alizadeh-Hamidi, B. Influence of non-circular cross section shapes on torsional vibration of a micro-rod based on modified couple stress theory. *Acta Astronaut.* **2021**, *178*, 805–812. [CrossRef]
37. Kumar, H.; Mukhopadhyay, S. Response of deflection and thermal moment of Timoshenko microbeams considering modified couple stress theory and dual-phase-lag heat conduction model. *Compos. Struct.* **2021**, *263*, 113620. [CrossRef]
38. Al-Shewailiah, D.M.R.; Al-Shujairi, M.A. Static bending of functionally graded single-walled carbon nanotube conjunction with modified couple stress theory. *Mater. Today Proc.* **2022**, *61*, 1023–1037. [CrossRef]
39. Jena, S.K.; Chakraverty, S.; Mahesh, V.; Harursampath, D. Application of Haar wavelet discretization and differential quadrature methods for free vibration of functionally graded micro-beam with porosity using modified couple stress theory. *Eng. Anal. Bound. Elem.* **2022**, *140*, 167–185. [CrossRef]
40. Abouelregal, A.E.; Ersoy, H.; Civalek, O. Solution of Moore–Gibson–Thompson equation of an unbounded medium with a cylindrical hole. *Mathematics* **2021**, *9*, 1536. [CrossRef]
41. Civalek, O.; Uzun, B.; Yaylı, M.Ö.; Akgöz, B. Size-dependent transverse and longitudinal vibrations of embedded carbon and silica carbide nanotubes by nonlocal finite element method. *Eur. Phys. J. Plus* **2020**, *135*, 381. [CrossRef]
42. Numanoğlu, H.M.; Ersoy, H.; Akgöz, B.; Civalek, O. A new eigenvalue problem solver for thermo-mechanical vibration of Timoshenko nanobeams by an innovative nonlocal finite element method. *Math. Methods Appl. Sci.* **2022**, *45*, 2592–2614. [CrossRef]
43. Yayli, M.Ö.; Yanik, F.; Kandemir, S.Y. Longitudinal vibration of nanorods embedded in an elastic medium with elastic restraints at both ends. *Micro Nano Lett.* **2015**, *10*, 641–644. [CrossRef]
44. Aydogdu, M. Axial vibration analysis of nanorods (carbon nanotubes) embedded in an elastic medium using nonlocal elasticity. *Mech. Res. Commun.* **2012**, *43*, 34–40. [CrossRef]
45. Uzun, B.; Kafkas, U.; Deliktaş, B.; Yaylı, M.Ö. Size-Dependent Vibration of Porous Bishop Nanorod with Arbitrary Boundary Conditions and Nonlocal Elasticity Effects. *J. Vib. Eng. Technol.* **2022**, 1–18. [CrossRef]
46. Alizadeh Hamidi, B.; Khosravi, F.; Hosseini, S.A.; Hassannejad, R. Closed form solution for dynamic analysis of rectangular nanorod based on nonlocal strain gradient. *Waves Random Complex Media* **2020**, *32*, 2067–2083. [CrossRef]
47. Babaei, A. Forced vibration analysis of non-local strain gradient rod subjected to harmonic excitations. *Microsyst. Technol.* **2021**, *27*, 821–831. [CrossRef]
48. Li, L.; Hu, Y.; Li, X. Longitudinal vibration of size-dependent rods via nonlocal strain gradient theory. *Int. J. Mech. Sci.* **2016**, *115*, 135–144. [CrossRef]
49. Zhu, X.; Li, L. Closed form solution for a nonlocal strain gradient rod in tension. *Int. J. Eng. Sci.* **2017**, *119*, 16–28. [CrossRef]
50. Akgöz, B.; Civalek, O. A size-dependent shear deformation beam model based on the strain gradient elasticity theory. *Int. J. Eng. Sci.* **2013**, *70*, 1–14. [CrossRef]
51. Akgöz, B.; Civalek, O. Bending analysis of embedded carbon nanotubes resting on an elastic foundation using strain gradient theory. *Acta Astronaut.* **2016**, *119*, 1–12. [CrossRef]
52. Güven, U. Love–Bishop rod solution based on strain gradient elasticity theory. *C. R. Méc.* **2014**, *342*, 8–16. [CrossRef]
53. Narendar, S.; Ravinder, S.; Gopalakrishnan, S. Strain gradient torsional vibration analysis of micro/nano rods. *Int. J. Nano Dimens.* **2012**, *3*, 1–17.
54. Thai, L.M.; Luat, D.T.; Phung, V.B.; Minh, P.V.; Thom, D.V. Finite element modeling of mechanical behaviors of piezoelectric nanoplates with flexoelectric effects. *Arch. Appl. Mech.* **2022**, *92*, 163–182. [CrossRef]
55. Duc, D.H.; Thom, D.V.; Cong, P.H.; Minh, P.V.; Nguyen, N.X. Vibration and static buckling behavior of variable thickness flexoelectric nanoplates. *Mech. Based Des. Struct. Mach.* **2022**, 1–29. [CrossRef]
56. Tho, N.C.; Ta, N.T.; Thom, D.V. New numerical results from simulations of beams and space frame systems with a tuned mass damper. *Materials* **2019**, *12*, 1329. [CrossRef]

57. Tuan, L.T.; Dung, N.T.; Van Thom, D.; Van Minh, P.; Zenkour, A.M. Propagation of non-stationary kinematic disturbances from a spherical cavity in the pseudo-elastic cosserat medium. *Eur. Phys. J. Plus* **2021**, *136*, 1199. [CrossRef]
58. Eringen, A.C. On differential equations of nonlocal elasticity and solutions of screw dislocation and surface waves. *J. Appl. Phys.* **1983**, *54*, 4703–4710. [CrossRef]
59. Halpin, J.C. *Effects of Environmental Factors on Composite Materials*; Air Force Materials Lab Wright-Patterson AFB: Dayton, OH, USA, 1969.
60. Agarwal, B.D.; Broutman, L.J. *Analysis and Performance of Fiber Composites*, 2nd ed.; Wiley: New York, NY, USA, 1990.

materials

Article

Bimetallic Thin-Walled Box Beam Thermal Buckling Response

Sandra Kvaternik Simonetti, Goran Turkalj, Damjan Banić and Domagoj Lanc *

Department of Engineering Mechanics, Faculty of Engineering, University of Rijeka, 51000 Rijeka, Croatia
* Correspondence: dlanc@riteh.hr

Abstract: A beam model for thermal buckling analysis of a bimetallic box beam is presented. The Euler–Bernoulli–Vlasov beam theory is employed considering large rotations but small strains. The nonlinear stability analysis is performed using an updated Lagrangian formulation. In order to account for the thermal effects of temperature-dependent (TD) and temperature-independent (TID) materials, a uniform temperature rise through beam wall thickness is considered. The numerical results for thin-walled box beams are presented to investigate the effects of different boundary conditions, beam lengths and material thickness ratios on the critical buckling temperature and post-buckling responses. The effectiveness and accuracy of the proposed model are verified by means of comparison with a shell model. It is revealed that all of the abovementioned effects are invaluable for buckling analysis of thin-walled beams under thermal load. Moreover, it is shown that the TD solutions give lower values than the TID one, emphasizing the importance of TD materials in beams.

Keywords: thin-walled; FEM; thermal buckling analysis

Citation: Simonetti, S.K.; Turkalj, G.; Banić, D.; Lanc, D. Bimetallic Thin-Walled Box Beam Thermal Buckling Response. *Materials* **2022**, *15*, 7537. https://doi.org/10.3390/ma15217537

Academic Editor: Haim Abramovich

Received: 28 September 2022
Accepted: 24 October 2022
Published: 27 October 2022

1. Introduction

Thin-walled beams and structures are increasingly used in engineering branches, in standalone forms and as a stiffeners for plate- and shell-like structures, due to their high strength and light weight. However, these structures show susceptibility to local buckling and buckling failure [1,2]. Buckling analysis and the post-buckling response of such weight-optimized structures have been the topic of many research papers, such as [3–6], especially in the field of composite materials [7–11].

If the thermal environment is considered, the stability of structures has received significant attention in recent years: Duan et al. [12] performed thermal analysis of a beam element, Saha and Ali [13] presented a post-buckling mathematical model of a slender road under uniform temperature rise, while Cui and Hu [14] analyzed the thermal buckling and vibration of a beam. Jeyaraj et al. [15] investigated experimental and theoretical non-uniform heating of an isotropic beam. Burgreen and Mannit [16] and Burgreen and Regal [17] analysed the thermal buckling of bimetallic beams. In the case of composite beams, Aydogdu [18] obtained critical buckling temperatures of composite beams, Luan et al. [19] presented an analytical solution for buckling and vibration of FG beams, Kiani and Eslami [20,21] investigated buckling analysis under different types of thermal loads, while Giunta [22] analyzed FG beams under thermal/mechanical load using the Carrera unified formulation. However, there are not many papers about thermal buckling analyses of thin-walled structures: Librescu [23] studied stability problems in a high-temperature environment and Ziane et al. [24] studied analytical methods for buckling and vibration responses of porous beams under thermomechanical loads.

In the present work, thermal buckling analysis of a thin-walled bimetallic box beam and frame structures is presented. The material is assumed to be linear, elastic and isotropic. The model is based on Euler–Bernoulli–Vlasov theory and on assumptions of large rotations and small strains. It is also assumed that the cross-section is not deformed in its own plane and that there are no shear strains in the middle surface. The nonlinear displacement field, which includes nonlinear displacement terms due to large rotation effects, is implemented.

Using the UL description, the element geometric stiffness is derived. As an incremental iterative solution scheme, the Newton–Raphson method is used. Furthermore, this paper is a continuation of the research in which thermal buckling analysis of temperature-independent materials was conducted [25], which has now been further expanded with temperature-dependent materials' properties. As far as the authors are aware, there is no beam model solution for thermal buckling analysis of thin-walled beam-type structures with temperature-dependent materials' properties.

The numerical results for thin-walled box beams are presented to investigate the effects of different boundary conditions, namely clamped–clamped, simply supported and clamped–simply supported, beam lengths and material thickness ratios on the critical buckling temperature and post buckling responses. In order to demonstrate the accuracy of the numerical algorithm, benchmark examples using shell FEM code were developed. Numerical results show that the abovementioned effects have a huge impact on the buckling analysis.

2. Materials and Methods

2.1. Kinematics

Two sets of coordinate systems related to the angle of orientation β are considered. The first one is a Cartesian (z, x, y) coordinate system where the z-axis coincides with the longitudinal beam that passes through the centroid O of each cross section, while the x- and y-axes are principal axes. The second one is a contour coordinate system where the s-axis is tangential to the middle surface directed along the contour line of the cross-section while the n-axis is perpendicular to the s-axis.

The field of incremental displacement measures of a cross section are defined as [8]:

$$\begin{aligned} w_0 = w_0(z);\ u_s = u_s(z);\ v_s = v_s(z);\ \varphi_z = \varphi_z(z); \\ \varphi_x = \varphi_x(z) = -\tfrac{dv_s}{dz};\ \varphi_y = \varphi_y(z) = \tfrac{du_s}{dz};\ \theta = \theta(z) = -\tfrac{d\varphi_z}{dz}(z), \end{aligned} \tag{1}$$

where w_0, u_s and v_s are the rigid-body translations of the cross-section centroid in the z-, x- and y-direction, respectively, while φ_z, φ_x and φ_y are the rigid-body rotations about the aforementioned axis; θ is a warping parameter of the cross-section.

In the case of small rotations, the incremental displacement field consists of the first-order displacement values:

$$\begin{aligned} u_z(z,x,y) &= w_0(z) - y\tfrac{dv_s}{dz}(z) - x\tfrac{du_s}{dz}(z) - \omega(x,y)\tfrac{d\varphi_z}{dz}(z), \\ u_x(z,x,y) &= u_s(z) - (y - y_s)\varphi_z(z) \\ u_y(z,x,y) &= v_s(z) + (x - x_s)\varphi_z(z), \end{aligned} \tag{2}$$

where u_z, u_x and u_y are the linear displacement increments of an arbitrary point on the cross-section defined by the x and y coordinates and the warping function $\omega(x,y)$. When the large rotations are considered, nonlinear displacement increments are expressed as follows:

$$\begin{aligned} \tilde{u}_z(z,x,y) &= 0.5\left[-(x - x_s)\varphi_z\varphi_x + (y - y_s)\varphi_z\varphi_y\right], \\ \tilde{u}_x(z,x,y) &= 0.5\left\{-\varphi_x\varphi_y y - \left[\varphi_z^2 + \varphi_y{}^2\right]x + x_s\varphi_z^2\right\}, \\ \tilde{u}_y(z,x,y) &= 0.5\left\{-\varphi_x\varphi_y x - \left[\varphi_z^2 + \varphi_x{}^2\right]y + y_s\varphi_z^2\right\}, \end{aligned} \tag{3}$$

and should be added to those from Equation (2).

Considering the nonlinear displacement field, the Green–Lagrange strain tensor components can be written as:

$$\begin{aligned} \varepsilon_{ij} = \tfrac{1}{2}\left[(u_i + \tilde{u}_i)_{,j} + (u_j + \tilde{u}_j)_{,i} + (u_k + \tilde{u}_k)_{i} + (u_k + \tilde{u}_k)_{,j}\right] &\cong e_{ij} + \eta_{ij} + \tilde{e}_{ij}, \\ 2e_{ij} &= u_{i,j} + u_{j,i} \\ 2\eta_{ij} &= u_{k,i} + u_{k,j} \\ 2\tilde{e}_{ij} &= \tilde{u}_{i,j} + \tilde{u}_{j,i} \end{aligned} \tag{4}$$

where e_{ij} and η_{ij} are the linear and nonlinear strain components corresponding to the linear displacement, while $\tilde{e}_{ij}$ is the linear strain component corresponding to the nonlinear displacement due to the large rotations.

The contour mid-line displacement $\overline{w}$, and $\overline{v}$ can be seen more detail in [26].

Due to the in-plane rigidity hypothesis of the cross-section, the non-zero strain components are [7]:

$$e_{zz} = \frac{\partial w}{\partial z},\ e_{zs} = \frac{\partial w}{\partial s} + \frac{\partial v}{\partial z}, \tag{5}$$

$$\begin{aligned} \eta_{zz} &= \frac{1}{2}\left[\left(\frac{\partial w}{\partial z}\right)^2 + \left(\frac{\partial u}{\partial z}\right)^2 + \left(\frac{\partial v}{\partial z}\right)^2\right], \\ \eta_{zs} &= \frac{\partial w}{\partial z}\frac{\partial w}{\partial s} + \frac{\partial u}{\partial z}\frac{\partial u}{\partial s} + \frac{\partial v}{\partial z}\frac{\partial v}{\partial s}, \end{aligned} \tag{6}$$

$$\tilde{e}_{zz} = \frac{\partial w}{\partial z},\ \tilde{e}_{zs} = \frac{\partial w}{\partial s} + \frac{\partial v}{\partial z}. \tag{7}$$

The stress resultants of the beam can be defined as:

$$\begin{aligned} F_z &= \int_A \sigma_z \mathrm{d}n\mathrm{d}s, \\ M_x = \int_A \sigma_z (y - n\cos\beta)\mathrm{d}n\mathrm{d}s,&\ M_y = \int_A \sigma_z (x + n\sin\beta)\mathrm{d}n\mathrm{d}s, \\ M_t = \int_A \tau_{sz}\left(n + \frac{F_s}{t}\right)\mathrm{d}n\mathrm{d}s,&\ M_\omega = \int_A \sigma_z(\omega - nq)\mathrm{d}n\mathrm{d}s, \end{aligned} \tag{8}$$

where F_z represents axial force, M_x and M_y are bending moments with respect to x- and y-axis, respectively, M_t is the torsion moment and M_ω is the warping moment (bimoment). t is the thickness of the closed section contour and F_s is the St. Venant circuit flow [26].

2.2. Constitutive Equations

Consider a bimetallic beam made of two different metals. A beam wall with a core thickness of Ti–6Al–4V λt on the outer surface and SUS304 in the inner part of cross-section beam wall is shown in Figure 1. It is assumed that the layers of materials are perfectly bonded.

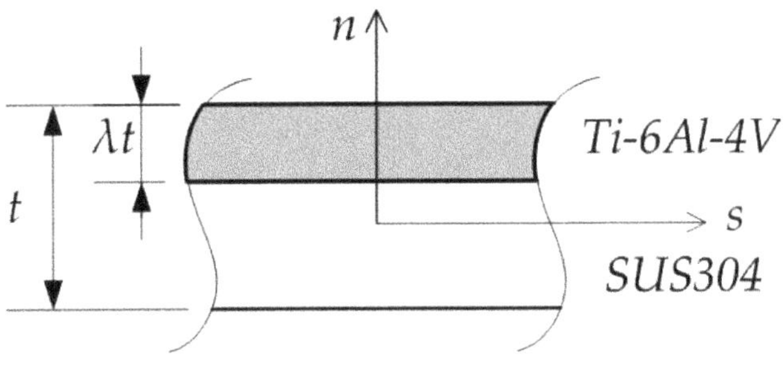

Figure 1. Bimetallic beam wall.

If the thermo-elastic material properties are considered as a function of temperature T, they can be calculated for each material, as described in [19,27]:

$$P(T) = P_0\left(1 + P_{-1}T^{-1} + P_1T + P_2T^2 + P_3T^3\right), \tag{9}$$

where P represents Young's modulus E and thermal expansion coefficient α, while P_0, P_{-1}, P_1, P_2 and P_3 are temperature-dependent coefficients listed in Table 1 for different metals [28]. For simplicity, Poisson's ratio ν is assumed to be constant, $\nu = 0.3$.

Table 1. Temperature dependent coefficients [19].

Material	Properties	P_0	P_1	P_1	P_2	P_3
Ti–6Al–4V	E (Pa)	122.56×10^9	0.0	-4.586×10^{-4}	0.0	0.0
	α (1/K)	7.5788×10^{-6}	0.0	6.638×10^{-4}	-3.147×10^{-6}	0.0
SUS304	E (Pa)	201.04×10^9	0.0	3.079×10^{-4}	-6.534×10^{-7}	0.0
	α (1/K)	12.330×10^{-6}	0.0	8.086×10^{-4}	0.0	0.0

It is assumed that the temperature of the whole beam is uniform and increased from the current ambient temperature T_0 to the critical value in incremental steps of 1 °C. If the axial displacements are prevented, the temperature at a point $T(n,z)$ may be raised to $T+\Delta T$, in the way that the beam buckles. ΔT is the temperature rise. The temperature that is read as the critical buckling temperature is the temperature difference compared to the ambient temperature. The process can be described as quasi-adiabatic since the heat exchange between the environment and the beam is neglected.

The stress–strain relations of the bimetallic beam can be written as:

$$\begin{pmatrix} \sigma_z \\ \tau_{sz} \end{pmatrix} = \begin{pmatrix} E(n,z,T) & 0 \\ 0 & G(n,z,T) \end{pmatrix} \cdot \begin{pmatrix} e_{zz} - \alpha(n,z,T)\Delta T \\ \gamma_{sz} \end{pmatrix}, \tag{10}$$

where

$$G(n,z) = \frac{E(n,z,T)}{2[1+\nu(n,z,T)]}. \tag{11}$$

Using Equations (5), (10) and (11), the beam forces can be expressed in a matrix form as:

$$\begin{Bmatrix} F_z \\ M_y \\ M_x \\ M_\omega \\ M_t \end{Bmatrix} = \begin{bmatrix} R_{11} & R_{12} & R_{13} & R_{14} & 0 \\ R_{21} & R_{22} & R_{23} & R_{24} & 0 \\ R_{31} & R_{32} & R_{33} & R_{34} & 0 \\ R_{41} & R_{42} & R_{43} & R_{44} & 0 \\ 0 & 0 & 0 & 0 & R_{55} \end{bmatrix} \begin{Bmatrix} dw_0/dz \\ -d^2u_s/dz^2 \\ -d^2v_s/dz^2 \\ -d^2\varphi_z/dz^2 \\ 2d\varphi_z/dz \end{Bmatrix} - \begin{Bmatrix} N_z^T \\ M_y^T \\ M_x^T \\ M_\omega^T \\ 0 \end{Bmatrix}, \tag{12}$$

where R_{ij} represents the thin-walled beam stiffness, as shown in Appendix A. N_z^T, M_x^T, M_y^T and M_ω^T are thermal force and thermal moments, respectively:

$$\begin{aligned} N_z^T &= \int_A E(n,z,T)\alpha(n,z,T)\Delta T \mathrm{d}n\mathrm{d}s; \\ M_x^T &= \int_A E(n,z,T)\alpha(n,z,T)\Delta T(y - n\cos\beta)\mathrm{d}n\mathrm{d}s \\ M_y^T &= \int_A E(n,z,T)\alpha(n,z,T)\Delta T(x + n\sin\beta)\mathrm{d}n\mathrm{d}s \\ M_\omega^T &= \int_A E(n,z,T)\alpha(n,z,T)\Delta T(\omega - nq)\mathrm{d}n\mathrm{d}s \end{aligned} \tag{13}$$

2.3. Finite Element Formulation

A two-nodded beam element with 14 degrees of freedom is shown in Figure 2. The nodal displacements and nodal force vectors are as follows:

$$(\mathrm{u}^e)^T = \{w_A\ u_A\ v_A\ \varphi_{zA}\ \varphi_{xA}\ \varphi_{yA}\ w_B\ u_B\ v_B\ \varphi_{zB}\ \varphi_{xB}\ \varphi_{yB}\ \theta_A\ \theta_B\} \tag{14}$$

$$(\mathrm{f}^e)^T = \{F_{zA}\ F_{xA}\ F_{yA}\ M_{zA}\ M_{xA}\ M_{yA}\ F_{zB}\ F_{xB}\ F_{yB}\ M_{zB}\ M_{xB}\ M_{yB}\ M_{\omega A}\ M_{\omega B}\} \tag{15}$$

where the superscript e denotes the eth finite element. It should be noted that nodal displacement w and nodal forces F_z, M_x and M_y are defined in the centroid O, while other nodal components are defined in the shear center.

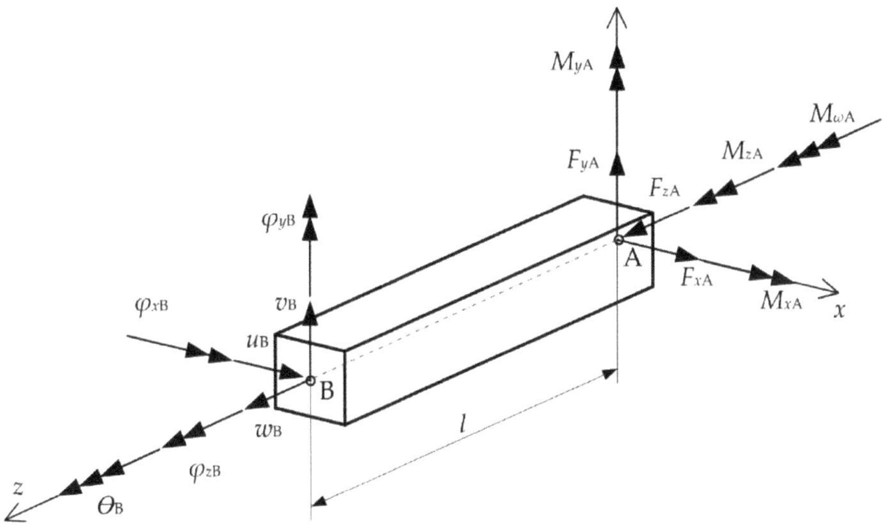

Figure 2. Nodal force vectors and displacements.

Applying the principle of the virtual work, the incremental equilibrium equations of a beam element in linearized form are:

$$\delta U_E + \delta U_G = \delta^2 W - \delta^1 W, \tag{16}$$

where the equations from the left side consist of incremental virtual elastic strain energy:

$$\delta U_E = \int_{1V} {}_1C_{ijkl}\;{}_1e_{kl}\;\delta_1\, e_{ij}{}^1\;\mathrm{d}V, \tag{17}$$

and the incremental virtual geometric potential:

$$\delta U_G = \int_{1V} {}^1S_{ij}\;\delta_1\,\eta_{ij}{}^1\;\mathrm{d}V + \int_V {}^1S_{ij}\;\delta_1\,\tilde{e}_{ij}{}^1\;\mathrm{d}V - \int_{{}^1A_\sigma} {}^1t_i\;\delta\tilde{u}_i{}^1\;\mathrm{d}A_\sigma. \tag{18}$$

On the right side of the equations, the terms represent the virtual work completed by external forces at the end and at the beginning of the present increment:

$$\begin{aligned} \delta^2 W &= \textstyle\int_{{}^1A_\sigma} {}^2t_i\;\delta u_i{}^1\;\mathrm{d}A_\sigma, \\ \delta^1 W &= \textstyle\int_{1V} {}^1S_{ij}\;\delta_1\, e_{ij}{}^1\;\mathrm{d}V = \int_{{}^1A_\sigma} {}^1t_i\;\delta u_i{}^1\;\mathrm{d}A_\sigma. \end{aligned} \tag{19}$$

In these equations, S_{ij} is the second Piola–Kirchoff stress tensor, t_i denotes the surface tractions, C_{ijkl} presents the stress–strain tensor and the symbol δ indicates virtual quantities. By applying the linear interpolation functions for w_0 displacement and cubic interpolations for w_s, u_s and v_s, one can obtain:

$$\delta U_E = \int_V \mathrm{S_{ij}}\delta \mathrm{e_{ij}}\mathrm{d}V = (\delta \mathrm{u}^e)^T \mathrm{k}_E^e \mathrm{u}^e, \tag{20}$$

$$\delta U_G = \int_V \mathrm{S_{ij}}(\delta\eta_{\mathrm{ij}} + \delta\tilde{\mathrm{e}}_{\mathrm{ij}})\mathrm{d}V - \int_{A_\sigma} \mathrm{t_i}\delta\tilde{\mathrm{u}}_{\mathrm{i}}\mathrm{d}A_\sigma = (\delta \mathrm{u}^e)^T \mathrm{k}_G^e \mathrm{u}^e \tag{21}$$

$$\delta W = \int_{A_\sigma} t_i \delta u_i \mathrm{d}A_\sigma = (\delta \mathrm{u}^e)^T \mathrm{f}^e \quad (22)$$

where f^e is the nodal force vector, k_E^e is the elastic stiffness matrix and k_G^e is the geometric stiffness matrix of the beam element. Nonlinear equilibrium equations are solved using the Newton–Raphson method as incremental iterative approach [29,30], and the explicit form of the terms given in nonlinear components were described previously in [31].

3. Results and Discussion

In numerical examples, the thin-walled box beam with height $h = 100$ mm, width $b = 150$ mm and thickness $t = 10$ mm is considered (Figure 3). For verification purposes, the critical buckling temperatures were obtained by shell FEM commercial code [32]. In order to simulate the bimetallic material, the beam walls were divided into two layers of different metals with a variable thickness ratio λ. Note that for $\lambda = 0$, the beam wall is fully SUS304, while as the index thickness ratio λ increases, the beam wall becomes fully Ti-6Al-4V.

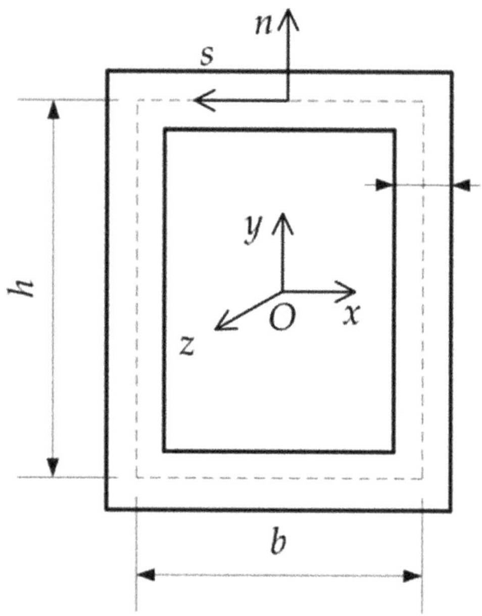

Figure 3. Box beam cross-section.

3.1. Box Beam

In the first example, the eigenvalue results of the box beam for different boundary conditions, which are clamped–clamped (C-C), clamped–simply supported (C-S) and simply supported (S-S), beam lengths of $L_1 = 6$ m, $L_2 = 8$ m and $L_3 = 10$ m and different material thickness ratios λ are given in Tables 2–4. Critical buckling temperatures are given for the first two flexural buckling modes. As the thickness ratio λ increases, the critical buckling temperatures increase as well due to the material properties of TI-6Al-4V. Furthermore, longer beams obtain lower critical buckling temperatures. As expected, clamped–clamped beams exhibit the highest eigenvalues for every beam length. Good agreement of the present results and solutions derived from the 2D model for both flexural buckling modes is achieved.

Table 2. Critical buckling temperatures of the beam $L_1 = 6$ m for different boundary conditions and material thickness ratios.

			λ						
BC	**Mode**	**Method**	**0**	**0.2**	**0.4**	**0.5**	**0.6**	**0.8**	**1**
C-C	Y	Present	128.87	135.26	145.83	153.64	164.09	199.19	284.51
		Shell	130.22	138.77	151.18	159.72	170.71	205.83	287.97
	X	Present	239.24	252.56	273.89	288.9	308.62	373.43	528.16
		Shell	238.95	254.69	277.45	293.11	313.24	377.6	527.52
C-S	Y	Present	66.58	69.91	75.39	79.44	84.84	102.97	146.98
		Shell	65.98	70.313	76.604	80.936	86.504	104.29	145.66
	X	Present	122.87	130.02	140.09	148.58	158.73	192.03	271.48
		Shell	120.46	128.39	139.88	147.78	157.93	190.37	265.92
S-S	Y	Present	32.72	34.368	37.07	39.06	41.72	50.62	72.24
		Shell	32.196	34.314	37.39	39.507	42.226	50.904	71.077
	X	Present	60.28	60.29	69.07	72.86	77.83	94.15	133.08
		Shell	58.725	62.6	68.208	72.063	77.015	92.827	129.64

Table 3. Critical buckling temperatures of the beam $L_2 = 8$ m for different boundary conditions and material thickness ratios.

			λ						
BC	**Mode**	**Method**	**0**	**0.2**	**0.4**	**0.5**	**0.6**	**0.8**	**1**
C-C	Y	Present	73.17	76.83	82.85	87.3	93.24	113.16	161.54
		Shell	73.542	78.368	85.375	90.2	96.405	116.24	162.35
	X	Present	135.52	142.96	154.88	163.38	174.53	211.15	298.53
		Shell	135.02	143.91	156.78	165.62	177	213.37	298.08
C-S	Y	Present	37.63	39.52	42.63	44.91	47.97	58.21	83.07
		Shell	37.323	39.774	43.33	45.783	48.933	58.997	82.396
	X	Present	69.35	73.33	79.45	83.81	89.53	108.31	153.09
		Shell	68.256	72.754	79.261	83.736	89.487	107.87	150.68
S-S	Y	Present	18.45	19.38	20.9	22.03	23.53	28.55	40.73
		Shell	18.211	19.409	21.149	22.346	23.885	28.793	40.204
	X	Present	33.95	35.9	38.9	41.04	43.84	53.03	74.95
		Shell	33.278	35.473	38.651	40.836	43.642	52.603	73.466

Table 4. Critical buckling temperatures of the beam $L_3 = 10$ m for different boundary conditions and material thickness ratios.

			λ						
BC	**Mode**	**Method**	**0**	**0.2**	**0.4**	**0.5**	**0.6**	**0.8**	**1**
C-C	Y	Present	47.03	49.39	53.27	56.13	59.95	72.75	103.83
		Shell	47.158	50.252	54.745	57.839	61.818	74.536	104.11
	X	Present	86.74	91.72	99.37	104.82	111.98	135.47	191.49
		Shell	86.605	92.308	100.56	106.23	113.53	136.86	191.19
C-S	Y	Present	24.14	25.35	27.35	28.81	30.77	37.34	53.28
		Shell	23.965	25.538	27.823	29.397	31.419	37.881	52.906
	X	Present	44.43	46.99	50.91	53.71	57.38	69.4	98.09
		Shell	43.872	46.763	50.945	53.822	57.518	69.335	96.854
S-S	Y	Present	11.82	12.42	13.39	14.11	15.07	18.29	26.09
		Shell	11.695	12.464	13.582	14.35	15.338	18.491	25.818
	X	Present	21.74	22.99	24.91	26.28	28.08	33.96	47.99
		Shell	21.396	22.807	24.85	26.255	28.059	33.82	47.234

In order to further perform the nonlinear stability analysis of the box beam, a perturbation force of intensity $\Delta F = 500$ N is introduced in the y-axis direction at the midpoint of the clamped–clamped beam. Temperature–displacement curves of the shortest beam with the comparison of temperature-dependent (TD) and temperature-independent (TID) materials are shown in Figures 4–6. The results are given for different thickness ratios: $\lambda = 0.2$, $\lambda = 0.5$ and $\lambda = 0.8$. It can be seen that curves match very well with the critical buckling temperatures achieved in the eigenvalue manner. As expected, TD solutions obtained lower critical buckling temperatures. In the case of $\lambda = 0.2$, the difference in critical temperature is around 4%, for $\lambda = 0.5$ the difference is 3.3%, and for $\lambda = 0.8$ it is 1.5%.

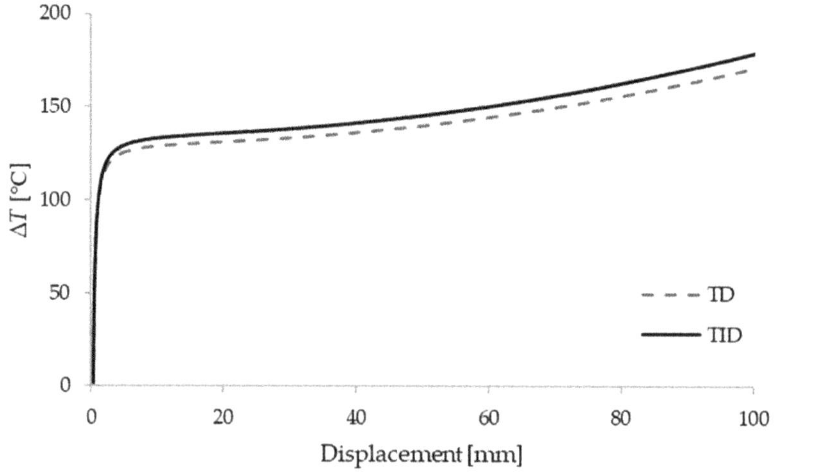

Figure 4. Critical temperatures vs. displacement for a clamped–clamped beam $L_1 = 6$ m with $\lambda = 0.2$.

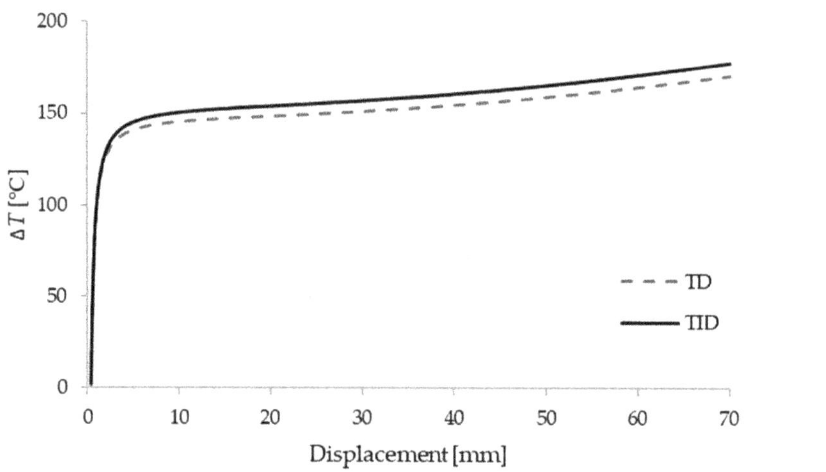

Figure 5. Critical temperatures vs. displacement for a clamped–clamped beam $L_1 = 6$ m with $\lambda = 0.5$.

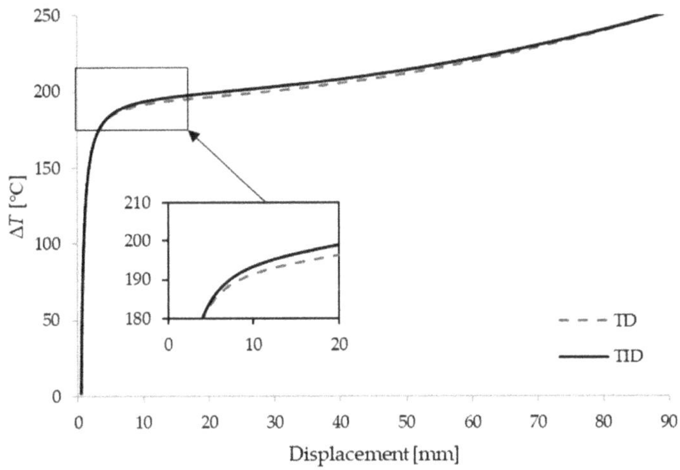

Figure 6. Critical temperatures vs. displacement for a clamped–clamped beam $L_1 = 6$ m with $\lambda = 0.8$.

3.2. L-Frame

Furthermore, the model is tested for thermal buckling analysis of an L-frame with the length of both legs being $L = 5$ m and cross-section described in the previous chapter (Figure 7). The frame is fixed at points A and C, while at point B, in-plane translations are prevented.

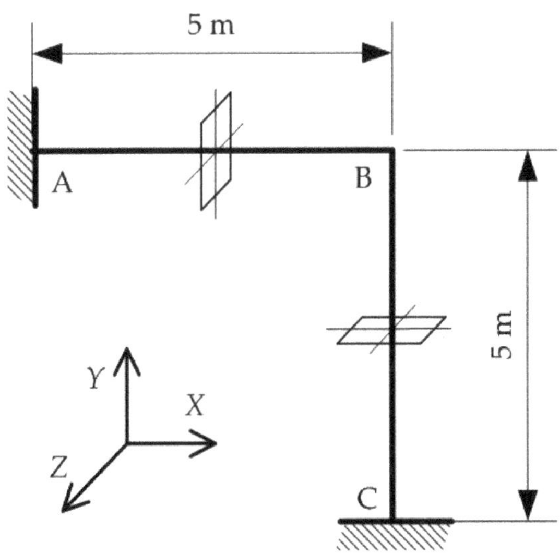

Figure 7. L-frame.

To verify the results, the critical buckling temperatures for the full SUS304, $\Delta T_{cr(SUS304)} = 16.09$ °C, and full Ti-6Al-4V $\Delta T_{cr(\text{Ti-6Al-4V})} = 35.52$ °C, sections are solved by a shell commercial code. To perform nonlinear analysis, a small perturbation force $\Delta F = 50$ N acting in the z-axis direction at point B is applied. The results are shown in Figure 8 for pure metals and for $\lambda = 0.2t$, $\lambda = 0.5t$ and $\lambda = 0.8t$. The good recognition of the critical values can be noted. It can be observed that with an increase in the proportion of Ti-6Al-4V material, higher critical temperatures are achieved.

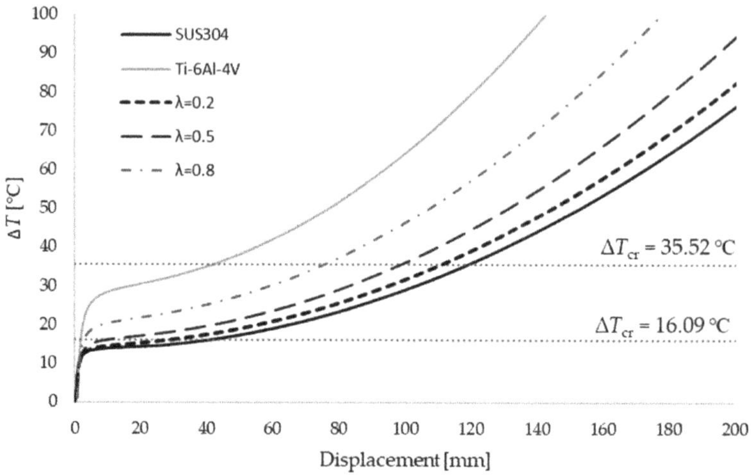

Figure 8. Displacements of the L-frame for several values of the material thickness ratio.

4. Conclusions

A thin-walled beam model capable of thermal buckling analysis has been presented. By means of the updated Lagrangian formulation, the incremental equilibrium equations have been developed using the nonlinear displacement field of the cross-section, taking into account the effects of large rotations. The reliability of the present model was verified by studying the benchmark examples, and the values obtained with the proposed model are in good agreement with those of the shell model. The effects of boundary conditions, the length of the beam and material thickness ratio on the critical buckling temperature and post-buckling response are of great importance. Additionally, it is shown that the TD solutions provide lower values than those of the TID solutions. The model was found to be efficient in predicting eigenvalues and nonlinear buckling behavior.

Author Contributions: Conceptualization, S.K.S. and D.L.; methodology, S.K.S., G.T. and D.L.; software, S.K.S. and D.B.; validation, S.K.S. and D.B.; formal analysis, S.K.S., G.T. and D.L.; investigation, S.K.S.; resources, D.L. and G.T.; data curation, S.K.S.; writing—original draft preparation, S.K.S.; writing—review and editing, S.K.S. and D.L.; visualization, S.K.S.; supervision D.L. and G.T.; project administration, D.L.; funding acquisition, D.L. and G.T. All authors have read and agreed to the published version of the manuscript.

Funding: This research was funded by the Croatian Science Foundation, grant number: IP-2019-04-8615 and University of Rijeka, grant numbers: uniri-tehnic-18-107 and uniri-tehnic-18-139.

Institutional Review Board Statement: Not applicable.

Informed Consent Statement: Not applicable.

Data Availability Statement: Not applicable.

Acknowledgments: The research presented in this paper was made possible by the financial support of Croatian Science Foundation.

Conflicts of Interest: The authors declare no conflict of interest.

Appendix A

$$R_{11} = \int\limits_A E(n,z,T)\mathrm{d}n\mathrm{d}s,$$
$$R_{12} = R_{21} = \int\limits_A E(n,z,T)(x + n\sin\beta)\mathrm{d}n\mathrm{d}s,$$
$$R_{13} = R_{31} = \int\limits_A E(n,z,T)(y - n\cos\beta)\mathrm{d}n\mathrm{d}s,$$
$$R_{14} = R_{41} = \int\limits_A E(n,z,T)(\omega - nq)\mathrm{d}n\mathrm{d}s,$$
$$R_{22} = \int\limits_A E(n,z,T)(x + n\sin\beta)^2\mathrm{d}n\mathrm{d}s,$$
$$R_{23} = R_{32} = \int\limits_A E(n,z,T)(x + n\sin\beta)(y - n\cos\beta)\mathrm{d}n\mathrm{d}s,$$
$$R_{23} = R_{32} = \int\limits_A E(n,z,T)(x + n\sin\beta)(y - n\cos\beta)\mathrm{d}n\mathrm{d}s,$$
$$R_{24} = R_{42} = \int\limits_A E(n,z,T)(x + n\sin\beta)(\omega - nq)\mathrm{d}n\mathrm{d}s,$$
$$R_{33} = \int\limits_A E(n,z,T)(y - n\cos\beta)^2\mathrm{d}n\mathrm{d}s,$$
$$R_{34} = R_{43} = \int\limits_A E(n,z,T)(y - n\cos\beta)(\omega - nq)\mathrm{d}n\mathrm{d}s,$$
$$R_{44} = \int\limits_A E(n,z,T)(\omega - nq)^2\mathrm{d}n\mathrm{d}s,$$
$$R_{55} = \int\limits_A G(n,z,T)\left(n + \tfrac{F_s}{t}\right)^2\mathrm{d}n\mathrm{d}s.$$

References

1. Trahair, N.S. *Flexural-Torsional Buckling of Structures*; CRC Press: Boca Raton, FL, USA, 1993.
2. Alfutov, N.A. *Stability of Elastic Structures*; Springer: Berlin/Heidelberg, Germany, 2000.
3. Bin Kamarudin, M.N.; Mohamed Ali, J.S.; Aabid, A.; Ibrahim, Y.E. Buckling Analysis of a Thin-Walled Structure Using Finite Element Method and Design of Experiments. *Aerospace* **2022**, *9*, 541. [CrossRef]
4. Turkalj, G.; Brnić, J. Nonlinear Stability Analysis of Thin-Walled Frames Using Ul–Esa Formulation. *Int. J. Struct. Stab. Dyn.* **2004**, *4*, 45–67. [CrossRef]
5. Nguyen, N.D.; Vo, T.P.; Nguyen, T.K. An Improved Shear Deformable Theory for Bending and Buckling Response of Thin-Walled FG Sandwich I-Beams Resting on the Elastic Foundation. *Compos. Struct.* **2020**, *254*, 112823. [CrossRef]
6. Pagani, A.; Augello, R.; Carrera, E. Frequency and Mode Change in the Large Deflection and Post-Buckling of Compact and Thin-Walled Beams. *J. Sound Vib.* **2018**, *432*, 88–104. [CrossRef]
7. Lanc, D.; Turkalj, G.; Vo, T.P.; Brnić, J. Nonlinear Buckling Behaviours of Thin-Walled Functionally Graded Open Section Beams. *Compos. Struct.* **2016**, *152*, 829–839. [CrossRef]
8. Lanc, D.; Vo, T.P.; Turkalj, G.; Lee, J. Buckling Analysis of Thin-Walled Functionally Graded Sandwich Box Beams. *Thin-Walled Struct.* **2015**, *86*, 148–156. [CrossRef]
9. Banat, D.; Mania, R.J.; Degenhardt, R. Stress State Failure Analysis of Thin-Walled GLARE Composite Members Subjected to Axial Loading in the Post-Buckling Range. *Compos. Struct.* **2022**, *289*, 115468. [CrossRef]
10. Nguyen, T.-T.; Kim, N.-I.; Lee, J. Analysis of Thin-Walled Open-Section Beams with Functionally Graded Materials. *Compos. Struct.* **2016**, *138*, 75–83. [CrossRef]
11. Nguyen, T.; Lee, J. Optimal Design of Thin-Walled Functionally Graded Beams for Buckling Problems. *Compos. Struct.* **2017**, *179*, 459–467. [CrossRef]
12. Jin, M.X.; Xiang, D.Z. Thermally-Induced Bending-Torsion Coupling Vibration of Large Scale Space Structures. *Comput. Mech.* **2007**, *40*, 707–723. [CrossRef]
13. Saha, S. Thermal Buckling and Postbuckling Characteristics of Extensional Slender Elastic Rods. *J. Mech. Eng.* **2009**, *40*, 1–8. [CrossRef]
14. Cui, D.F.; Hu, H.Y. Thermal Buckling and Natural Vibration of the Beam with an Axial Stick—Slip—Stop Boundary. *J. Sound Vib.* **2014**, *333*, 2271–2282. [CrossRef]
15. George, N.; Jeyaraj, P.; Murigendrappa, S.M. Thin-Walled Structures Buckling of Non-Uniformly Heated Isotropic Beam: Experimental and Theoretical Investigations. *Thin Walled Struct.* **2016**, *108*, 245–255. [CrossRef]
16. Burgreen, D.; Manitt, P.J. Thermal Buckling of a Bimetallic Beams. *J. Eng. Mech. Div.* **1969**, *95*, 421–432. [CrossRef]
17. Burgreen, D.; Regal, D. Higher Mode Buckling of Bimetallic Beam. *J. Eng. Mech. Div.* **1971**, *97*, 1045–1056. [CrossRef]

18. Aydogdu, M. Thermal Buckling Analysis of Cross-Ply Laminated Composite Beams with General Boundary Conditions. *Compos. Sci. Technol.* **2007**, *67*, 1096–1104. [CrossRef]
19. Trinh, L.C.; Vo, T.P.; Thai, H.T.; Nguyen, T.K. An Analytical Method for the Vibration and Buckling of Functionally Graded Beams under Mechanical and Thermal Loads. *Compos. Part B Eng.* **2016**, *100*, 152–163. [CrossRef]
20. Kiani, Y.; Eslami, M.R. Thermal Buckling Analysis of Functionally Graded Material Beams. *Int. J. Mech. Mater. Des.* **2010**, *6*, 229–238. [CrossRef]
21. Kiani, Y.; Eslami, M. Thermomechanical Buckling Dependent FGM Beams Of. *Lat. Am. J. Solids Struct.* **2013**, *10*, 223–246. [CrossRef]
22. Giunta, G.; Crisafulli, D.; Belouettar, S.; Carrera, E. A Thermo-Mechanical Analysis of Functionally Graded Beams via Hierarchical Modelling. *Compos. Struct.* **2013**, *95*, 676–690. [CrossRef]
23. Librescu, L.; Oh, S.Y.; Song, O. Thin-Walled Beams Made of Functionally Graded Materials and Operating in a High Temperature Environment: Vibration and Stability. *J. Therm. Stress.* **2005**, *28*, 649–712. [CrossRef]
24. Ziane, N.; Meftah, S.A.; Ruta, G.; Tounsi, A. Thermal Effects on the Instabilities of Porous FGM Box Beams. *Eng. Struct.* **2017**, *134*, 150–158. [CrossRef]
25. Kvaternik Simonetti, S.; Turkalj, G.; Lanc, D. Thin-Walled Structures Thermal Buckling Analysis of Thin-Walled Closed Section FG Beam-Type Structures. *Thin-Walled Struct.* **2022**, *181*, 110075. [CrossRef]
26. Gjelsvik, A. *The Theory of Thin-Walled Bars*; Wiley: New York, NY, USA, 1981.
27. Shen, H.S.; Wang, Z.X. Nonlinear Analysis of Shear Deformable FGM Beams Resting on Elastic Foundations in Thermal Environments. *Int. J. Mech. Sci.* **2014**, *81*, 195–206. [CrossRef]
28. Reddy, J.N.; Chin, C.D. Thermomechanical Analysis of Functionally Graded Cylinders and Plates. *J. Therm. Stress.* **1998**, *21*, 593–626. [CrossRef]
29. Turkalj, G.; Brnic, J.; Kravanja, S. A Beam Model for Large Displacement Analysis of Flexibly Connected Thin-Walled Beam-Type Structures. *Thin-Walled Struct.* **2011**, *49*, 1007–1016. [CrossRef]
30. Bathe, K.J. *Finite Element Procedure*; Klaus-Jurgen Bathe, 2007. Available online: https://www.amazon.com/Finite-Element-Procedures-K-J-Bathe/dp/097900490X (accessed on 27 September 2022).
31. Turkalj, G.; Brnic, J.; Prpic-Orsic, J. Large Rotation Analysis of Elastic Thin-Walled Beam-Type Structures Using ESA Approach. *Comput. Struct.* **2003**, *81*, 1851–1864. [CrossRef]
32. *Ansys Mechanical*; Release 18.1; Help System. ANSYS, Inc.: Canonsburg, PA, USA. Available online: https://www.ansys.com/academic/terms-and-conditions (accessed on 27 September 2022).

materials

Article

The Effect of Different Morphologies of WO_3/GO Nanocomposite on Photocatalytic Performance

Banu Esencan Türkaslan [1,*], Aziz Kerim Çelik [1], Ayça Dalbeyler [1] and Nicholas Fantuzzi [2]

1 Department of Chemical Engineering, Faculty of Engineering, University of Süleyman Demirel, Isparta 32260, Turkey
2 Department of Civil, Chemical, Environmental, and Materials Engineering, University of Bologna, 40126 Bologna, Italy
* Correspondence: banuturkaslan@sdu.edu.tr

Abstract: Tungsten trioxide/graphene oxide (WO_3/GO) nanocomposites have been successfully synthesized using in situ and ex situ chemical approaches. Graphite and tungsten carbide (WC) were employed to perform in situ synthesis, and WO_3 and GO were employed to perform the ex situ synthesis of WO_3/GO nanocomposites. GO, which was required for ex situ synthesis, is synthesized via the modified and improved Hummers method. XRD, SEM/EDS, and FTIR are used for the characterization of the nanocomposite. From the XRD of the WO_3/GO nanocomposites, it was observed that WO_3 distributed uniformly on graphene oxide sheets or was incorporated between the sheets. The photocatalytic activities of WO_3/GO nanocomposites were evaluated by methylene blue (MB) adsorption and visible light photocatalytic degradation activities by UV-vis spectroscopy. The results showed that the efficiency of the photocatalytic activity of the nanocomposite depends on different synthesis methods and the morphology resulting from the changed method. WO_3/GO nanocomposites synthesized by both methods exhibited much higher photocatalytic efficiencies than pure WO_3, and the best degradation efficiencies for MB was 96.30% for the WO_3/GO in situ synthesis nanocomposite.

Keywords: tungsten trioxide; graphene oxide; photocatalysis; nanocomposite

Citation: Esencan Türkaslan, B.; Çelik, A.K.; Dalbeyler, A.; Fantuzzi, N. The Effect of Different Morphologies of WO_3/GO Nanocomposite on Photocatalytic Performance. *Materials* **2022**, *15*, 8019. https://doi.org/10.3390/ma15228019

Academic Editor: Michela Alfè

Received: 7 October 2022
Accepted: 9 November 2022
Published: 14 November 2022

1. Introduction

In recent years, due to a notable increase in industrial activities that produce wastewater, the development of easily accessible adsorbent materials for cleaning water resources gained great momentum. Organic dyes have a dangerous and toxic effect on humans, which are found in the wastewater of various industries [1].

Among the different methods, degradation by photocatalysis has become a promising technique as it is simple and sustainable, and it enables the conversion of dyes to non-hazardous waste.

Nanocomposite materials, including metal oxide nanoparticles, with their advanced physicochemical properties, are common photoactive semiconductor materials that make them suitable for use in photocatalysis [2,3].

Tungsten oxide (WO_3), a nanostructured metal oxide, is widely used in the fields of gas detection [4,5], lithium-ion batteries [6,7], smart windows [8], and photocatalysis [9,10] practices. However, minimal photonic efficiency, the fast recombination rate of the charge carriers (photogenerated hole–electron), and low absorption ranges restrict the performance of WO_3 as an efficient photocatalyst [11–14]. The most important features that distinguish an ideal adsorbent from others are high adsorption capacity, rapid adsorption rate, and high selectivity. In this sense, doping WO_3 with other elements or compounds to improve its photocatalytic activity is deemed as an effective approach [15–17]. This may help improve the morphology of WO_3 material as well as the affected band levels and characteristics of the charge carriers [18].

In particular, graphene and its derivatives are frequently preferred in improving the performance of metal oxide nanostructures due to their extraordinary properties [19–21].

Contrary to graphene, the use of hydrophilic GO, which contains various proportions of carbon, oxygen, and hydrogen in its structure, is expanding day by day due to its easy dispersion in solutions, its dielectric properties, transparency and adjustable electronic properties [22,23]. However, it is difficult to separate GO from water after adsorption. To overcome this drawback, the hybridization of GO with other inorganic or organic materials is an alternative [24].

Hummers and Offeman synthesized GO via the method of oxidizing graphite in H_2SO_4, $NaNO_3$, and $KMnO_4$ atmospheres in 1958 [25]. Hummers' methodology was widely accepted, yet many disadvantages of this method have been reported, such as the production of toxic gas (NO_2 and N_2O_4), the residual nitrate, and low efficiency levels. In the last 20 years, alternative methods have been attempted relative to Hummer's method, including the addition of a peroxidation phase before $KMnO_4$ oxidation (without $NaNO_3$), increasing the amount of $KMnO_4$ rather than $NaNO_3$, and replacing $KMnO_4$ with K_2FeO_4 when $NaNO_3$ is extracted [26,27].

Studies conducted on WO_3/GO show that while the photocatalytic degradation activities of composites formed by combining WO_3 and GO structures that are synthesized separately via the ex situ method [28–32], there are no studies conducted regarding the photocatalytic degradation activities of nanocomposite structures, which are synthesized via in situ methods. It is thought that the composite structures synthesized using different methods are in the form of nanoparticles and nanowires [33–36], and this will affect the photocatalytic efficiency level.

In this study, highly efficient photocatalysts were developed, which is a method of producing renewable energy. To this end, both ex situ and in situ syntheses of WO_3/GO composites were performed, and the photocatalytic activity of the formed structure was examined. GO was synthesized via the modified and improved Hummers method without using $NaNO_3$. Given that there is no other WO_3/GO study conducted to examine the photocatalytic degradation activity of nanocomposite structures synthesized via in situ methods, our study bears the characteristics of the first study conducted in this field, and it is also the first study that compares photocatalytic activities of ex-situ and in-situ synthesized WO_3/GO composite structures.

2. Materials and Methods

Graphite flake (≥75% min), sulfuric acid (H_2SO_4, 98%), potassium permanganate ($KMnO_4$, 99%), hydrogen peroxide (H_2O_2, 30%), hydrochloric acid (HCI, 37%), and tungsten (VI) oxide (WO_3, <100 nm) were obtained from Sigma-Aldrich, and tungsten carbide (WC 45 nm, 99%) was obtained for the synthesis process from Nanokar company (İstanbul, Turkey).

2.1. Graphene Oxide (GO) Synthesis

GO was synthesized from layered graphite via the modified and improved Hummers method. Firstly, graphite (2 g) and then $KMnO_4$ (6 g) were gradually added into H_2SO_4 within the ice bath and mixed. Later, 300 mL of deionized water was added to the mixture. In order to stop the oxidation process and remove the impurities in the structure, the mixture was filtered by adding H_2O_2 and HCl, respectively. Synthesized graphite oxide measuring 1 g was taken into 350 mL of pure water and dispersed for 3 h. After two hours of sonication in order to facilitate the exfoliation of clumped graphite oxide layers on GO layers, the mixture was centrifuged, and GO was produced.

2.2. In Situ and Ex Situ Synthesis of WO_3/GO Composites

A total of 10 mg WO_3 and 30 mg GO were separately mixed for 2 h in 10 mL and 30 mL water, respectively, for the ex situ synthesis of WO_3/GO nanocomposites. Later, the two solutions were combined and mixed for another 2 h in a magnetic stirrer. The achieved solution was centrifuged and finally dried at 60 °C for 24 h to obtain the WO_3/GO composite.

WC powder measuring 1 g and 10 mL H_2SO_4 were mixed in ice bath for the in situ synthesis of WO_3/GO nanocomposite. $KMnO_4$ measuring 3 g was gradually added into this mixture by continuously stirring. After adding $KMnO_4$, the mixture was stirred for an additional 2 h; then, 15 mL H_2O_2 (%30 *w*/*w*) was added, and it was observed that the color of the mixture turned bright yellow. The achieved solution was centrifuged and finally dried at 60 °C for 24 h.

2.3. Photocatalytic Activities of WO_3/GO Composites

Methylene blue (MB) was used as typical pollutants to study the photocatalytic activity of the synthesized WO_3/GO composites, which are synthesized with in situ and ex situ methods. In order to simulate the coloring agent, 75 mL of 20 ppm MB solutions was prepared. WO_3 was added to one of the solutions, and 15 mg of in situ WO_3/GO and 15 mg of ex situ WO_3/GO were added to other solutions as catalysts. Then, the solution was deposited into tubes in equal amounts in order to be able to make measurements at different time intervals.

After mixing for 30 min in the dark to ensure an adsorption–desorption balance, the solution tubes that were placed in the UV cabinet were exposed to a total of 2 xenon lamps, each of which was 150 Watt (Figure 1).

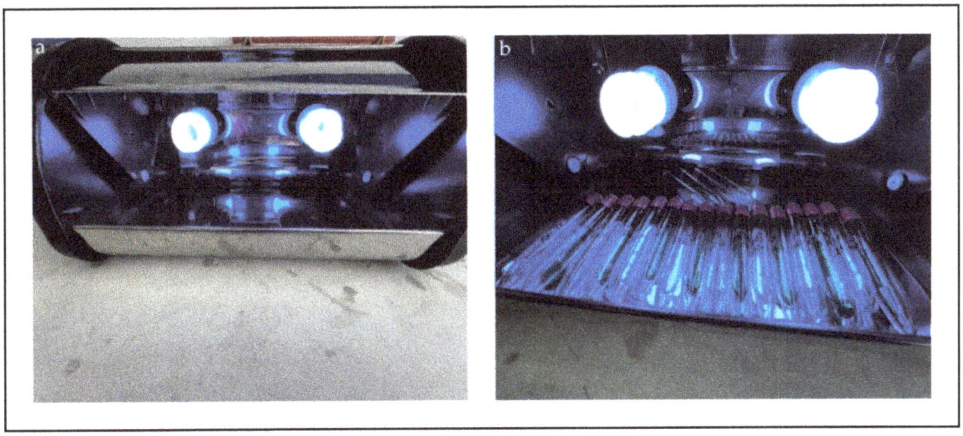

Figure 1. Designed UV cabinet (**a**) and tubes prepared for measurements to be made with different time intervals (**b**).

The distance between the lamp and the center of the tubes was measured as 8 cm. Samples were then taken at regular intervals to observe the degradation of methylene blue at 660 nm. The first measurement was made in the 15th minute; the next measurements were set to be made every 30 min, and regular measurements were made.

3. Results

The GO characterization was performed with X-ray diffraction (XRD) and the scanning electron microscopy (SEM/EDS) technique. WO_3/GO composites were also evaluated by using a scanning electron microscope (SEM, Quanta Feg 250; FEI, Eindhoven, the Netherlands). WO_3/GO composites were examined with a low-vacuum at 20.00 kV and 12.7–13.2 mm working distance at 5000× and 10000× magnifications, respectively. The elemental analysis of nanocomposites was carried out using an SEM microscope equipped with an energy-dispersive X-ray spectroscopy (EDX, Quanta Feg 250; FEI, Eindhoven, The Netherlands). The distribution and atomic composition of WO_3/GO was examined using elemental mappings at an accelerating voltage of 20 kV. The crystalline phase of WO_3/GO were examined by X-ray diffraction (XRD, Bruker D8 Advance Twin-Twin; Bruker, Karlsruhe, Germany) at 40 kV, 40 mA, and 1600 watts. In order to determine the photocatalytic capacity of WO_3/GO nanocomposites, a UV-Vis spectrophotometer (UV-Vis Carry 60) device was used.

3.1. Characterization of GO

When the XRD spectrum of the GO structure was examined, it was observed that the peak formed at $2\theta = 11.52°$ for GO, which is consistent with the results in the literature and it shows that the GO structure is obtained properly (Figure 2) [37].

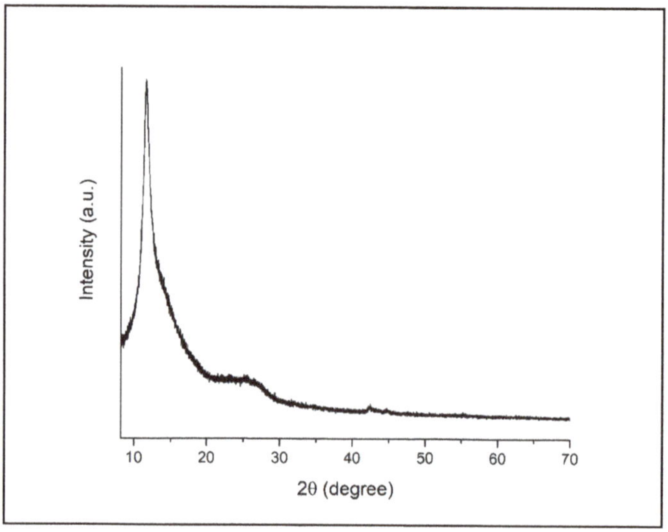

Figure 2. XRD patterns of GO.

SEM analysis results showed that the GO structure was formed by layered wavy structures piled on top of each other (Figure 3).

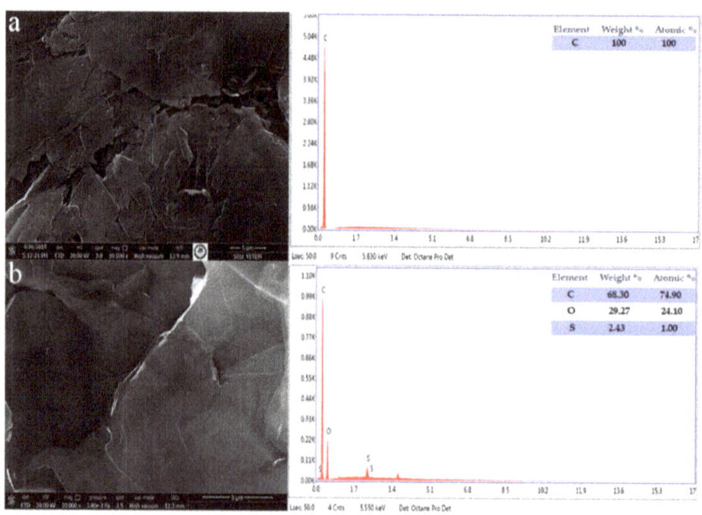

Figure 3. SEM/EDS image of (**a**) graphite and (**b**) GO.

3.2. Characterization of WO_3/GO Composites

Figure 4 provides the comparative FTIR spectra of GO and WO_3/GO composites. The peaks of <1000 cm^{-1} within the composite structures show the existence of pure WO_3 [28].

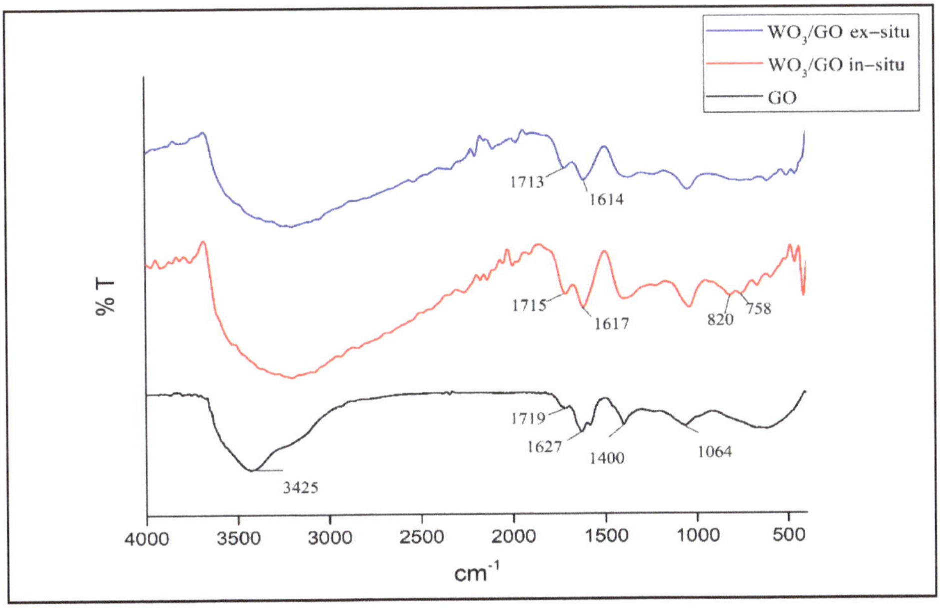

Figure 4. FTIR Spectra of GO and WO_3/GO Composites.

The FTIR spectrum of GO shows that the hydroxyl bond (-OH) is at 3425 cm^{-1}, the carbonyl bond (C=O) is at 1719 cm^{-1}, the aromatic bond (C=C) is at 1627cm^{-1}, the epoxy bond is at 1627 cm^{-1}, (C-O) is at 1400 cm^{-1}, and the alkoxy bond (C-O) is at 1064 cm^{-1}. The bands at 1715 cm^{-1} and 1617 cm^{-1} in the in situ synthesized composite structure and the bands at 1713 cm^{-1} and 1614 cm^{-1} in the ex situ synthesized composite structure belong to C=O and C=O vibrations, indicating the presence of GO in the composite's structure.

Peaks under <1000 cm^{-1} observed in composites that are not observed in GO structure are caused by O-W-O stretch vibrations and show that nanoparticles bind to GO nanolayers strongly [38]. O-W-O vibrations observed approximately at 820 and 758 cm^{-1} in the in situ synthesized composite structure were observed less in ex situ structures.

SEM micrographs were used to identify the morphology of the synthesized composites and the location of the metal oxide in the carbon matrix. It was observed that while the pure WO_3 structure comprised spongy structures of various sizes placed in such a way as to form spaces between them, WC consisted of randomly distributed and irregularly shaped coarse grains (Figure 5).

It is observed that the wavy interlayer spaces in GO structures are randomly dispersed by some spherical WO_3 particles to form smoother surfaces in the WO_3/GO ex situ synthesized nanocomposite (Figure 5e,f) structures. In WO_3/GO insitu synthesis morphologies, however, it is seen that a single-phase homogeneous composite morphology formed with good interfacial interaction between GO and WC (Figure 5g,h). The homogeneous coating of the GO surface with WO_3 as a result of good interfacial interaction in the in situ synthesis structure shows parallelisms with the O-W-O vibration bands observed in the FTIR results, while the lesser observation of these bands in the ex situ synthesis also supports the WO_3 particle structure observed between the GO layers.

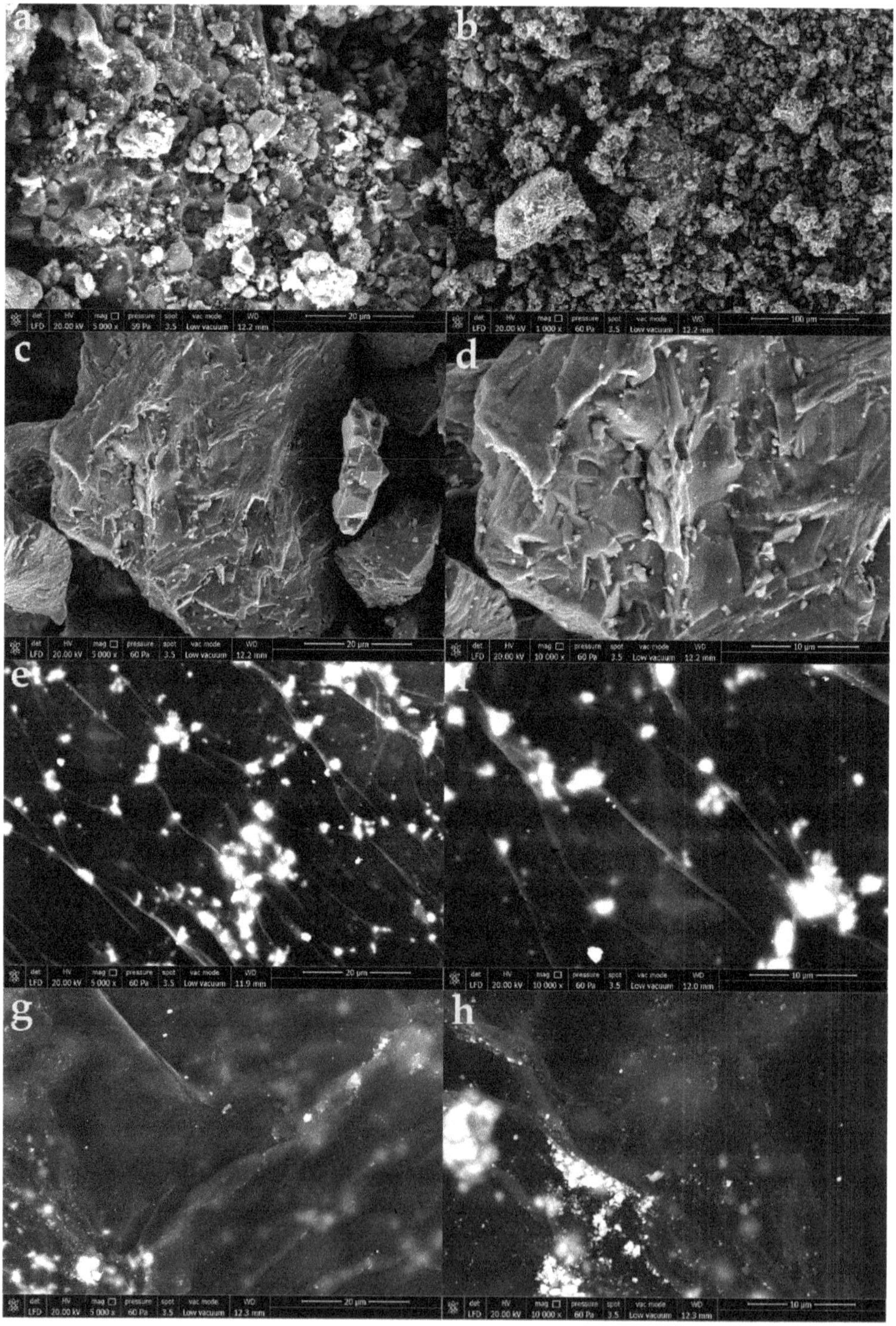

Figure 5. SEM images of WO_3 (**a**,**b**), WC (**c**,**d**), WO_3/GO ex situ synthesis (**e**,**f**), and WO_3/GO in situ synthesis (**g**,**h**).

The EDS analysis of WO_3 powders shows a tungsten atomic percentage of 79.35% and an oxygen atomic percentage of 20.65%. WC powders show tungsten at 90.34% and a carbon atomic percentage of 9.66%. After the synthesis of nanocomposites, a notable decrease in the atomic percentages of W elements was observed. Furthermore, the presence of GO was confirmed via EDS analyses, which showed additional carbon elements and oxygen elements in WO_3/GO ex situ and in situ, respectively (Table 1).

Table 1. Elemental composition of WO_3, WC and WO_3/GO in ex situ synthesis and in WO_3/GO in situ synthesis.

Samples	% at.		
	Oxygen	Tungsten	Carbon
WO_3	20.65	79.35	-
WC	-	90.34	9.66
WO_3/GO ex situ	41.10	8.62	50.28
WO_3/GO in situ	26.97	16.71	56.32

Figure 6 shows the XRD patterns of the ex situ and in situ WO_3/GO, WO_3/GO nanocomposites, which confirms the presence of both WO_3 and GO.

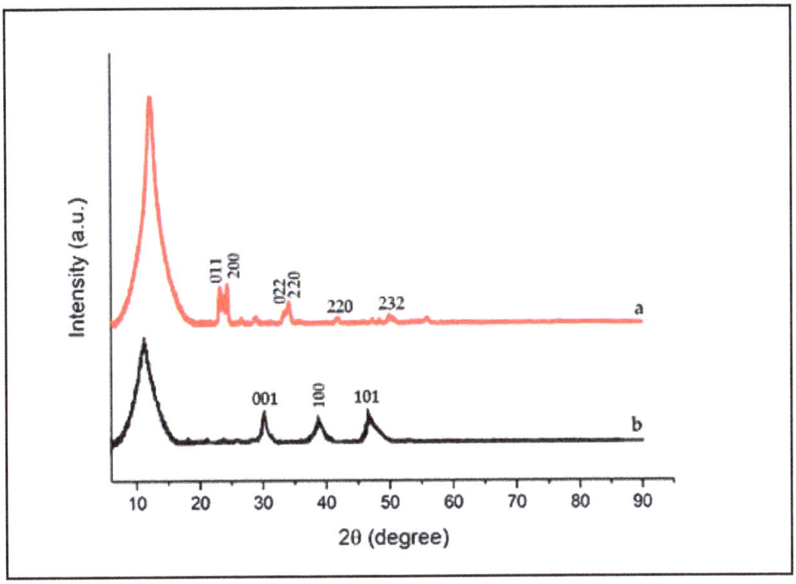

Figure 6. XRD patterns of WO_3/GO ex situ synthesis (a) and WO_3/GO in situ synthesis (b).

A low intensity peak at 10.8° indicates the formation of GO sheets in the in situ synthesis of WO_3/GO, which is due to the poor crystalline nature of carbon. The other peaks at 2θ values of 23.3, 24.5, 34.2, 42.1, 47.49, and 50.22 confirm the presence of WO_3 particles [39].

The WC diffraction spectrum shows three major intense peaks located at 2θ = 30.42°, 38.98°, and 47.03°, which correspond well to the crystallographic planes (001, 100, and 101) of WC, respectively [40].

3.3. Photocatalytic Degradation

Figure 7 shows the adsorption capacities (q_e) of WO_3 and WO_3/GO ex situ and WO_3/GO in situ nanocomposites, which were calculated using the following formula:

$$q_e = \frac{C_0 - C_e}{m} * V$$

where C_0 (mg/L) and C_e (mg/L) refer, respectively, to the initial concentration of the coloring agent and the concentration of the coloring agent remaining in the solution after adsorption, m (g) refers to the amount of adsorbents, and V (mL) represents the volume of the solution. Accordingly, WO_3 showed adsorption capacities in the range of 12.65–15.59, while WO_3/GO ex situ and WO_3/GO in situ nanocomposites showed adsorption capacities in the range of 15.18–19.84 and 19.46–23.91, respectively. During the experiment, the adsorption maximum capacity was determined with WO_3/GO in situ nanocomposites at t = 270, and the lowest was determined with WO_3 at t = 0. Within the increasing time intervals, the adsorption capacity of WO_3/GO in situ nanocomposites showed a significant increase after 3 h. This shows that this situation can be associated with the surface area of the nanocomposite, thus leading to the understanding that the photocatalytic effect increases with time.

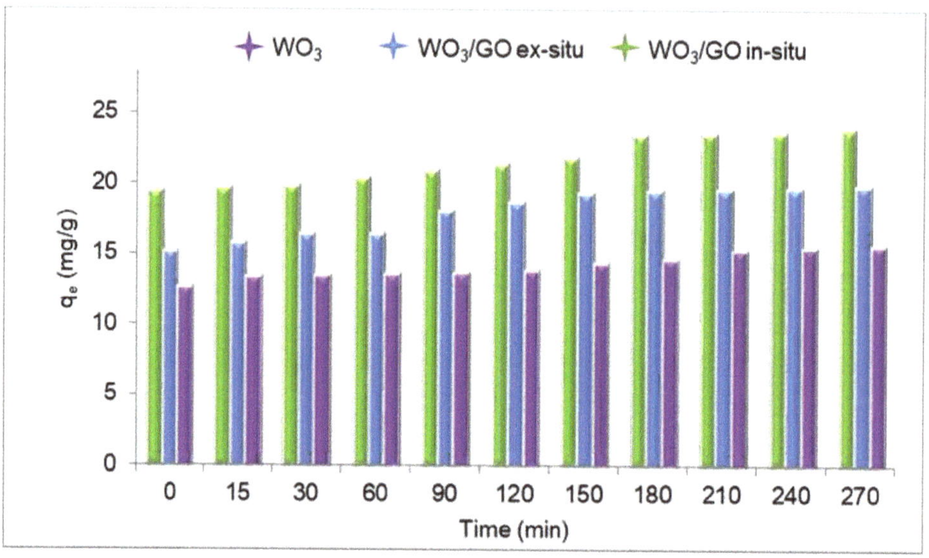

Figure 7. Relationship between adsorption capacity and time.

The degradation efficiency of was calculated using the following formula:

$$\eta\% = \left(1 - \frac{C}{C_0}\right) \times 100$$

where C_0 is the absorption maximum at t = 0, and C is the absorption maximum after complete degradation.

WO_3 and WO_3/GO nanocomposites synthesized ex situ/in situ showed the maximum degradation of 75.79%, 90.52%, and 96.30% respectively (Figure 8). It was observed that the degradation amount of the WO_3/GO in situ catalyst was higher than the one of ex situ synthesis, and it was concluded that this difference depends on the synthesis method. When GO is added to the matrix, it increased the photocatalytic effect, and in parallel with this, the synthesized nanocomposite structures reach higher percentage degradation efficiency values in a shorter time compared to WO_3.

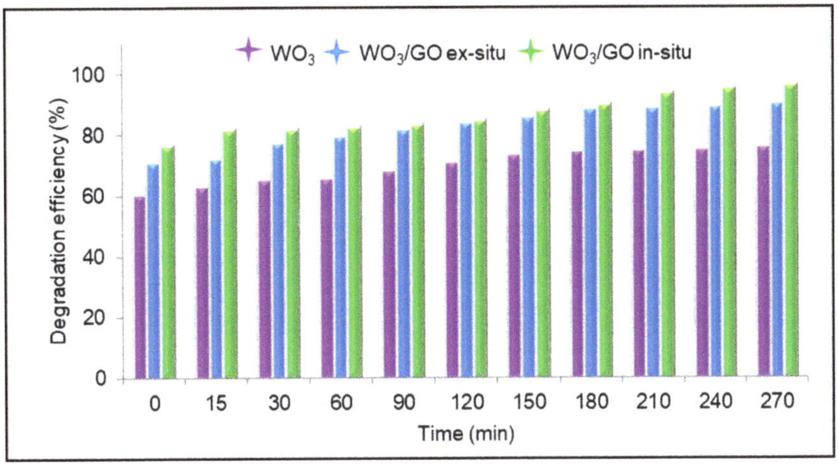

Figure 8. Degradation efficiency of MB.

The chemical structure of MB has cationic atoms and aromatic rings. The degradation mechanism starts with the MB dye adsorption on the nanocomposite's surface followed by its photodegradation, which can be summarized in the following steps. Figure 9 illustrates the mechanism of the photocatalytic degradation of MB via the WO_3/GO nanocomposite's UV irradiation. First, visible light radiation allows the transfer of electrons in in the valence band WO_3 to the conduction band of GO. Therefore, holes (h) and electrons (e-) are formed on the surface of the WO_3 photocatalyst. GO behaves as an electron acceptor via electrostatic and $\pi-\pi$ stacking interactions. Then, while the holes react with the hydroxide ion, the electrons react with dissolved oxygen to produce OH^-, which degrades MB dyes into non-toxic gases such as carbon dioxide and water. In addition, hydrogen peroxide reacts with electrons to produce more OH- to increase the degradation of the dye.

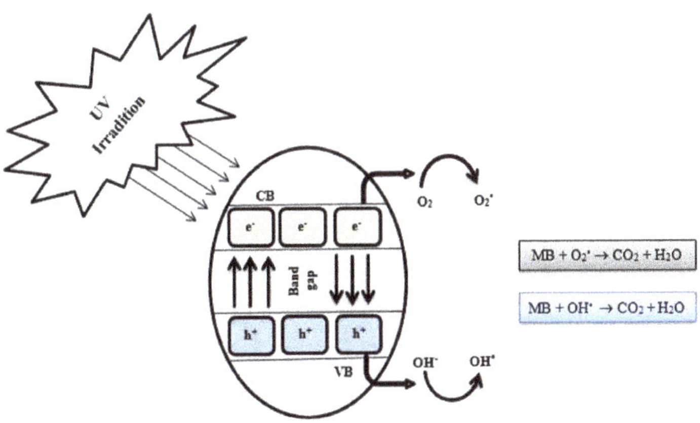

Figure 9. Schematic representation of photocatalytic degradation of methylene blue by WO_3/GO nanocomposites under UV irradiation illumination.

As shown in Table 2, prepared photocatalysts were compared with other WO_3-based nanocomposites. Their photocatalytic efficiency is usually at 80–97%. In this work,

WO_3/GO nanocomposites prepared by in situ synthesis were higher than that of the tungsten-oxide-based counterparts. In addition, the method is simple and does not require synthesizing GO separately; the in situ oxidation of graphite is provided while the nanocomposite is formed.

Table 2. Comparison of photocatalytic performance between this work and reported references.

Photocatalyst	Methods of Synthesis	Photodecomposition	Photocatalytic Effciency	References
WO_3/GO	In situ ex situ chemical oxidation	MB	96.30% 90.52%	Current work
WO_3/GO	Ultrasonication Method	MB	97.03%	[28]
WO_3/GO	Sol-gel method	MB	82%	[31]
WO_3/GO	Photo-reduction method	MO	92.7%	[32]
WO_3/GR	Hydrothermal method	MB	83%	[39]
WO_3/rGO	In situ slvothermal method	MB	94%	[41]

4. Discussion

The photocatalytic activities of the WO_3 and WO_3/GO nanocomposites were evaluated by the degradation of MB in aqueous solutions. Compared with WO_3, the photocatalytic activity of the WO_3/GO nanocomposites was enhanced, and the best degradation efficiencies for MB was 96.30% for the WO_3/GO in situ synthesis nanocomposite. It was attributed to the large surface area of GO, which served as an acceptor of the electrons generated in the WO_3 and effectively decreased the recombination. To overcome the rapid recombination and slow migration of charge carriers, different morphologies have been developed, such as nanoplates, nanotubes, and nano-sheets. The homogeneous nanoplate structure formed in in situ syntheses showed higher photocatalytic effects due to its large surface area than the nanorod-like structure formed as a result of ex situ syntheses.

Although there are various different pollutant sources in the environment [42], the application of composite-based photocatalysts is limited to water treatments. Expanding the application areas of GO-based photocatalysts with different studies is necessary. Laboratory equipment is mostly used for the degradation processes in the laboratory. To prepare photocatalysts on a commercial scale, cost studies should be carried out for the large-scale degradation of pollutants, and systems should be modified with appropriate strategies. It is seen that suitable morphologies can be obtained by adjusting different methods and reaction conditions, and morphology control is an important parameter for photocatalytic activities.

5. Conclusions

In summary, we reported a simple chemical in situ and ex situ synthesis process and the physical properties and photocatalytic activities of WO_3/GO composite structures. We observed that different synthesis methods affect WO_3/GO's morphology, while different WO_3/GO morphologies affect photocatalytic performances. The ex situ preparation of the composite leads to the formation of well-dispersed WO_3 with smoother surface in the WO_3/GO. However, the in situ-prepared WO_3 nanostructures have showed that single-phase homogeneous composite morphologies formed with good interfacial interactions between GO and WC. The prepared WO_3 and its nanocomposite with GO was evidenced for the dye degradation of MB. The best degradation efficiencies for MB were 96.30% for the WO_3/GO in situ synthesis nanocomposite, which are much better than that of WO_3. The results showed that WO_3/GO composites exhibited an enhanced WO_3 photocatalysis efficiency in visible light. This study gave a new perspective for applications of WO_3/GO nanocomposite photocatalysts for various areas.

Author Contributions: Conceptualization, B.E.T. and A.K.Ç.; methodology B.E.T., A.K.Ç. and N.F.; validation, B.E.T., A.D. and A.K.Ç.; formal analysis, N.F.; investigation, B.E.T. and A.D.; resources, B.E.T. and A.D.; data curation, B.E.T. and A.D.; writing—original draft preparation, B.E.T., A.D. and N.F.; writing—review and editing, B.E.T. and N.F.; visualization, B.E.T. and A.D.; supervision, B.E.T. and N.F.; project administration, B.E.T., A.K.Ç., A.D. and N.F. All authors have read and agreed to the published version of the manuscript.

Funding: This research was funded by grants from the Scientific and Technological Research Council of Turkey (TUBITAK, project code: 1919B012001762).

Institutional Review Board Statement: Not applicable.

Informed Consent Statement: Not applicable.

Data Availability Statement: Not applicable.

Conflicts of Interest: The authors declare no conflict of interest.

References

1. Alam, M.W.; Aamir, M.; Farhan, M.; Albuhulayqah, M.; Ahmad, M.M.; Ravikumar, C.; Dileep Kumar, V.; Ananda Murthy, H. Green Synthesis of Ni-Cu-Zn Based Nanosized Metal Oxides for Photocatalytic and Sensor Applications. *Crystals* **2021**, *11*, 1467. [CrossRef]
2. Hossain, S.; Chu, W.S.; Lee, C.S.; Ahn, S.H.; Chun, D.M. Photocatalytic performance of few-layer Graphene/WO_3 thin films prepared by a nano-particle deposition system. *Mater. Chem. Phys.* **2019**, *226*, 141–150. [CrossRef]
3. Alam, M.W.; Al Qahtani, H.S.; Souayeh, B.; Ahmed, W.; Albalawi, H.; Farhan, M.; Abuzir, A.; Naeem, S. Novel Copper-Zinc-Manganese Ternary Metal Oxide Nanocomposite as Heterogeneous Catalyst for Glucose Sensor and Antibacterial Activity. *Antioxidants* **2022**, *11*, 1064. [CrossRef] [PubMed]
4. Chatterjee, S.G.; Chatterjee, S.; Ray, A.K.; Chakraborty, A.K. Graphene–metal oxide nanohybrids for toxic gas sensor. *Sens. Actuators B Chem.* **2015**, *221*, 1170–1181. [CrossRef]
5. Chang, X.; Zhou, Q.; Sun, S.; Shao, C.; Lei, Y.; Liu, T.; Yin, Y. Graphene-tungsten oxide nanocomposites with highly enhanced gas-sensing performance. *J. Alloys Compd.* **2017**, *705*, 659–667. [CrossRef]
6. Tian, W.; Liu, X.; Yu, W. Research Progress of Gas Sensor Based on Graphene and Its Derivatives. *Appl. Sci.* **2018**, *8*, 1118. [CrossRef]
7. Zhang, J.; Liu, X.; Neri, G.; Pinna, N. Nanostructured materials for room-temperature gas sensors. *Adv. Mater.* **2016**, *28*, 795–831. [CrossRef]
8. Le, X.V.; Luu, T.L.A.; Nguyen, H.L.; Nguyen, C.T. Synergistic enhancement of ammonia gassensing properties at low temperature by compositing carbon nanotubes with tungsten oxide nanobricks. *Vacuum* **2019**, *168*, 108861. [CrossRef]
9. Korotcenkov, G.; Brinzari, V.; Ivanov, M.; Cerneavschi, A.; Rodríguez, J.R.; Cirera, A.; Cornet, A.; Morante, J.R. Structural stability of indium oxide films deposited by spray pyrolysis during thermal annealing. *Thin Solid Films* **2005**, *479*, 38–51. [CrossRef]
10. Righettoni, M.; Amann, A.; Pratsinis, S.E. Breath analysis by nanostructured metal oxides as chemoresistive gas sensors. *Mater. Today* **2015**, *18*, 163–171. [CrossRef]
11. Sayama, K.; Hayashi, H.; Arai, T.; Yanagida, M.; Gunji, T.; Sugihara, H. Highly active WO_3 semiconductor photocatalyst prepared from amorphous peroxo-tungstic acid for the degradation of various organic compounds. *Appl. Catal. B Environ.* **2010**, *94*, 150–157. [CrossRef]
12. Guo, Y.; Quan, X.; Lu, N.; Zhao, H.; Chen, S. High photocatalytic capability of self-assembled nanoporous WO_3 with preferential orientation of (002) planes. *Environ. Sci. Technol.* **2007**, *41*, 4422–4427. [CrossRef] [PubMed]
13. Sonia, S.; Kumar, P.S.; Mangalaraj, D.; Ponpandian, N.; Viswanathan, C. Influence of growth and photocatalytic properties of copper selenide (CuSe) nanoparticles using reflux condensation method. *Appl. Surf. Sci.* **2013**, *283*, 802–807. [CrossRef]
14. Deng, F.; Pei, X.; Luo, Y.; Luo, X.; Dionysiou, D.D.; Wu, S.; Luo, S. Fabrication of hierarchically porous reduced graphene oxide/SnIn4S8 composites by a low-temperature coprecipitation strategy and their excellent visible-light photocatalytic mineralization performance. *Catalysts* **2016**, *6*, 113. [CrossRef]
15. Sun, S.; Wang, W.; Zeng, S.; Shang, M.; Zhang, L. Preparation of ordered mesoporous Ag/WO_3 and its highly efficient degradation of acetaldehyde under visible-light irradiation. *J. Hazard. Mater.* **2010**, *178*, 427–433. [CrossRef]
16. Zheng, H.; Mathe, M. Hydrogen evolution reaction on single crystal WO_3/C nanoparticles supported on carbon in acid and alkaline solution. *Int. J. Hydrogen Energy* **2011**, *36*, 1960–1964. [CrossRef]
17. Cao, L.; Yuan, J.; Chen, M.; Shangguan, W. Photocatalytic energy storage ability of TiO2-WO_3 composite prepared by wet-chemical technique. *Res. J. Environ. Sci.* **2010**, *22*, 454–459. [CrossRef]
18. Kalanur, S.S.; Seo, H. Aligned nanotriangles of tantalum doped tungsten oxide for improved photoelectrochemical water splitting. *J. Alloys Compd.* **2019**, *785*, 1097–1105. [CrossRef]
19. Chen, H.; Chen, Z.; Yang, H.; Wen, L.; Yi, Z.; Zhou, Z.; Dai, B.; Zhang, J.; Wu, X.; Wu, P. Multi-mode surface plasmon resonance absorber based on dart-type single-layer graphene. *RSC Adv.* **2022**, *12*, 7821–7829. [CrossRef]

20. Zhang, Z.; Cai, R.; Long, F.; Wang, J. Development and application of tetrabromobisphenol A imprinted electrochemical sensor based on graphene/carbon nanotubes three-dimensional nanocomposites modified carbon electrode. *Talanta* **2015**, *134*, 435–442. [CrossRef]
21. Shangguan, Q.; Chen, Z.; Yang, H.; Cheng, S.; Yang, W.; Yi, Z.; Wu, X.; Wang, S.; Yi, Y.; Wu, P. Design of Ultra-Narrow Band Graphene Refractive Index Sensor. *Sensors* **2022**, *22*, 6483. [CrossRef] [PubMed]
22. Alam, S.N.; Sharma, N.; Kumar, L. Synthesis of graphene oxide (GO) by modified hummers method and its thermal reduction to obtain reduced graphene oxide (rGO). *Graphene* **2017**, *6*, 1–18. [CrossRef]
23. Khan, Z.U.; Kausar, A.; Ullah, H.; Badshah, A.; Khan, W.U. A review of graphene oxide, graphene buckypaper, and polymer/graphene composites: Properties and fabrication techniques. *J. Plast. Film Sheeting* **2016**, *32*, 336–379. [CrossRef]
24. Cheng, Z.; Liao, J.; He, B.; Zhang, F.; Zhang, F.; Huang, X.; Zhou, L. One-Step Fabrication of Graphene Oxide Enhanced Magnetic Composite Gel for Highly Efficient Dye Adsorption and Catalysis. *ACS Sustain. Chem. Eng.* **2015**, *3*, 1677–1685. [CrossRef]
25. Hummers, W.S., Jr.; Offeman, R.E. Preparation of graphitic oxide. *J. Am. Chem. Soc.* **1958**, *80*, 1339. [CrossRef]
26. Yu, H.; Zhang, B.; Bulin, C.; Li, R.; Xing, R. High-efficient synthesis of graphene oxide based on improved hummers method. *Sci. Rep.* **2016**, *6*, 36143. [CrossRef] [PubMed]
27. Yazıcı, M.; Tiyek, İ.; Ersoy, M.S.; Alma, M.H.; Dönmez, U.; Yıldırım, B.; Salan, T.; Karataş, Ş.; Uruş, S.; Karteri, İ.; et al. Modifiye Hummers Yöntemiyle Grafen Oksit (GO) Sentezi ve Karakterizasyonu. *GU J. Sci. Part C* **2016**, *4*, 43–50.
28. Jeevitha, G.; Abhinayaa, R.; Mangalaraj, D.; Ponpandian, N. Tungsten oxide-graphene oxide (WO_3-GO) nanocomposite as an efficient photocatalyst, antibacterial and anticancer agent. *J. Phys. Chem. Solids* **2018**, *116*, 137–147. [CrossRef]
29. Fu, L.; Xia, T.; Zheng, Y.; Yang, J.; Wang, A.; Wang, Z. Preparation of WO_3-reduced graphene oxide nanocomposites with enhanced photocatalytic property. *Ceram. Int.* **2015**, *41*, 5903–5908. [CrossRef]
30. Kofuji, Y.; Isobe, Y.; Shiraishi, Y.; Sakamoto, H.; Ichikawa, S.; Tanaka, S.; Hirai, T. Hydrogen peroxide production on a carbon nitride–boron nitride-reduced graphene oxide hybrid photocatalyst under visible light. *ChemCatChem* **2018**, *10*, 2070–2077. [CrossRef]
31. Ismail, A.A.; Faisal, M.; Al-Haddad, A. Mesoporous WO_3-graphene photocatalyst for photocatalytic degradation of Methylene Blue dye under visible light illumination. *Res. J. Environ. Sci.* **2018**, *66*, 328–337. [CrossRef]
32. Zhou, M.; Yan, J.; Cui, P. Synthesis and enhanced photocatalytic performance of WO_3 nanorods@ graphene nanocomposites. *Mater. Lett.* **2012**, *89*, 258–261. [CrossRef]
33. Bragaglia, M.; Paleari, L.; Lamastra, F.R.; Puglia, D.; Fabbrocino, F.; Nanni, F. Graphene nanoplatelet, multiwall carbon nanotube, and hybrid multiwall carbon nanotube–graphene nanoplatelet epoxy nanocomposites as strain sensing coatings. *J. Reinf. Plast. Compos.* **2021**, *40*, 632–643. [CrossRef]
34. Pittella, E.; D'Alvia, L.; Palermo, E.; Piuzzi, E. Microwave Characterization of 3D Printed PLA and PLA/CNT Composites. In Proceedings of the 2021 IEEE 6th International Forum on Research and Technology for Society and Industry (RTSI), Rome, Italy, 7–9 June 2021. [CrossRef]
35. Paleari, L.; Bragaglia, M.; Fabbrocino, F.; Nanni, F. Structural Monitoring of Glass Fiber/Epoxy Laminates by Means of Carbon Nanotubes and Carbon Black Self-Monitoring Plies. *Nanomaterials* **2021**, *11*, 1543. [CrossRef]
36. Izadi, R.; Tuna, M.; Trovalusci, P.; Ghavanloo, E. Torsional Characteristics of Carbon Nanotubes: Micropolar Elasticity Models and Molecular Dynamics Simulation. *Nanomaterials* **2021**, *11*, 453. [CrossRef] [PubMed]
37. Hanifah, M.F.R.; Jaafar, J.; Aziz, M.; Ismail, A.F.; Rahman, M.A.; Othman, M.H.D. Synthesis of kaphene oxide nanosheets via modified hummers method and ıts physicochemical properties. *J. Technol.* **2015**, *74*, 189–192. [CrossRef]
38. Murugan Vadivel, A.; Muraliganth, T.; Manthiram, A. Rapid, Facile Microwave-Solvothermal Synthesis of Graphene Nanosheets and Their Polyaniline Nanocomposites for Energy Strorage. *Chem. Mater.* **2009**, *21*, 5004–5006. [CrossRef]
39. Hu, X.; Xu, P.; Gong, H.; Yin, G. Synthesis and Characterization of WO_3/Graphene Nanocomposites for Enhanced Photo-catalytic Activities by One-Step In-Situ Hydrothermal Reaction. *Materials* **2018**, *11*, 147. [CrossRef]
40. Vakhrushev, A.V. Synthesis of WO_3 Nanostructures and Their Nanocomposites with Graphene Derivatives via Novel Chemical Approach. In *Nanomechanics—Theory and Application*; IntechOpen: London, UK, 2021; 146p. [CrossRef]
41. Tie, L.; Yu, C.; Zhao, Y.; Chen, H.; Yang, S.; Sun, J.; Dong, S.; Sun, J. Fabrication of WO_3 na-norods on reduced graphene oxide sheets with augmented visible light photocatalytic ac-tivity for efficient mineralization of dye. *J. Alloys Compd.* **2018**, *769*, 83–91. [CrossRef]
42. Derradji, M.; Mehelli, O.; Liu, W.; Fantuzzi, N. Sustainable and Ecofriendly Chemical Design of High Performance Bio-Based Thermosets for Advanced Applications. *Front. Chem.* **2021**, *9*, 691117. [CrossRef]

materials MDPI

Article

An Experimental and Numerical Study of the Influence of Temperature on Mode II Fracture of a T800/Epoxy Unidirectional Laminate

Yu Gong [1,2,3], Linfei Jiang [4], Linkang Li [1] and Jian Zhao [5,*]

1 Chongqing Key Laboratory of Heterogeneous Material Mechanics, College of Aerospace Engineering, Chongqing University, Chongqing 400044, China
2 State Key Laboratory of Coal Mine Disaster Dynamics and Control, Chongqing University, Chongqing 400044, China
3 State Key Laboratory of Automotive Simulation and Control, Jilin University, Changchun 130022, China
4 School of Civil Engineering, Chongqing University, Chongqing 400044, China
5 School of Aerospace Engineering and Applied Mechanics, Tongji University, Shanghai 200092, China
* Correspondence: zhaojian@tongji.edu.cn

Abstract: Studies on mode II fracture have promoted the establishment of the delamination theory for unidirectional composite laminates at room temperature. However, under thermal conditions, the fracture behavior of composite laminates will exhibit certain differences. The delamination theory should be extended to consider the temperature effect. To achieve this goal, in this study, the mode II static delamination growth behavior of an aerospace-grade T800/epoxy composite is investigated at 23 °C, 80 °C and 130 °C. The mode II fracture resistance curve (*R*-curve) is experimentally determined. A fractographic study on the fracture surface is performed using a scanning electron microscope (SEM), in order to reveal the failure mechanism. In addition, a numerical framework based on the cohesive zone model with a bilinear constitutive law is established for simulating the mode II delamination growth behavior at the thermal condition. The effects of the interfacial parameters on the simulations are investigated and a suitable value set for the interfacial parameters is determined. Good agreements between the experimental and numerical load–displacement responses illustrate the applicability of the numerical model. The research results provide helpful guidance for the design of composite laminates and an effective numerical method for the simulation of mode II delamination growth behavior.

Keywords: composite laminate; mode-II delamination; *R*-curve; temperature; cohesive law

Citation: Gong, Y.; Jiang, L.; Li, L.; Zhao, J. An Experimental and Numerical Study of the Influence of Temperature on Mode II Fracture of a T800/Epoxy Unidirectional Laminate. *Materials* **2022**, *15*, 8108. https://doi.org/10.3390/ma15228108

Academic Editors: Abbas S. Milani, Michele Bacciocchi and Panagiotis G. Asteris

Received: 28 September 2022
Accepted: 11 November 2022
Published: 16 November 2022

1. Introduction

Composite materials with high specific strength and specific stiffness have been widely applied in automobile, aerospace, and other fields [1–4]. Composite laminates are one of the most commonly used configuration types. However, the interlaminar properties of composite laminates are much lower than the in-plane properties because of the lack of reinforcement in the thickness direction. At the same time, owing to the mismatch of Poisson's ratio between the laminates and the anisotropy of the coefficient of thermal expansion, serious stress concentration will be produced at locations with variable thickness, free edges, and hole edges. This weak interlaminar performance and high interlaminar stress can easily lead to interlaminar failure occurring at the interface between adjacent layers in the form of interlayer debonding, which is called delamination. Delamination is one of the most common and important failure modes in composite structures [5–7]. The occurrence and growth of delamination will lead to a significant reduction in the strength and stiffness of composite structures and may even cause unpredictable catastrophic failure [8–10]. This is extremely unfavorable to the integrity, life, and safety of the structures and will seriously restrict the application of composite materials in engineering structures. Therefore, the studies on delamination behavior are of great significance and practical value and are also the focus of damage tolerance design and analysis of composite structures [11,12].

Moreover, aerospace composite structures may be exposed to thermal conditions [13–15]. Resin matrix composites are very sensitive to high temperature. The variation in temperature will change the mechanical properties of components and the internal stress between composite layers and then affect the delamination propagation behavior of composite materials. In order to ensure the reliable application of carbon fiber-reinforced composites under service conditions, it is necessary to quantitatively characterize the temperature effect on the delamination behavior and failure mechanism of carbon fiber-reinforced composites. However, the importance of the degradation of the fracture toughness caused by the thermal environment has been ignored, based on the relatively small number of published studies, especially for the mode II delamination. More efforts should be made to study mode II delamination to promote the establishment of test standards.

Limited studies have shown that the relationship between the measured mode II fracture toughness and temperature is still not clear. Hashemi et al. [16] found that the mode II fracture toughness of AS4/PEEK laminates increased with the temperature increasing from 20 °C to 130 °C. A similar conclusion was also made by Machado et al. [17]. They measured the mode II fracture toughness of laminates fabricated by the carbon/epoxy composite (SEAL® Texipreg HS 160 RM) at temperatures of 20 °C, 80 °C and −30 °C. Mode II fracture toughness of AS4/3501-6 was also found to increase with temperatures in the range of −50 °C to 100 °C [18], owing to increasing matrix ductility. However, Davies and Charentenay [19] found that there was no temperature effect on the mode II fracture toughness of T300/914C laminates evaluated in the range from 30 °C to 120 °C. More researchers found that the mode II fracture toughness decreased with the increase in temperature [20–22]. The results of Sjögren and Asp [23,24] showed that the mode II fracture toughness of HTA/6376C laminates decreased with the increase in temperature from −50 °C to 100 °C. However, the fracture surface characteristics at different temperatures were similar to each other. It was observed that the crack jumped between the upper and lower boundaries of the fiber when the temperature was 100 °C. However, at room temperature, the delamination grew along the upper boundary of the fiber. Cowley and Beaumont [25] performed end-notched flexure (ENF) tests on composite laminates fabricated by IM8/APC and IM8/954-2 materials. They found that the mode II fracture toughness decreased more obviously with the increase in temperature, especially when close to the glass transition temperature of the material. Davidson et al. [26] found that the matrix ductility increased with the increase in temperature, resulting in the decrease in mode II fracture toughness of T800H/3900-2 laminates. Boni et al. [15] measured the mode II fracture toughness of Hexcel 913C-HTA laminates evaluated at room temperature and 70 °C and found that it decreased at a higher temperature. Zulkifli and Azari [27] studied the mode II delamination growth behavior of silk fiber/epoxy laminates at 20 °C, 23 °C, 26 °C, 38 °C, 50 °C and 75 °C. It was shown that the sliding cracking of matrix and fiber appeared during the experiment, and the mode II fracture toughness at 75 °C showed a 71% decrease compared to that at room temperature.

The aerospace field has shown much interest in studying the effect of temperature on the mode II delamination behavior of widely used composites for design and verification purposes. To the authors' knowledge, the mode II resistance curve (*R*-curve), which relates the strain energy release rate (SERR) to the delamination growth length, is still lacking in-depth investigation, especially in the thermal environment. This information is important and valuable if a damage tolerance philosophy is to be applied to the design of composite structures [28]. In addition, accurate simulation of the mode II delamination is required. Therefore, this study aims to investigate the temperature dependence of the mode II delamination behavior and establish a numerical framework for the delamination simulation based on the cohesive zone model.

2. Delamination Tests on the ENF Specimen

2.1. Specimen and Test Set-Up

Specimens were fabricated by carbon fiber T800/epoxy prepregs, which were provided by the Cytec Corporation. The layer-up panel was cured according to the supplier's suggested procedures. The basic material properties are listed in Table 1, where E_1, E_2, and E_3 are the Young's moduli in the 1-direction (main orthotropic material axe), 2-direction (perpendicular orthotropic material axe) and 3-direction, respectively; G_{12}, G_{13} and G_{23} are the shear moduli in the 1-2 plane, 1-3 plane, and 2-3 plane, respectively; ν_{12}, ν_{13} and ν_{23} are the Poisson's ratios in the 1-2 plane, 1-3 plane, and 2-3 plane, respectively. X_T and X_C are the tensile and compressive strengths in the 1-direction, respectively; Y_T and Y_C are the tensile and compressive strengths in the 2-direction, respectively. Z_T and Z_C are the tensile and compressive strengths in the 3-direction, respectively; S_{12}, S_{13}, and S_{23} are the shear strengths in the 1-2 plane, 1-3 plane, and 2-3 plane, respectively. The cured panel was cut into specimens with uniform geometrical dimensions of 180 mm length, 25 mm width and 4.8 mm thickness. A 40 mm long and 25 mm wide Teflon film was prearranged between the center plies to create an artificial pre-crack for the delamination. Unidirectional laminates with a stacking sequence of $[0/0]_{12}$ were studied here.

Table 1. Basic mechanical properties of T800/epoxy material.

E_1 (Gpa)	E_2, E_3 (Gpa)	G_{12}, G_{13} (Gpa)	G_{23} (Gpa)	ν_{12}, ν_{13}	ν_{23}
163.5	9.0	4.14	3.08	0.32	0.46
X_T (MPa)	X_C (MPa)	Y_T, Z_T (MPa)	Y_C, Z_C (MPa)	S_{12}, S_{13} (MPa)	S_{23} (MPa)
2992	1183	70.6	278	172	105

The edges of specimens were painted with a thin layer of white correction fluid to facilitate the identification of the crack tip. The edge of the specimen was monitored by a traveling microscope, enabling continuous reading of the position of the crack tip with a precision of at least 0.1 mm. An insulated test chamber was used to keep the specimens at the designated temperature by electrical heating. Interior fans in the chamber were used to circulate the air and ensure a uniform temperature environment. The test was performed when the specimen remained at the stabilized temperature for about half an hour to allow the specimen to arrive at the equilibrium temperature. The chamber had one large, outfitted glass window and inside lighting, which enabled the observation of the position of the crack tip and the measurement of the delamination length during the tests.

2.2. Test Procedure

Mode II delamination was achieved by the end-notched flexure (ENF) test. The ENF tests were conducted following the ASTM standard D7905/D7905M-14 [29] on a computer-controlled MTS 858 testing machine. Displacement controlling mode was adopted. The load was applied through a loading roller as shown in Figure 1. The total span between the external rollers was set at 100 mm. The ratio of the initial delamination length to the half span, a_0/L, was set at 0.7 [30]. In this case, the delamination could be initiated in a relatively stable manner, followed by continued and relatively stable delamination growth over a relatively long distance before reaching the maximum load.

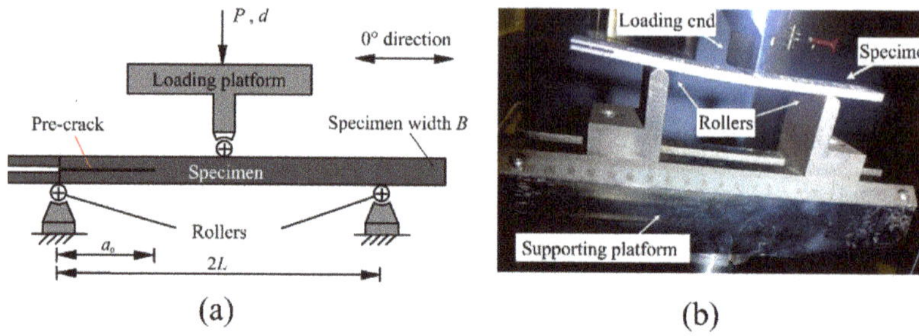

Figure 1. (**a**) Schematic diagram and (**b**) picture of the ENF test set-up.

Before conducting the mode II delamination test, it was necessary to calibrate the specimen compliance in order to determine the relationship between the compliance C and the delamination length a. The derivative of the specimen compliance with respect to the delamination length $\partial C/\partial a$ was used to calculate the mode II fracture toughness. By moving the position of the specimen on the support roller to achieve different delamination lengths, the compliance calibration was applied to obtain five data points. In order to facilitate the compliance calibration and positioning, the specimen edges were marked at 15, 20, 25, 30, and 35 mm away from the end of the pre-crack. At the same time, a ruler was set on the support platform, and the center line of the support roller was marked for accurate placement of the specimen. When placing the specimen, the marking line on the roller was aligned with the scale line on the ruler of the support platform to ensure an exact delamination length during the compliance calibration and the delamination test. According to ASTM standard D7905/D7905M-14 [29], the maximum force during the compliance calibration should be less than 50% of the ultimate load. In this work, the load applied during the compliance calibration was in the linear elastic stage of the material. The maximum applied load was 800 N and then unloaded. After each compliance test, the specimen was fully unloaded, and then the procedure of "adjusting the distance-holding heating-reloading-unloading" was repeated until the compliance data corresponding to five different delamination lengths were obtained. The loading speed during delamination tests was 0.1 mm/min in order to obtain slow delamination growth and maximize the number of data points obtained. The unloading speed was 0.5 mm/min. It should be noted that once the incubator was opened to adjust the delamination length, re-heating was required for more than 10 min to ensure the stability of the tested temperature.

2.3. *Date Reduction Method*

The mode II fracture toughness can be determined by linear elastic fracture mechanics. Three different calculation methods are available [31], including the Experimental Compliance Method (ECM) [29], the Simple Beam Theory Method (SBT), and the Corrected Beam Theory with Effective Crack Length Method (CBTE) [32]. All of them can be expressed as the classical Irwin–Kies equation:

$$G_{\mathrm{II}} = \frac{P^2}{2b}\frac{\partial C}{\partial a} \tag{1}$$

where P is the applied load, C is the specimen compliance, a and b are the delamination length and the specimen width, respectively.

2.3.1. ECM Method

The ECM is established based on the assumption of a cubic relationship between the specimen compliance and the delamination length:

$$C = A + m{\cdot}a^3 \tag{2}$$

where C is the specimen compliance (the ratio between the displacement and the load, δ/P), the parameters m and A are the slope and intercept, respectively, of the fitting curve obtained by the least squares linear regression analysis. Substituting Equation (2) into Equation (1), the expression for mode II fracture toughness can be derived.

$$G_{II} = \frac{3P^2a^2m}{2b} \quad (3)$$

2.3.2. SBT Method

According to the Simple Beam Theory, the compliance of the ENF specimen is as follows.

$$C = \frac{\delta}{P} = \frac{3a^3 + 2L^3}{8bh^3E_1} \quad (4)$$

Substituting Equation (4) into Equation (1), the expression for the fracture toughness is:

$$G_{II} = \frac{9P^2a^2}{16b^2h^3E_1} \quad (5)$$

where L is the half span, δ is the applied displacement, P is the applied load, a is the delamination length, b is the specimen width, h is a half of the specimen thickness, E_1 is the flexural modulus. The limitation of the SBT method is that the contribution of shear deformation of the specimen to the energy release rate is neglected. Zero compliance at the crack root is assumed in Equation (4). However, in an actual case, there is some deflection and rotation at the crack tip.

2.3.3. CBTE

The above ECM and SBT methods must visually observe the position of the crack tip, which is usually not easy to achieve in the mode II delamination. This is because there is no clear opening during the delamination growth [33]. Even with the help of a microscope with high magnification, it is difficult to distinguish between the damage and the delamination growth [34]. Therefore, the CBTE method without the requirement for visual observation of the crack tip is proposed. The compliance of the ENF specimen is defined as Equation (6).

$$C = \frac{\delta}{P} = \frac{3(a_e)^3 + 2L^3}{8bh^3E_1} \quad (6)$$

Substituting Equation (6) into Equation (1), the equation for the fracture toughness is given by:

$$G_{II} = \frac{9P^2{a_e}^2}{16b^2h^3E_1} \quad (7)$$

where a_e is the effective delamination length and can be obtained by Equation (8).

$$a_e = \left(\frac{1}{3}(8bCh^3E_1 - 2L^3)\right)^{\frac{1}{3}} \quad (8)$$

3. Test Results

3.1. Compliance Calibration Results

For the ECM method, the load–displacement curves of specimens with different delamination lengths were used to obtain the relationship between the compliance and the crack length. Figure 2a shows the obtained load–displacement curve of the specimen in the compliance calibration process. As shown in Figure 2b, the linear sections of the curves are abstracted for linear fitting in order to determine the specimen compliance. In this way, the plots of the compliance against the cubic of crack length at RT, 80 °C, and 130 °C are presented in Figure 3, which are fitted by Equation (2). For all temperatures, it can be seen that a good linear relation (with the R^2 coefficient equal to 0.99) was obtained between the compliance and the cubic of crack length. The detailed fitting formulas are given in Figure 3. The values of the m at RT, 80 °C, and 130 °C were 0.006, 0.0089, and 0.0078, respectively. These values were used in Equation (3) for the calculation of fracture toughness.

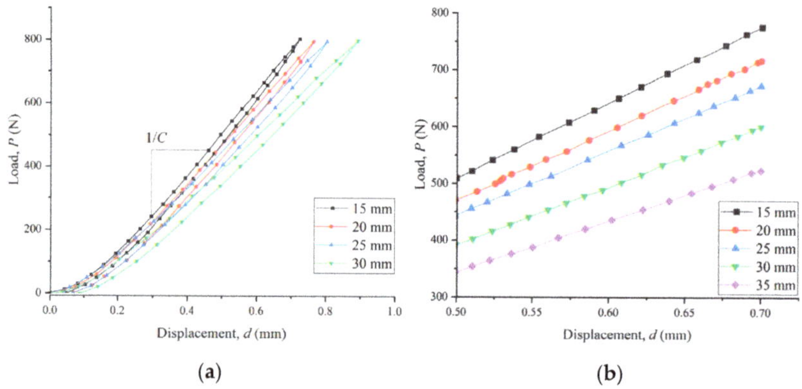

Figure 2. *P-d* curves from the compliance calibration of ENF specimens (**a**) original curve and (**b**) linear elastic section for the determination of compliance.

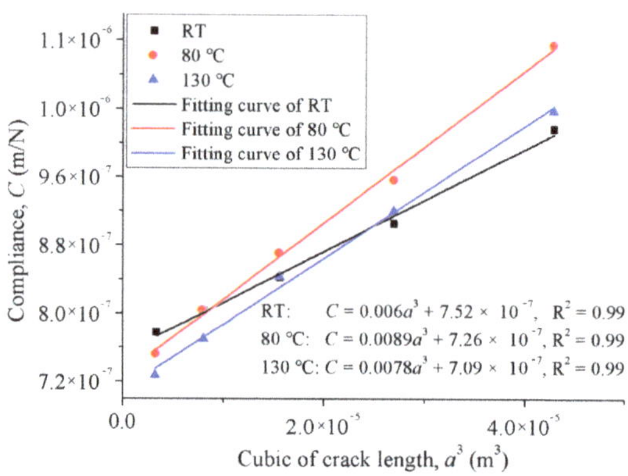

Figure 3. Compliance calibration of ENF specimens evaluated at different temperatures.

3.2. Load–Displacement Curve

Figure 4 shows the load–displacement curves of the ENF specimens under various temperature conditions. From this figure, it can be seen that all specimens exhibit a short nonlinear behavior at the initial stage of loading when the specimen first contacted with the loading nose. Then, the load–displacement curve exhibits a linear part, and the load gradually increases until reaching the maximum value. Before arriving at the final load drop, the curve exhibits a certain degree of non-linear response owing to the formation of a process zone of the matrix deformation and micro cracks at the crack tip, and delamination failure. Subsequently, the load decreased with further delamination growth. Comparing the load–displacement curves at different temperatures, both the specimen stiffness and the maximum load decreased with the increase in temperature. At room temperature, the load dropped rapidly after reaching the maximum value, which illustrated the relatively unstable delamination growth. However, at high temperatures, the load drop process tended to be gentler, especially when the temperature was 130 °C. More significant nonlinear behavior appeared, and the specimen carried the load in a relatively stable manner. This phenomenon was consistent with the experimental results in [35].

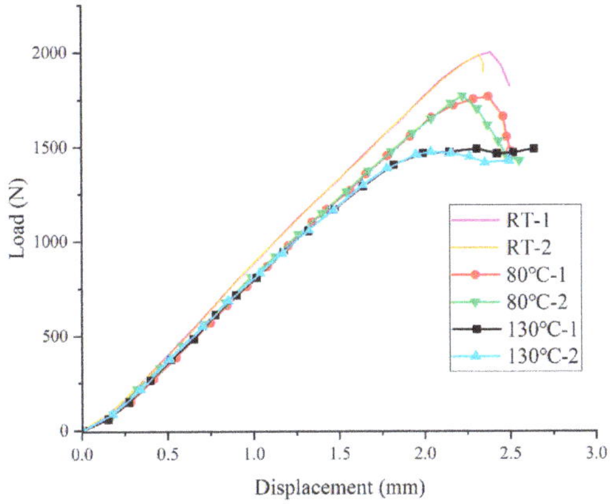

Figure 4. Load–displacement curves of ENF specimens evaluated at different temperatures.

3.3. Fracture Toughness

Based on the recorded load, deflection, and delamination length, the mode II fracture toughness under various temperature conditions was calculated using the three data reduction methods given in Section 2.3. The obtained results are shown in Figure 5. It can be seen that the ECM and SBT produced similar results. However, At RT and 80 °C, the fracture toughness values obtained by the CBTE method were obviously higher than those of the other two methods. Only at 130 °C was there no obvious difference between the results calculated by the three methods. Detailed values of the mode II fracture toughness are listed in Table 2. The mean value and standard deviation are also listed in Table 2. In this work, the ENF tests were conducted referring to the ASTM standard D7905/D7905M-14 and the ECM is the only one method covered in this ASTM standard. Therefore, the mode II fracture toughness calculated by the ECM method ECM was compared. It was determined that the fracture toughness decreased with the increase in temperature. The fracture toughness values at 80 °C and 130 °C decreased by 13.88% and 29.47%, respectively, compared with that at room temperature. With the increase in temperature, the decrease in fracture toughness may be caused by the degradation of the matrix, which led to the decrease in shear strength and stiffness and the increase in resin ductility [22,36,37].

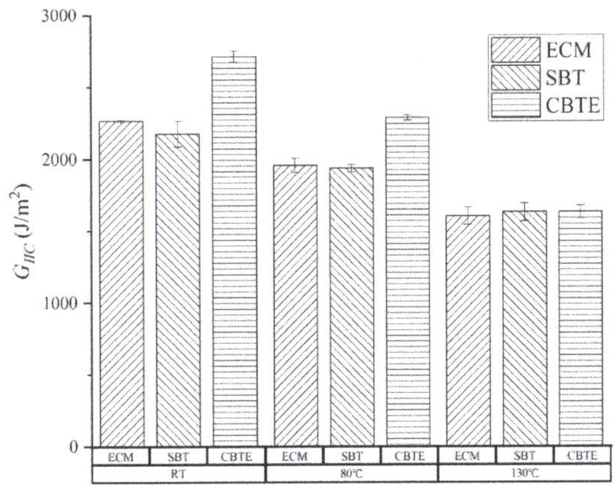

Figure 5. Comparison of mode II fracture toughness calculated by different data reduction methods.

Table 2. Mode II fracture toughness of ENF specimens evaluated at different temperatures. (Unit: J/m^2).

Temperature	Method	Test 1	Test 2	Mean Value	S.D.	CoV (%)
RT	ECM	2269.68	2259.09	2264.39	5.29	0.2
	SBT	2237.93	2110.97	2174.45	63.48	2.9
	CBTE	2738.67	2684.17	2711.42	27.25	1.0
80 °C	ECM	1920.93	1989.62	1955.28	34.34	1.8
	SBT	1916.81	1953.56	1935.19	18.38	0.9
	CBTE	2273.08	2300.07	2286.58	13.50	0.6
130 °C	ECM	1644.45	1558.94	1601.70	42.78	2.7
	SBT	1584.24	1671.48	1627.86	43.63	2.7
	CBTE	1660.58	1595.40	1627.99	32.59	2.0

An obvious *R*-curve was observed over the entire evaluated temperature range, and the *R*-curves are shown in Figure 6. The *R*-curve behavior was also presented in mode I delamination [38,39] and mixed-mode I/II delamination [40–42] of composite laminates. Wang et al. [38,39] carried out mode I delamination tests on six kinds of T800/X850 laminates with different interfaces of 0°/0°, +22.5°/−22.5°, +45°/−45°, 90°/90°, 0°/45°, and 0°/90°. An interface-dependent analytical model was proposed for the plateau fracture toughness. Gong et al. [40] experimentally determined the mixed-mode I/II *R*-curves of the unidirectional and multidirectional laminates made from carbon/epoxy composites by using the mixed-mode bending tests. It was found that the initial and steady-state fracture toughness values were strongly influenced by the interfacial ply and the mode mixity. The fiber bridging length of mode I delamination was much higher than that of mode II delamination. However, the fiber bridging length in the mixed-mode I/II delamination was approximately the same for unidirectional laminates and multidirectional laminates with a +22.5°/−22.5° interface. Furthermore, mixed-mode I/II delamination tests were performed on multidirectional T700/QY9511 laminates with a +45°/−45° interface [41,42]. Mixed-mode I/II *R*-curves were obtained, while the fracture toughness increased with the increase in delamination growth length while there was no plateau value. They also found a similar *R*-curve behavior of fatigue delamination resistance, which is also mainly resulted by the bridging fibers [43–47]. However, the resulting damage status in fatigue delamination is usually different from that in static delamination. For the mode II delamination studied here, the *R*-curve primarily arose from an increasing amount of crack-tip plasticity developing as the crack grew [16]. At RT and 80 °C, the fracture toughness gradually increased with the delamination growth until reaching a relatively stable value after a certain delamination growth length. At 130 °C, the fracture toughness linearly increased with the delamination growth, and no stable value was exhibited. With the increase in temperature, a longer delamination growth length is required to achieve the saturation state of the formed crack tip plasticity. However, because of the limitation of the effective crack length in the ENF test, the fracture toughness corresponding to the delamination length higher than 50 mm could not be obtained.

3.4. Analysis on the Fracture Surface

A fractographic study was conducted on the ENF specimens in order to analyze the failure mechanisms at different temperatures. The tested specimens were opened and the exposed fracture surfaces were gold sputter-coated and then examined with a scanning electron microscope (SEM) at a magnification of 2000. It was noted that the main micro characteristics of the fracture surfaces were shear cusps or hackles. At room temperature, a brittle failure mode with relatively little matrix deformation was observed on the fracture surface. The low fiber/matrix debonding strength resulted in a large amount of exposed fibers, as shown in Figure 7a. For the delamination tests at high temperatures, an increase in the resin ductility was observed with the increase in temperature, a feature that was highlighted by the higher degree of matrix deformation and the amount of matrix adhered to the fibers on the fracture surfaces, as shown in Figure 7b,c. The decrease in the amount of unstable delamination growth from room temperature to the elevated high temperature also reflected this transition from brittle to ductile behavior. This was consistent with the discussion in Section 3.3.

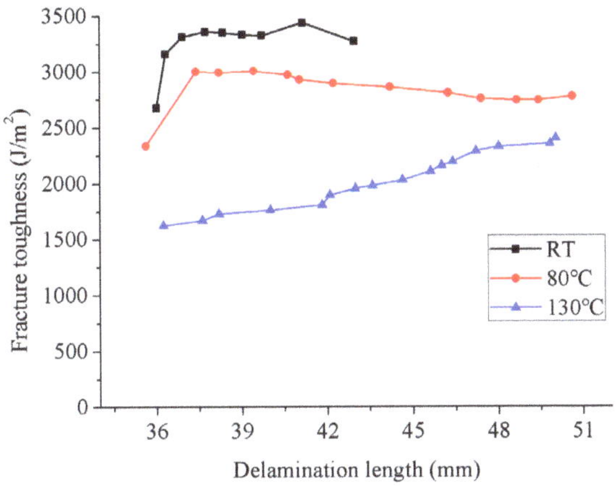

Figure 6. *R*-curves of ENF specimens evaluated at different temperatures.

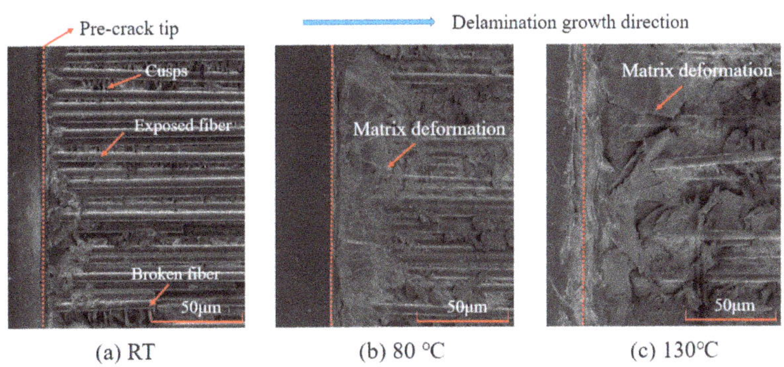

Figure 7. SEM images of fracture surfaces from the specimens tested at different temperatures.

4. Numerical Simulation for the Mode II Delamination in ENF Specimens

In this section, a numerical framework is established for the mode II delamination growth behavior at the thermal conditions. Influences of temperature on the basic material mechanical parameters were considered using a material model, which can quantitatively characterize the temperature-dependent effect on the mechanical properties of composite materials based on exponential functions of the temperature. The delamination growth behavior was modeled using the cohesive elements with a bilinear constitutive law. The effects of interfacial parameters on the simulations were investigated, and a suitable value set for the interfacial parameters was determined.

4.1. A Model for the Thermal-Affected Material Properties

The constitutive law of a transverse isotropic material can be expressed as $\boldsymbol{\varepsilon} = \boldsymbol{S} \bullet \boldsymbol{\sigma}$, where $\boldsymbol{\varepsilon}$, $\boldsymbol{S}$, and $\boldsymbol{\sigma}$ are the matrixes of engineering strain, compliance, and stress, respectively. The element S_{ij} is a function of engineering constants E_{ij}, ν_{ij}, G_{ij} (i, j = 1, 2, 3) of the composite material [48]. In order to include the temperature effect on the material mechanical properties, the compliance matrix $\boldsymbol{S}$ is modified, as shown in Equation (9):

$$S^{\text{temp}} = \begin{bmatrix} 1/E_{11}^{\text{temp}} & -\nu_{21}^{\text{temp}}/E_{22}^{\text{temp}} & -\nu_{31}^{\text{temp}}/E_{33}^{\text{temp}} & 0 & 0 & 0 \\ & 1/E_{22}^{\text{temp}} & -\nu_{32}^{\text{temp}}/E_{33}^{\text{temp}} & 0 & 0 & 0 \\ & & 1/E_{33}^{\text{temp}} & 0 & 0 & 0 \\ & & & 1/G_{23}^{\text{temp}} & 0 & 0 \\ & & & & 1/G_{31}^{\text{temp}} & 0 \\ \text{symm.} & & & & & 1/G_{12}^{\text{temp}} \end{bmatrix} \tag{9}$$

where "temp" denotes the thermal condition. A non-dimensional temperature T^* as Equation (10) is introduced to evaluate the temperature-dependent characteristics of the composite material, which has also been adopted by other researchers [49,50].

$$T^* = \frac{T_g - T_{\text{opr}}}{T_g - T_{\text{rm}}} \tag{10}$$

Here, T_g is the glass transition temperature, T_{opr} is the operation temperature, and T_{rm} is the room temperature. A united model [51], such as Equation (11), can be used to quantitatively characterize the temperature-dependent effect on the mechanical property of composite materials. The matrix and fiber stiffness and strength data are empirically fitted by exponential formulas as a function of T^*. This model was first proposed for the hygrothermal environment. When there is no moisture absorption, this model degrades to the case with only temperature effect and is also applicable.

$$\begin{array}{c} \frac{E_{11}^{\text{temp}}}{E_{11}^{0}} = (T^*)^a, \frac{E_{22}^{\text{temp}}}{E_{22}^{0}} = (T^*)^b, \frac{G_{12}^{\text{temp}}}{G_{12}^{0}} = (T^*)^c \\ \frac{X_T^{\text{temp}}}{X_T^{0}} = (T^*)^d, \frac{X_C^{\text{temp}}}{X_C^{0}} = (T^*)^e, \frac{Y_T^{\text{temp}}}{Y_T^{0}} = (T^*)^f \\ \frac{Y_C^{\text{temp}}}{Y_C^{0}} = (T^*)^g, \frac{S_{12}^{\text{temp}}}{S_{12}^{0}} = (T^*)^h, \frac{\nu_{12}^{\text{temp}}}{\nu_{12}^{0}} = 1 \end{array} \tag{11}$$

Here, the symbols a–h denote empirical constants. X_T, X_C, Y_T, Y_C and S_{12} are the longitudinal tensile and compressive, transverse tensile and compressive, and in-planar shear strengths, respectively. The superscript "0" denotes the room temperature. Values of the exponents in Equation (11) can be determined by a micro-level based method [50,52,53]. As shown in Equation (12), mechanical properties E_{33}, G_{13}, ν_{13}, G_{23}, Z_T, Z_C and S_{13} at the thermal condition are determined based on the transverse isotropic assumption. The ν_{23} and S_{23} are determined based on the method proposed by Christensen [54] and Pinho et al. [55]. The values of the exponents in Equation (11) are listed in Table 3 from which the basic mechanical properties of the composite material at 80 °C and 130 °C can be determined and are also listed in Table 3.

$$\begin{array}{c} E_{33}^{\text{temp}} = E_{22}^{\text{temp}}, G_{13}^{\text{temp}} = G_{12}^{\text{temp}}, \nu_{13}^{\text{temp}} = \nu_{12}^{\text{temp}}, \\ \nu_{23}^{\text{temp}} = \nu_{12}^{\text{temp}}\left(1 - \nu_{12}^{\text{temp}} E_{22}^{\text{temp}}/E_{11}^{\text{temp}}\right)/\left(1 - \nu_{12}^{\text{temp}}\right), \\ G_{23}^{\text{temp}} = E_{22}^{\text{temp}}/2\left(1 + \nu_{23}^{\text{temp}}\right), \\ Z_T^{\text{temp}} = Y_T^{\text{temp}}, Z_C^{\text{temp}} = Y_C^{\text{temp}}, S_{13}^{\text{temp}} = S_{12}^{\text{temp}}, \\ S_{23}^{\text{temp}} = \frac{Y_C^{\text{temp}}}{2\tan(53°)}. \end{array} \tag{12}$$

Table 3. Values of the exponents for the engineering constant calculations and the determined mechanical properties of T800/epoxy composites at high temperatures.

Item	*a*	*b*	*c*	*d*	*e*
Value	0.04	0.5	0.5	0.04	0.54
Item	*f*	*g*	*h*	T_g/°C	T_{rm}/°C
Value	0.50	0.50	0.50	185	23
Elastic property	80 °C	130 °C	Strength (MPa)	80 °C	130 °C
E_1 (GPa)	160.7	156.6	X_T	2941	2866
E_2, E_3 (GPa)	7.245	5.24	X_C	936	661
G_{12}, G_{13} (GPa)	3.33	2.41	Y_T, Z_T	56.8	41.2
G_{23} (GPa)	2.44	1.76	Y_C, Z_C	224	162
ν_{12}, ν_{13}	0.33	0.33	S_{12}, S_{13}	138	100
ν_{23}	0.485	0.487	S_{23}	84	61

4.2. Cohesive Zone Model with a Bilinear Constitutive Law

Different methods have been proposed to simulate the delamination growth behavior, such as the virtual fracture closure technique (VCCT) [56], extended finite element method (XFEM) [57,58], and cohesive zone model (CZM) [7,59]. The XFEM can effectively model the delamination migration [60,61] in composite structures [62]. In addition, Reddy and co-workers [63,64] illustrated that delamination can be conveniently modeled using an element deletion/failure method in conjunction with a non-local fracture criterion. They developed a novel three-dimensional, rate form-based finite-deformation constitutive theory to describe the damage and fracture in viscoelastic solids. The CZM is a prevailing method which combines strength-based analysis to predict delamination initiation and fracture mechanics to predict delamination growth. The constitutive law relates cohesive traction (σ) to crack opening/shearing displacement (δ) at the interface. Different constitutive laws have been proposed for different material systems and structural forms, such as bilinear, exponential, trapezoidal, and multi-linear [65]. To deal with the effect of fiber bridging on the delamination behavior, modified delamination growth criteria and modified cohesive zone model were proposed. In the modified delamination growth criteria, the *R*-curve was integrated into the traditional cohesive zone model with bilinear constitutive law [41,42]. Gong et al. [65] proposed a physical-based three-linear CZM superposed by two bilinear CZMs that represent two different phenomena including the quasi-brittle matrix fracture characterized by a higher peak stress and a shorter relative opening displacement, and the fiber bridging characterized by a lower peak stress and a longer relative opening displacement. Furthermore, a novel four-linear cohesive law was developed [66], which was established based on the realistic failure mechanism during the delamination growth. All the necessary cohesive parameters can be obtained by tests except for the fracture toughness ratio, which is introduced to characterize the non-linear softening behavior of the bridging fibers in a simple way. In addition, the implementation of the four-linear cohesive law in FEM is easier and more efficient than the non-linear. In order to determine the bridging law that can be applied in the multi-linear cohesive law, different semi-analytical methods have been developed [67–69] based on the beam theory. Only experimental load and displacement data are required and the visual observation for the delamination is avoided, which makes the semi-analytical methods time saving and cost effective. A simple procedure was also established for determining the bridging law [70]. The crack opening displacement at the initial pre-crack tip was obtained by a numerical method. The bridging law was determined via the *J*-integral method. The obtained results agreed well with the experimental ones. In this study, the bilinear cohesive law, which is the most widely used because of its simplicity and intuitive physical meaning [71], was adopted. Figure 8 shows the constitutive behavior of pure mode II delamination in an ENF specimen.

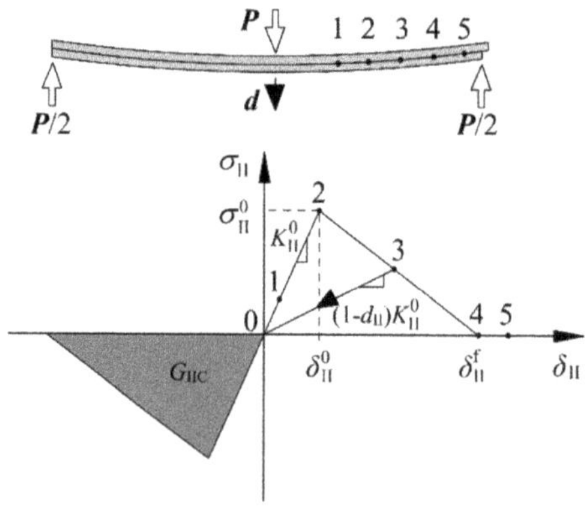

Figure 8. Bilinear constitutive law for the mode II delamination in an ENF specimen.

Interface stiffness K_{II}^0 is used to characterize the mechanical behavior of the cohesive element in the linear elastic range (point 1 in Figure 8). When the interface stress reaches the shear strength σ_{II}^0 (point 2), the damage onset in cohesive elements occurs. The relative displacement at this point can be expressed as: $\delta_{\text{II}}^0 = \sigma_{\text{II}}^0 / K_{\text{II}}^0$. Point 3 indicates that the element is in the softening stage and has been partially damaged. The interface stiffness is degraded to $K_{\text{II}} = (1 - d_{\text{II}})\, K_{\text{II}}^0$, where $d_{\text{II}} \in [0,1]$ is the damage variable. The element completely fails and loses its load-bearing capacity when the interface stiffness reduces to zero (point 4). According to Griffith's fracture theory, the area under the traction-relative displacement relationship is equal to the fracture toughness G_{IIC} as shown in Equation (13):

$$\int_0^{\delta_{\text{II}}^f} \sigma_{\text{II}} d\delta_{\text{II}} = G_{\text{IIC}} \tag{13}$$

where δ_{II}^f is the relative displacement when the cohesive element just loses the load-bearing capacity. The detailed constitutive law is as follows:

$$\sigma_{\text{II}} = \begin{cases} K_{\text{II}}\delta_{\text{II}}, & \delta_{\text{II}}^{max} \leq \delta_{\text{II}}^0 \\ (1 - d_{\text{II}})K_{\text{II}}\delta_{\text{II}}, & \delta_{\text{II}}^0 < \delta_{\text{II}}^{max} \leq \delta_{\text{II}}^f, \\ 0, & \delta_{\text{II}}^{max} > \delta_{\text{II}}^f \end{cases} \tag{14}$$

$$d_{\text{II}} = \frac{\delta_{\text{II}}^f (\delta_{\text{II}}^{max} - \delta_{\text{II}}^0)}{\delta_{\text{II}}^{max} \left(\delta_{\text{II}}^f - \delta_{\text{II}}^0\right)}, \; d_{\text{II}} \in [0,1] \tag{15}$$

where δ_{II}^{max} is the maximum relative sliding displacement and is defined as:

$$\delta_{\text{II}}^{max} = max\{\delta_{\text{II}}^{max}, |\delta_{\text{II}}|\}. \tag{16}$$

4.3. Numerical Model of the ENF Specimen

Considering the transverse isotropy of the unidirectional laminate studied here, a two-dimensional finite element model of the ENF specimen was established in the Abaqus commercial FE software, as shown in Figure 9. A low computational cost was achieved without reducing the prediction accuracy, as shown in the following simulation. The specimen was modeled as the upper and lower arms tied with a cohesive layer in the uncracked region, while contact interactions were present in the pre-cracked region to prevent the penetration between the arm surfaces. The two arms were meshed with four-node plane stress elements with a coarse mesh along the specimen length. Four-node two-dimensional cohesive elements (COH2D4) were pre-arranged along the middle plane of the specimen to simulate the delamination growth [72]. In the thickness direction, one element for each composite layer was used. A refined mesh was adopted for the cohesive layer to capture accurate stress/strain distributions in the cohesive zone and ensure the independence of simulation results on the mesh size. To reproduce the ENF test, the specimen was sustained by two semi-cylindrical rigid supports (B and C) with a span of 100 mm and loaded by a central nose (A). Boundary conditions were assigned to the two supports through their reference points, which were constrained in translation and rotation. A displacement boundary condition was applied to the central semi-cylinder A, and the horizontal translation was inhibited, such as the in-plane rotation. In this work, the delamination growth was evaluated by comparing the SERR of the cohesive element with the fracture toughness. The fracture toughness was obtained from the tests as presented in Section 3.3. The critical interface parameters of the cohesive element were numerically determined, as shown in the following, Section 4.4.

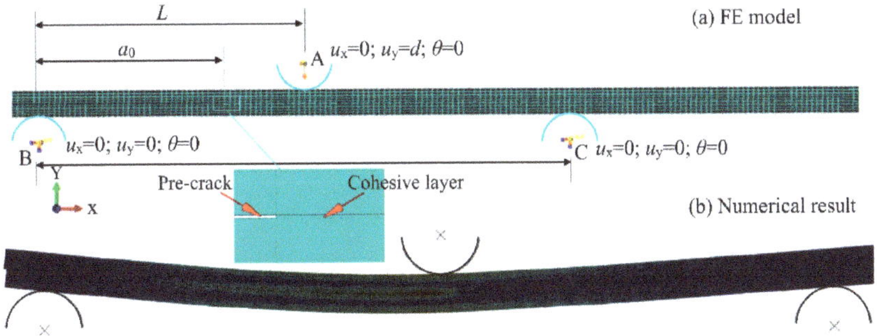

Figure 9. Simulation of the ENF specimen (**a**) FE model and (**b**) numerical result.

4.4. Simulated Results

When using the cohesive zone model for the delamination modeling, the interface stiffness, the viscosity coefficient of the cohesive elements, the mesh size, and the interface strength are critical parameters. The interface stiffness is a non-physical parameter used to maintain the rigid connection between the undamaged cohesive elements and the laminates. Ideally, the interface stiffness should be infinite, so that the overall stiffness of the structure is accurately modeled. However, it will result in a converging problem and spurious oscillation. Therefore, a reasonable value of the interface stiffness should be adopted. The viscosity is introduced in the cohesive elements in order to improve the convergence of calculation. The viscous regularization taken by the constitutive law is realized by permitting the stresses beyond the interface strength, and it adopts a relatively small characteristic time increment. Computational costs and influences on predictions should be considered when determining the optimal viscosity coefficient. Reasonable mesh size of the cohesive element is particularly important for the accuracy of simulation results. The selection of mesh size depends on the length of the cohesive zone and the number of cohesive elements in this zone. The length of the cohesive zone is related to the interface strength [73]. In order to accurately capture the stress/strain field in the cohesive zone, a sufficiently fine meshing for the cohesive elements is required. At present, there is still no universal rule for determining the values of those parameters. In this work, the following procedure was adopted based on the trial-and-error method. First, the mesh of the cohesive element was fine to ensure enough elements in the cohesive zone, and the effect of the interface stiffness on the elastic stage of the load–displacement response was investigated. Then, the effect of the viscosity coefficient of cohesive elements was studied. Based on the determined values of interface stiffness and viscosity coefficient, the influence of mesh size and interface strength was finally studied. From this procedure, the suitable value set of these critical parameters were obtained.

Three kinds of interfacial stiffness (10^{14} N/m^3, 10^{15} N/m^3 and 10^{16} N/m^3) were studied to reveal its influence on the initial slope of the predicted load–displacement curve. The predictions are shown in Figure 10a. 'Test 1' and 'Test 2' represented the experimental results of two tested specimens. For better comparisons, the initial nonlinear stage of the experimental results was removed. It indicated that the influence of interface stiffness was trivial and the predictions had good agreement with the experimental ones. Considering that a larger value of interface stiffness can ensure less structural stiffness loss, its value was reasonably chosen as 10^{15} N/m^3 for the specimens evaluated at 80 °C. The influence of the viscosity coefficient (10^{-6}, 10^{-5}, and 10^{-4}) on the predicted load–displacement curves is shown in Figure 10b and compared with the experimental ones. It can be seen that the viscosity coefficient had no effect on the linear and elastic stage, while it had an obvious effect on the subsequent load-drop stage. Consistent results were obtained when the viscosity coefficient was chosen as 10^{-6} and 10^{-5} because a higher viscosity coefficient usually results in a better convergence, which means fewer analysis steps and less computational time are required. Thus, the viscosity coefficient was chosen as 10^{-5} for the specimens evaluated at 80 °C. The predicted load–displacement curves with different mesh sizes (0.25, 0.5, and 1 mm) are shown in Figure 10c. It can be seen that the effect of the studied mesh size on the predictions was also trivial. Because a finer mesh size will increase the computational cost, the mesh size was chosen as 0.5 mm. The final numerical results are compared with experimental results in Figure 10d, and good agreement was achieved between them.

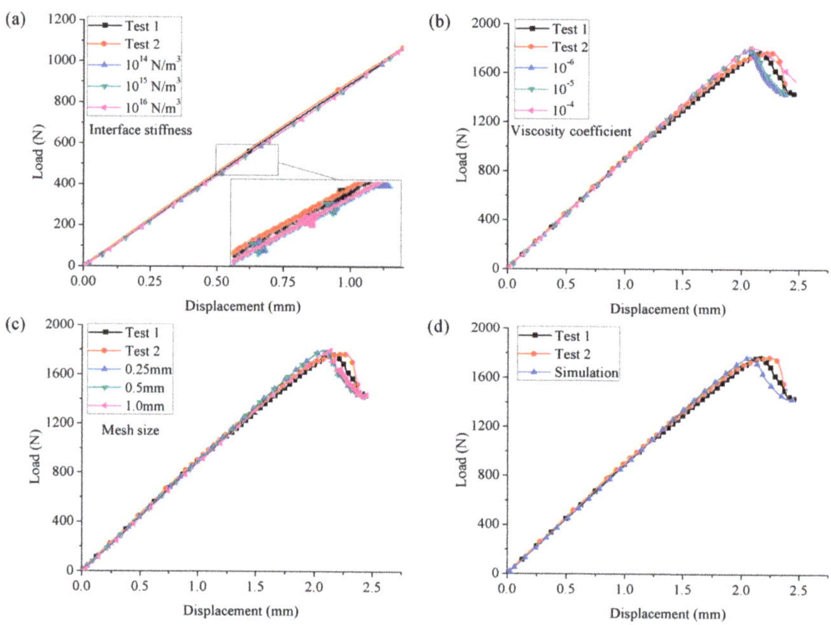

Figure 10. Influence of (**a**) interface stiffness, (**b**) viscosity coefficient, (**c**) mesh size on the predicted load–displacement response and (**d**) numerical results of the ENF test at 80 °C.

The reasonable interfacial parameters for modeling the ENF test at 130 °C were determined by following the above method. Figure 11 shows the effects of interface stiffness (10^{14}, 10^{15}, and 10^{16} N/m^3), viscosity coefficient (10^{-5}, 10^{-4} and 10^{-3}) and mesh size (0.25 and 0.5 mm) on the numerical results. It can be seen that in the studied range, the interface stiffness and mesh size did not affect the predictions. For different viscosity coefficients, the predicted linear stages were the same. However, only when its value was 10^{-5} was a satisfactory agreement between the predictions and the experimental results obtained for the load-drop stage. Therefore, the interfacial parameters of the cohesive elements suitable for ENF tests evaluated at 130 °C were as follows: interface stiffness of 10^{15} N/m^3, viscosity coefficient of 10^{-5}, and mesh size of 0.5 mm. Based on the suitable interfacial parameter set, the numerical model for simulating the mode II delamination of composite laminates was established. The final numerical results were obtained and are presented in Figure 11d, and they agreed well with the experimental results.

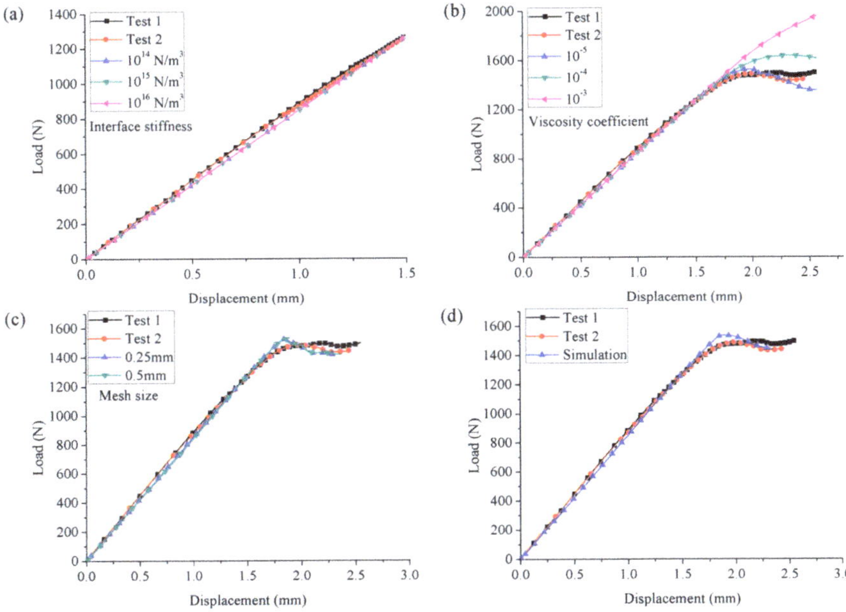

Figure 11. Influence of (**a**) interface stiffness, (**b**) viscosity coefficient, (**c**) mesh size on the predicted load–displacement response, and (**d**) numerical results of the ENF test at 130 °C.

5. Conclusions

Mode II delamination tests were conducted on CFRP laminates using the ENF set-up to investigate the temperature dependence of delamination behavior. The fracture toughness at 80 °C and 130 °C decreased by 13.88% and 29.47%, respectively, compared with that at room temperature. This observation was consistent with most published results. The decrease in fracture toughness with the increase in temperature may be caused by the degradation of matrix, which led to the decrease in shear strength and stiffness and the increase in resin ductility. In addition, obvious *R*-curve behaviors were experimentally observed in the specimens tested at different temperatures, which suggested that the fracture toughness increased with the increase in temperature. The *R*-curve was mainly caused by the increasing amount of crack-tip plasticity. At RT and 80 °C, the fracture toughness value was relatively stable after a certain delamination growth length. However, no obvious stable value was observed for the specimen tested at 130 °C, owing to the limitation of the effective crack length in the work. It may need a longer delamination growth length to achieve the stable value, and this requires further validation from future work. Micrographs of the fracture surfaces obtained via scanning electron microscope showed that a brittle failure mode with relatively little matrix deformation was observed in specimens evaluated at room temperature. However, with the increase in temperature, an increase in the matrix ductility was observed, as confirmed by the higher degree of matrix deformation and the amount of matrix adhered to the fibers on the fracture surfaces. The decrease in the amount of unstable delamination growth from the room temperature to the higher temperature also reflected this transition from brittle to ductile behavior.

Based on the cohesive zone model with a bilinear constitutive law, a numerical framework was established for the delamination simulation of ENF tests. Temperature effects on the mechanical parameters of the composite material, including the elastic and strength parameters, were considered. Detailed values of these mechanical parameters were determined by a power exponential model formulated as a function of a non-dimensional temperature. The basic elastic parameters and strengths decreased with the increase in temperature. The effects of critical cohesive parameters, including interface stiffness, viscosity, and mesh size, were numerically investigated. A suitable value set for the critical cohesive parameters was identified and applied to the delamination modeling. It indicated that the interfacial parameters of the cohesive elements suitable for ENF tests evaluated at 80 °C and 130 °C were as follows: interface stiffness of 10^{15} N/m^3, viscosity coefficient of 10^{-5}, and mesh size of 0.5 mm. The temperature did not seem to affect the suitable value of the interfacial parameter. The predicted load–displacement responses showed satisfactory agreement with the experimental, which illustrated the applicability and accuracy of the established numerical model. It is worthwhile to point out that the aerospace composite structures are usually exposed to thermal cycles. Further studies are required on the delamination behavior of those structures exposed to thermal cycles, which should be an interesting topic.

Author Contributions: Conceptualization, investigation, supervision, funding acquisition, writing—original draft, review and editing, project administration, Y.G.; methodology, investigation, writing—original draft, review and editing, L.J.; investigation, writing—review and editing, L.L.; formal analysis, resources, writing—review and editing, funding acquisition, J.Z. All authors have read and agreed to the published version of the manuscript.

Funding: The research work is supported by the National Natural Science Foundation of China (Project nos. 12172067, 11902054, 52175144), the Fundamental Research Funds for the Central Universities (Project no. 2022CDJKYJH011), the Chongqing Natural Science Foundation (Project nos. CSTB2022NSCQ-MSX0308, CSTB2022TFII-OFX0004), the Young Elite Scientists Sponsorship Program by CAST (Project no. 2020QNRC001), the Chongqing Talent Plan (Project no. cstc2022ycjh-bgzxm0144) and the Foundation of the State Key Laboratory of Automotive Simulation and Control (Project no. 20201202).

Institutional Review Board Statement: Not applicable.

Informed Consent Statement: Not applicable.

Data Availability Statement: Not applicable.

Conflicts of Interest: The authors declare no conflict of interest.

References

1. Li, B.; Gong, Y.; Gao, Y.; Hou, M.; Li, L. Failure analysis of hat-stringer-stiffened aircraft composite panels under four-point bending loading. *Materials* **2022**, *15*, 2430. [CrossRef]
2. Wang, Y.; Gong, Y.; Zhang, Q.; He, Y.; Liu, Z.; Hu, N. Fatigue behavior of 2.5D woven composites based on the first-order bending vibration tests. *Compos. Struct.* **2022**, *284*, 115218. [CrossRef]
3. Li, X.; Gong, Y.; Jiang, L.; Zhao, Y.; Hu, N. A parametric study on the failure strength of multi-bolt composite repairs with different configurations. *Theor. Appl. Fract. Mech.* **2022**, *121*, 103459. [CrossRef]
4. Li, B.; Gong, Y.; Xiao, H.; Gao, Y.; Liang, E. A two-dimensional model for pin-load distribution and failure analysis of composite bolted joints. *Materials* **2021**, *14*, 3646. [CrossRef]
5. Hosseini, M.R.; Taheri-Behrooz, F.; Salamat-talab, M. Mode II interlaminar fracture toughness of woven E-glass/epoxy composites in the presence of mat interleaves. *Int. J. Adhes. Adhes.* **2020**, *98*, 102523. [CrossRef]
6. Herráez, M.; Pichler, N.; Pappas, G.A.; Blondeau, C.; Botsis, J. Experiments and numerical modelling on angle-ply laminates under remote mode II loading. *Compos. Part A Appl. Sci. Manuf.* **2020**, *134*, 105886. [CrossRef]
7. Heidari-Rarani, M.; Sayedain, M. Finite element modeling strategies for 2D and 3D delamination propagation in composite DCB specimens using VCCT, CZM and XFEM approaches. *Theor. Appl. Fract. Mech.* **2019**, *103*, 102246. [CrossRef]

8. Gong, Y.; Chen, X.; Li, W.; Zhao, L.; Tao, J.; Zhang, J.; Hu, N. Delamination in carbon fiber epoxy DCB laminates with different stacking sequences: R-curve behavior and bridging traction-separation relation. *Compos. Struct.* **2021**, *262*, 113605. [CrossRef]
9. Gong, Y.; Tao, J.; Chen, X.; Zhao, J.; Hu, N. A semi-analytical method for determining mode-II fracture toughness and bridging law of composite laminates. *Eng. Fract. Mech.* **2022**, *265*, 108371. [CrossRef]
10. Zhang, Y.; Jiang, L.; Gong, Y.; Liu, H.; Zhao, J.; Ren, S. Failure prediction of marine non-planar composite π joints based on combined 2D material model and progressive damage method. *Ocean Eng.* **2021**, *222*, 108581. [CrossRef]
11. Gong, Y.; Tian, D.; Cao, T.; Zhao, L.; Zhang, J.; Hu, N.; Zhang, C. An R-curve effect-included delamination growth criterion for mixed-mode I/II delamination predictions of composite laminates. *Compos. Struct.* **2022**, *295*, 115846. [CrossRef]
12. Liu, C.; Gong, Y.; Gong, Y.; Li, W.; Liu, Z.; Hu, N. Mode II fatigue delamination behaviour of composite multidirectional laminates and the stress ratio effect. *Eng. Fract. Mech.* **2022**, *264*, 108321. [CrossRef]
13. Abdel-Monsef, S.; Renart, J.; Carreras, L.; Maimí, P.; Turon, A. Effect of environmental conditioning on pure mode I fracture behaviour of adhesively bonded joints. *Theor. Appl. Fract. Mech.* **2020**, *110*, 102826. [CrossRef]
14. Gong, Y.; Jiang, L.; Li, L.; Ren, S.; Zhao, Y.; Wang, Z.; Hu, N. Temperature effect on the static mode I delamination behavior of aerospace-grade composite laminates: Experimental and numerical study. *Fatigue Fract. Eng. Mater. Struct.* **2022**, *45*, 2827–2844. [CrossRef]
15. Boni, L.; Fanteria, D.; Lazzeri, L.; Panettieri, E.; Mariani, U.; Rigamonti, M. Influence of environment conditioning on the interlaminar fracture toughness of a graphite/epoxy unidirectional material. *Compos. Part B Eng.* **2018**, *153*, 97–107. [CrossRef]
16. Hashemi, S.; Kinloch, A.J.; Williams, J.G. The effects of geometry, rate and temperature on the mode I, mode II and mixed-mode I/II interlaminar fracture of carbon-fibre/poly (ether-ether ketone) composites. *J. Compos. Mater.* **1990**, *24*, 918–956. [CrossRef]
17. Machado, J.J.M.; Marques, E.A.S.; Campilho, R.D.S.G.; da Silva, L.F.M. Mode II fracture toughness of CFRP as a function of temperature and strain rate. *Compos. Part B Eng.* **2017**, *114*, 311–318. [CrossRef]
18. Chapman, T.; Smiley, A.; Pipes, R. Rate and temperature effects on mode II interlaminar fracture toughness in composite materials. Deformation and fracture of laminated composites. In Proceedings of the 6th International Conference on Composite Materials (ICCM-6), London, UK, 20–24 July 1987; pp. 3.295–3.304.
19. Davies, P.; de Charentenay, F. Effect of temperature on the interlaminar fracture of tough composites. Deformation and fracture of laminated composites. In Proceedings of the 6th International Conference on Composite Materials (ICCM-6), London, UK, 20–24 July 1987; pp. 3.284–3.294.
20. Russell, A.; Street, K. *Moisture and Temperature effects on the Mixed-Mode Delamination Fracture of Unidirectional Graphite/Epoxy*; Delamination and Debonding of Materials; ASTM International: West Conshohocken, PA, USA, 1985.
21. Garg, A.C. Intralaminar and interlaminar fracture in graphite/epoxy laminates. *Eng. Fract. Mech.* **1986**, *23*, 719–733. [CrossRef]
22. Kim, H.; Wang, W.; Takao, Y.; Ben, G. Effects of temperature on mode II fracture toughness of multidirectional CFRP laminates. *Int. J. Mod. Phys. B* **2003**, *17*, 1717–1723. [CrossRef]
23. Asp, L.E. The effects of moisture and temperature on the interlaminar delamination toughness of a carbon/epoxy composite. *Compos. Sci. Technol.* **1998**, *58*, 967–977. [CrossRef]
24. Sjögren, A.; Asp, L.E. Effects of temperature on delamination growth in a carbon/epoxy composite under fatigue loading. *Int. J. Fatigue* **2002**, *24*, 179–184. [CrossRef]
25. Cowley, K.D.; Beaumont, P.W.R. The interlaminar and intralaminar fracture toughness of carbon-fibre/polymer composites: The effect of temperature. *Compos. Sci. Technol.* **1997**, *57*, 1433–1444. [CrossRef]
26. Davidson, B.D.; Kumar, M.; Soffa, M.A. Influence of mode ratio and hygrothermal condition on the delamination toughness of a thermoplastic particulate interlayered carbon/epoxy composite. *Compos. Part A Appl. Sci. Manuf.* **2009**, *40*, 67–79. [CrossRef]
27. Zulkifli, R.; Azhari, C.H. Mode II interlaminar fracture properties of treated silk fibre/ epoxy composites at low and high temperature range. *Int. J. Eng. Technol.* **2018**, *7*, 129–132. [CrossRef]
28. Landry, B.; LaPlante, G.; LeBlanc, L.R. Environmental effects on mode II fatigue delamination growth in an aerospace grade carbon/epoxy composite. *Compos. Part A Appl. Sci. Manuf.* **2012**, *43*, 475–485. [CrossRef]
29. *D7905/D7905M-14*; Standard Test Method for Determination of the Mode II Interlaminar Fracture Toughness of Unidirectional Fiber-Reinforced Polymer Matrix Composites. ASTM International: West Conshohocken, PA, USA, 2014.
30. Carlsson, L.A.; Gillespie, J.W.; Pipes, R.B. On the analysis and design of the end notched flexure (ENF) specimen for mode II testing. *J. Compos. Mater.* **1986**, *20*, 594–604. [CrossRef]
31. Wilk, J. Applicability of mode II interlaminar fracture toughness testing methods for characterization of thermoplastic laminates with woven fabric reinforcements. *Eng. Fract. Mech.* **2019**, *216*, 106533. [CrossRef]
32. *ISO 15114*; Fibre-Reinforced Plastic Composites-Determination of the Mode II Fracture Resistance for Unidirectionally Reinforced Materials Using the Calibrated End-Loaded Split (C-ELS) Test and an Effective Crack Length Approach. International Standards Organization: Geneva, Switzerland, 2014.
33. Gong, Y.; Xia, K.; Wang, Y.; Zhao, L.; Zhang, J.; Hu, N. A semi-analytical model for the mode II fracture toughness of multidirectional composite laminates. *Thin-Walled Struct.* **2023**, *182*, 110235. [CrossRef]
34. Blackman, B.; Brunner, A.J.; Williams, J.G. Mode II fracture testing of composites: A new look at an old problem. *Eng. Fract. Mech.* **2006**, *73*, 2443–2455. [CrossRef]
35. Czabaj, M.W.; Davidson, B.D. Determination of the mode I, mode II, and mixed-mode I–II delamination toughness of a graphite/polyimide composite at room and elevated temperatures. *J. Compos. Mater.* **2015**, *50*, 2235–2253. [CrossRef]

36. Zou, L.; Gong, Y.; Tian, D.; Zhao, L.; Zhang, J.; Hu, N. Effect of interface angle on mode I delamination damage behavior of multidirectional fully isotropic laminates with the same global stiffness. *Thin-Walled Struct.* **2023**, *182*, 110211. [CrossRef]
37. Lee, S.M. Mode II delamination failure mechanisms of polymer matrix composites. *J. Mater. Sci.* **1997**, *32*, 1287–1295. [CrossRef]
38. Zhao, L.; Wang, Y.; Zhang, J.; Gong, Y.; Lu, Z.; Hu, N.; Xu, J. An interface-dependent model of plateau fracture toughness in multidirectional CFRP laminates under mode I loading. *Compos. Part B Eng.* **2017**, *131*, 196–208. [CrossRef]
39. Zhao, L.; Wang, Y.; Zhang, J.; Gong, Y.; Hu, N.; Li, N. XFEM-based model for simulating zigzag delamination growth in laminated composites under mode I loading. *Compos. Struct.* **2017**, *160*, 1155–1162. [CrossRef]
40. Gong, Y.; Zhang, B.; Zhao, L.; Zhang, J.; Hu, N.; Zhang, C. R-curve behaviour of the mixed-mode I/II delamination in carbon/epoxy laminates with unidirectional and multidirectional interfaces. *Compos. Struct.* **2019**, *223*, 110949. [CrossRef]
41. Gong, Y.; Zhao, L.; Zhang, J.; Wang, Y.; Hu, N. Delamination propagation criterion including the effect of fiber bridging for mixed-mode I/II delamination in CFRP multidirectional laminates. *Compos. Sci. Technol.* **2017**, *151*, 302–309. [CrossRef]
42. Gong, Y.; Zhao, L.; Zhang, J.; Hu, N. An improved power law criterion for the delamination propagation with the effect of large-scale fiber bridging in composite multidirectional laminates. *Compos. Struct.* **2018**, *184*, 961–968. [CrossRef]
43. Zhao, L.; Gong, Y.; Zhang, J.; Wang, Y.; Lu, Z.; Peng, L.; Hu, N. A novel interpretation of fatigue delamination growth behavior in CFRP multidirectional laminates. *Compos. Sci. Technol.* **2016**, *133*, 79–88. [CrossRef]
44. Gong, Y.; Zhao, L.; Zhang, J.; Hu, N. A novel model for determining the fatigue delamination resistance in composite laminates from a viewpoint of energy. *Compos. Sci. Technol.* **2018**, *167*, 489–496. [CrossRef]
45. Gong, Y.; Li, W.; Liu, H.; Yuan, S.; Wu, Z.; Zhang, C. A novel understanding of the normalized fatigue delamination model for composite multidirectional laminates. *Compos. Struct.* **2019**, *229*, 111395. [CrossRef]
46. Gong, Y.; Zhao, L.; Zhang, J.; Hu, N.; Zhang, C. An insight into three approaches for determining fatigue delamination resistance in DCB tests on composite laminates. *Compos. Part B-Eng.* **2019**, *176*, 107206. [CrossRef]
47. Gong, Y.; Zhao, L.; Zhang, J.; Hu, N. Crack closure in the fatigue delamination of composite multidirectional DCB laminates with large-scale fiber bridging. *Compos. Struct.* **2020**, *244*, 112220. [CrossRef]
48. Kaw, A.K. *Mechanics of Composite Materials*, 2nd ed.; CRC Press: Boca Raton, FL, USA, 2005.
49. Zhang, J.; Qi, D.; Zhou, L.; Zhao, L.; Hu, N. A progressive failure analysis model for composite structures in hygrothermal environments. *Compos. Struct.* **2015**, *133*, 331–342. [CrossRef]
50. Benkhedda, A.; Tounsi, A.; Bedia, E.A.A. Effect of temperature and humidity on transient hygrothermal stresses during moisture desorption in laminated composite plates. *Compos. Struct.* **2008**, *82*, 629–635. [CrossRef]
51. Shan, M.; Zhao, L.; Hong, H.; Liu, F.; Zhang, J. A progressive fatigue damage model for composite structures in hygrothermal environments. *Int. J. Fatigue* **2018**, *111*, 299–307. [CrossRef]
52. Megueni, A.; Tounsi, A.; Bedia, E.A. Evolution of the stress intensity factor for patched crack with bonded hygrothermal aged composite repair. *Mater. Design.* **2007**, *28*, 287–293. [CrossRef]
53. Chang, F.; Shahid, I.; Engdahl, R.A. Predicting Moduli and Strengths Reduction of Unidirectional Graphite/Epoxy Composites Due to Hygrothermal Effects. *J. Reinf. Plast. Compos.* **1989**, *8*, 106–132. [CrossRef]
54. Christense, R.M. Tensor transformations and failure criteria for the analysis of fiber composite materials. *J. Compos. Mater.* **1988**, *22*, 874–897. [CrossRef]
55. Pinho, S.T.; Iannucci, L.; Robinson, P. Physically-based failure models and criteria for laminated fibre-reinforced composites with emphasis on fibre kinking: Part I: Development. *Compos. Part A Appl. Sci. Manuf.* **2006**, *37*, 63–73. [CrossRef]
56. Krueger, R. Virtual crack closure technique: History, approach, and applications. *Appl. Mech. Rev.* **2004**, *57*, 109–143. [CrossRef]
57. Belytschko, T.; Moës, N.; Usui, S.; Parimi, C. Arbitrary discontinuities in finite elements. *Int. J. Numer. Meth. Eng.* **2001**, *50*, 993–1013. [CrossRef]
58. Abdullah, N.A.; Curiel-Sosa, J.L.; Taylor, Z.A.; Tafazzolimoghaddam, B.; Vicente, J.L.M.; Zhang, C. Transversal crack and delamination of laminates using XFEM. *Compos. Struct.* **2017**, *173*, 78–85. [CrossRef]
59. Camanho, P.P.; Davila, C.G.; de Moura, M.F. Numerical simulation of mixed-mode progressive delamination in composite materials. *J. Compos. Mater.* **2003**, *37*, 1415–1438. [CrossRef]
60. Gong, Y.; Zhang, B.; Mukhopadhyay, S.; Hallett, S.R. Experimental study on delamination migration in multidirectional laminates under mode II static and fatigue loading, with comparison to mode I. *Compos. Struct.* **2018**, *201*, 683–698. [CrossRef]
61. Gong, Y.; Zhang, B.; Hallett, S.R. Delamination migration in multidirectional composite laminates under mode I quasi-static and fatigue loading. *Compos. Struct.* **2018**, *189*, 160–176. [CrossRef]
62. Hu, X.F.; Chen, B.Y.; Tirvaudey, M.; Tan, V.; Tay, T.E. Integrated XFEM-CE analysis of delamination migration in multi-directional composite laminates. *Compos. Part A Appl. Sci. Manuf.* **2016**, *90*, 161–173. [CrossRef]
63. Thamburaja, P.; Sarah, K.; Srinivasa, A.; Reddy, J.N. Fracture of viscoelastic materials: FEM implementation of a non-local & rate form-based finite-deformation constitutive theory. *Comput. Methods Appl. Mech. Eng.* **2019**, *354*, 871–903.
64. Sarah, K.; Thamburaja, P.; Srinivasa, A.; Reddy, J.N. Numerical simulations of damage and fracture in viscoelastic solids using a nonlocal fracture criterion. *Mech. Adv. Mater. Struct.* **2020**, *27*, 1085–1097. [CrossRef]
65. Gong, Y.; Hou, Y.; Zhao, L.; Li, W.; Zhang, J.; Hu, N. A modified mode I cohesive zone model for the delamination growth in DCB laminates with the effect of fiber bridging. *Int. J. Mech. Sci.* **2020**, *176*, 105514. [CrossRef]
66. Yin, S.; Gong, Y.; Li, W.; Zhao, L.; Zhang, J.; Hu, N. A novel four-linear cohesive law for the delamination simulation in composite DCB laminates. *Compos. Part B Eng.* **2020**, *180*, 107526. [CrossRef]

67. Ye, J.; Gong, Y.; Tao, J.; Cao, T.; Zhao, L.; Zhang, J.; Hu, N. Efficiently determining the R-curve and bridging traction-separation relation of mode I delamination in a simple way. *Compos. Struct.* **2022**, *288*, 115388. [CrossRef]
68. Cao, T.; Zhao, L.; Gong, Y.; Gong, X.; Zhang, J. An enhanced beam theory based semi-analytical method to determine the DCB mode I bridging-traction law. *Compos. Struct.* **2020**, *245*, 112306. [CrossRef]
69. Cao, T.; Zhao, L.; Wang, L.; Wang, K.; Gong, Y.; Zhang, J. An efficient semi-analytical method to extract the mode II bridging-traction law in ENF tests directly from the experimental load displacement data. *Compos. Struct.* **2022**, *285*, 115229. [CrossRef]
70. Gong, Y.; Chen, X.; Tao, J.; Zhao, L.; Zhang, J.; Hu, N. A simple procedure for determining the mode I bridging stress of composite DCB laminates without measuring the crack opening displacement. *Compos. Struct.* **2020**, *243*, 112147. [CrossRef]
71. Zhao, L.; Gong, Y.; Zhang, J.; Chen, Y.; Fei, B. Simulation of delamination growth in multidirectional laminates under mode I and mixed mode I/II loadings using cohesive elements. *Compos. Struct.* **2014**, *116*, 509–522. [CrossRef]
72. Gong, Y.; Chen, X.; Zou, L.; Li, X.; Zhao, L.; Zhang, J.; Hu, N. Experimental and numerical investigations on the mode I delamination growth behavior of laminated composites with different z-pin fiber reinforcements. *Compos. Struct.* **2022**, *287*, 115370. [CrossRef]
73. Hillerborg, A.; Modeer, M.; Petersson, P.E. Analysis of crack formation and crack growth in concrete by means of fracture mechanics and finite elements. *Cem. Concr. Res.* **1976**, *6*, 773–781. [CrossRef]

materials

MDPI

Article

Viscoelasticity in Large Deformation Analysis of Hyperelastic Structures

Shahriar Dastjerdi [1], Bekir Akgöz [1,*] and Ömer Civalek [1,2]

1 Division of Mechanics, Civil Engineering Department, Akdeniz University, Antalya 07070, Turkey
2 Department of Medical Research, China Medical University Hospital, China Medical University, Taichung 404, Taiwan
* Correspondence: bekirakgoz@akdeniz.edu.tr

Abstract: In this paper, an annular/circular plate made of hyperelastic material and considering the viscoelastic property was investigated based on a novel nonlinear elasticity theory. A new approach for hyperelastic materials in conjunction with the Kelvin–Voigt scheme is employed to obtain the structure's large deformation under uniform transverse loading. The constitutive equations were extracted using the energy method. The derived partial differential time-dependent equations have been solved via the semi-analytical polynomial method (SAPM). The obtained results have been validated by ABAQUS software and the available paper. In consequence, a good agreement between the results was observed. Finally, several affecting parameters on the analysis have been attended to and studied, such as the nonlinear elasticity analysis, the boundary conditions, loading, and the material's viscosity. It can be possible to obtain the needed time for achieving the final deformation of the structure based on the applied analysis in this research.

Keywords: nonlinear hyperelastic material; viscoelasticity; semi-analytical polynomial method; large deformation

Citation: Dastjerdi, S.; Akgöz, B.; Civalek, Ö. Viscoelasticity in Large Deformation Analysis of Hyperelastic Structures. *Materials* **2022**, *15*, 8425. https://doi.org/10.3390/ma15238425

Academic Editors: Abbas S. Milani, Michele Bacciocchi and Rui Miranda Guedes

Received: 12 October 2022
Accepted: 23 November 2022
Published: 26 November 2022

1. Introduction

Various materials can be found in nature that demonstrate different mechanical behaviors under applied loads, such as metals displaying linear elastic behavior until yielding. At the same time, irreversible deformation is then observed up to failure. On the other hand, brittle materials exhibit a slight elastic deformation linearly and then fail without irreversible deformation. However, it is notable that all materials do not exhibit linear elasticity. Hyperelastic materials will exhibit elastic deformations excessively before failure without irreversible deformation. These materials exhibit an extremely nonlinear stress-strain behavior that ascends monotonically up to fracture. It is well-known that linear elastic materials can be defined with two material constants: modulus of elasticity and Poisson's ratio. On the contrary, hyperelastic materials can be defined by a strain-energy density function. The stored strain energy remains constant when a sample is subjected to constant strain. Therefore, hyperelastic materials are modeled in the context of strain energy potentials such as Arruda-Boyce [1], Gent [2], Neo-Hookean [3], and Yeoh [4].

Some recent studies on the mechanical responses of hyperelastic structures are summarized below. Erchiqui et al. [5] implemented a dynamic finite element method to model the visco-hyperelastic behaviors of a thin, isotropic, and incompressible thermoplastic membrane based on the Ogden and Mooney-Rivlin models. Kocatürk and Akbaş [6] examined the geometrically nonlinear static response of a hyperelastic simply supported beam under a non-follower load by the finite element and Newton–Raphson iteration methods. Li et al. [7] perused the dynamic behavior of the visco-hyperelastic dielectric elastomer structures based on the Gent hyperelastic model. Alibakhshi et al. [8] analyzed the nonlinear vibration of dielectric elastomer microbeam resonators based on a hyperelastic Cosserat model by the Runge-Kutta time domain method. Almasi et al. [9] studied the thermomechanical analysis of a hyperelastic thick cylindrical pressure vessel by giving analytical and numerical solutions. Asgari and Hashemi [10] developed an efficient visco-hyperelastic constitutive model for a hollow cylinder elastomer under dynamic and impact loadings. Pascon [11,12] implemented a finite element formulation based on a two-dimensional beam element to investigate various viscoelastic functionally graded materials and beams made of functionally graded hyperelastic material. The constitutive equations are extracted based on the neo-Hookean model. Gharooni and Ghannad [13] studied the nonlinear analysis of functionally graded tapered hyperelastic cylindrical pressure vessels subjected to non-uniform pressure load.

Hosseini and Rahimi [14] conducted the nonlinear bending analysis of a neo-Hookean hyperelastic plate based on the Mindlin plate theory. Xu et al. [15] introduced the plate element formulation by quadratic interpolation. They carried out static and dynamic analyses of incompressible hyperelastic silicone plates. Dadgar-Rad and Firouzi [16] presented a nonlinear finite element formulation for the viscoelastic deformation of hyperelastic structures under several loadings and boundary conditions. Ansari et al. [17] developed a numerical approach to survey the deformations of hyperelastic Mindlin rectangular plates in compressible and nearly incompressible regimes based on the Neo-Hookean model. Tashiro et al. [18] analyzed blood clots using a nonlinear viscoelastic and hyperelastic model. They employed the visco-hyperelastic finite element method to estimate the mechanical behavior of blood clots. Shariyat and Abadi [19] studied the nonlinear dynamic and impact responses of incompressible neo-Hookean hyperelastic plates with stiff elastic reinforcing particles. Runge-Kutta time integration and penalty methods are used in the solution. Karimi et al. [20] conducted the nonlinear dynamic analysis of an embedded neo-Hookean hyperelastic membrane under a uniformly distributed hydrostatic pressure. Alibakhshi and Heidari [21] carried out the nonlinear vibration of a dielectric elastomer balloon via the Gent hyperelastic model. A time integration-based solver is utilized to solve the resulting equations. Alibakhshi et al. [22] elaborated on a suitable detection mechanism for scanning the surface profile of a micro-sample by AFM, considering that the probe is made of a hyperelastic material. Falope et al. [23] proposed a finite element-based theoretical model to study the bending behavior of a hyperelastic solid.

Hosseini et al. [24] perused the nonlinear static behavior of functionally graded hyperelastic plates based on FSDT. The potential energy function is formulated according to the neo-Hookean model and the Cauchy–Green tensor. Coda et al. [25] elaborated a finite element formulation to analyze the laminated and functionally graded hyperelastic one-dimensional structure with transverse shear stress distribution. Dastjerdi et al. [26] introduced a comprehensive theoretical method to examine the mechanical responses of hyperelastic structures. The proposed method can analyze the geometrically and physically nonlinear hyperelastic materials. Zhao et al. [27] investigated the nonlinear dynamic responses of the visco-hyperelastic spherical shells under uniform radial loads. Additionally, Zhao et al. [28] perused dynamic loads and structural damping influences for incompressible hyperelastic spherical shells. They considered the Yeoh strain energy function in their study. Bacciocchi and Tarantino [29] examined the finite bending of hyperelastic beams based on the compressible Mooney-Rivlin model. More information about the nonlinear dynamics of hyperelastic structures can be found in the recent review article [30]. Additionally, many studies have been carried out on analyzing viscoelastic structures [31–45]. Li et al. [46] presented a perturbation approach for the lateral vibration analysis of viscoelastic microstructures. The clamped microbeam has been subjected to external harmonic excitation. Recently, some researchers developed a structure-preserving approach to solve the macroscopic/microscopic coupling dynamic problems with large deformation [47]. Additionally, a large deformation analysis of a nano-sized structure has been attended by Yan et al. [48].

As stated before, many recent studies have been performed on the mechanical analysis of structures made of hyperelastic materials based on various strain energy functions, especially the neo-Hookean model. However, there is no study on the static response of an annular circular visco-hyperelastic plate. The new approach for hyperelastic materials simulates the annular circular visco-hyperelastic plate in conjunction with the Kelvin–Voigt scheme. The general formulations are derived according to the first-order shear deformation theory (FSDT). The constitutive equations and boundary conditions are extracted by the energy method. Then, the semi-analytical polynomial method is utilized for solving the partial differential time-dependent equations. Finally, the effects of various parameters on the bending analysis of the annular circular visco-hyperelastic plate are investigated and discussed in detail.

2. Geometry of the Structure

First, the geometry of the problem is discussed. The geometry of the analyzed structure will play a significant role in obtaining the governing equations. The more comprehensive the geometry of the problem, the more different structures can be modeled only by considering a single formulation. Analysis of the structures made of hyperelastic materials will make the simulation more complicated. Therefore, an attempt has been made to focus more on the effect of viscosity on hyperelastic materials. Hence, an annular–circular structure made of visco-hyperelastic material has been assumed. The geometry of the analyzed problem is shown in Figure 1, along with the considered coordinate system, which is the cylindrical coordinate system.

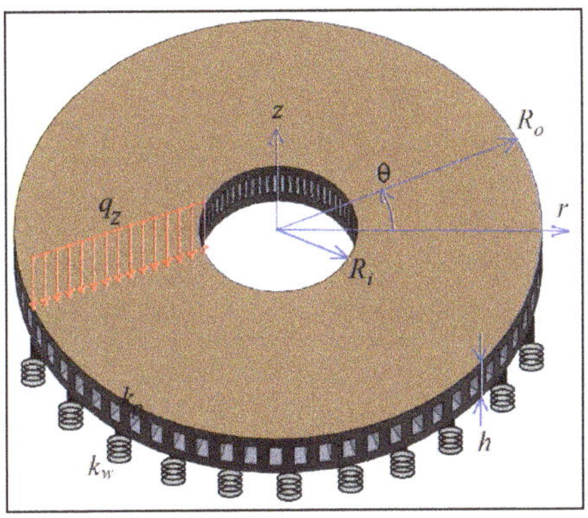

Figure 1. Schematic view of the structure.

As can be seen, the inner radius of the structure is R_i, and its outer radius is R_o. Additionally, the thickness of the structure is constant and equal to h. The structure is placed on an elastic foundation with two components, the Winkler (k_w) and Pasternak (k_p). The external load on the structure is considered a uniform distributed load in the z direction equal to q_z.

3. Viscoelastic Property

In this research, an attempt has been made to perform two nonlinear elastic properties simultaneously with the changes in the deformation of the material with respect to the strain rate. In other words, large strains against the strain rate are considered. The structure's material is considered in such a way that the deformation created in it due to the applied load is different during the duration of the load application. On the other hand, the final deformation of the structure does not happen as soon as the load is applied, and it takes place over time. The Kelvin–Voigt model is used to consider the viscoelastic properties of the material [43]. Additionally, other models such as Zener and mixing linear viscoelasticity and nonlinear elasticity (as has been done in this paper) have been used by other researchers [49,50]. In this type of time-dependent modeling for stress–strain, a spring and a damper are placed in parallel, which can be seen in Figure 2. As shown in

Figure 2, the viscous property of the material (time-dependent strain) is simulated by a damper. Additionally, the material's elasticity is modeled by a spring (E represents Young's elasticity modulus). In linear elastic materials, the value of E is constant and changes linearly. However, this value is not fixed or variable in this research. Therefore, the model (as explained in the next part) uses linear viscoelasticity according to the Kelvin–Voigt scheme. However, the principle of elasticity has been considered nonlinearly.

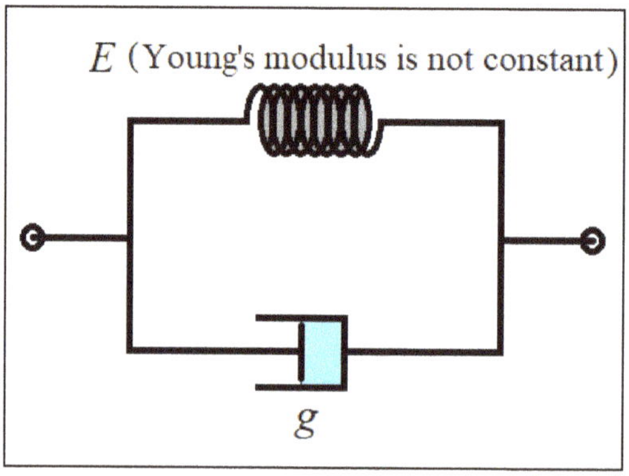

Figure 2. The Kelvin–Voigt model for viscoelastic property.

Stress and strain will be written according to the following equations based on the Kelvin–Voigt simulation.

$$\sigma_{Total} = \sigma_s + \sigma_D \left(\sigma_s = E\varepsilon(r,\theta,z,t), \sigma_D = g\frac{\partial \varepsilon(r,\theta,z,t)}{\partial t} \right) \tag{1}$$

It can be seen in Figure 2 that E is the elasticity modulus, and g is the viscosity of the material. The amount of σ_s is linear here according to Equation (1). However, it is not constant in this study due to the nonlinear elasticity analysis. This issue will be discussed in detail in the next part.

4. Nonlinear Elastic Material and Governing Equations

In a linear elastic material, it can be observed that the changes in stress and strain are linear. However, in a structure with nonlinear elastic property, the stress increases nonlinearly with the increase of strain. In this research, the material with nonlinear elastic properties is assumed. Therefore, calculations based on nonlinear elasticity theory are considered. Various mathematical simulations and theories have been presented regarding the analysis of nonlinear elastic or hyperelastic structures [2,9–12]. A new method has been presented previously for analyzing the mechanical behavior of nonlinear elastic structures [26]. According to this theory, the stress-strain diagram of a nonlinear elastic material obtained from the experiment is approximated by a polynomial function ($\sigma(\varepsilon) = \sum_{i=1}^{n} E_i \varepsilon^i$). According to the polynomial degree, a higher accuracy can be obtained by choosing more n. Consequently, the expressed viscous stress can be formulated as $\sigma_s(\varepsilon) = \sum_{i=1}^{n} E_i \varepsilon(r,\theta,z,t)^i$. The calculations related to the strain field and the governing equations can be obtained according to the following equations by considering the viscoelastic property.

$$\overleftrightarrow{\varepsilon}_{ij} = \begin{bmatrix} \varepsilon_{rr} & \varepsilon_{r\theta} & \varepsilon_{rz} \\ \varepsilon_{\theta r} & \varepsilon_{\theta\theta} & \varepsilon_{\theta z} \\ \varepsilon_{zr} & \varepsilon_{z\theta} & \varepsilon_{zz} \end{bmatrix}$$

$$\varepsilon_{rr} = \left(1+g\tfrac{\partial}{\partial t}\right)\ \left(\tfrac{\partial U_r}{\partial r}+\tfrac{1}{2}\left(\tfrac{\partial U_z}{\partial r}\right)^2\right); 2\varepsilon_{r\theta} = 2\varepsilon_{\theta r} = 0;\ 2\varepsilon_{rz} = 2\varepsilon_{zr} = \left(1+g\tfrac{\partial}{\partial t}\right)\left(\tfrac{\partial U_r}{\partial z}+\tfrac{\partial U_z}{\partial r}\right);\ \varepsilon_{\theta\theta} = \left(1+g\tfrac{\partial}{\partial t}\right)\left(\tfrac{1}{r}\left(\tfrac{\partial U_\theta}{\partial \theta}+U_r\right)\right);\ 2\varepsilon_{\theta z} = 2\varepsilon_{z\theta} = 0;\ \varepsilon_{zz} = 0 \quad (2)$$

The strain field is assumed based on FSDT as

$$\begin{cases} U_r(r,z,t) = u_0(r,t) + z\varphi(r,t) \\ U_\theta(r,z) = 0 \\ U_z(r,z,t) = w_0(r,t) \end{cases} \quad (3)$$

$$\sigma_{rr} = \sum_{i=1}^{n} \frac{E_i}{(1-\nu^2)^i}[\varepsilon_{rr}+\nu\varepsilon_{\theta\theta}]^i; \sigma_{\theta\theta} = \sum_{i=1}^{n} \frac{E_i}{(1-\nu^2)^i}[\varepsilon_{\theta\theta}+\nu\varepsilon_{rr}]^i;$$
$$\sigma_{rz} = \sum_{i=1}^{n} \frac{E_i}{[2(1+\nu)]^i}\varepsilon_{rz}^i \quad (4)$$

In the above equations, u_0 and w_0 are transport displacements, and φ is the rotation function around the θ axes. Now, using the energy method [43–45] ($\delta P_{Total} = \iiint_V \sigma_{ij}\delta\varepsilon_{ij} + \iint_A (q_z - k_W w_0 + k_p\nabla^2 w_0)\delta w_0\ (i,j = r,\theta,z)$), the mathematical description of boundary conditions and governing equations can be introduced as the following equations.

$$\delta u_0 : \frac{\partial N_{rr}}{\partial r} + \frac{1}{r}(N_{rr} - N_{\theta\theta}) = 0 \quad (5)$$

$$\delta w_0 : \frac{\partial N_{rz}}{\partial r} + \frac{N_{rz}}{r} + N_{rr}\frac{\partial^2 w_0}{\partial r^2} + \frac{N_{rr}}{r}\frac{\partial w_0}{\partial r} + \frac{\partial N_{rr}}{\partial r}\frac{\partial w_0}{\partial r} + q_z - k_w w_0 + k_p\nabla^2(w_0) = 0 \quad (6)$$

$$\delta\varphi : \frac{\partial M_{rr}}{\partial r} + \frac{1}{r}(M_{rr} - M_{\theta\theta}) - N_{rz} = 0 \quad (7)$$

$$(N_{rr}, N_{\theta\theta}, N_{rz}) = \int_{-\frac{h}{2}}^{\frac{h}{2}} (\sigma_{rr}, \sigma_{\theta\theta}, \sigma_{rz})dz;\ (M_{rr}, M_{\theta\theta}) = \int_{-\frac{h}{2}}^{\frac{h}{2}} (\sigma_{rr}, \sigma_{\theta\theta},)zdz \quad (8)$$

Considering the uniform transverse load on the structure, the problem is symmetrical, and there will be no changes in the θ direction. The only independent variables of the problem are r and t, where t represents the time duration. The mathematical definition of the different types of boundary conditions is introduced according to the following relations: Clamped (C), Simply supported (S), and Free (F) at the edges of $r = R_i,\ R_o$.

$$\begin{cases} Clamped:\ u_0 = w_0 = \varphi = 0\ (r = R_i,\ R_o; t = 0, t_N) \\ Simply\ supported:\ u_0 = w_0 = M_{rr} = 0\ (r = R_i,\ R_o; t = 0, t_N) \\ Free:\ N_{rr} = N_{rz} = M_{rr} = 0\ (r = R_i,\ R_o; t = 0, t_N) \end{cases} \quad (9)$$

5. Solution Method

This research uses the semi-analytical method based on polynomials (SAPM) to solve the governing equations [43,44]. According to this method, the displacement functions (u_0, w_0, and φ) are formulated as comprehensive polynomials based on the number of nodes (N and M) in each direction of the independent variable of the problem (r and z).

$$u_0(r,z) = \sum_{i=1}^{N}\sum_{j=1}^{M} a_{(j+M(i-1))} r^{(i-1)} t^{(j-1)} \quad (10)$$

$$w_0(r,z) = \sum_{i=1}^{N}\sum_{j=1}^{M} a_{(j+M(i-1)+M\cdot N)} r^{(i-1)} t^{(j-1)} \quad (11)$$

$$\varphi(r,z) = \sum_{i=1}^{N}\sum_{j=1}^{M} a_{(j+M(i-1)+2M\cdot N)} r^{(i-1)} t^{(j-1)} \quad (12)$$

Now, a system of algebraic equations can be obtained by inserting the functions (Equations (10)–(12)) in the mathematical definition of the boundary conditions at the edges and constitutive equations. Finally, by solving the obtained algebraic equations, the unknown functions in the displacement field and, accordingly, other unknowns, including stresses and strains, will be obtained for the visco-hyperelastic annular–circular sheet under uniform transverse load.

6. Discussion

6.1. Validation

6.1.1. The Solving Method (SAPM)

First, the accuracy of the used solution method is checked. Figure 3 shows the background changes in terms of time for an annular/circular sheet with the following specifications for different values of the number of nodes in the direction of r and t.

$$R_i = 0.1m, R_o = 0.5m, h = 0.03m, E_1 = 2.1467 \times 10^6, E_2 = -6.4659 \times 10^5, E_3 = -69779, E_4 = 5.66989 \times 10^5, v = 0.3, q_z = 1000Pa, \quad (13)$$

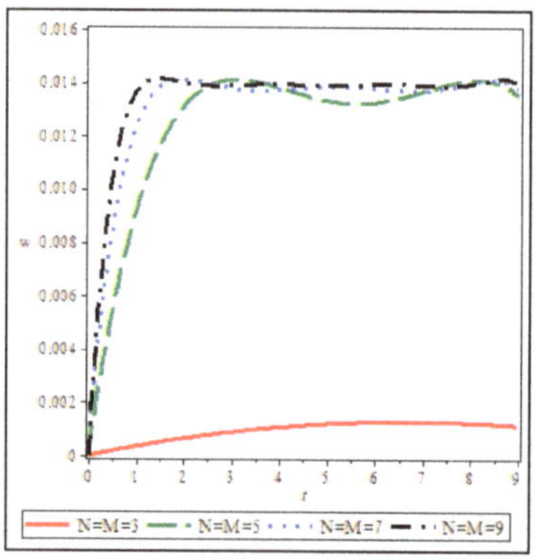

Figure 3. Investigation of the result's convergence.

As can be seen, the accuracy of the results increases versus the increase of nodes in the calculations. A sudden jump is observed between the results of $N = M = 3$ and $N = M = 5$, so the used solution method has a very high convergence. It can be seen that the results of seven and nine nodes are very close, and therefore, the results obtained from seven or nine nodes can be used with reasonable confidence. Increasing the number of nodes in the calculations will significantly increase the time to obtain the results. Therefore, the optimized number of nodes in the network for solving the problem is an important point that should be considered. In the first form, by choosing only nine nodes in each direction (r and t), you can get the results with the desired accuracy for the subsequent analysis.

6.1.2. Comparisons

Figure 4a,b and Table 1 are provided to validate the results. Figure 4 shows the maximum deflection between the results of (a) the present study and (b) the obtained results of ABAQUS software for an annular/circular plate. It is demonstrated that the results agree well. Finding relevant papers in visco-hyperelasticity to fit with the present work is hard. For example, some papers can be found. However, their modeled geometry might be different from the present study. Another assessment was done, and the results can be observed in Figure 5 for the visco-hyperelastic structure. As the tested model [51] is a thick circular sheet, we considered a quasi-three-dimensional analysis [52]. Therefore, the captured results have been monitored into a single plot in conjunction with the experimental results [51]. The compared results in Figure 5 show that an acceptable difference is available between the obtained results. The obtained results in this paper are near to the ABAQUS results, according to the originally depicted figure in [51]. Eventually, the applied theory and solving method are reliable. Additionally, according to Table 1, the same conclusion can be made for different boundary conditions.

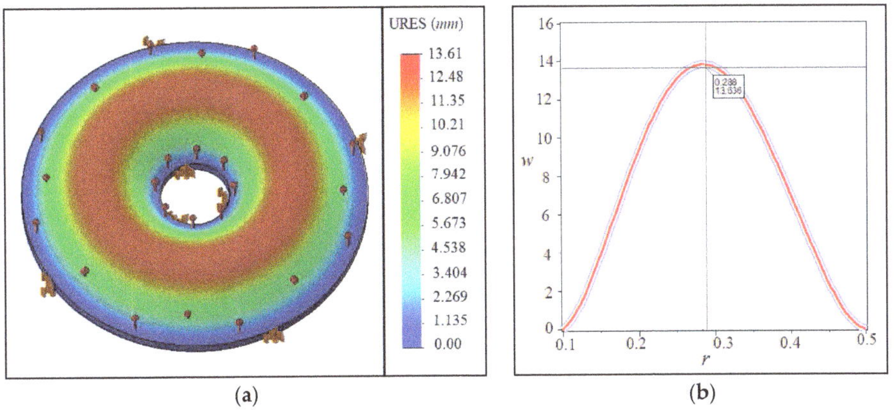

Figure 4. Deflection results of (**a**) ABAQUS software and (**b**) the present study.

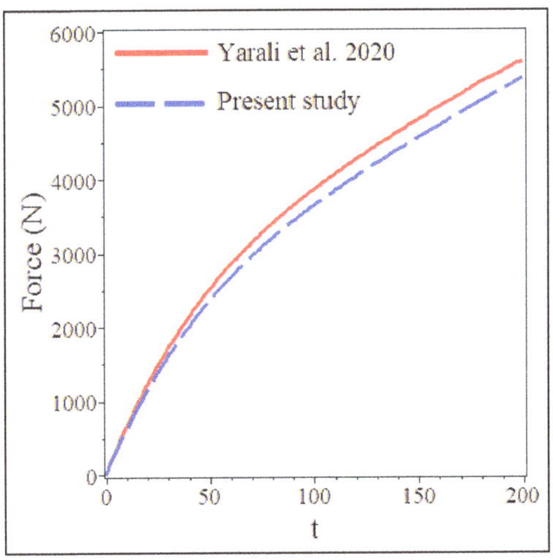

Figure 5. Comparison between the results of this study and [51] for the visco-hyperelastic analysis.

Table 1. Comparison between the maximum deflection (mm) results of the present study and [53] for aluminum circular plates.

Boundary Conditions	Maximum Deflection (mm)							
	$q_z = 5\ MPa$		$q_z = 20\ MPa$		$q_z = 50\ MPa$		$q_z = 100\ MPa$	
	Present	[53]	Present	[53]	Present	[53]	Present	[53]
SS	3.789	3.883	8.232	8.333	12.11	12.15	15.56	15.68
CC	1.261	1.263	4.559	4.562	8.819	8.820	12.93	12.93

6.2. Numerical Results

To investigate different boundary conditions on nonlinear-elastic analysis, Figure 6 depicts a circular sheet with the same characteristics as Figure 3. As can be seen, the maximum deflection increases with the increase of transverse load on the structure. The increase in the maximum deflection is completely nonlinear and will be accompanied by a decreasing slope. In other words, as the transverse load on the sheet increases, it is observed that the slope of the changes decreases. The importance of this matter is that if the linear analysis were considered, the obtained results would increase linearly. These linear results would be acceptable only for low load values. However, because the nonlinear analysis of the sheet made of material with nonlinear elastic properties has been studied in this research, the simulation results can be used with appropriate accuracy. Another point (according to Figure 6, which is drawn for two boundary conditions, *CC* and *SS*, at the edges of R_i and R_o) is that the decreasing trend of the results at the beginning of loading is more for the *CC* boundary conditions than SS. However, with increasing load on the plate, the changes are almost the same, with a slight difference for the two boundary conditions, *CC* and *SS*. In other words, by increasing the load on the structure, the effects related to the boundary conditions are reduced, and the behavior of the sheet against the applied load will be less affected by the applied boundary conditions.

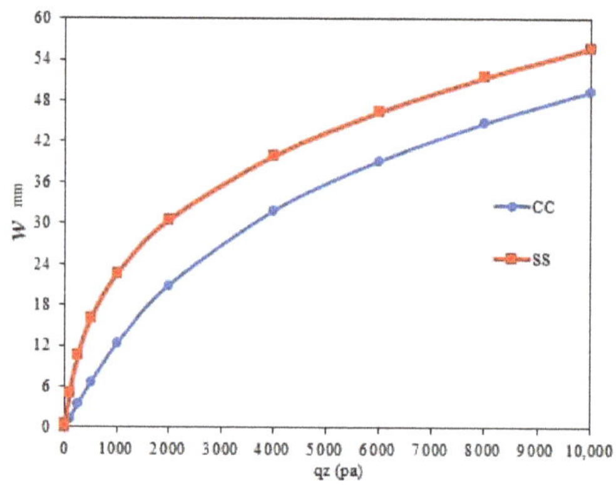

Figure 6. Deflection changes versus the applied transverse loading (q_z) for different types of boundary conditions.

The viscoelastic property in nonlinear elasticity analysis is a significant issue considered in this research. Therefore, Figures 7 and 8 show the effect of viscosity (changes in deformation over time) on the results of the maximum deflection of the circular sheet (specifications of Figure 3). Viscosity $g = 0$ means that, when the load is applied on the sheet, the maximum deflection will occur, and the structure will reach the absolute limit of deformation, which can be seen in Figure 7. Of course, it can be observed that the mentioned result will be for $t = 1s$. This calculation error is due to the limitation in choosing the number of nodes in time when solving the governing equations. It can be observed that with the rise of the viscosity value (g), the needed time for the sheet to reach its final deformation will increase. In other words, with the increase in viscosity, the slope variations will decrease over time. Of course, the effect of viscosity on the results is not linear. In other words, the distance between the results from $g = 0$ to $g = 5$ is more remarkable than from $g = 5$ to $g = 10$. As the viscosity of the structure material increases, the intensity of its impact on the results decreases. Figure 8 is one of the results of the curves in Figure 7 ($g = 5$), which here show the changes in the deflection in two directions: r and t, simultaneously. According to Figure 8, the maximum deflection on the sheet will occur in its middle ($r = R_i,\ R_o$), according to the *CC* boundary conditions.

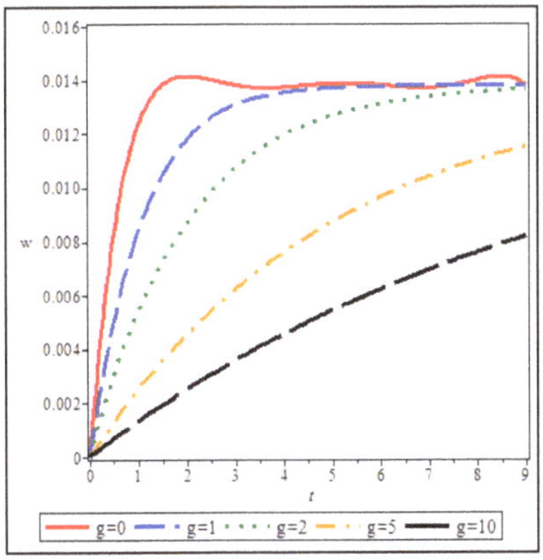

Figure 7. Viscoelastic effects on the results.

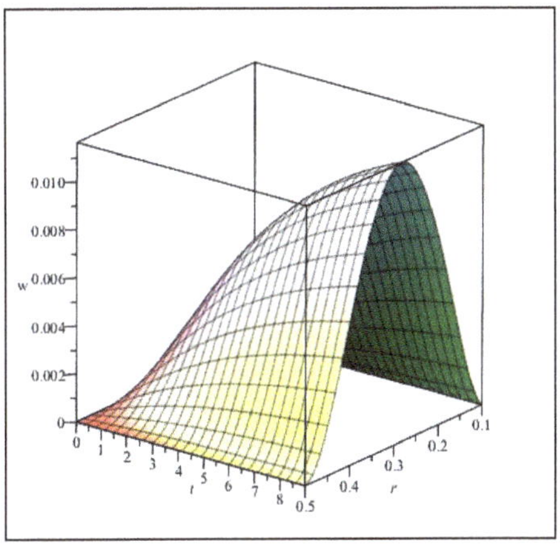

Figure 8. Three-dimensional view of deflection changes versus r and time (t) for $g = 5$.

In Figures 7 and 8, the effect of viscosity on the nonlinear elastic structure is investigated. As seen, with the increase in viscosity, the time for the structure to reach its final shape increases. Now, this topic, how much time is required for the final deflection (considering a specific viscosity), is further investigated. Figure 9a is drawn for viscosity $g = 5$, and Figure 9b is drawn for $g = 10$. Here, the criterion for drawing two shapes is to reach the absolute limit of the deformation and be equal to a specific number. In other words, the final deflection for both Figure 9a,b is equal. It can be seen that the final deflection in Figure 9a is for $t = 25s$. However, this result in Figure 9b is equal to $t = 50s$. This analysis is crucial, because it is observed that, with the doubling of the viscosity of the structure, the time it takes to reach the final shape change also doubles. Although the variation rate is nonlinear, the deformation of the structure at the final time has a linear and direct relationship with the value of the viscosity of the structure. Of course, this conclusion has been reached for this particular problem. Such a result may not be obtained by considering other assumptions, including environmental factors such as temperature, humidity, or electric fields. Investigating the impact of the mentioned factors can also be considered by other researchers who work in this field.

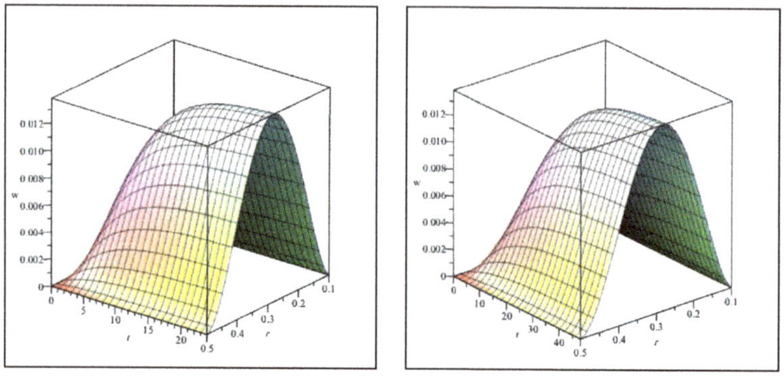

(**a**) (**b**)

Figure 9. Deflection changes versus r and time for (**a**) $g = 5$ and (**b**) $g = 10$.

7. Conclusions

The present study investigated the viscoelastic analysis of nonlinear elastic (hyperelastic) materials. The governing partial differential equations and mathematical definitions of boundary conditions were derived based on a new method and were solved by the semi-analytical method based on polynomials (SAPM). The viscoelastic property of the structural material is assumed by Kelvin–Voigt modeling. In general, the significant results obtained from the research can be categorized as follows:

1. The new theory of hyperelastic structures can be used with appropriate confidence for viscoelastic properties.
2. Structures made of nonlinear elastic material are sensitive to changes in applied transverse loads, and the changes are entirely nonlinear, even with the low-load application.
3. For low loads, it has a significant impact on the deformation by the boundary conditions. However, as the load increases, these effects decrease.
4. As the viscosity increases, the duration of the final deformation increases, which has a direct relationship with the viscosity of the material.

Author Contributions: Conceptualization, S.D. and B.A.; methodology, S.D.; software, S.D.; validation, S.D.; formal analysis, S.D.; investigation, S.D.; resources, B.A.; data curation, S.D.; writing—original draft preparation, B.A.; writing—review and editing, B.A. and Ö.C; visualization, B.A.; supervision, Ö.C.; project administration, Ö.C.; and funding acquisition, Ö.C. All authors have read and agreed to the published version of the manuscript.

Funding: This research received no external funding.

Institutional Review Board Statement: Not applicable.

Informed Consent Statement: Not applicable.

Data Availability Statement: There is no data for sharing, and all data are available within the paper.

Conflicts of Interest: The authors declare no conflict of interest.

References

1. Boyce, M.C.; Arruda, E.M. Constitutive models of rubber elasticity: A review. *Rubber Chem. Technol.* **2000**, *73*, 504–523. [CrossRef]
2. Gent, A.N. A new constitutive relation for rubber. *Rubber Chem. Technol.* **1996**, *69*, 59–61. [CrossRef]
3. Anssari-Benam, A.; Bucchi, A. A generalised neo-Hookean strain energy function for application to the finite deformation of elastomers. *Int. J. Non-Linear Mech.* **2021**, *128*, 103626. [CrossRef]
4. Yeoh, O.H. Characterization of Elastic Properties of Carbon-Black-Filled Rubber Vulcanizates. *Rubber Chem. Technol.* **1990**, *63*, 792–805. [CrossRef]
5. Erchiqui, F.; Gakwaya, A.; Rachik, M. Dynamic finite element analysis of nonlinear isotropic hyperelastic and viscoelastic materials for thermoforming applications. *Polym. Eng. Sci.* **2005**, *45*, 125–134. [CrossRef]

6. Kocaturk, T.; Akbas, S.D. Geometrically nonlinear static analysis of a simply supported beam made of hyperelastic material. *Struct. Eng. Mech.* **2010**, *35*, 677–697. [CrossRef]
7. Li, Y.L.; Oh, I.; Chen, J.H.; Zhang, H.H.; Hu, Y.H. Nonlinear dynamic analysis and active control of visco-hyperelastic dielectric elastomer membrane. *Int. J. Solids Struct.* **2018**, *152*, 28–38. [CrossRef]
8. Alibakhshi, A.; Dastjerdi, S.; Akgoz, B.; Civalek, O. Parametric vibration of a dielectric elastomer microbeam resonator based on a hyperelastic cosserat continuum model. *Compos. Struct.* **2022**, *287*, 115386. [CrossRef]
9. Almasi, A.; Baghani, M.; Moallemi, A. Thermomechanical analysis of hyperelastic thick-walled cylindrical pressure vessels, analytical solutions and FEM. *Int. J. Mech. Sci.* **2017**, *130*, 426–436. [CrossRef]
10. Asgari, M.; Hashemi, S.S. Dynamic visco-hyperelastic behavior of elastomeric hollow cylinder by developing a constitutive equation. *Struct. Eng. Mech.* **2016**, *59*, 601–619. [CrossRef]
11. Pascon, J.P. Finite element analysis of functionally graded hyperelastic beams under plane stress. *Eng. Comput.* **2020**, *36*, 1265–1288. [CrossRef]
12. Pascon, J.P. Large deformation analysis of functionally graded visco-hyperelastic materials. *Comput. Struct.* **2018**, *206*, 90–108. [CrossRef]
13. Gharooni, H.; Ghannad, M. Nonlinear analysis of radially functionally graded hyperelastic cylindrical shells with axially-varying thickness and non-uniform pressure loads based on perturbation theory. *J. Comput. Appl. Mech.* **2019**, *50*, 324–340.
14. Hosseini, S.; Rahimi, G. Nonlinear Bending Analysis of Hyperelastic Plates Using FSDT and Meshless Collocation Method Based on Radial Basis Function. *Int J Appl Mech* **2021**, *13*, 2150007. [CrossRef]
15. Xu, Q.P.; Liu, J.Y.; Qu, L.Z. A Higher-Order Plate Element Formulation for Dynamic Analysis of Hyperelastic Silicone Plate. *J. Mech.* **2019**, *35*, 795–808. [CrossRef]
16. Dadgar-Rad, F.; Firouzi, N. Large deformation analysis of two-dimensional visco-hyperelastic beams and frames. *Arch. Appl. Mech.* **2021**, *91*, 4279–4301. [CrossRef]
17. Ansari, R.; Hassani, R.; Oskouie, M.F.; Rouhi, H. Nonlinear bending analysis of hyperelastic Mindlin plates: A numerical approach. *Acta Mech.* **2021**, *232*, 741–760. [CrossRef]
18. Tashiro, K.; Shobayashi, Y.; Ota, I.; Hotta, A. Finite element analysis of blood clots based on the nonlinear visco-hyperelastic model. *Biophys. J.* **2021**, *120*, 4547–4556. [CrossRef] [PubMed]
19. Shariyat, M.; Abedi, S. An accurate hyperelasticity-based plate theory and nonlinear energy-based micromechanics for impact and shock analyses of compliant particle-reinforced FG hyperelastic plates. *Zamm-Z. Angew. Math. Mech.* **2022**, *102*, e202100099. [CrossRef]
20. Karimi, S.; Ahmadi, H.; Foroutan, K. Nonlinear vibration and resonance analysis of a rectangular hyperelastic membrane resting on a Winkler-Pasternak elastic medium under hydrostatic pressure. *J. Vib. Control* **2022**. [CrossRef]
21. Alibakhshi, A.; Heidari, H. Nonlinear dynamics of dielectric elastomer balloons based on the Gent-Gent hyperelastic model. *Eur. J. Mech. A-Solid* **2020**, *82*, 103986. [CrossRef]
22. Alibakhshi, A.; Rahmanian, S.; Dastjerdi, S.; Malikan, M.; Karami, B.; Akgoz, B.; Civalek, O. Hyperelastic Microcantilever AFM: Efficient Detection Mechanism Based on Principal Parametric Resonance. *Nanomaterials* **2022**, *12*, 2598. [CrossRef] [PubMed]
23. Falope, F.O.; Lanzoni, L.; Tarantino, A.M. FE Analyses of Hyperelastic Solids under Large Bending: The Role of the Searle Parameter and Eulerian Slenderness. *Materials* **2020**, *13*, 1597. [CrossRef] [PubMed]
24. Hosseini, S.; Rahimi, G.; Shahgholian-Ghahfarokhi, D. A meshless collocation method on nonlinear analysis of functionally graded hyperelastic plates using radial basis function. *Zamm-Z. Angew. Math. Mech.* **2022**, *102*, e202100216. [CrossRef]
25. Coda, H.B.; Bernardo, C.C.L.G.; Paccola, R.R. A FEM formulation for the analysis of laminated and functionally graded hyperelastic beams with continuous transverse shear stresses. *Compos. Struct.* **2022**, *292*, 115606. [CrossRef]
26. Dastjerdi, S.; Alibakhshi, A.; Akgoz, B.; Civalek, O. A Novel Nonlinear Elasticity Approach for Analysis of Nonlinear and Hyperelastic Structures. *Eng. Anal. Bound. Elem.* **2022**, *143*, 219–236. [CrossRef]
27. Zhao, Z.T.; Niu, D.T.; Zhang, H.W.; Yuan, X.G. Nonlinear dynamics of loaded visco-hyperelastic spherical shells. *Nonlinear Dyn.* **2020**, *101*, 911–933. [CrossRef]
28. Zhao, Z.T.; Yuan, X.G.; Zhang, W.Z.; Niu, D.T.; Zhang, H.W. Dynamical modeling and analysis of hyperelastic spherical shells under dynamic loads and structural damping. *Appl. Math. Model.* **2021**, *95*, 468–483. [CrossRef]
29. Bacciocchi, M.; Tarantino, A.M. Bending of hyperelastic beams made of transversely isotropic material in finite elasticity. *Appl. Math. Model.* **2021**, *100*, 55–76. [CrossRef]
30. Khaniki, H.B.; Ghayesh, M.H.; Chin, R.; Amabili, M. A review on the nonlinear dynamics of hyperelastic structures. *Nonlinear Dyn.* **2022**, *110*, 963–994. [CrossRef]
31. Zenkour, A.M. Nonlocal thermal vibrations of embedded nanoplates in a viscoelastic medium. *Struct. Eng. Mech.* **2022**, *82*, 701–711.
32. Yuan, Y.; Niu, Z.Q.; Smitt, J. Magneto-hygro-thermal vibration analysis of the viscoelastic nanobeams reinforcedwith carbon nanotubes resting on Kerr's elastic foundation based on NSGT. *Adv. Compos. Mater.* **2022**. [CrossRef]
33. Soleimani-Javid, Z.; Arshid, E.; Amir, S.; Bodaghi, M. On the higher-order thermal vibrations of FG saturated porous cylindrical micro-shells integrated with nanocomposite skins in viscoelastic medium. *Def. Technol.* **2022**, *18*, 1416–1434. [CrossRef]

34. Moayeri, M.; Darabi, B.; Sianaki, A.H.; Adamian, A. Third order nonlinear vibration of viscoelastic circular microplate based on softening and hardening nonlinear viscoelastic foundation under thermal loading. *Eur. J. Mech. A-Solid* **2022**, *95*, 104644. [CrossRef]
35. Li, Y.; Liu, B. Thermal buckling and free vibration of viscoelastic functionally graded sandwich shells with tunable auxetic honeycomb core. *Appl. Math. Model.* **2022**, *108*, 685–700. [CrossRef]
36. Dang, R.Q.; Yang, A.M.; Chen, Y.M.; Wei, Y.Q.; Yu, C.X. Vibration analysis of variable fractional viscoelastic plate based on shifted Chebyshev wavelets algorithm. *Comput. Math. Appl.* **2022**, *119*, 149–158. [CrossRef]
37. Alizadeh, A.; Shishehsaz, M.; Shahrooi, S.; Reza, A. Free vibration characteristics of viscoelastic nano-disks based on modified couple stress theory. *J. Strain Anal. Eng.* **2022**. [CrossRef]
38. Ghobadi, E.; Shutov, A.; Steeb, H. Parameter Identification and Validation of Shape-Memory Polymers within the Framework of Finite Strain Viscoelasticity. *Materials* **2021**, *14*, 2049. [CrossRef]
39. Dacol, V.; Caetano, E.; Correia, J.R. A New Viscoelasticity Dynamic Fitting Method Applied for Polymeric and Polymer-Based Composite Materials. *Materials* **2020**, *13*, 5213. [CrossRef]
40. Chang, J.J.; Li, Y.Y.; Zeng, X.F.; Zhong, H.Y.; Wan, T.L.; Lu, C. Study on the Viscoelasticity Measurement of Materials Based on Surface Reflected Waves. *Materials* **2019**, *12*, 1875. [CrossRef] [PubMed]
41. Wang, D.Z.; de Boer, G.; Ghanbarzadeh, A. A Numerical Model for Investigating the Effect of Viscoelasticity on the Partial Slip Solution. *Materials* **2022**, *15*, 5182. [CrossRef] [PubMed]
42. Itou, H.; Kovtunenko, V.A.; Rajagopal, K.R. The Boussinesq flat-punch indentation problem within the context of linearized viscoelasticity. *Int. J. Eng. Sci.* **2020**, *151*, 103272. [CrossRef]
43. Dastjerdi, S.; Akgoz, B.; Civalek, O. On the effect of viscoelasticity on behavior of gyroscopes. *Int. J. Eng. Sci.* **2020**, *149*, 103236. [CrossRef]
44. Dastjerdi, S.; Akgoz, B.; Civalek, O. On the shell model for human eye in Glaucoma disease. *Int. J. Eng. Sci.* **2021**, *158*, 103414. [CrossRef]
45. Dastjerdi, S.; Malikan, M.; Akgoz, B.; Civalek, O.; Wiczenbach, T.; Eremeyev, V.A. On the deformation and frequency analyses of SARS-CoV-2 at nanoscale. *Int. J. Eng. Sci.* **2022**, *170*, 103604. [CrossRef]
46. Li, C.; Zhu, C.; Sui, S.; Yan, J. A Perturbation Approach for Lateral Excited Vibrations of a Beam-like Viscoelastic Microstructure Using the Nonlocal Theory. *Appl. Sci.* **2022**, *12*, 40. [CrossRef]
47. Hu, W.; Xu, M.; Song, J.; Gao, Q.; Deng, Z. Coupling dynamic behaviors of flexible stretching hub-beam system. *Mech. Syst. Signal Process.* **2021**, *151*, 107389. [CrossRef]
48. Yan, J.W.; Lai, S.K.; He, L.H. Nonlinear dynamic behavior of single-layer graphene under uniformly distributed loads. *Compos. B Eng.* **2019**, *165*, 473–490. [CrossRef]
49. Pascon, J.P.; Coda, H.B. Finite deformation analysis of visco-hyperelastic materials via solid tetrahedral finite elements. *Finite Elem. Anal. Des.* **2017**, *133*, 25–41. [CrossRef]
50. López-Campos, J.A.; Segade, A.; Fernández, J.R.; Casarejos, E.; Vilán, J.A. Behavior characterization of visco-hyperelastic models for rubber-like materials using genetic algorithms. *Appl. Math. Model.* **2019**, *66*, 241–255. [CrossRef]
51. Yarali, E.; Baniasadi, M.; Bodaghi, M.; Bodaghi, M. 3D constitutive modeling of electro-magneto-viscohyperelastic elastomers: A semi-analytical solution for cylinders under large torsion–extension deformation. *Smart Mater. Struct.* **2020**, *29*, 085031. [CrossRef]
52. Dastjerdi, S.; Tadi Beni, Y.; Malikan, M. A comprehensive study on nonlinear hygro-thermo-mechanical analysis of thick functionally graded porous rotating disk based on two quasi-three-dimensional theories. *Mech Based Des Struct* **2022**, *50*, 3596–3625. [CrossRef]
53. Zhang, D.G.; Zhou, H.M. Nonlinear bending analysis of FGM circular plates based on physical neutral surface and higher-order shear deformation theory. *Aerosp. Sci. Technol.* **2015**, *41*, 90–98. [CrossRef]

materials

MDPI

Article

Application of the $ps-$Version of the Finite Element Method to the Analysis of Laminated Shells

Cheng Angelo Yan and Riccardo Vescovini *

Dipartimento di Scienze e Tecnologie Aerospaziali, Politecnico di Milano, Via La Masa 34, 20156 Milano, Italy; chengangelo.yan@polimi.it
* Correspondence: riccardo.vescovini@polimi.it; Tel.: +39-02-2399-8332

Abstract: The development of accurate and efficient numerical methods is of crucial importance for the analysis and design of composite structures. This is even more true in the presence of variable stiffness (VS) configurations, where intricate load paths can be responsible for complex and localized stress profiles. In this work, we present the $ps-$version of the finite elements method ($ps-$FEM), a novel FE approach which can perform global/local analysis through different refinement strategies efficiently and easily. Within this framework, the global behavior is captured through a $p-$refinement by increasing the polynomial order of the elements. For the local one, a mesh−superposition technique, called $s-$refinement, is used to improve locally the solution by defining a local/fine mesh overlaid to the global/coarse one. The combination of $p-$ and $s-$refinements enables us to achieve excellent accuracy−to−cost ratios. This paper aims to present the numerical formulation and the implementation aspects of this novel approach to VS composite shell analysis. Numerical tests are reported to illustrate the potential of the method. The results provide a clear insight of its potential to guarantee fast convergence and easy mesh refinement where needed.

Keywords: finite element method; numerical methods; thin shells; variable−stiffness structures

Citation: Yan, C.A.; Vescovini, R. Application of the $ps-$Version of the Finite Element Method to the Analysis of Laminated Shells. *Materials* **2023**, *16*, 1395. https://doi.org/10.3390/ma16041395

Academic Editors: Michele Bacciocchi and Abbas S. Milani

Received: 13 January 2023
Revised: 2 February 2023
Accepted: 3 February 2023
Published: 7 February 2023

1. Introduction

The anisotropic nature of composite laminates is responsible for structural responses characterized by complex couplings between extension, bending, torsion, and shearing modes [1]. Such effects are particularly relevant when laminates with curvilinear reinforcements [2–5] are of concern.

These effects require appropriate numerical tools to be available to conduct accurate and reliable analyses. This is even more true as composite thin shells are increasingly used in primary load−carrying structural components. Hence, the interest of the designers is not restricted to the global response, but includes also local effects, such as stress profiles in critical regions due to geometric discontinuities, material interphases, and areas affected by load introduction. In this framework, the finite element method (FEM) is the most commonly used tool for the analysis and design of composite structures. The main advantages can be found in the high flexibility of modeling complex domains, the possibility of considering arbitrary material properties, and handling different loading/boundary conditions.

Over the years, different strategies have been developed to maximize the accuracy of the FE solution, while keeping the computational cost at a minimum. The classical approaches to improve accuracy are represented by the $h-$ and $p-$refinements [6]. In the former, accuracy is increased by reducing the mesh size h, while the interpolation order of the elements p is fixed [7]. In the latter, the polynomial degree p is increased, while the mesh resolution h is left unchanged [8]. The $h-$ and $p-$refinements provide two systematic ways to improve the precision of the FE solution and yield excellent results when global quantities are of interest. Different applications of these strategies to advanced composite shells and plates can be found in the literature, such as sandwich plates [9], functionally graded laminates [10,11], shell structures with orthogonal periodic configurations [12], cylindrical shells made of metamaterials with spatially variable elastic properties [13], and variable stiffness composites [14–17].

Local refinement strategies are needed whenever focus is required on the response in specific regions. Adaptive $h-$refinement procedures have been proposed in the literature, e.g., [18–20]. In these works, a posteriori error estimators are coupled with the $h-$refinement strategy to generate adaptable meshes locally refined in the most critical zones. The $hp-$refinement approach [21] represents an extension of the adaptive $h-$refinement where an increase of the interpolation order of elements ($p-$refinement) is added on top of a suitable mesh adaptation ($h-$refinement). A simultaneous increase of the order p combined with a graded increase of the mesh resolution h yields exponential convergence rate for non−smooth problems, where steep gradients or singular points are present [22,23].

One of the challenges associated with simultaneous $h-$ and $p-$refinements relates to the complexity in generating an $h-$adaptable mesh. Typical procedures to cope with this problem involve the adoption of transition elements [24–26] or sophisticated multi-constraints approaches [27].

An alternative and simpler way to perform local mesh refinements is given by mesh superposition techniques. One of the first works in this field was by Motc [28], who employed the Ritz method to capture the global deflections of the structure, while a superposed local FE model was proposed for capturing stress concentrations.

The spectral overlay finite element method (SOFEM) proposed by Belytschko [29] is another example of an FE scheme based on solution superposition, where local refinements are performed by a spectral overlay on an FE mesh. A generalization of SOFEM is represented by the superposition−version, or $s-$version, of the finite element method ($s-$FEM) proposed by Fish [30]. In the $s-$FEM the solution accuracy is locally enhanced by superposing one or more high−order local meshes onto a global one. This extension of the FEM has been applied successfully for linear static, frequency, and buckling analysis of laminated composite plates and shells [31–34], stress analysis of laminated smart structures [35], microscopic stress analysis of heterogeneous materials [36], and dynamic crack analysis in steel structures [37].

A further superposition refinement approach is implemented in the $hp-d-$version of the finite element method ($hp-d-$FEM) developed by Rank [38]. The $hp-d-$FEM improves the solution efficiency by representing the smooth part of the solution with a coarse/high−order global mesh, while the non−smooth features are resolved by one or more [39] fine/low−order local meshes. The method is applied for solving linear and nonlinear elasticity problems with singularities in [39,40]. The multi−level $hp-$method introduced by Zander et al. [41] recovers the idea of the $hp-d-$FEM where the superposition meshes used for the local refinement employ high−order instead of low−order elements. This approach permits to achieve much faster convergence rates, as shown in [42,43], where the method is used for solving fracture mechanics, linear elastodynamics, and shell problems.

All the previous works have shown the potential of mesh superposition techniques to achieve refined solutions while keeping at minimum the implementation efforts. The present work aims at recovering such techniques to realize an efficient global/local formulation of the FEM, here denoted as the $ps-$version of the finite element method ($ps-$FEM), for the analysis of laminated composite shells. More specifically, the FE framework presented herein exploits a $p-$refinement strategy, based on the hierarchic polynomial space first formulated by [8], to capture the global response of the laminate. For the resolution of the local solution features, the $p-$refinement is combined with the mesh superposition refinement approach, or $s-$refinement, developed by [30,41].

Differently from existing works in the literature, superposition techniques are employed here for the first time for the analysis of composite shells, including straight−fiber and VS configurations.

The outline of the paper is as follows: the shell mathematical model is presented in Section 2; Section 3 discusses the formulation and implementation aspects of the proposed FEM; in Section 4, the method is applied to several test cases involving laminated shells. Finally, the main findings of this study are summarized in Section 5.

2. Theoretical Model

This section illustrates the theoretical framework for the analysis of laminated composite shells. In the first part, the shell geometry is discussed by presenting the relevant equations employed next in the finite element approximation. Then, the constitutive law for handling the case of curvilinear fiber paths is presented. Finally, the shell governing equations are derived in the context of a total Lagrangian description.

2.1. Geometry Description

The geometric description of shells is developed under the theory of surfaces, which provides a general framework to define panels with arbitrary configurations and curvatures.

The reference surface Ω is assumed to be coinciding with the shell's midsurface and is parametrized by two arc−length coordinates $\xi_1 \in [0, a]$ and $\xi_2 \in [0, b]$, where a and b are the curvilinear lengths of the shell edges. The normal direction to Ω is parametrized by the coordinate $\zeta \in [-t/2, +t/2]$, where t is the thickness of the shell. It is assumed that the triad (ξ_1, ξ_2, ζ) forms an orthogonal curvilinear coordinate system such that ξ_1 and ξ_2 are aligned with the principal lines of curvature on the midsurface, as shown in Figure 1.

Considering a global reference system (O, x, y, z), any point on Ω can be described by the position vector $\boldsymbol{r}$:

$$\boldsymbol{r}(\xi_1, \xi_2) = r_x(\xi_1, \xi_2)\boldsymbol{e}_x + r_y(\xi_1, \xi_2)\boldsymbol{e}_y + r_z(\xi_1, \xi_2)\boldsymbol{e}_z \tag{1}$$

where r_x, r_y and r_z are definite, continuous, single-valued functions of ξ_1 and ξ_2 representing the components of $\boldsymbol{r}$ on the global reference system with unit directors $\boldsymbol{e}_x$, $\boldsymbol{e}_y$ and $\boldsymbol{e}_z$.

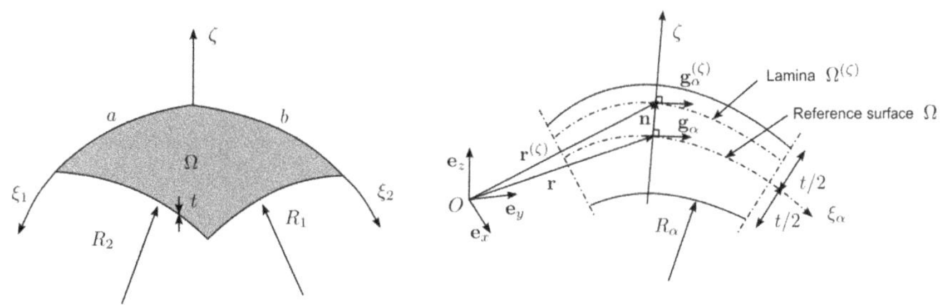

Figure 1. Shell geometry definition.

The tangent vectors on Ω to the two coordinate lines, ξ_1 and ξ_2, are obtained by differentiation of Equation (1):

$$\boldsymbol{g}_\alpha = \frac{\partial \boldsymbol{r}}{\partial \xi_\alpha} = \boldsymbol{r}_{,\alpha} \quad \text{for} \quad \alpha = 1, 2 \tag{2}$$

where the comma denotes differentiation with respect to the coordinate ξ_α.

Accordingly, the unit normal to the reference surface is defined as:

$$\boldsymbol{n} = \frac{\boldsymbol{g}_1 \times \boldsymbol{g}_2}{|\boldsymbol{g}_1 \times \boldsymbol{g}_2|} \tag{3}$$

where $\times$ denotes the cross product and $|\cdot|$ the Euclidean norm. The unit normal in Equation (3) is assumed to be common to all the shell laminas $\Omega^{(\zeta)}$, see Figure 1.

Considering a lamina at a distance ζ from Ω, the position vector of the generic point on $\Omega^{(\zeta)}$ is expressed as:

$$\boldsymbol{r}^{(\zeta)}(\xi_1,\xi_2,\zeta) = \boldsymbol{r}(\xi_1,\xi_2) + \zeta\boldsymbol{n}(\xi_1,\xi_2) \tag{4}$$

while the corresponding tangent vectors along the curvilinear lines ξ_1 and ξ_2 are:

$$\boldsymbol{g}_\alpha^{(\zeta)} = \boldsymbol{r}_{,\alpha}^{(\zeta)} = \boldsymbol{r}_{,\alpha} + \zeta\boldsymbol{n}_{,\alpha} \quad \text{for} \quad \alpha = 1,2 \tag{5}$$

By application of the Weingarten–Gauss relations [1]:

$$\boldsymbol{n}_{,\alpha} = \frac{\boldsymbol{g}_\alpha}{R_\alpha} \tag{6}$$

the following relation between the tangent vectors on $\Omega^{(\zeta)}$ and Ω holds:

$$\boldsymbol{g}_\alpha^{(\zeta)} = \left(1 + \frac{\zeta}{R_\alpha}\right)\boldsymbol{g}_\alpha \quad \text{for} \quad \alpha = 1,2 \tag{7}$$

where R_α are the principal radii of curvature of the reference surface and are defined as [44]:

$$R_\alpha = -\frac{\boldsymbol{g}_\alpha \cdot \boldsymbol{g}_\alpha}{\boldsymbol{g}_{\alpha,\alpha} \cdot \boldsymbol{n}} \quad \text{for} \quad \alpha = 1,2 \tag{8}$$

where $\cdot$ denotes the dot product. In general, the two principal radii of curvatures R_1 and R_2 are not constant and can be a function of the coordinate lines ξ_1 and ξ_2.

Referring to Equation (4), the differential position vector is given as:

$$d\boldsymbol{r}^{(\zeta)} = \boldsymbol{r}_{,1}^{(\zeta)}d\xi_1 + \boldsymbol{r}_{,2}^{(\zeta)}d\xi_2 + \boldsymbol{r}_{,\zeta}^{(\zeta)}d\zeta = \boldsymbol{g}_1^{(\zeta)}d\xi_1 + \boldsymbol{g}_2^{(\zeta)}d\xi_2 + \boldsymbol{n}d\zeta \tag{9}$$

From Equation (9), the square distance dl between two arbitrary points in the shell volume $P(\xi_1,\xi_2,\zeta)$ and $P'(\xi_1 + d\xi_1, \xi_2 + d\xi_2, \zeta + d\zeta)$ is computed as:

$$(dl)^2 = d\boldsymbol{r}^{(\zeta)} \cdot d\boldsymbol{r}^{(\zeta)} = \left(a_1^{(\zeta)}d\xi_1\right)^2 + \left(a_2^{(\zeta)}d\xi_2\right)^2 + (d\zeta)^2 \tag{10}$$

where $a_\alpha^{(\zeta)}$ are the Lamè parameters defined as:

$$a_\alpha^{(\zeta)} = \sqrt{\boldsymbol{g}_\alpha^{(\zeta)} \cdot \boldsymbol{g}_\alpha^{(\zeta)}} = \left(1 + \frac{\zeta}{R_\alpha}\right)a_\alpha \quad \text{for} \quad \alpha = 1,2 \tag{11}$$

with:

$$a_\alpha = \sqrt{\boldsymbol{g}_\alpha \cdot \boldsymbol{g}_\alpha} \quad \text{for} \quad \alpha = 1,2 \tag{12}$$

For the development of the shell mathematical model, the geometric measures are expressed in terms of reference surface Ω quantities. Accordingly, the element lines and cross section areas along the coordinates ξ_1 and ξ_2 are given as:

$$\begin{aligned} dl_\alpha &= a_\alpha^{(\zeta)}d\xi_\alpha = \left(1 + \frac{\zeta}{R_\alpha}\right)a_\alpha d\xi_\alpha \\ dA_\alpha &= dl_\alpha d\zeta = \left(1 + \frac{\zeta}{R_\alpha}\right)a_\alpha d\xi_\alpha d\zeta \quad \text{for} \quad \alpha = 1,2 \end{aligned} \tag{13}$$

while the element surface and volume are given by:

$$\begin{aligned} d\Omega &= \left|\boldsymbol{r}_{,1}^{(\zeta)}d\xi_1 \times \boldsymbol{r}_{,2}^{(\zeta)}d\xi_2\right| = a_1a_2\left(1 + \frac{\zeta}{R_1}\right)\left(1 + \frac{\zeta}{R_2}\right)d\xi_1d\xi_2 \\ dV &= \left(\boldsymbol{r}_{,1}^{(\zeta)}d\xi_1 \times \boldsymbol{r}_{,2}^{(\zeta)}d\xi_2\right) \cdot \boldsymbol{r}_{,\zeta}^{(\zeta)}d\zeta = a_1a_2\left(1 + \frac{\zeta}{R_1}\right)\left(1 + \frac{\zeta}{R_2}\right)d\xi_1d\xi_2d\zeta \end{aligned} \tag{14}$$

The geometric formulation presented here furnishes the background to develop the finite elements discussed next, where plates and shells with variable curvature can be taken into account.

2.2. Curvilinear Fiber Path Description

The shell elements considered here can be applied to the analysis of isotropic and composite structures. For generality purposes, the model is developed to allow non−uniform elastic properties to be considered along the surface. Hence, new configurations such as variable stiffness (VS) laminates can be studied within the proposed framework. Specifically, a laminate with VS properties is achieved by stacking together plies with curvilinear reinforcing fibers [2]. Different strategies have been proposed in the literature for describing the path of the fibers. Without loss of generality, the approach considered here relies on the Lagrange polynomials introduced in [4]. According to this mathematical formulation, the curvilinear paths are described by a set of parameters T_{mn} according to:

$$\theta(\xi_1, \xi_2) = \sum_{m=0}^{M-1} \sum_{n=0}^{N-1} T_{mn} \prod_{n \neq i} \frac{\left(\xi_1 - \xi_1^{(i)}\right)}{\left(\xi_1^{(n)} - \xi_1^{(i)}\right)} \cdot \prod_{m \neq j} \frac{\left(\xi_2 - \xi_2^{(j)}\right)}{\left(\xi_2^{(m)} - \xi_2^{(j)}\right)} \tag{15}$$

where $\left(\xi_1^{(n)}, \xi_2^{(m)}\right)$ are pre−selected control points over a uniform grid $N \times M$ in $[0, a] \times [0, b]$. From Equation (15), it is possible to understand the parameters T_{mn} as the fiber orientation angle at the control points, as one can see that $\theta\left(\xi_1^{(n)}, \xi_2^{(m)}\right) = T_{mn}$. The information of fiber orientation on a the generic VS ply can thus be defined through a matrix $\boldsymbol{T} \in \mathbb{R}^{M \times N}$. This enables us to specify VS lamination sequences very concisely as $[\boldsymbol{T}_1/\boldsymbol{T}_2/\ldots/\boldsymbol{T}_{N_p}]$, where N_p stands for the number of plies.

An alternative formulation is also possible where the control points are defined at a quarter of the laminate and the resulting fiber orientation distribution is reflected to the rest of the domain. In this case, the expression of Equation (15) modifies as:

$$\theta(\xi_1, \xi_2) = \sum_{m=0}^{M-1} \sum_{n=0}^{N-1} T_{mn} \prod_{n \neq i} \frac{\left(|\tilde{\xi}_1| - \tilde{\xi}_1^{(i)}\right)}{\left(\tilde{\xi}_1^{(n)} - \tilde{\xi}_1^{(i)}\right)} \cdot \prod_{m \neq j} \frac{\left(|\tilde{\xi}_2| - \tilde{\xi}_2^{(j)}\right)}{\left(\tilde{\xi}_2^{(m)} - \tilde{\xi}_2^{(j)}\right)} \tag{16}$$

where the new coordinates are $\tilde{\xi}_1 = \xi_1 - a/2$ and $\tilde{\xi}_2 = \xi_2 - b/2$, while the control points $\left(\tilde{\xi}_1^{(n)}, \tilde{\xi}_2^{(m)}\right)$ are now defined over a grid $N \times M$ in $[0, a/2] \times [0, b/2]$.

2.3. Shell Description

The displacements of an arbitrary point on the shell domain Ω are denoted by u_1, u_2 and u_3. These components correspond to the displacements along the ξ_1, ξ_2 and ζ directions, respectively. The component u_3 is taken positive in the outward direction from the center of the smallest radius of curvature. Considering only von Kàrmàn nonlinearities, the components of the Green–Lagrange strain tensor in the orthogonal curvilinear coordinate system (ξ_1, ξ_2, ζ) are [45]:

$$
\begin{aligned}
\epsilon_{11} &= \frac{1}{1+\zeta/R_1}\left(\frac{u_{1,1}}{a_1}+\frac{a_{1,2}}{a_1a_2}u_2+\frac{u_3}{R_1}\right)+\frac{1}{2}\left(\frac{1}{1+\zeta/R_1}\right)^2\left(\frac{u_{3,1}}{a_1}\right)^2 \\
\epsilon_{22} &= \frac{1}{1+\zeta/R_2}\left(\frac{u_{2,2}}{a_2}+\frac{a_{2,1}}{a_1a_2}u_1+\frac{u_3}{R_2}\right)+\frac{1}{2}\left(\frac{1}{1+\zeta/R_2}\right)^2\left(\frac{u_{3,2}}{a_2}\right)^2 \\
\gamma_{12} &= \frac{1}{1+\zeta/R_1}\left(\frac{u_{2,1}}{a_1}-\frac{a_{1,2}}{a_1a_2}u_1\right)+\frac{1}{1+\zeta/R_2}\left(\frac{u_{1,2}}{a_2}-\frac{a_{2,1}}{a_1a_2}u_2\right)+\frac{1}{(1+\zeta/R_1)(1+\zeta/R_2)}\left(\frac{u_{3,1}}{a_1}\frac{u_{3,2}}{a_2}\right) \\
\gamma_{13} &= u_{1,\zeta}+\frac{1}{1+\zeta/R_1}\left(\frac{u_{3,1}}{a_1}-\frac{u_1}{R_1}\right) \\
\gamma_{23} &= u_{2,\zeta}+\frac{1}{1+\zeta/R_2}\left(\frac{u_{3,2}}{a_2}-\frac{u_2}{R_2}\right) \\
\epsilon_{33} &= u_{3,\zeta}
\end{aligned}
\tag{17}
$$

where $\epsilon_{\alpha\alpha}$ and $\gamma_{\alpha\beta}$ are the normal and shearing strains, respectively.

The shell kinematics is described according to the first shear deformation theory (FSDT), so:

$$
\begin{aligned}
u_1(\xi_1,\xi_2,\zeta,t) &= u(\xi_1,\xi_2)+\zeta\phi_1(\xi_1,\xi_2,t) \\
u_2(\xi_1,\xi_2,\zeta,t) &= v(\xi_1,\xi_2)+\zeta\phi_2(\xi_1,\xi_2,t) \\
u_3(\xi_1,\xi_2,\zeta,t) &= w(\xi_1,\xi_2)
\end{aligned}
\tag{18}
$$

where u, v and w are the displacement components of an arbitrary point on Ω along ξ_1, ξ_2 and ζ, respectively, while ϕ_1 and ϕ_2 are the rotations. The choice of FSDT is due to the easier FE implementation compared with the classical Kirchhoff theory (CLT). At the same time, FSDT allows shear deformability to be taken into account with a small number of unknown fields. Higher–order theories are not considered here, but can be of practical interest to guarantee improved description of the thickness–wise response, while avoiding the use of the shear factor. The strains of the kinematic model at hand can be obtained upon the substitution of Equation (18) into Equation (17):

$$
\boldsymbol{\epsilon} = \boldsymbol{\epsilon}^0+\zeta\boldsymbol{k}, \quad \boldsymbol{\gamma} = \boldsymbol{\gamma}^0 \tag{19}
$$

where $\boldsymbol{\epsilon} = \{\epsilon_{11}\ \epsilon_{22}\ \gamma_{12}\}^{\mathrm{T}}$ and $\boldsymbol{\gamma} = \{\gamma_{13}\ \gamma_{23}\}^{\mathrm{T}}$. The definition of $\boldsymbol{\epsilon}^0 = \{\epsilon^0_{11}\ \epsilon^0_{22}\ \gamma^0_{12}\}^{\mathrm{T}}$, $\boldsymbol{k} = \{k_{11}\ k_{22}\ k_{12}\}^{\mathrm{T}}$ and $\boldsymbol{\gamma}^0 = \{\gamma^0_{13}\ \gamma^0_{23}\}^{\mathrm{T}}$ is available in the Appendix A.

The membrane resultant on the cross section dA_2, i.e., the section normal to ξ_1, is:

$$
\int_{-t/2}^{t/2}\sigma_{11}dA_2 = \left[\int_{-t/2}^{t/2}\sigma_{11}\left(1+\frac{\zeta}{R_2}\right)d\zeta\right]a_2d\xi_2 = N_{11}a_2d\xi_2 \tag{20}
$$

where the definition of the resultant N_{11} follows from the equation above. The other resultants are derived in a similar manner, leading to:

$$
\begin{aligned}
&\begin{Bmatrix} N_{11} \\ N_{22} \\ N_{12} \\ N_{21} \end{Bmatrix} = \int_{-t/2}^{t/2}\begin{Bmatrix} \sigma_{11}\left(1+\frac{\zeta}{R_2}\right) \\ \sigma_{22}\left(1+\frac{\zeta}{R_1}\right) \\ \sigma_{12}\left(1+\frac{\zeta}{R_2}\right) \\ \sigma_{21}\left(1+\frac{\zeta}{R_1}\right) \end{Bmatrix}d\zeta, \quad
\begin{Bmatrix} M_{11} \\ M_{22} \\ M_{12} \\ M_{21} \end{Bmatrix} = \int_{-t/2}^{t/2}\zeta\begin{Bmatrix} \sigma_{11}\left(1+\frac{\zeta}{R_2}\right) \\ \sigma_{22}\left(1+\frac{\zeta}{R_1}\right) \\ \sigma_{12}\left(1+\frac{\zeta}{R_2}\right) \\ \sigma_{21}\left(1+\frac{\zeta}{R_1}\right) \end{Bmatrix}d\zeta, \\
&\begin{Bmatrix} Q_1 \\ Q_2 \end{Bmatrix} = K_s\int_{-t/2}^{t/2}\begin{Bmatrix} \sigma_{13}\left(1+\frac{\zeta}{R_2}\right) \\ \sigma_{23}\left(1+\frac{\zeta}{R_1}\right) \end{Bmatrix}d\zeta
\end{aligned}
\tag{21}
$$

where K_s is the shear correction factor, which is taken as equal to $5/6$. The shear stress resultants, N_{12} and N_{21}, and the twisting moments, M_{12} and M_{21}, are different due to the curvature terms. For shallow and moderately thick shells, i.e., $h/R_\alpha < 1/20$, the simplifying assumption $N_{12} \approx N_{21}$ and $M_{12} \approx M_{21}$ can be introduced.

Based on the assumption above, the constitutive law reads:

$$\begin{Bmatrix} \boldsymbol{N} \\ \boldsymbol{M} \\ \boldsymbol{Q} \end{Bmatrix} = \begin{bmatrix} \boldsymbol{A}(\xi_1,\xi_2) & \boldsymbol{B}(\xi_1,\xi_2) & 0 \\ \boldsymbol{B}(\xi_1,\xi_2) & \boldsymbol{D}(\xi_1,\xi_2) & 0 \\ 0 & 0 & \overline{\boldsymbol{A}}(\xi_1,\xi_2) \end{bmatrix} \begin{Bmatrix} \boldsymbol{\epsilon}^0 \\ \boldsymbol{k} \\ \boldsymbol{\gamma}^0 \end{Bmatrix} \tag{22}$$

where $\boldsymbol{N} = \{N_{11}\ N_{22}\ N_{12}\}^{\mathrm{T}}$, $\boldsymbol{M} = \{M_{11}\ M_{22}\ M_{12}\}^{\mathrm{T}}$, $\boldsymbol{Q} = \{Q_1\ Q_2\}^{\mathrm{T}}$ are the vectors collecting the force and moment resultants, whereas $\boldsymbol{A}$, $\boldsymbol{D}$, $\boldsymbol{B}$ and $\overline{\boldsymbol{A}}$ are the membrane, bending, membrane–bending coupling and shear stiffness matrices, respectively.

The equations of motion are derived by referring to Hamilton's principle:

$$\int_{t_1}^{t_2} \delta\Pi \mathrm{d}t = \int_{t_1}^{t_2} \delta[K - (U + V)]\mathrm{d}t = 0 \tag{23}$$

where K, U and V are the kinetic and elastic energies and the potential of the applied loads, respectively. For the shell model developed in this work, these energy quantities are defined as follows:

$$\begin{aligned} K =& \frac{1}{2}\int_\Omega \left[I_0\left(\dot{u}^2 + \dot{v}^2 + \dot{w}^2\right) + 2I_1(\dot{\phi}_1\dot{u} + \dot{\phi}_2\dot{v}) + I_2\left(\dot{\phi}_1{}^2 + \dot{\phi}_2{}^2\right)\right] a_1 a_2 d\xi_1 d\xi_2 \\ U =& \frac{1}{2}\int_\Omega \left[\boldsymbol{\epsilon}^{0\mathrm{T}} \boldsymbol{A}\boldsymbol{\epsilon}^0 + \boldsymbol{k}^{\mathrm{T}}\boldsymbol{D}\boldsymbol{k} + 2\boldsymbol{\epsilon}^{0\mathrm{T}}\boldsymbol{B}\boldsymbol{k} + \boldsymbol{\gamma}^{0\mathrm{T}}\overline{\boldsymbol{A}}\boldsymbol{\gamma}^0\right] a_1 a_2 d\xi_1 d\xi_2 \\ V =& -\int_\Omega \left[w\left(1 + \frac{h}{2R_1}\right)\left(1 + \frac{h}{2R_2}\right) q^+ \right] a_1 a_2 d\xi_1 d\xi_2 \\ &- \int_{\partial\Omega_1} (u\overline{N}_{11} + v\overline{N}_{12}) a_2 d\xi_2 + \int_{\partial\Omega_2} (v\overline{N}_{22} + u\overline{N}_{12}) a_1 d\xi_1 \end{aligned} \tag{24}$$

where the dot denotes the time derivative, q^+ is the pressure applied on the upper surface ($\zeta = h/2$), $\partial\Omega_1$ and $\partial\Omega_2$ are the boundaries at $\xi_1 = \text{const}$ and $\xi_2 = \text{const}$, respectively. The terms I_α are the moment of inertia defined as:

$$I_\alpha = \int_A \rho\left(1 + \frac{\zeta}{R_1}\right)\left(1 + \frac{\zeta}{R_2}\right)(\zeta)^\alpha d\zeta \quad \alpha = 0, 1, 2 \tag{25}$$

where ρ is the mass density.

3. The $ps-$Version of the Finite Element Method

The finite element method (FEM) developed in this work is a combination of two FE schemes, i.e., the $p-$version of the finite element method ($p-$FEM) and the $s-$version ($s-$FEM). The former is an extension of the conventional FEM, or $h-$version ($h-$FEM), where the accuracy is increased by increasing the interpolation order ($p-$refinement). The latter is an FE scheme where arbitrary local improvements of the mesh resolution are possible through the adoption of advanced mesh superposition techniques ($s-$refinement). The two features of these numerical approaches are put together into an extended FE framework, called the $ps-$version of the finite element method ($ps-$FEM), where simultaneous $p-$ and $s-$refinement ($ps-$refinement) can be performed to adaptively adjust the interpolation order p and the size h of the elements. Within the $ps-$FEM, the construction of the polynomial space S^p is inspired from the $p-$FEM, while the design of the mesh Δ follows the ideas implemented in the $s-$FEM, see Figure 2. In the following, the two main pillars of the $ps-$FEM, i.e., $p-$FEM and $s-$FEM, are presented. Then, the $ps-$FEM is introduced and applied to composite shell problems by referring to the equilibrium equations of Section 2.3.

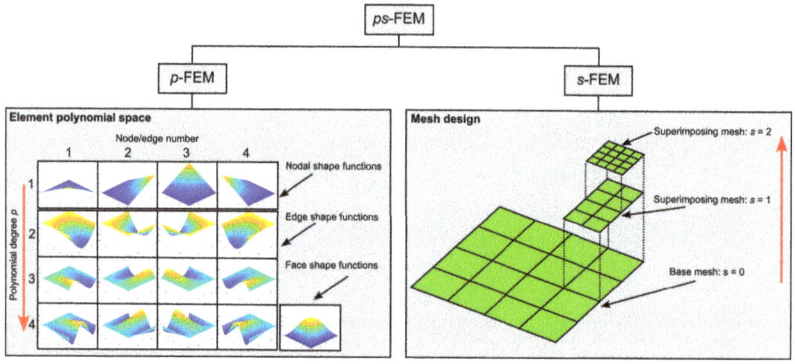

Figure 2. Building blocks of the $ps-$FEM framework.

3.1. *p−Refinement*

The core idea of the $p-$refinement strategy is to progressively increase the polynomial order p of the shape functions to improve the quality of the numerical solution. One crucial aspect regards the worsening of the conditionality of the global stiffness matrix as p is increased. In the $p-$FEM, this issue can be mitigated by constructing the set of shape functions starting from Legendre polynomials. The properties of these functions enable better numerical conditioning in comparison to the Lagrange polynomials commonly used in the $h-$FEM. High−order polynomials can be considered in the $p-$FEM—even beyond $p = 3$—, without encountering numerical issues.

Considering a generic two−dimensional solution field ϕ depending on x and y, the FE discretization in the elemental domain $\Omega^{(k)}$ is:

$$\phi^{(k)}(x,y) = \phi^{(k)}(\chi^{(k)}) = \sum_{i=1}^{p+1}\sum_{j=1}^{p+1} c_{ij} f_i(\xi) f_j(\eta) \quad \text{in} \quad \Omega^{(k)} \tag{26}$$

where $-1 < \{\xi,\eta\} < 1$ are the nondimensional coordinates in the computational domain Ω_{st}, $\chi^{(k)}$ is an array of functions mapping the standard element in the $k-$th element of the mesh Δ, see Figure 3, the coefficients p and c_{ij} are the element interpolation order and degrees of freedom, respectively, while f_i and f_j are polynomial shape functions. In the $p-$FEM the polynomial space S^p is constructed as:

$$\begin{aligned} f_1(\xi) &= \frac{1}{2}(1+\xi), \quad f_2(\xi) = \frac{1}{2}(1-\xi) \\ f_{i+1}(\xi) &= \sqrt{\frac{2i-1}{2}} \int_{-1}^{\xi} P_{i-1}(\xi) d\xi \quad \text{for} \quad i = 2,3,\ldots p \end{aligned} \tag{27}$$

where $P_n(\xi)$ is the Legendre polynomial functions of order n:

$$\begin{aligned} P_0(\xi) &= 1, \quad P_1(\xi) = \xi \\ P_n(\xi) &= \frac{1}{n}[\xi(2n-1)P_{n-1}(\xi) - (n-1)P_{n-2}(\xi)] \quad \text{for} \quad n = 2,3,\ldots p \end{aligned} \tag{28}$$

From Equation (27), it is possible to distinguish two families of functions describing different deformation modes, i.e., nodal and internal ones. Nodal modes are represented by $f_1(\xi)$ and $f_2(\xi)$, which are Lagrange linear interpolation polynomials. Internal modes are reproduced by the higher−order terms, $f_\alpha(\xi)$ (for $\alpha \geq 3$), which are based on the integrals of Legendre polynomials. The combination of nodal and internal modes leads to different types of two−dimensional shape functions, i.e., nodal (two nodal modes), edge (one nodal mode and one internal mode), and face (two internal modes) shape functions, see Figure 4.

The series expansion in Equation (27) offers several advantages over the classical Lagrangian polynomial base. The first one is that it inherits the orthogonal properties of the original Legendre polynomials, indeed:

$$\int_{-1}^{+1} \frac{\partial f_i}{\partial \xi} \frac{\partial f_j}{\partial \xi} d\xi = \delta_{ij} \quad \text{for} \quad i \geq 3, j \geq 1 \quad \text{or} \quad \text{for} \quad j \geq 3, i \geq 1 \tag{29}$$

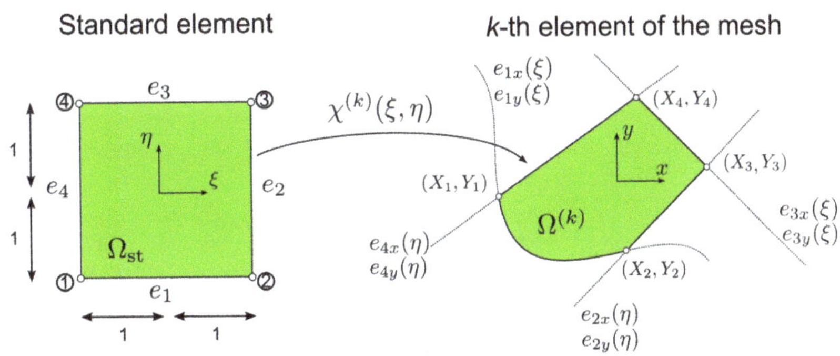

Figure 3. Element mapping procedure.

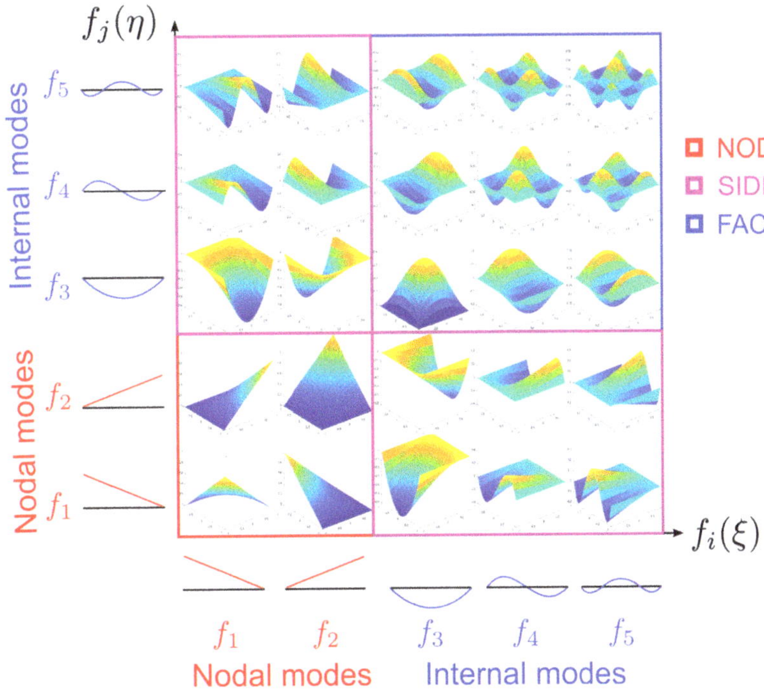

Figure 4. Two−dimensional shape functions.

This property provides the stiffness matrix with a quasi−diagonal structure, which significantly eases the solution process and enables the use of shape functions with a very high polynomial order. The second feature regards the hierarchical nature of these functions: p−refinements implies adding new higher−order terms in Equation (27), while the low−order ones remain unchanged. As a consequence, the final stiffness matrix is obtained by successively adding rows and columns associated with new levels of $p-$refinements.

Relatively few elements are in most cases sufficient to meet the desired levels of accuracy. At the same time, the method should be capable of appropriately representing the geometry of the structure with a limited number of elements. For this purpose, blending functions [46,47] are usually combined with the $p-$FEM. The mapping procedure for the transformation of the standard element in Ω_{st} to the $k-$th element of the mesh in $\Omega^{(k)}$ is:

$$\begin{aligned} x = Q_x^{(k)}(\xi,\eta) =& f_1(\eta)[e_{x1}(\xi) - f_1(\xi)X_1] + f_2(\xi)[e_{x2}(\eta) - f_1(\eta)X_2] \\ & f_2(\eta)[e_{x3}(\xi) - f_2(\xi)X_3] + f_1(\xi)[e_{x4}(\eta) - f_2(\eta)X_4] \\ y = Q_y^{(k)}(\xi,\eta) =& f_1(\eta)\left[e_{y1}(\xi) - f_1(\xi)Y_1\right] + f_2(\xi)\left[e_{y2}(\eta) - f_1(\eta)Y_2\right] \\ & f_2(\eta)\left[e_{y3}(\xi) - f_2(\xi)Y_3\right] + f_1(\xi)\left[e_{y4}(\eta) - f_2(\eta)Y_4\right] \end{aligned} \tag{30}$$

where $Q_x^{(k)}(\xi,\eta)$ and $Q_y^{(k)}(\xi,\eta)$ are general nonlinear mapping functions, (X_α, Y_α) are the nodal coordinates of the element, while $e_{\alpha x}$ and $e_{\alpha y}$ with $\alpha = 1,\ldots,4$ are the functions defining the curves of the element edge, as illustrated in Figure 3.

From the mapping of Equation (30), the derivatives with respect to the physical coordinates x and y are computed as:

$$\begin{Bmatrix} \frac{\partial}{\partial x} \\ \frac{\partial}{\partial y} \end{Bmatrix} = \begin{bmatrix} \frac{\partial Q_x^{(k)}}{\partial \xi} & \frac{\partial Q_y^{(k)}}{\partial \xi} \\ \frac{\partial Q_x^{(k)}}{\partial \eta} & \frac{\partial Q_y^{(k)}}{\partial \eta} \end{bmatrix}^{-1} \begin{Bmatrix} \frac{\partial}{\partial \xi} \\ \frac{\partial}{\partial \eta} \end{Bmatrix} = \begin{bmatrix} J_{11}^{(k)} & J_{12}^{(k)} \\ J_{21}^{(k)} & J_{22}^{(k)} \end{bmatrix}^{-1} \begin{Bmatrix} \frac{\partial}{\partial \xi} \\ \frac{\partial}{\partial \eta} \end{Bmatrix} \tag{31}$$

where $J_{\alpha\beta}^{(k)}$ are the components of the Jacobian matrix $\boldsymbol{J}^{(k)}$.
Line integrals at the four sides of the element read:

$$\begin{aligned} \int_{e_\alpha} F(x,y)dx &= \int_{-1}^{1} \mathcal{F}(\xi,\pm 1)\sqrt{J_{11}^{(k)^2} + J_{12}^{(k)^2}}\, d\xi \quad \text{for} \quad \alpha = 1,3 \\ \int_{e_\alpha} F(x,y)dy &= \int_{-1}^{1} \mathcal{F}(\pm 1,\eta)\sqrt{J_{21}^{(k)^2} + J_{22}^{(k)^2}}\, d\eta \quad \text{for} \quad \alpha = 2,4 \end{aligned} \tag{32}$$

while the surface ones are:

$$\int_{\Omega^{(k)}} F(x,y)dxdy = \int_{-1}^{1}\int_{-1}^{1} \mathcal{F}(\xi,\eta)Jd\xi d\eta \tag{33}$$

where F is the generic integrand, $\mathcal{F}$ is obtained from F by replacing x and y with the mapping functions in Equation (30), while $J = \det\boldsymbol{J}^{(k)}$ is the determinant of the Jacobian matrix.

3.2. *s−Refinement*

The main problem in performing local mesh refinements relates to the need to generate a transition between refined and unrefined regions. Different approaches are available in the literature to address this issue, for example transition elements or multi−point constraints approaches [24–27]. The $s-$refinement strategy offers the advantage of simplifying this process by allowing an element size reduction in the desired regions only. This result is achieved through the definition of an independent local/fine mesh which is superimposed to a global/coarse one [30].

Using this idea, the final FE approximation ϕ can be represented as:

$$\phi = \begin{cases} \phi_G & \text{in} \quad \Omega - \Omega_L \\ \phi_G + \phi_L & \text{in} \quad \Omega_L \end{cases} \tag{34}$$

where ϕ_G is the global mesh solution defined in Ω, while ϕ_L is the local mesh solution defined in $\Omega_L \subset \Omega$, see Figure 5. Note that the mesh superposition technique allows for incompatible discretization between global and local mesh. This gives an extremely high level of flexibility when performing local $h-$refinements, as no transition regions [24–26] or multi-point constraints [27] are required.

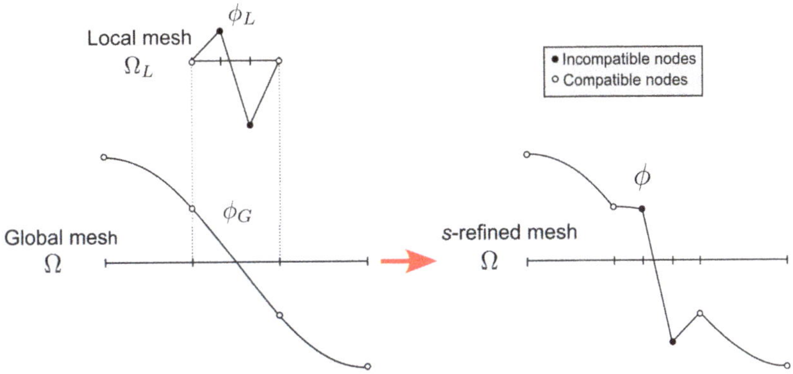

Figure 5. Conceptual idea of mesh superposition.

This concept of solution superposition can be extended to multi−level refinements [41]. In this case, the elements of the local mesh can be further refined by superposing one over the other multiple levels of overlaid meshes. In this case, the final FE solution ϕ becomes:

$$\phi = \begin{cases} \phi_G & \text{in} \quad \Omega - \Omega_L^{(1)} \\ \phi_G + \phi_L^{(1)} & \text{in} \quad \Omega_L^{(1)} - \Omega_L^{(2)} \\ \cdots \\ \phi_G + \phi_L^{(1)} + \cdots + \phi_L^{(s)} + \cdots + \phi_L^{(N_s)} & \text{in} \quad \Omega_L^{(N_s)} \end{cases} \tag{35}$$

where $\phi_L^{(s)}$ is the local solution given by the mesh at level s covering the domain $\Omega_L^{(s)} \subset \Omega_L^{(s-1)} \subset \cdots \subset \Omega$. Note that the solution on the global ϕ_G and local meshes $\phi_L^{(s)}$ can be represented by any FE scheme, such as the $p-$FEM presented in Section 3.1.

Two conditions, i.e., compatibility of the basis functions and their linear independency, are required to apply this multi−level decomposition of the solution field ϕ.

The first condition implies C^1-continuity within each elements and C^0-continuity across the element boundaries. The C^1-continuity is satisfied by construction. On the contrary, the inter−element continuity is not guaranteed and needs to the imposed. This is achieved by enforcing homogeneous Dirichlet boundary conditions on the boundary of the overlaid meshes, as depicted in Figure 6a.

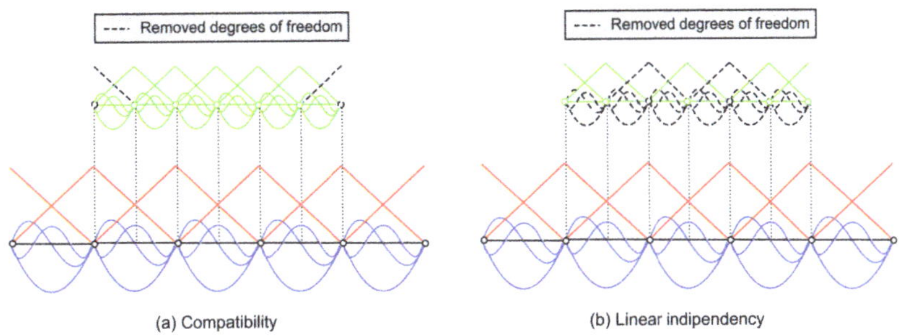

Figure 6. Conditions to satisfy when performing $s-$refinements: (**a**) compatibility and (**b**) linear indipendency.

The second condition on the linear independency is required to avoid singularities in the stiffness matrix. In general, the redundant degrees−of−freedom can be removed during the factorization process by elimination of the equations with zero pivots [30]. If the $p-$FEM is employed for the global and local meshes, this is avoided by ensuring that shape functions of the same type (nodal, side, face) and polynomial order p appear only once in regions with multiple meshes. This idea is graphically illustrated in Figure 6b.

3.3. *ps−Refinement*

The $ps-$refinement procedure is a combination of the $p-$ and $s-$FEM approaches. In this framework, the order of interpolation p can be increased with no numerical issues owing to the properties of Legendre polynomials; at the same time, the elements size h can be adaptively reduced by overlaying different levels of superposition meshes s.

In this work, the $ps-$refinement is used to develop an advanced FEM, called the $ps-$FEM, which is exploited to perform global/local analysis of laminated shells. In particular, the $p-$refinement strategy is exploited to capture the smooth features of the solution at global scale, such as the global deformation field, while the $s-$refinement approach is employed to capture local effects, such as stress concentrations.

The refinement process is outlined in Figure 7. First, a global/base mesh Δ_G is defined. The resolution of Δ_G is chosen to guarantee appropriate definition of the loading and boundary conditions, and to avoid the description of the geometry using distorted elements. Then, a $p-$refinement is performed until the required accuracy on the global response is reached. Finally, the solution is $s-$refined with local meshes in the regions where enhanced interpolation capability is required.

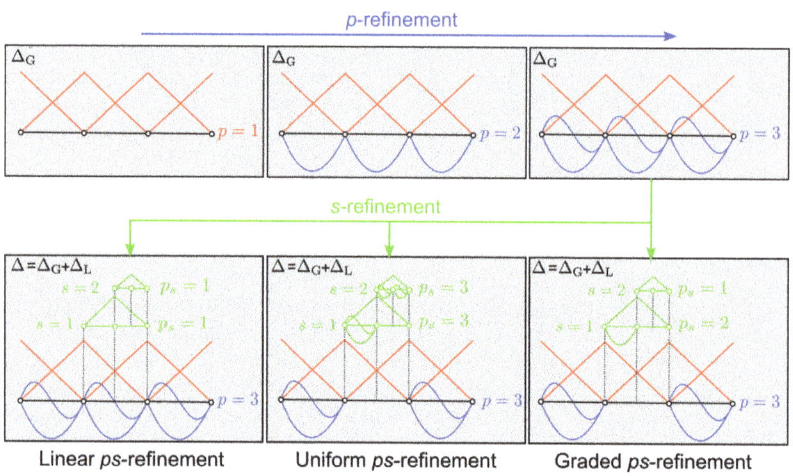

Figure 7. Refinement process in the $ps-$FEM.

The $ps-$FEM shares similar convergence features with the $hp-$FEM. Indeed, by increasing the order of the polynomial expansion in combination with overlaid meshes, it is possible achieve an exponential decaying global approximation error. This is true even when the solution presents steep gradients or singular points. More insights on convergence properties can be found in the work of [42].

In the proposed framework, three different combinations of $p-$ and $s-$refinement are allowed, namely linear, uniform, and graded $ps-$refinements. In the linear case [39], the global solution is represented by a coarse high−order mesh, while locally a multi−level $s-$refinement is performed with low−order meshes. The uniform and graded $ps-$refinement strategies [41] are an extension of the previous one, where high−order local meshes are employed. In the first case, the local meshes have the same order p of the global one, while in the second case p is different from level to level.

In principle, $s-$refinement can be carried out considering local meshes with arbitrary resolution and orientation, as originally done in [30]. A simplified approach is implemented here, inspired by the work of [41]. In particular, the elements of the refined region are divided to half of the original size. These elements are then used to define the superposed mesh at the upper level. This procedure allows for a quicker refinement, with no need to iterate to correlate the generic coordinates of the local and global element domain [30].

In the $ps-$FEM, the shell generalized displacements for the $k-$th element with interpolation order p at the $s-$level mesh are:

$$
\begin{aligned}
u(\xi_1,\xi_2,t) &= u(\chi^{(k,s)},t) = \sum_{p=1}^{p_1+1}\sum_{q=1}^{p_1+1} c_{pq}^{(1)}(t) f_p(\xi) f_q(\eta) \\
v(\xi_1,\xi_2,t) &= v(\chi^{(k,s)},t) = \sum_{r=1}^{p_2+1}\sum_{s=1}^{p_2+1} c_{rs}^{(2)}(t) f_r(\xi) f_s(\eta) \\
w(\xi_1,\xi_2,t) &= w(\chi^{(k,s)},t) = \sum_{m=1}^{p_3+1}\sum_{n=1}^{p_3+1} c_{mn}^{(3)}(t) f_m(\xi) f_n(\eta) \\
\phi_1(\xi_1,\xi_2,t) &= \phi_1(\chi^{(k,s)},t) = \sum_{i=1}^{p_4+1}\sum_{j=1}^{p_4+1} c_{ij}^{(4)}(t) f_i(\xi) f_j(\eta) \\
\phi_2(\xi_1,\xi_2,t) &= \phi_2(\chi^{(k,s)},t) = \sum_{k=1}^{p_5+1}\sum_{l=1}^{p_5+1} c_{kl}^{(5)}(t) f_k(\xi) f_l(\eta) \quad \text{in} \quad \Omega^{(k)}
\end{aligned} \tag{36}
$$

where p_α and $c^{(\alpha)}$ $(\alpha = 1,2,3,4,5)$ are the element expansion order and unknown amplitude coefficients for the displacement fields, while $\chi^{(k,s)}$ is the vector collecting the mapping functions between the element computational space $\Omega_{\text{st}}^{(k,s)}$ and element physical space $\Omega^{(k,s)}$. This mapping procedure is illustrated in Figure 8. For the global mesh elements (s=0), the vector of mapping functions is defined by Equation (30). For the superimposed mesh elements ($s > 0$), the following sequence of mapping is operated:

$$
\chi^{(k,s)} = \chi^{(k,0)} \circ \mathbf{\Psi}^{(k,1)} \circ \quad \dots \quad \circ \mathbf{\Psi}^{(k,s-1)} \circ \mathbf{\Psi}^{(k,s)} \tag{37}
$$

where $\mathbf{\Psi}^{(k,s)}$ is the vector collecting the mapping functions between the computational domain of the element and the underlying one, as shown in Figure 8. Each element is therefore provided with two set of mapping functions, the global one $\chi^{(k,s)}$ and the local one $\mathbf{\Psi}^{(k,s)}$. These functions are required for defining the derivatives and integrating at element level.

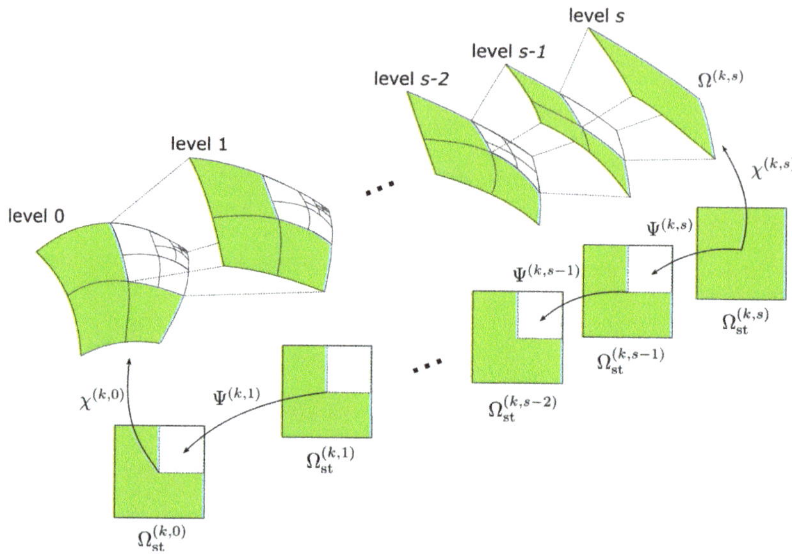

Figure 8. Mapping procedure between local meshes and global one.

It is noted that the $ps-$FEM can be applied for the numerical solution of any mathematical model, such as higher−order shear deformation theories, as well as three−dimensional elasticity theories. The effectiveness of FE techniques relying on mesh superposition has been discussed in the literature for such problems. Examples can be found in [42], where the $hp-d-$FEM is employed to solve 3D linear elastodynamic problems, and in [32,35] where the $s-$FEM is employed in the context of multiple model methods.

Using the approximation of Equation (36), the energy contributions of Equation (24) are:

$$
\begin{aligned}
K^{(k,s)} =& \frac{1}{2}\Big[M^{(11)}_{(pq)(\overline{pq})}\dot{c}^{(1)}_{pq}\dot{c}^{(1)}_{\overline{pq}} + M^{(22)}_{(rs)(\overline{rs})}\dot{c}^{(2)}_{rs}\dot{c}^{(2)}_{\overline{rs}} + M^{(33)}_{(mn)(\overline{mn})}\dot{c}^{(3)}_{mn}\dot{c}^{(3)}_{\overline{mn}} + M^{(44)}_{(ij)(\overline{ij})}\dot{c}^{(4)}_{ij}\dot{c}^{(4)}_{\overline{ij}} + M^{(55)}_{(kl)(\overline{kl})}\dot{c}^{(5)}_{kl}\dot{c}^{(5)}_{\overline{kl}} \\
&+2M^{(14)}_{(pq)(mn)}\dot{c}^{(1)}_{pq}\dot{c}^{(4)}_{mn} + 2M^{(25)}_{(pq)(mn)}\dot{c}^{(2)}_{pq}\dot{c}^{(5)}_{mn}\Big] \\
U^{(k,s)} =& \frac{1}{2}\Big[K^{(11)}_{(pq)(\overline{pq})}c^{(1)}_{pq}c^{(1)}_{\overline{pq}} + 2K^{(12)}_{(pq)(rs)}c^{(1)}_{pq}c^{(2)}_{rs} + 2K^{(13)}_{(pq)(mn)}c^{(1)}_{pq}c^{(3)}_{mn} + 2K^{(14)}_{(pq)(ij)}c^{(1)}_{pq}c^{(4)}_{ij} + 2K^{(15)}_{(pq)(kl)}c^{(1)}_{pq}c^{(5)}_{kl} \\
&+K^{(22)}_{(rs)(\overline{rs})}c^{(2)}_{rs}c^{(2)}_{\overline{rs}} + 2K^{(23)}_{(rs)(mn)}c^{(2)}_{rs}c^{(3)}_{mn} + 2K^{(24)}_{(rs)(ij)}c^{(2)}_{rs}c^{(4)}_{ij} + 2K^{(25)}_{(rs)(kl)}c^{(2)}_{rs}c^{(5)}_{kl} \\
&+K^{(33)}_{(mn)(\overline{mn})}c^{(3)}_{mn}c^{(3)}_{\overline{mn}} + 2K^{(34)}_{(mn)(ij)}c^{(3)}_{pq}c^{(4)}_{ij} + 2K^{(35)}_{(mn)(kl)}c^{(3)}_{pq}c^{(5)}_{kl} \\
&+K^{(44)}_{(ij)(\overline{ij})}c^{(4)}_{ij}c^{(4)}_{\overline{ij}} + 2K^{(45)}_{(ij)(kl)}c^{(4)}_{ij}c^{(5)}_{kl} \\
&+K^{(55)}_{(kl)(\overline{kl})}c^{(5)}_{kl}c^{(5)}_{\overline{kl}} \\
&+\mathcal{N}^{(133)}_{(pq)(mn)(\overline{mn})}c^{(1)}_{pq}c^{(3)}_{mn}c^{(3)}_{\overline{mn}} + \mathcal{N}^{(233)}_{(rs)(mn)(\overline{mn})}c^{(2)}_{rs}c^{(3)}_{mn}c^{(3)}_{\overline{mn}} + \mathcal{N}^{(433)}_{(ij)(mn)(\overline{mn})}c^{(4)}_{ij}c^{(3)}_{mn}c^{(3)}_{\overline{mn}} \\
&+\mathcal{N}^{(533)}_{(kl)(mn)(\overline{mn})}c^{(5)}_{kl}c^{(3)}_{mn}c^{(3)}_{\overline{mn}} + \mathcal{N}^{(333)}_{(mn)(\overline{mn})(\tilde{mn})}c^{(3)}_{mn}c^{(3)}_{\overline{mn}}c^{(3)}_{\tilde{mn}} + \mathbb{N}^{(3333)}_{(mn)(\overline{mn})(\tilde{mn})(\hat{mn})}c^{(3)}_{mn}c^{(3)}_{\overline{mn}}c^{(3)}_{\tilde{mn}}c^{(3)}_{\hat{mn}}\Big] \\
V^{(k,s)} =& P^{(1)}_{pq}c^{(1)}_{pq} + P^{(2)}_{rs}c^{(2)}_{rs} + P^{(3)}_{mn}c^{(3)}_{mn}
\end{aligned}
\tag{38}
$$

where use is made of the index notation presented in [48]. According to this notation, a_{pq} represents a vector, $A_{(pq)(rs)}$, $\mathcal{A}_{(pq)(rs)(mn)}$, $\mathbb{A}_{(pq)(rs)(mn)(ij)}$ are second−, third− and fourth−order fourth−order arrays.

By application of Hamilton's principle to the generic element, the following kernel is obtained:

$$\int_{t_1}^{t_2} \delta\Pi^{(k,s)} dt = \delta \boldsymbol{c}^{(k,s)^T} \boldsymbol{M}^{(k,s)} \ddot{\boldsymbol{c}}^{(k,s)} + \delta \boldsymbol{c}^{(k,s)^T} \boldsymbol{K}^{(k,s)} \boldsymbol{c}^{(k,s)} + \delta \boldsymbol{c}^{(k,s)} \boldsymbol{P}^{(k,s)} = 0 \quad (39)$$

where $\boldsymbol{u}^{(k,s)}$ is the vector collecting the element degrees of freedom, $\boldsymbol{M}^{(k,s)}$, $\boldsymbol{K}^{(k,s)}$, $\boldsymbol{P}^{(k,s)}$ are the element stiffness, mass matrices, and load vector.

Upon assembly of the contributions of all elements, the governing equations are obtained as follows:

$$\boldsymbol{M}\ddot{\boldsymbol{c}} + \boldsymbol{K}\boldsymbol{c} + \boldsymbol{N}_2\boldsymbol{c} + \boldsymbol{N}_3\boldsymbol{c} = \boldsymbol{P} \quad (40)$$

where $\boldsymbol{M}, \boldsymbol{K}$ and $\boldsymbol{P}$ are the global mass, linear stiffness matrices and vector of external loads, respectively, while $\boldsymbol{N}_2$ and $\boldsymbol{N}_3$ are the nonlinear contributions due to cubic and quartic terms in the elastic energy, respectively. The set of ODEs reported in Equation (40) describe the nonlinear dynamics of the shell and can be solved via implicit or explicit integration techniques. In this work, the discussion is restricted to the quasi−static case, so the relevant equations simplify to:

$$\boldsymbol{K}\boldsymbol{c} + \boldsymbol{N}_2\boldsymbol{c} + \boldsymbol{N}_3\boldsymbol{c} = \boldsymbol{P} \quad (41)$$

for which the solution is computed by referring to the Newton–Raphson method.

4. Results

This section presents applications of $ps-$FEM to laminated shell problems. The method has been implemented in a Matlab environment, while all analysis have been run on a laptop with the following characteristics: 1.4 GHz Intel Core i7 processor, 16 GB 1867 MHz LPDDR3 memory, Intel HD Graphics 6,151,536 MB.

In the first part of the section, a series of validation studies is proposed. Then, the numerical technique proposed herein is exploited to solve exemplary test cases for laminated shell problems. The potential of the $ps-$refinement is illustrated against classical $h-$ and $p-$refinements for linear and geometrically nonlinear problems.

4.1. Validation

4.1.1. Test Case 1: Vibrations of Elliptical Shells

The first study aims at validating the shell model implemented and, for this purpose, the vibration response is studied for three elliptical cylindrical shells. The test case is taken from the literature [49] and covers the case of shells characterized by different eccentricities e. The position vector of the reference surface is expressed as:

$$\boldsymbol{r}(\xi_1, \xi_2) = \xi_1 \boldsymbol{e}_x + R_A \sin(\xi_2) \boldsymbol{e}_y + R_B \cos(\xi_2) \boldsymbol{e}_z \quad \text{for} \quad \xi_1 \in [0, L] \quad \text{and} \quad \xi_2 \in [0, 2\pi] \quad (42)$$

where R_A and R_B are the semi−major and semi−minor axis, respectively, while L is the length. A sketch of the shell is provided in Figure 9, while the geometric configurations considered in this study are summarized in Table 1. The shells are made of an aluminum whose properties are $E = 68,950$ MPa, $\nu = 0.3$, and $\rho = 2766$ kgm/^{3}. Clamped−free boundary conditions are considered.

Table 1. Shell geometry.

	L (mm)	R_A (mm)	R_B (mm)	e (-)
Shell 1	594.9188	304.8000	304.8000	0.0000
Shell 2	595.3250	328.9300	279.6540	0.5265
Shell 3	595.3760	365.5060	237.4900	0.7601

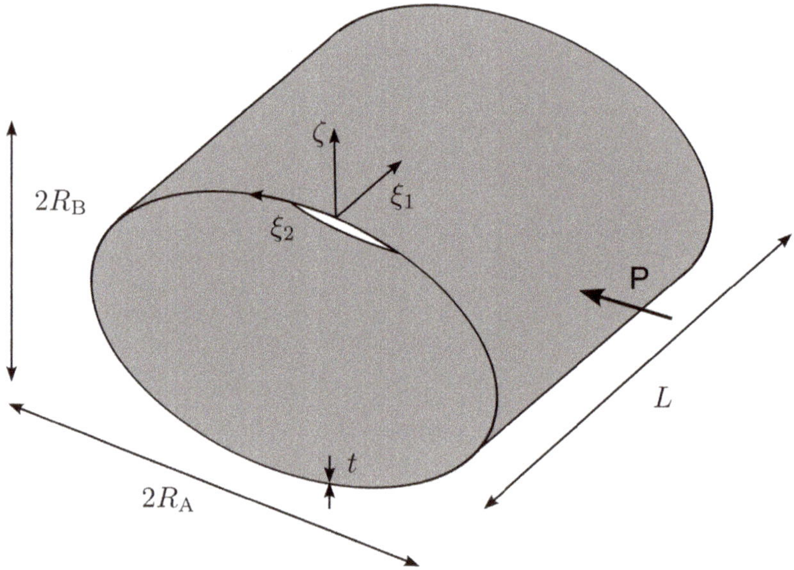

Figure 9. Cylindrical elliptical shell: geometry.

The numerical model consists of a coarse mesh with five elements along the circumferential direction and two along the axial one, see Figure 10. The polynomial order is $p = 10$.

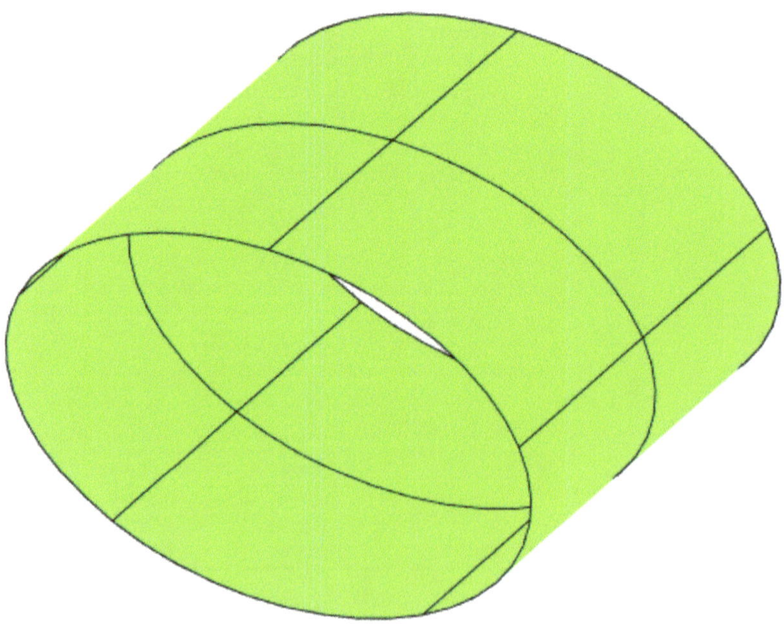

Figure 10. Cylindrical elliptical shell: mesh.

The results are reported in Table 2 for the first ten natural frequencies. A comparison is illustrated against the reference results taken from [49], where the same shell kinematics and geometric formulation were adopted. In [49] the shell governing equations are expressed in strong−form and are solved using the generalized differential quadrature method. Excellent matching is achieved for the reported frequencies, where the error is evaluated as $\mathbb{E}_{\%} = |\omega - \omega_{ref}|/\omega_{ref} \times 100$. For the reported modes, the percentage errors are of the order of $\mathbb{E}_{\%} = 10^{-2} - 10^{-3}$. The almost perfect agreement gives evidence of the correct implementation of the shell geometry and kinematics.

Table 2. Natural frequencies ω (rad/s) for elliptical shells with different eccentricities e. Subscript: percent difference against ref. [49].

Mode N°	$e = 0.0000$	$e = 0.5265$	$e = 0.7601$
1	$611.2500^{(0.0031)}$	$597.7019^{(0.0033)}$	$539.5629^{(0.0041)}$
2	$611.2500^{(0.0031)}$	$597.7020^{(0.0034)}$	$539.5683^{(0.0059)}$
3	$643.1510^{(0.0003)}$	$618.4085^{(0.0016)}$	$542.8651^{(0.0025)}$
4	$643.1510^{(0.0003)}$	$618.4086^{(0.0015)}$	$542.8681^{(0.0023)}$
5	$701.4954^{(0.0042)}$	$712.9803^{(0.0025)}$	$715.4793^{(0.0009)}$
6	$701.4954^{(0.0042)}$	$712.9810^{(0.0025)}$	$715.5105^{(0.0049)}$
7	$857.9875^{(0.0014)}$	$847.2782^{(0.0010)}$	$785.3787^{(0.0042)}$
8	$857.9875^{(0.0014)}$	$847.2840^{(0.0014)}$	$785.5405^{(0.0080)}$
9	$864.4224^{(0.0025)}$	$870.6317^{(0.0001)}$	$892.8773^{(0.0072)}$
10	$864.4224^{(0.0025)}$	$870.6408^{(0.0003)}$	$893.0180^{(0.0147)}$

4.1.2. Test Case 2: Vibration of Variable Stiffness Plate

The second study aims at validating the ability of the proposed numerical code to handle the case of structures with non−uniform stiffness. In this regards, the example presented herein provides a validation of the correctness of the constitutive law implemented [50].

A free vibration analysis is conducted by considering a variable stiffness (VS) rectangular laminate, as illustrated in Figure 11. In particular, the geometry is defined by $a = b = 300$ mm and $t = 1.2$ mm. The material is characterized by the following elastic properties: $E_{11} = 10{,}000$ MPa, $E_{22} = 9000$ MPa, $G_{12} = G_{13} = G_{23} = 5000$ MPa, $\nu_{12} = \nu_{13} = \nu_{23} = 0.3$. The stacking sequence is described in compact manner as $[\pm T]_{2s}$, where the entries of the the matrix $\boldsymbol{T}$ are:

$$\boldsymbol{T} = \langle T_{11} | T_{12} \rangle \quad \text{with} \quad T_{11} = -45 \quad \text{and} \quad T_{12} = \{45, 30, 15, 0, -15, -30, -45\} \tag{43}$$

where different values of T_{12} define different fiber paths. Starting from the configurations defined by Equation (43), the local fiber orientation is obtained by referring to Equation (16).

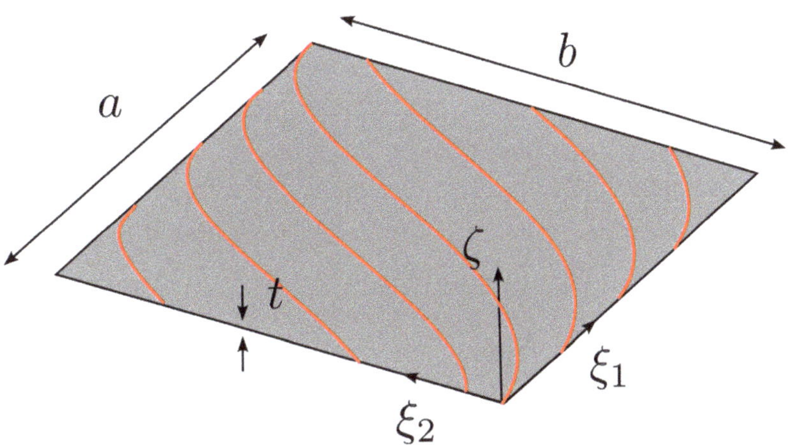

Figure 11. Variable stiffness plate—Geometry.

The VS plate is clamped at the four edges and is modeled with one single element of polynomial order p. The results are presented in terms of the first nondimensional frequency $\Omega = \omega a \sqrt{\rho / E_{22}}$ in Table 3. The comparison against the reference results is provided by considering different degrees of $p-$refinement. As shown, the FE results converge quickly upon $p-$refinement with frequencies converging from above. This behavior is motivated by the hierarchical properties of the Legendre expansion [6]. It is important to note that different FE discretizations may imply different distributions of the elastic properties leading to nonmonothonic convergence of the solution. The comparison of the converged frequencies with the reference ones [50] provides differences below 1%, thus illustrating the possibility of successfully handling structures with non−uniform elastic properties within the present framework.

Table 3. Natural frequencies (rad/s) of rectangular plate with different variable stiffness layups. Subscript: percent difference against ref. [50].

p	$\langle -45 \vert 45 \rangle$	$\langle -45 \vert 30 \rangle$	$\langle -45 \vert 15 \rangle$	$\langle -45 \vert 0 \rangle$	$\langle -45 \vert -15 \rangle$	$\langle -45 \vert -30 \rangle$	$\langle -45 \vert -45 \rangle$
4	$0.0894^{(3.95)}$	$0.0966^{(1.68)}$	$0.1028^{(1.78)}$	$0.1058^{(2.72)}$	$0.1049^{(0.87)}$	$0.1010^{(1.00)}$	$0.0962^{(0.21)}$
6	$0.0882^{(2.56)}$	$0.0957^{(0.74)}$	$0.1016^{(0.59)}$	$0.1044^{(1.36)}$	$0.1037^{(-0.29)}$	$0.1005^{(0.50)}$	$0.0960^{(0.00)}$
8	$0.0868^{(0.93)}$	$0.0948^{(-0.21)}$	$0.1011^{(0.10)}$	$0.1040^{(0.97)}$	$0.1035^{(-0.48)}$	$0.1004^{(0.40)}$	$0.0959^{(-0.10)}$
10	$0.0866^{(0.70)}$	$0.0946^{(-0.42)}$	$0.1009^{(-0.10)}$	$0.1039^{(0.87)}$	$0.1035^{(-0.48)}$	$0.1004^{(0.40)}$	$0.0959^{(-0.10)}$

4.1.3. Test Case 3: Static Analysis of Plate with Cutout

This test case deals with the static analysis of plate with circular cutout loaded in traction. The presence of the cutout determines the onset of stress concentrations, thus providing a useful mean for investigating the validity of the proposed $ps-$refinement strategies.

The plate, an illustration of which is reported in Figure 12, is characterized by dimensions $a = 75$ mm, $b = 50$ mm and $r = 10$ mm. The elastic properties considered for the material are: $E_{11} = 1{,}500{,}00$ MPa, $E_{22} = 9000$ MPa, $G_{12} = G_{13} = G_{23} = 5000$ MPa, $\nu_{12} = \nu_{13} = \nu_{23} = 0.32$. A quasi−isotropic layup $[\pm 45, 90, 0]_s$ is considered and each ply has thickness $t_{ply} = 1$ mm. The traction load is prescribed at two parallel edges with magnitude $\overline{N}_{xx} = 100$ N/mm. The FE model is restricted to a quarter of the full structure owing to the double symmetry of the problem.

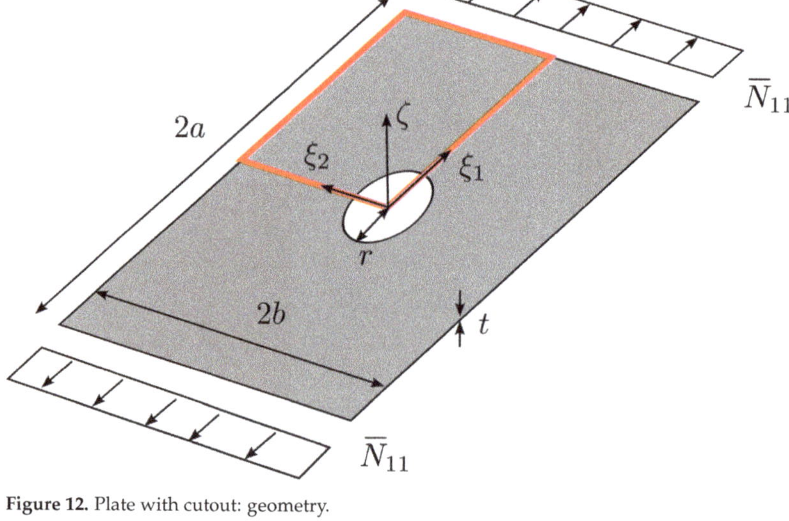

Figure 12. Plate with cutout: geometry.

The $s-$refinement strategy is considered for designing the mesh of the numerical model. In particular, a base mesh having five elements with polynomial order p is superposed with $s = 5$ levels of overlaying meshes of decreasing size, as illustrated in Figure 13. In the validation, the order p is increased until convergence of the global solution is reached. The number of overlaying layers s is selected to have a proper representation of the local response, while the resolution of the base mesh (number/size of elements) is chosen to guarantee the appropriate subdivision of the domain for performing $s-$refinements in the regions of interest. Specifically, the overlaying meshes are placed around the corners of the cutout, where stress gradients are expected.

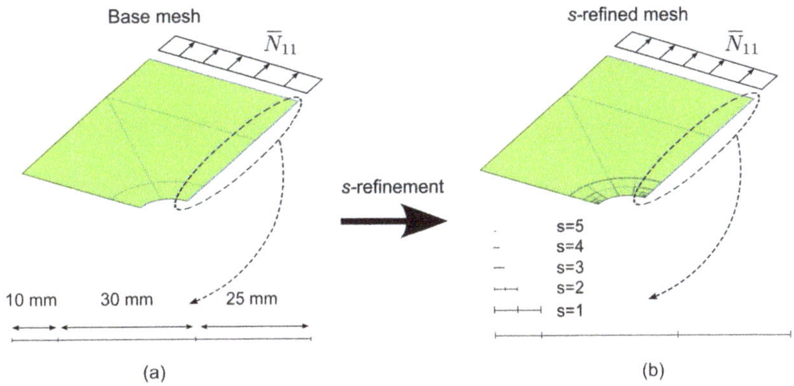

Figure 13. Plate with cutout — static analysis: (**a**) Base mesh, (**b**) $s-$refined mesh.

Three FE models employing the $ps-$refinement strategies illustrated in Figure 7 are set up for the validation study. The interpolation order of the superimposing elements are $p_s = 1$ and $p_s = p$ for the linear and uniform $ps-$refinement, respectively. Regarding the graded $ps-$refinement, the polynomial order p_s is set according to the following law:

$$p_s = p - \text{floor}\left(\frac{s}{m}\right) \tag{44}$$

where m is a constant that controls the rate of decrease of p for increasing levels s. The law of Equation (44) will be adopted in the rest of this work for the graded $ps-$refinement strategy, unless otherwise specified. In the present study, the $ps-$refinements are performed fixing the mesh and increasing the polynomial degree of the elements on the global mesh p, while the order of the overlaying elements p_s is adjusted according to the specific refinement strategy, i.e., linear, uniform, or graded.

The validation is carried out using a FE model realized in Abaqus with S4 shell elements. A preliminary convergence study was carried out with this model. Based on this study, the total number of elements has been taken equal to 350,000 elements ($h-$refinement) to guarantee convergence. The corresponding number of membrane degrees of freedom is 7×10^5, approximately. Results are shown in Table 4 for different types and levels of $ps-$refinements, and are presented in terms of nondimensional elastic energy:

$$\overline{U} = U\frac{E_{11}}{\overline{N}_{xx}^2 r^2 t^3} \tag{45}$$

From Table 4, it can be seen that the nondimensional elastic energies of the three FE models approach the reference one as p is increased. In particular, one can note a quick monotonic convergence from below, i.e., the strain energy increases as the FE models are more refined. This is an expected behaviour as the numerical model becomes more flexible with an increasing number of degrees of freedom. For the linear and graded $ps-$refinement strategies, convergence is reached with $p = 5$. Regarding the uniform $ps-$refinement, convergence is achieved with a slightly lower polynomial order ($p = 4$). This is due to the use of high−order elements in the upper mesh levels which enable us to obtain a higher approximation capability.

Table 4. Convergence of nondimensional energy for different ps−refinement strategies.

		Linear ps-Refinement	**Uniform ps-Refinement**	**Graded ps-Refinement ($m = 1$)**
		$p_s = 1$	$p_s = p$	$p_s = p - \text{floor}(\frac{s}{m})$
$p = 1$		52.4124	52.4570	52.3428
$p = 2$		52.5307	52.5348	52.5254
$p = 3$		52.5433	52.5439	52.5433
$p = 4$		52.5445	52.5446	52.5445
$p = 5$		52.5446	52.5446	52.5446
$p = 6$		52.5446	52.5446	52.5446
Reference	52.5446			

In Figure 14, the convergence curves are presented for the stress resultant N_{xx} measured at $(\xi_1, \xi_2) = (0, r)$ for the three FE models considered for the validation. In all cases, one can note a strong and non−monotonic convergence of N_{xx} to the reference value of 378.26 MPa. The curves in Figure 14 provide insights on the local convergence rate of the ps−refinement strategies. For the present problem, convergence of the stress measures is reached for approximately 10^3 degrees of freedom, which is much lower than the one required by the reference model where a h−refinement was employed.

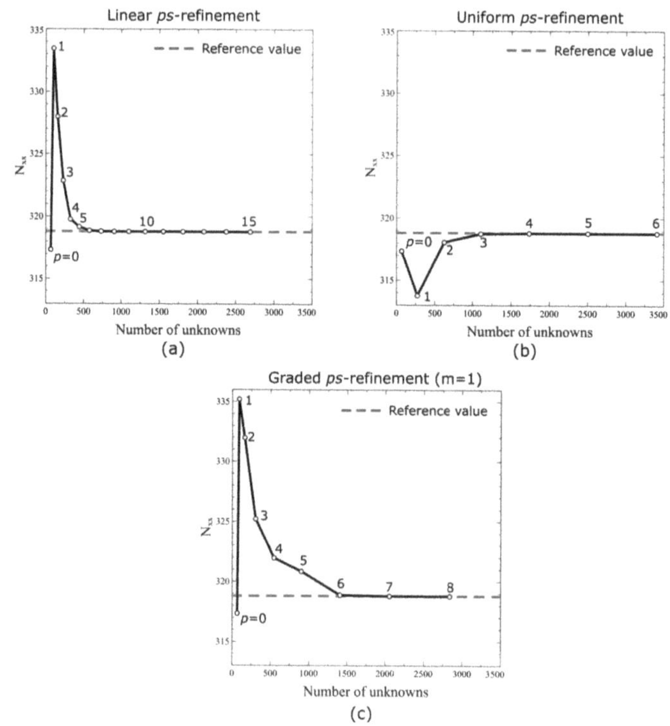

Figure 14. Computed values of N_{xx} at $(\xi_1, \xi_2) = (0, r)$ for different levels and types of ps−refinement: (**a**) linear, (**b**) uniform, and (**c**) graded.

The consistency of these results, both in terms of global and local quantities, provide a validation for the $ps-$refinement strategies implemented in this work.

4.2. Applications

4.2.1. Example 1: Vibration and Buckling of a Highly Anisotropic Laminated Plate

The first application of the $ps-$FEM is an example taken from [51], and relates to the vibration and buckling analysis of a highly anisotropic composite plate. The following data are considered for the analysis: $a/b = 1$ and $a/t = 100{,}000$, see Figure 11. The material has elastic properties given by: $E_{11} = 393{,}000$ MPa, $E_{22} = 5030$ MPa, $G_{12} = G_{13} = G_{23} = 5240$ MPa, $\nu_{12} = \nu_{13} = \nu_{23} = 0.31$, $\rho = 1500$ kg$/m^3$. The plate is a laminate with a single ply oriented at $\theta = 45$. Simply supported boundary conditions are considered along the four edges, while uniform edge−shortening conditions are considered for the buckling analysis.

This problem has been studied in previous efforts in the literature due to its challenging convergence features. The eigenvalues of the problem, both for free vibrations and buckling, tend to convergence with a slow rate as a consequence of the high orthotropy ratio $E_{11}/E_{22} = 73.36$, whose effects are exacerbated by the 45−oriented ply. Indeed, this orientation promotes strong bending/twisting elastic couplings with highly localized gradients in the modal shapes.

Two meshes are realized for conducting the numerical tests. The first one is a mesh with 5×5 quadrilateral elements, see Figure 15a. The second one is a $s-$refined mesh, whose features are outlined in Figure 15b; the $s-$refinement is obtained starting from the base mesh 5×5 and adding 15 layers of overlaying meshes. The refinement is conducted to allow local effects arising from vibration and buckling modes to be accurately captured.

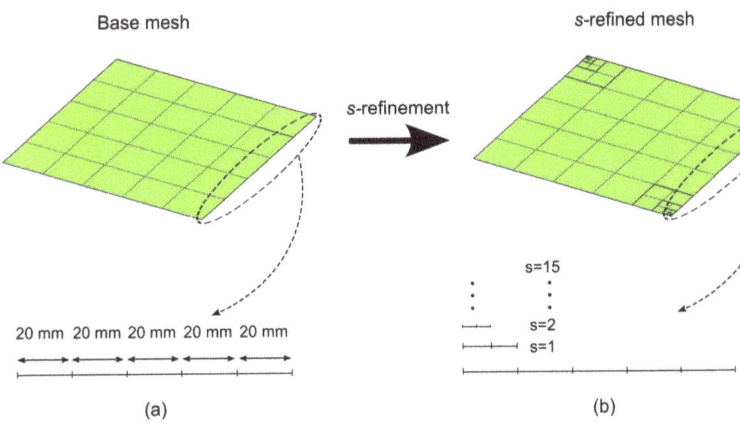

Figure 15. Highlyanisotropic plate – vibration and buckling analysis: (**a**) Base mesh, (**b**) $s-$refined mesh.

Starting from these two meshes, five FE models are developed to illustrate the effect of different refinement strategies. The first two models, $h-$ and $p-$models, employ the mesh in Figure 15a and implement the $h-$ and $p-$refinement strategies, respectively. The other three models, $ps-$L−, $ps-$U−, and $ps-$G−models, are constructed from the $s-$refined mesh in Figure 15b and adopt the linear (L), uniform (U), and graded (G) $ps-$refinement strategies, respectively. A summary of the FE models is reported in Table 5, whose nomenclature will be used for the other examples reported next.

Table 5. Nomenclature for finite element models.

	Refinement Strategy	Mesh Resolution (h)	Polynomial Order (p)	Superposition Levels (s)
$h-$model	h	Increased	Fixed	-
$p-$model	p	Fixed	Increased	-
$ps-$L$-$model	Linear ps	Fixed	Increased	Increased, $p_s = 1$
$ps-$U$-$model	Uniform ps	Fixed	Increased	Increased, $p_s = p$
$ps-$G$-$model	Graded ps	Fixed	Increased	Increased, p_s set with Equation (44)

In absence of available exact solutions, the first step consists of obtaining highly accurate results to be used as a reference for comparing the five models later. The $ps-$U$-$model is used for this scope. The results are summarized in Table 6 in terms of nondimensional frequency and buckling load, defined as:

$$\hat{\omega} = \omega \frac{a^2}{t} \sqrt{\frac{\rho}{E_{22}}}, \quad \hat{N}_{xx} = \overline{N}_{xx} \frac{a^2}{E_{22} t^3} \tag{46}$$

where $\overline{N}_{xx}$ is the force per unit length on the loaded edges.

Table 6. Nondimensional frequency $\hat{\omega}$ and buckling load $\hat{N}_{xx}$ for a SSSS anisotropic plate using a uniform $ps-$refinement strategy.

	Number of Unknowns	$\hat{\omega}$	$\hat{N}_{xx}$
$p = 1$	1119	24.9319	52.5977
$p = 2$	2714	22.0819	31.6995
$p = 3$	4999	21.9341	30.2107
$p = 4$	7974	21.9289	30.1487
$p = 5$	11,639	21.9267	30.1416
$p = 6$	15,994	21.9264	30.1407
$p = 7$	21,039	21.9263	30.1402

As shown, a value of $p = 7$ guarantees convergence up to the third digit for both vibration and buckling parameters. These values are retained as references for computing the errors obtained with different refinement strategies.

The vibration and buckling mode shapes are illustrated in Figures 16 and 17. Both modes are characterized by one single$-$skew half$-$wave, where stretching occurs in the direction of fibers, i.e., $\theta = 45$. From Figures 16a and 17a, one can note that the deflected shape w is smooth in most of the domain apart from the two corners. Here, it is possible to note highly localized effects, as evident from the plots of the twisting moments M_{xy} in Figures 16b and 17b where strong stress concentrations are observed. This will have drastic consequences on the convergence of the fundamental frequency and critical buckling load.

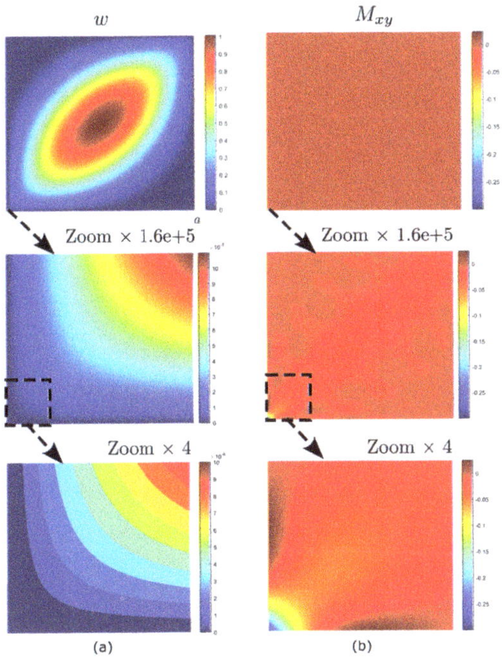

Figure 16. Highly anisotropic plate: mode shapes for free vibration analysis: (**a**) deflection shape and (**b**) twisting moment.

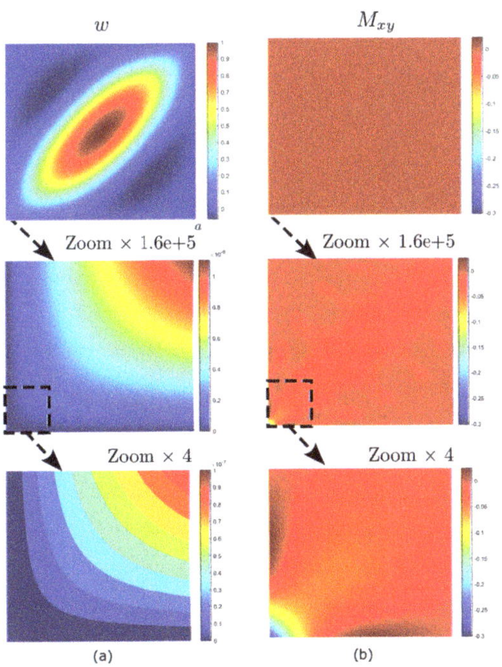

Figure 17. Highly anisotropic plate: mode shapes for buckling analysis: (**a**) deflection shape and (**b**) twisting moment.

The relative errors percentage from the numerical models of Table 5 are summarized in Figure 18, where an additional curve in the results of [51], i.e., Ritz $p-$refinement, is reported. The plots report the logarithmic relative errors percentage evaluated against the reference solutions.

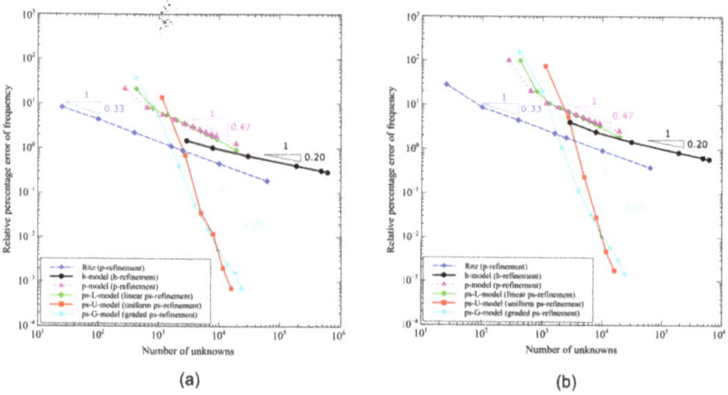

Figure 18. Highly anisotropic plate: convergence for different refinement strategies: (**a**) nondimensional fundamental frequency and (**b**) nondimensional buckling load.

Despite quantitative distinctions, similar trends are observed for the frequency and the buckling results, the latter requiring more degrees of freedom for a fixed corresponding error. From Figure 18, the fast convergence rate achieved with the uniform and graded $ps-$refinements is clearly visible. In particular, the $ps-$U$-$ and $ps-$G$-$refinements outperform FE models implementing standard $h-$ and $p-$refinement approaches. This effect is clear from the error$-$to$-$degrees$-$of$-$freedom ratio of the nondimensional frequency and buckling load in Figure 18, with the $ps-$U$-$ and $ps-$G$-$models presenting the curves with the steepest slopes. The excellent convergence properties just shown are made possible by the flexibility of the approach presented here: the degrees of freedom can be employed to refine specific areas of the domain with no need of generating transition regions, where refinement is not useful in terms of solution accuracy, but is a mean to generate compatible meshes.

The plots of Figure 18 allow us to distinguish regions where the adoption of a refinement strategy is advantageous over the others. For instance, percentage errors $\mathbb{E}_{\%}$ of the order of 0.1–1% suggest the adoption of a $p-$refinement strategy. On the contrary, if stricter requirements are set, the $ps-$refinements provide the best mean for solving the problem, with uniform and graded strategies as the most effective ones.

Benchmark results

The previous section demonstrated the advantages of the $ps-$FEM in deriving highly accurate solutions with relatively low computational costs. The proposed $ps-$FEM framework is then exploited to provide reference results to be used for comparison purposes in future studies in the field.

The same plate considered earlier are studied, but the investigation is now extended to different orientation angles, $\theta = [30, 45, 60]$, and orthotropic ratios, $E_{11}/E_{22} = [73.36, 40, 20, 10]$. By combining these values, twelve plates with different levels of anisotropy are obtained.

The analyses are performed using the $ps-$U$-$model considered in Table 6. The mesh is illustrated in Figure 15b, and the corresponding total number of degrees of freedom is approximately 2×10^4.

A summary of the results is provided in Table 7 in terms of the nondimensional eigenvalues according to Equation (46). The results are compared with the predictions obtained using the Ritz method proposed in [51]. Specifically, the Ritz predictions are obtained by expanding the unknowns with 250 $\times$ 250 Ritz functions, corresponding to a total of 6×10^4 degrees of freedom, i.e., six times higher than the present model.

As shown, the eigenvalues obtained via $ps-\text{U}-$model are always smaller than the Ritz ones. Owing to the convergence properties of the solution (the eigenvalues converge from above) the present results are thus more accurate. The maximum differences are reached for the plate with $E_{11}/E_{22} = 73.36$ and $\theta = 45$, which corresponds to the largest degree of anisotropy. Note, for the buckling load, the difference is as high as 0.40%, approximately, with respect to an already refined solution. The difference between the present and reference results becomes smaller as the degree of anisotropy is reduced either by reducing the off−axis angle or the orthotropy ratio.

Table 7. Nondimensional frequencies $\hat{\omega}$ and buckling load $\hat{N}_{xx}$ for SSSS plates with different degrees of anisotropy.

	E_{11}/E_{22}	$\hat{\omega}$			$\hat{N}_{xx}$		
		***ps*-FEM**	**Ritz Method**	**%diff**	***ps*-FEM**	**Ritz Method**	**%diff**
$\theta = 30$	73.64	22.6796	22.6929	0.0584	38.7882	38.8328	0.1148
	40	17.9948	17.9913	0.0195	26.7974	26.7858	0.0432
	20	14.0520	14.0524	0.0032	17.7050	17.7064	0.0078
	10	11.1987	11.1987	0.0004	11.8977	11.8978	0.0007
$\theta = 45$	73.64	21.9263	21.9674	0.1873	30.1402	30.2538	0.3754
	40	17.6666	17.6557	0.0619	22.9941	22.9633	0.1339
	20	13.9480	13.9496	0.0115	16.4120	16.4163	0.0262
	10	11.1537	11.1538	0.0010	11.4625	11.4628	0.0024
$\theta = 60$	73.64	22.6796	22.6929	0.0584	24.1328	24.1571	0.1005
	40	17.9948	17.9913	0.0195	19.3018	19.2959	0.0307
	20	14.0520	14.0524	0.0032	15.0556	15.0563	0.0050
	10	11.1987	11.1987	0.0004	11.7175	11.7176	0.0010

4.2.2. Example 2: Stress Analysis of Variable Stiffness Elliptical Shell

The response of a cylindrical laminated shell is investigated under the effect of a static point load. The shell has an elliptical cross−section with semi−axes $R_A = 1000$ mm and $R_B = 500$ mm, total thickness $t = 10$ mm, and length $L = 2000$ mm. The sketch of the structure is reported in Figure 9. The material has elastic properties $E_{11} = 150{,}000$ MPa, $E_{22} = 9000$ MPa, $G_{12} = G_{13} = G_{23} = 5000$ MPa, $\nu_{12} = \nu_{13} = \nu_{23} = 0.32$, while the stacking sequence is given as:

$$[\pm T]_s \quad \text{with} \quad T = [30, 45, 30] \tag{47}$$

where the fiber path is obtained via Equation (15). The shell is clamped at both ends and is loaded with a transverse point load with magnitude $P = 100$ N and applied at $(\xi_1, \xi_2) = (L/2, 3/2\pi)$. This loading condition is of interest due to the localized phenomena associated with the concentrated force and finds practical application in tests such as the single perturbation load analysis (SPLA) [52], used for the assessment of shell imperfection sensitivity.

Two FE models are developed and compared. The two models share the same number of degrees of freedom, 10^4, approximately, but have different features for the meshing strategy. The first model relies upon a $p-$refinement strategy ($p-$model) and is presented in Figure 19a. Its mesh has 8×8 elements, i.e., four elements along the circumferential and the axial directions, with a polynomial order $p = 11$.

The second model adopts the $s-$refined mesh illustrated in Figure 19b. The base mesh is unaltered with respect to the $p-$model. A local refinement is introduced with $s = 5$ layers of overlaying meshes in the surroundings of the loaded region. The uniform $ps-$refinement strategy ($ps-\text{U}-$model) is considered, with the polynomial order of the base elements set to $p = 5$.

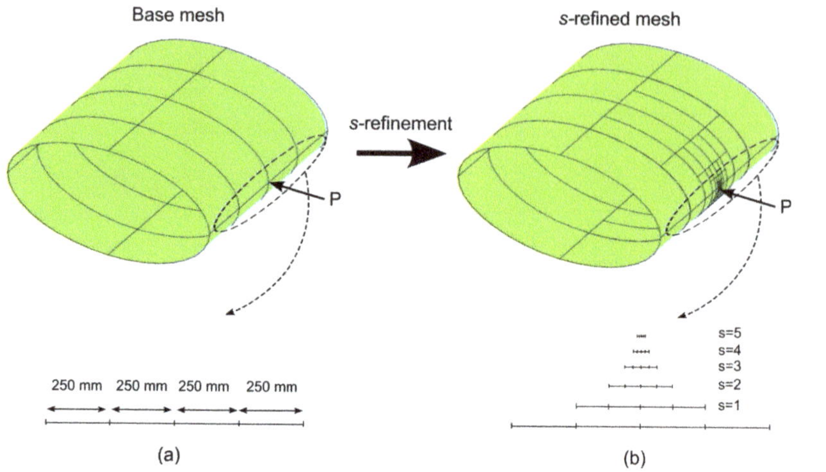

Figure 19. Variable Stiffness cylindrical elliptical shell – static analysis: (**a**) Base mesh, (**b**) $s-$refined mesh.

The plots of the out−of−plane displacement w and the stress component σ_{22} on the outmost ply at $\xi_2 = 3/2\pi$ are illustrated in Figure 20. Similar results are obtained for the two models in terms of deflections, as shown in Figure 20a.

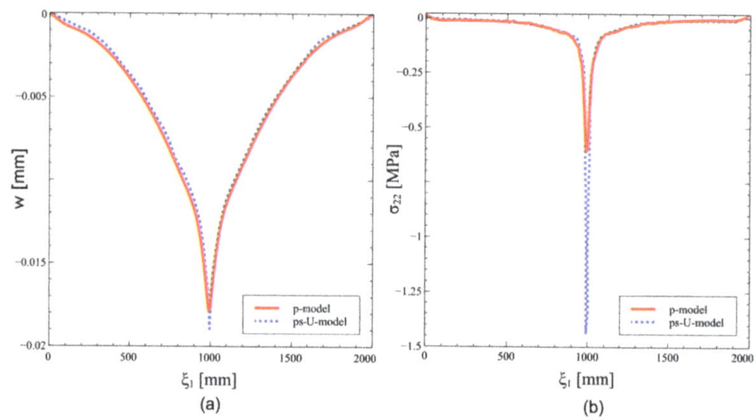

Figure 20. Variable stiffness cylindrical elliptical shell: comparison between $p-$ and $ps-$U$-$models: (**a**) displacement and (**b**) normal stress field.

The advantages of the $ps-$refinement are clear when addressing the predicted stress σ_{22} in Figure 20b, where the discrepancies between the two models are noticeable. An excellent prediction of the stress gradient, which is particularly exacerbated close in the loaded region, is achieved via $ps-$refinement. The comparison between the peak values of the two models reveals an underprediction of the $p-$model that is close to 60%.

The contour of the stress is reported in Figure 21. As shown, the $p-$model solution presents severe oscillations in the surroundings of the stress peak (Figure 21a). This phenomenon is typically observed in high−order approximations when the solution displays steep gradients or discontinuities. Oscillations are greatly attenuated in the case of the $ps-$refined solution, as revealed by Figure 21b. These results provide a clear insight into the advantages of the proposed strategy. By recalling that the two models rely upon the same number of unknowns, evidence is given on the superiority of the $ps-$refinement strategy for problem characterized by strong gradients. As a matter of fact, the combination of $p-$ and $s-$refinements allows us to combine the advantages of both strategies. The ability of high−order approximations is exploited to capture effectively the global response, while the superposed mesh provides an embedded approach suitable for detecting local effects.

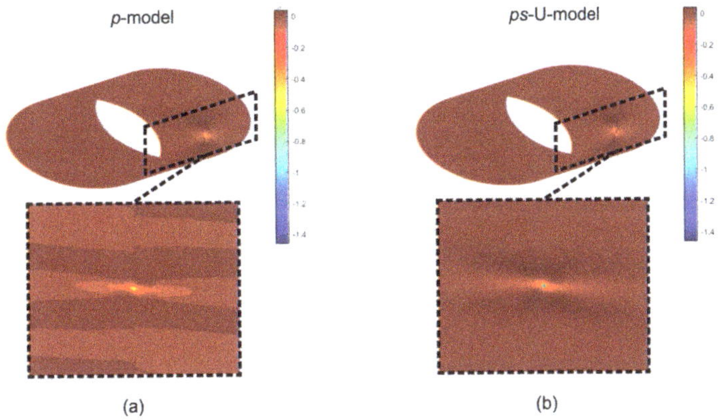

Figure 21. Variable stiffness cylindrical elliptical shell: stress field σ_{22}: (**a**) p−model and (**b**) ps−U−model.

4.2.3. Example 3: Snap−back of Cylindrical Panel

The static nonlinear response of a shell is discussed in this closing example. The test case is taken from [53] and considers an isotropic cylindrical panel loaded by a point force applied at its center and directed inward. The panel is illustrated in Figure 22, where the following parameters are considered: $a = b = 50.8$ mm, $R = 254$ mm, $t = 0.635$ mm. An isotropic material is considered with Young modulus $E = 310.125$ MPa and Poisson ratio $\nu = 0.3$. The panel is under simply−supported boundary conditions along the straight edges, $\xi_2 = 0, b$, while free conditions are imposed along the curved ones, $\xi_1 = 0, a$.

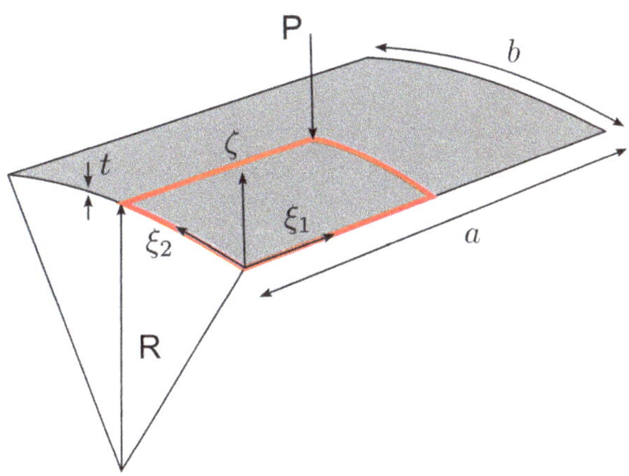

Figure 22. Geometry of cylindrical panel.

The results are presented by comparing the h−model, p−model and ps−L−model of Table 5. In all cases, the symmetry of the problem is exploited by modeling one quarter of the structure and imposing symmetry boundary conditions.

A convergence study is performed by considering two levels of refinement for each model, as displayed in Figure 23. The h−refinement procedure is performed for a fixed polynomial order $p = 1$, with increased mesh resolution, as reported in Figure 23a. The p−refinement strategy is carried out by increasing the polynomial order, $p = 1,3,6$, on a mesh of 2×2 element, see Figure 23b. The ps−refinement is performed by increasing both the polynomial order p and the degree of $s-$refinement of the mesh s. More specifically, the $p-$ and $s-$orders are increased simultaneously with steps of one ($p = 1, 2, 3$) and five ($s = 0, 5, 10$), respectively, see Figure 23c. For clarity, a summary of the FE refinement parameters is presented in Table 8 along with the number of degrees of freedom.

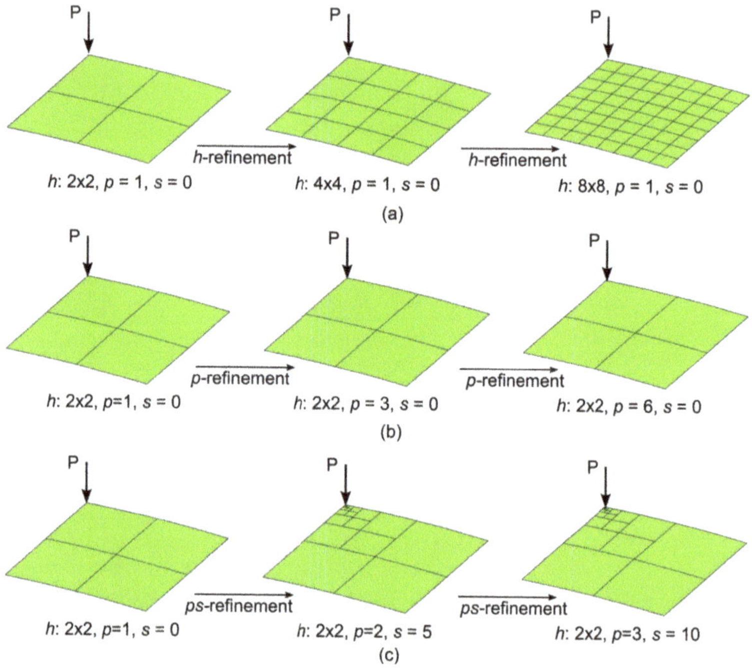

Figure 23. Cylindrical panel: refinement strategies: (**a**) h, (**b**) p, (**c**) $ps-$linear.

Table 8. Summary of the refinement parameters of the FE models used for solving the application example 3.

	Refinement	Number of Unknowns	h	p	s
$h-$model		91	2×2	1	-
	1	343	4×4	1	-
	2	1327	8×8	1	-
$p-$model		91	2×2	1	-
	1	343	2×2	3	-
	2	1021	2×2	6	-
$ps-$L$-$model		91	2×2	1	0
	1	252	2×2	2	5
	2	453	2×2	3	10

The results are presented in terms of load$-$deflection curves in Figure 24. An arc$-$length continuation procedure has been implemented for capturing the snap$-$back. As shown, all the solutions converge to the reference one provided sufficient refinement steps are performed. In particular, the results of Figure 24a demonstrate that 2 steps of refinement, leading to a total of 1327 dofs, are required for the $h-$model. On the contrary, one single step of refinement is needed for the $p-$model and the $ps-$L$-$model, as clear from the plots of Figure 24b,c. In these cases, the number of degrees of freedom is one order of magnitude lower for the $p-$model (343) and $ps-$L$-$model (252). These results demonstrate the advantage implied by the use of appropriate meshing and refinement strategies. Through a comparison of the $p-$ and $ps-$refinement strategies with a classical $h-$refinement approach, the quality of the predictions is preserved, but the number of degrees of freedom can be as low as 1/5. This saving becomes even more important in the context of nonlinear analyses, where the overall process requires repeated matrix factorizations and linear systems to be solved.

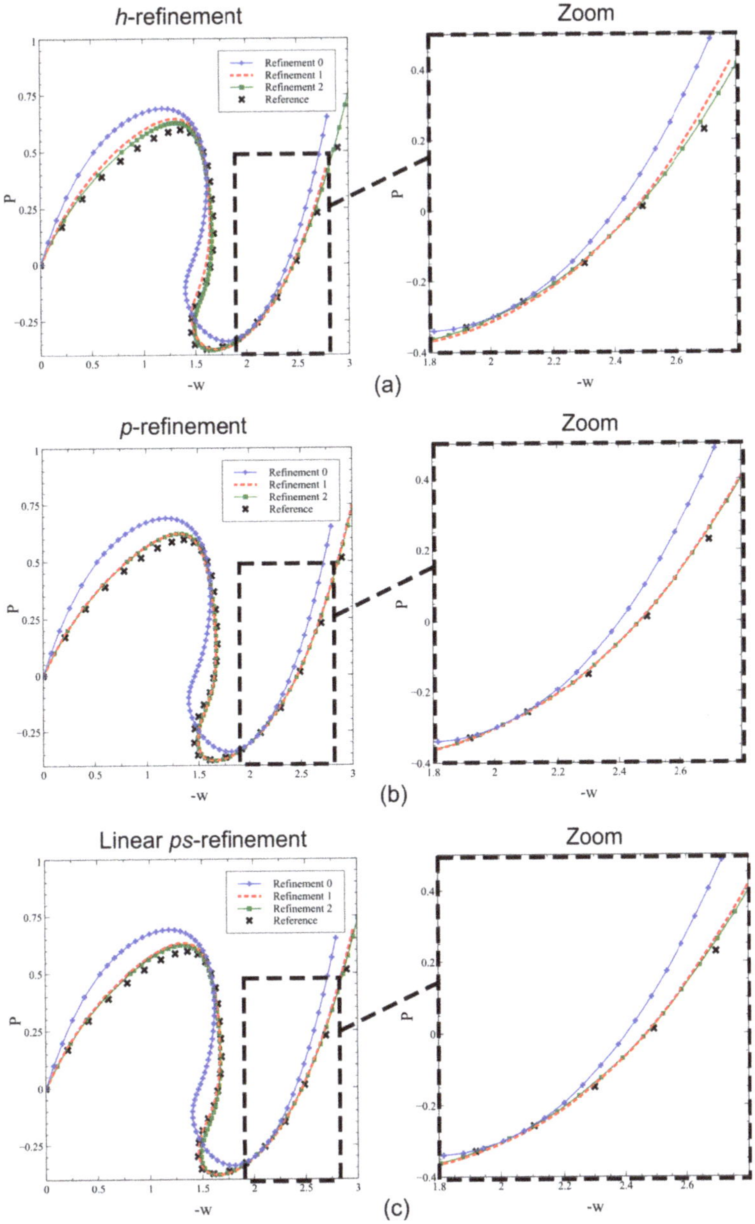

Figure 24. Cylindrical panel: load−deflection curves for different levels of (**a**) $h-$, (**b**) $p-$ and (**c**) $ps-$refinements.

5. Conclusions

This paper dealt with the application of the $ps-$version of the finite element method to the analysis of composite thin shells. The approach relies on the possibility of superposing multiple local meshes in arbitrary regions to refine the solution only where actually needed. No transition meshes are required nor special−purpose elements to connect areas with different grid densities.

The implementation proposed herein allows a wide class of structures to be analyzed, including shells with arbitrary curvature and innovative materials with variable in−plane elastic properties. Both geometrically linear and nonlinear features are introduced. Three different combinations of $p-$ and $s-$refinements were considered, although other strategies could be easily developed and implemented.

The proposed advanced refinement technique is useful under several circumstances in the analysis of composite shells. The results illustrate this aspect for the case of local stress concentrations, where the possibility of superposing multiple layers of refinement can be exploited to achieve accurate stress predictions with reduced effort. For 2D laminated structures, advanced refinement can be of interest even in case of global responses, such as for the eigenanalysis of highly anisotropic plates. Indeed, owing to complex elastic couplings, the ability to capture local effects can have a drastic influence also on the predictions at a global level. For these problems, conventional methods based on high−order global approximations can be unsuitable if refined predictions are of concern. Among the different combinations of $p-$ and $s-$refinements considered in the various test cases, no optimal choice can be established ex ante. The choice is, in general, problem−dependent.

The inherently hierarchical nature of the proposed tool suggests its use at different stages of the design process of composite shells. In the early design phases, simplified models with one or few macro elements with high−order interpolation can be the most suitable way for obtaining quick estimates. In more advanced phases, when increased detail is needed, the models can be easily improved by exploiting the local refinement capabilities to maximize the ratio between accuracy and number of degrees of freedom.

Author Contributions: Conceptualization, methodology, software, validation, investigation, writing—original draft, writing—review & editing, C.A.Y.; conceptualization, methodology, review & editing, supervision, R.V. All authors have read and agreed to the published version of the manuscript.

Funding: This research was funded by Ministero dell'Istruzione, dell'Università e della Ricerca, under the PRIN 2017 program.

Acknowledgments: The authors would like to thank Ministero dell'Istruzione, dell'Università e della Ricerca for funding this research under the PRIN 2017 program.

Conflicts of Interest: The authors declare no conflict of interest.

Appendix A

The components of the vectors of generalized strains $\boldsymbol{\epsilon}^0$, $\boldsymbol{k}$ and γ^0 are:

$$
\begin{aligned}
\epsilon_{11}^0 &= \frac{1}{1+\zeta/R_1}\left(\frac{u_{,1}}{a_1}+\frac{a_{1,2}}{a_1a_2}v+\frac{w}{R_1}\right)+\frac{1}{2}\left(\frac{1}{1+\zeta/R_1}\right)^2\left(\frac{w_{,1}}{a_1}\right)^2\\
\epsilon_{22}^0 &= \frac{1}{1+\zeta/R_2}\left(\frac{v_{,2}}{a_2}+\frac{a_{2,1}}{a_1a_2}u+\frac{w}{R_2}\right)+\frac{1}{2}\left(\frac{1}{1+\zeta/R_2}\right)^2\left(\frac{w_{,2}}{a_2}\right)^2\\
\gamma_{12}^0 &= \frac{1}{1+\zeta/R_2}\left(\frac{u_{,2}}{a_2}-\frac{a_{2,1}}{a_1a_2}v\right)+\frac{1}{1+\zeta/R_1}\left(\frac{v_{,1}}{a_1}-\frac{a_{1,2}}{a_1a_2}u\right)+\left(\frac{1}{1+\zeta/R_1}\right)\left(\frac{1}{1+\zeta/R_2}\right)\frac{w_{,1}w_{,2}}{a_1a_2}\\
\gamma_{13}^0 &= \phi_1+\frac{1}{1+\zeta/R_1}\left(\frac{w_{,1}}{a_1}-\frac{u}{R_1}\right)\\
\gamma_{23}^0 &= \phi_2+\frac{1}{1+\zeta/R_2}\left(\frac{w_{,2}}{a_2}-\frac{v}{R_2}\right)\\
k_{11} &= \frac{1}{1+\zeta/R_1}\left(\frac{\phi_{1,1}}{a_1}+\frac{a_{1,2}}{a_1a_2}\phi_2\right)\\
k_{22} &= \frac{1}{1+\zeta/R_2}\left(\frac{\phi_{2,2}}{a_2}+\frac{a_{2,1}}{a_1a_2}\phi_1\right)\\
k_{12} &= \frac{1}{1+\zeta/R_2}\left(\frac{\phi_{1,2}}{a_2}-\frac{a_{2,1}}{a_1a_2}\phi_2\right)+\frac{1}{1+\zeta/R_1}\left(\frac{\phi_{2,1}}{a_1}-\frac{a_{1,2}}{a_1a_2}\phi_1\right)
\end{aligned}
\tag{A1}
$$

References

1. Reddy, J. *Mechanics of Laminated Composite Plates and Shells: Theory and Analysis*; CRC Press: Boca Raton, CA, USA, 2004.
2. Gürdal, Z.; Olmedo, R. Composite laminates with spatially varying fiber orientations: Variable stiffness panel concept. In Proceedings of the 33rd AIAA/ASME/ASCE/AHS/ASC Structures, Structural Dynamics and Material Conference, Dallas, TX, USA, 13–15 April 1992.
3. Gürdal, Z.; Tatting, B.; Wu, C. Variable stiffness composite panels: Effects of stiffness variation on the in-plane and buckling response. *Compos. Part A Appl. Sci. Manuf.* **2008**, *39*, 911–922. [CrossRef]
4. Wu, Z.; Weaver, P.; Raju, G.; Kim, B. Buckling analysis and optimisation of variable angle tow composite plates. *Thin-Walled Struct.* **2012**, *60*, 163–172. [CrossRef]
5. Raju, G.; Wu, Z.; Weaver, P. Postbuckling analysis of variable angle tow plates using differential quadrature method. *Compos. Struct.* **2013**, *106*, 74–84. [CrossRef]
6. Szabó, B.; Babuška, I.B. *Introduction to Finite Element Analysis: Formulation, Verification and Validation*; John Wiley & Sons Ltd.: Chichester, UK, 2011.
7. Bathe, K. *Finite Element Procedures*; Prentice Hall: Hoboken, NJ, USA, 2006.
8. Babuška, I.; Szabó, B.; Katz, I. The p-version of the finite element method. *SIAM J. Numer. Anal.* **1981**, *18*, 515–545. [CrossRef]
9. Bacciocchi, M.; Luciano, R.; Majorana, C.; Tarantino, A. Free vibrations of sandwich plates with damaged soft-core and non-uniform mechanical properties: Modeling and finite element analysis. *Materials* **2019**, *12*, 2444. [CrossRef]
10. Bacciocchi, M.; Tarantino, A. Natural frequency analysis of functionally graded orthotropic cross-ply plates based on the finite element method. *Math. Comput. Appl.* **2019**, *24*, 52. [CrossRef]
11. Duc, N.; Trinh, T.; Do, T.V.; Doan, D. On the buckling behavior of multi-cracked FGM plates. In Proceedings of the International Conference on Advances in Computational Mechanics 2017: ACOME 2017, Phu Quoc, Vietnam, 2–4 August 2018.
12. Dong, H.; Zheng, X.; Cui, J.; Nie, Y.; Yang, Z.; Ma, Q. Multi-scale computational method for dynamic thermo-mechanical performance of heterogeneous shell structures with orthogonal periodic configurations. *Comput. Methods Appl. Mech. Eng.* **2019**, *354*, 143–180. [CrossRef]
13. Tsapetis, D.; Sotiropoulos, G.; Stavroulakis, G.; Papadopoulos, V.; Papadrakakis, M. A stochastic multiscale formulation for isogeometric composite Kirchhoff-Love shells. *Comput. Methods Appl. Mech. Eng.* **2021**, *373*, 113541. [CrossRef]
14. Cao, Z.; Guo, D.; Fu, H.; Han, Z. Mechanical simulation of thermoplastic composite fiber variable-angle laminates. *Materials* **2020**, *13*, 3374. [CrossRef]
15. Sanchez-Majano, A.; Pagani, A.; Petrolo, M.; Zhang, C. Buckling sensitivity of tow-steered plates subjected to multiscale defects by high-order finite elements and polynomial chaos expansion. *Materials* **2021**, *14*, 2706. [CrossRef]
16. Akhavan, H.; Ribeiro, P.; Moura, M.D. Large deflection and stresses in variable stiffness composite laminates with curvilinear fibres. *Int. J. Mech. Sci.* **2013**, *73*, 14–26. [CrossRef]
17. Yazdani, S.; Ribeiro, P. A layerwise p-version finite element formulation for free vibration analysis of thick composite laminates with curvilinear fibres. *Compos. Struct.* **2015**, *120*, 531–542. [CrossRef]

18. Bank, R.; Sherman, A.; Weiser, A. Some refinement algorithms and data structures for regular local mesh refinement. *Sci. Comput. Appl. Math. Comput. Phys. Sci.* **1983**, *1*, 3–17.
19. Zhu, J.; Zienkiewicz, O. Adaptive techniques in the finite element method. *Commun. Appl. Numer. Methods* **1988**, *4*, 197–204. [CrossRef]
20. Zienkiewicz, O.; Zhu, J. A simple error estimator and adaptive procedure for practical engineerng analysis. *Int. J. Numer. Methods Eng.* **1987**, *24*, 337–357. [CrossRef]
21. BabuÅ¡ka, I.; Dorr, M. Error estimates for the combined h and p versions of the finite element method. *Numer. Math.* **1981**, *37*, 257–277.
22. Guo, B.; Babuška, I. The hp version of the finite element method. *Comput. Mech.* **1986**, *1*, 21–41. [CrossRef]
23. Gui, W.; Babuška, I. The h, p and hp versions of the finite element method in 1 dimension. Part 1. In *The Error Analysis of the p-Version*; TN BN-1036; Laboratory for Numerical Analysis, University of Maryland: College Park, MD, USA, 1985.
24. Babuška, I.; Rank, E. An expert-system-like feedback approach in the hp-version of the finite element method. *Finite Elem. Anal. Des.* **1987**, *3*, 127–147. [CrossRef]
25. Verfürth, R. A review of a posteriori error estimation techniques for elasticity problems. *Comput. Methods Appl. Mech. Eng.* **1999**, *176*, 419–440. [CrossRef]
26. Eibner, T.; Melenk, J. An adaptive strategy for hp-FEM based on testing for analyticity. *Comput. Mech.* **2007**, *39*, 575–595. [CrossRef]
27. Bern, M.; Flaherty, J.; Luskin, M. *Grid generation and adaptive algorithms. In The IMA Volumes in Mathematics and its Applications*; Springer: New York, NY, USA, 2012.
28. Mote, C. Global-local finite element. *Int. J. Numer. Methods Eng.* **1971**, *3*, 565–574. [CrossRef]
29. Belytschko, T.; Fish, J.; Bayliss, A. The spectral overlay on finite elements for problems with high gradients. *Comput. Methods Appl. Mech. Eng.* **1990**, *81*, 71–89. [CrossRef]
30. Fish, J. The s-version of the finite element method. *Comput. Struct.* **1992**, *43*, 539–547. [CrossRef]
31. Fish, J.; Markolefas, S. The s-version of the finite element method for multilayer laminates. *Int. J. Numer. Methods Eng.* **1992**, *33*, 1081–1105. [CrossRef]
32. Fish, J.; Guttal, R. The s-version of finite element method for laminated composites. *Int. J. Numer. Methods Eng.* **1996**, *39*, 3641–3662. [CrossRef]
33. Fish, J.; Suvorov, A.; Belsky, V. Hierarchical composite grid method for global-local analysis of laminated composite shells. *Appl. Numer. Math.* **1997**, *23*, 241–258. [CrossRef]
34. Park, J.; Hwang, J.; Kim, Y. Efficient finite element analysis using mesh superposition technique. *Finite Elem. Anal. Des.* **2003**, *39*, 619–638. [CrossRef]
35. Reddy, J.; Robbins, D. Theories and Computational Models for Composite Laminates. *Appl. Mech. Rev.* **1994**, *47*, 147. [CrossRef]
36. Sakata, S.; Chan, Y.; Arai, Y. On accuracy improvement of microscopic stress/stress sensitivity analysis with the mesh superposition method for heterogeneous materials considering geometrical variation of inclusions. *Int. J. Numer. Methods Eng.* **2020**, *121*, 534–559. [CrossRef]
37. Kishi, K.; Takeoka, Y.; Fukui, T.; Matsumoto, T.; Suzuki, K.; Shibanuma, K. Dynamic crack propagation analysis based on the s-version of the finite element method. *Comput. Methods Appl. Mech. Eng.* **2020**, *366*, 113091. [CrossRef]
38. Rank, E. Adaptive remeshing and hp domain decomposition. *Comput. Methods Appl. Mech. Eng.* **1992**, *101*, 299–313. [CrossRef]
39. Schillinger, D.; Düster, A.; Rank, E. The hp-d-adaptive finite cell method for geometrically nonlinear problems of solid mechanics. *Int. J. Numer. Methods Eng.* **2012**, *89*, 1171–1202. [CrossRef]
40. Krause, R.; Rank, E. Multiscale computations with a combination of the h-and p-versions of the finite-element method. *Comput. Methods Appl. Mech. Eng.* **2003**, *192*, 3959–3983. [CrossRef]
41. Zander, N.; Bog, T.; Kollmannsberger, S.; Schillinger, D.; Rank, E. Multi-level hp-adaptivity: High-order mesh adaptivity without the difficulties of constraining hanging nodes. *Comput. Mech.* **2015**, *55*, 499–517. [CrossRef]
42. Zander, N.; Bog, T.; Elhaddad, M.; Frischmann, F.; Kollmannsberger, S.; Rank, E. The multi-level hp-method for three-dimensional problems: Dynamically changing high-order mesh refinement with arbitrary hanging nodes. *Comput. Methods Appl. Mech. Eng.* **2016**, *310*, 252–277. [CrossRef]
43. Zander, N.; Ruess, M.; Bog, T.; Kollmannsberger, S.; Rank, E. Multi-level hp-adaptivity for cohesive fracture modeling. *Int. J. Numer. Methods Eng.* **2017**, *109*, 1723–1755. [CrossRef]
44. Tornabene, F.; Fantuzzi, N.; Bacciocchi, M.; Dimitri, R. Dynamic analysis of thick and thin elliptic shell structures made of laminated composite materials. *Compos. Struct.* **2015**, *133*, 278–299. [CrossRef]
45. Amabili, M. *Nonlinear Vibrations and Stability of Shells and Plates*; Cambridge University Press: Cambrige, UK, 2008.
46. Gordon, W.; Hall, C. Construction of curvilinear coordinate systems and applications to mesh generation. *Int. J. Numer. Methods Eng.* **1973**, *7*, 461–477. [CrossRef]
47. Gordon, W.; Hall, C. Transfinite element methods: Blending-function interpolation over arbitrary curved element domains. *Numer. Math.* **1973**, *21*, 109–129. [CrossRef]
48. Vescovini, R.; Spigarolo, E.; Dozio, L. Efficient post-buckling analysis of variable-stiffness plates using a perturbation approach. *Thin-Walled Struct.* **2019**, *143*, 106211. [CrossRef]

49. Tornabene, F.; Fantuzzi, N.; Bacciocchi, M.; Dimitri, R. Free vibrations of composite oval and elliptic cylinders by the generalized differential quadrature method. *Thin-Walled Struct.* **2015**, *97*, 114–129. [CrossRef]
50. Houmat, A. Nonlinear free vibration of laminated composite rectangular plates with curvilinear fibers. *Compos. Struct.* **2013**, *106*, 211–224. [CrossRef]
51. Vescovini, R.; Dozio, L.; d'Ottavio, M.; Polit, O. On the application of the Ritz method to free vibration and buckling analysis of highly anisotropic plates. *Compos. Struct.* **2018**, *192*, 460–474. [CrossRef]
52. Kriegesmann, B.; Jansen, E.; Rolfes, R. Design of cylindrical shells using the single perturbation load approach potentials and application limits. *Thin-Walled Struct.* **2016**, *108*, 369–380. [CrossRef]
53. Wisniewski, K. Finite rotation shells. In *Basic Equations and Finite Elements for Reissner Kinematics;* Springer: Berlin, Germany, 2010.

MDPI
St. Alban-Anlage 66
4052 Basel
Switzerland
Tel. +41 61 683 77 34
Fax +41 61 302 89 18
www.mdpi.com

Materials Editorial Office
E-mail: materials@mdpi.com
www.mdpi.com/journal/materials

www.ingramcontent.com/pod-product-compliance
Lightning Source LLC
Chambersburg PA
CBHW061800210926
51850CB00006B/607